Single-Molecule Microscopy and Spectroscopy

London, UK

14–16 September 2015

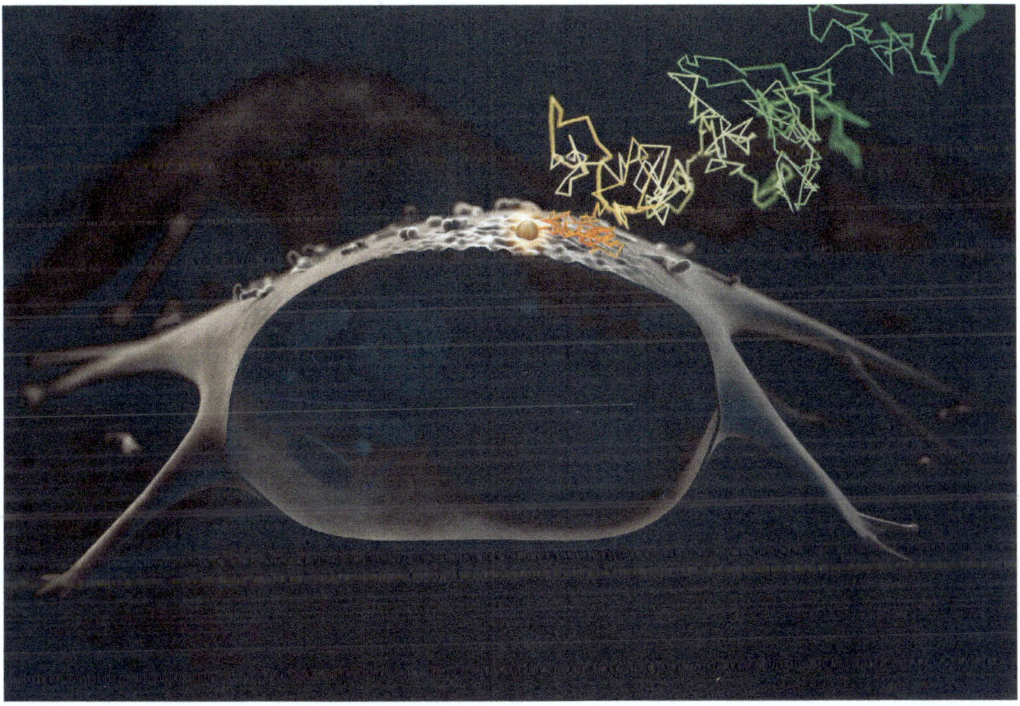

FARADAY DISCUSSIONS

Volume 184, 2015

ROYAL SOCIETY OF CHEMISTRY

The Faraday Division of the Royal Society of Chemistry, previously the Faraday Society, founded in 1903 to promote the study of sciences lying between Chemistry, Physics and Biology.

Faraday Discussions are unique international discussion meetings that focus on rapidly developing areas of chemistry and its interfaces with other scientific disciplines.

Faraday Discussions (Print ISSN 1359-6640, Electronic ISSN 1364-5498) is published 9 times a year by the Royal Society of Chemistry, Thomas Graham House, Science Park, Milton Road, Cambridge, UK CB4 0WF.

Volume 184 ISBN-13: 978-1-78262-461-5

2015 annual subscription price: print+electronic £877, US $1635; electronic only £833, US $1553. Customers in Canada will be subject to a surcharge to cover GST. Customers in the EU subscribing to the electronic version only will be charged VAT. All orders, with cheques made payable to the Royal Society of Chemistry, should be sent to RSC Order Department, Royal Society of Chemistry, Thomas Graham House, Science Park, Milton Road, Cambridge, CB4 0WF, UK. Tel +44 (0) 1223 432398; E-mail **www.orders@rsc.org**

If you take an institutional subscription to any RSC journal you are entitled to free, site-wide web access to that journal. You can arrange access via Internet Protocol (IP) address at **www.rsc.org/ip**. Customers should make payments by cheque in sterling payable on a UK clearing bank or in US dollars payable on a US clearing bank.

Printed in the UK

Single-Molecule Microscopy and Spectroscopy

Faraday Discussions

www.rsc.org/faraday_d

A General Discussion on Single-Molecule Microscopy and Spectroscopy was held in London, UK on the 14th, 15th and 16th of September 2015.

CONTENTS

ISSN 1359-6640; ISBN: 978-1-78262-461-5

Cover
Front cover image reproduced by permission of Prof. Haw Yang, FRSC.

INTRODUCTORY LECTURE

PAPERS AND DISCUSSIONS

SPONSOR

Cogent skills
for science industries

SUPPORTING ORGANISATIONS

British Biophysical Society

EPIC
European Photonics Industry Consortium

EXHIBITORS

PicoQuant

HAMAMATSU
PHOTON IS OUR BUSINESS

fianium
ultrafast fiber lasers

Laser Quantum
INNOVATIVE RELIABLE INTELLIGENT

IOP Publishing

CONCLUDING REMARKS

ADDITIONAL INFORMATION

Faraday Discussions

PAPER

Single-molecule spectroscopy and imaging over the decades

W. E. Moerner,* Yoav Shechtman and Quan Wang

Received 30th October 2015, Accepted 10th November 2015
DOI: 10.1039/c5fd00149h

As of 2015, it has been 26 years since the first optical detection and spectroscopy of single molecules in condensed matter. This area of science has expanded far beyond the early low temperature studies in crystals to include single molecules in cells, polymers, and in solution. The early steps relied upon high-resolution spectroscopy of inhomogeneously broadened optical absorption profiles of molecular impurities in solids at low temperatures. Spectral fine structure arising directly from the position-dependent fluctuations of the number of molecules in resonance led to the attainment of the single-molecule limit in 1989 using frequency-modulation laser spectroscopy. In the early 1990s, a variety of fascinating physical effects were observed for individual molecules, including imaging of the light from single molecules as well as observations of spectral diffusion, optical switching and the ability to select different single molecules in the same focal volume simply by tuning the pumping laser frequency. In the room temperature regime, researchers showed that bursts of light from single molecules could be detected in solution, leading to imaging and microscopy by a variety of methods. Studies of single copies of the green fluorescent protein also uncovered surprises, especially the blinking and photoinduced recovery of emitters, which stimulated further development of photoswitchable fluorescent protein labels. All of these early steps provided important fundamentals underpinning the development of super-resolution microscopy based on single-molecule localization and active control of emitting concentration. Current thrust areas include extensions to three-dimensional imaging with high precision, orientational analysis of single molecules, and direct measurements of photodynamics and transport properties for single molecules trapped in solution by suppression of Brownian motion. Without question, a huge variety of studies of single molecules performed by many talented scientists all over the world have extended our knowledge of the nanoscale and many microscopic mechanisms previously hidden by ensemble averaging.

1. Introduction

It has now been more than 25 years since the first experiment demonstrating optical detection and spectroscopy of single molecules in a condensed phase.[1]

Department of Chemistry, Stanford University, Stanford, California 94305, USA. E-mail: wmoerner@stanford.edu

Single-molecule spectroscopy (SMS) allows *exactly one*, molecule hidden deep within a crystal, polymer, or cell, to be observed *via* optical excitation of the molecule of interest (Fig. 1a). This represents detection and spectroscopy at the ultimate sensitivity level of $\sim 1.66 \times 10^{-24}$ moles of the molecule of interest (1.66 yoctomoles), or a quantity of moles equal to the inverse of Avogadro's number. Detection of the single molecule of interest must be done in the presence of billions to trillions of solvent or host molecules. To achieve this, a light beam (typically a laser) is used to pump an electronic transition of the one molecule resonant with the optical wavelength (Fig. 1b), and it is the interaction of this optical radiation with the molecule that allows the single molecule to be detected. Successful experiments must meet the requirements of (a) guaranteeing that only one molecule is in resonance in the volume probed by the laser, and (b) providing a signal-to-noise ratio (SNR) for the single-molecule signal that is greater than unity for a reasonable averaging time.

Why are single-molecule methods now regarded as a critical part of modern physical chemistry, chemical physics, and biophysics? By removing ensemble averaging, it becomes possible to directly measure distributions of behavior to explore hidden heterogeneity, a property that would be expected in complex environments. In the time domain, the ability to optically sense internal states of one molecule and the transitions among them allows measurement of hidden kinetic pathways and the detection of rare intermediates. Because typical single-molecule labels behave like tiny light sources, roughly 1–2 nm in size, and can report on their immediate local environment, single-molecule studies provide a new window into the nanoscale with intrinsic access to time-dependent changes.

The basic principles of single-molecule optical spectroscopy and imaging have been the subject of many reviews[2–12] and books.[13–16] In this introductory

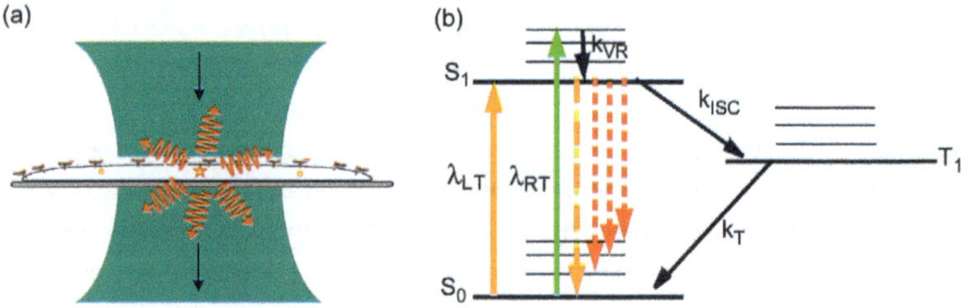

Fig. 1 (a) Schematic of a focused optical beam pumping a single resonant molecule in a cell or other condensed phase sample. The molecule may emit fluorescence or its presence may be detected by carefully measuring the transmitted beam. (b) Typical energy level scheme for single-molecule spectroscopy showing the interaction with the pumping light. S_0, ground singlet state; S_1, first excited singlet; T_1, lowest triplet state or other intermediate state. For each electronic state, several levels in the vibrational progression are shown. Typical low-temperature studies use wavelength λ_{LT} to pump the dipole-allowed (0–0) transition, while at room temperature shorter wavelengths λ_{RT} which pump vibronic sidebands are more common. Fluorescence emission, shown as dotted lines, originates from S_1 and terminates on various vibrationally excited levels of S_0 or S_0 itself. Molecules are typically chosen to minimize entry into dark states such as the triplet state (illustrated), although this or other dark processes can lead to blinking which is useful in super-resolution microscopy. The intersystem crossing or intermediate production rate is k_{ISC}, and the triplet decay rate is k_T.

paper, key ideas are summarized, starting from the early steps at low temperatures arising from explorations of spectral hole-burning as a method to achieve frequency domain optical storage. Some of the surprising physical effects first observed at low temperatures will then be described, followed by a short overview of the key early steps to achieve room temperature single-molecule detection, but the extremely wide range of applications which have been demonstrated can only be presented in a list. One relatively new advance, super-resolution optical microscopy, can be achieved by sequential single-molecule imaging and localization in situations where the emitting concentration is held to low levels.[17–19] Finally, selected recent research thrusts in single-molecule imaging and tracking, as well as trapping in solution, will be summarized.

1.1 Initial steps using low-temperature, high-resolution spectroscopy in solids

In the 1970s and 1980s, high-resolution optical spectroscopy of impurities and defects in solids was a major field of chemical and physical research. One goal of this work was to measure the true homogeneous width of the zero-phonon, purely electronic transition of molecules in solids. However, these zero-phonon lines were so narrow in frequency space that they became hidden in the broad distribution of resonance frequencies caused by the random strains and stresses present in solids, and the absorption lines were known to be inhomogeneously broadened.[20–22] It is for this reason that much research in the 1970s and 1980s was devoted to methods like fluorescence line narrowing (FLN)[23,24] and transient spectroscopies such as free induction decay, optical nutation, and photon echoes.[25–27] While these were all powerful methods with advantages and disadvantages, there was another method to assess the homogeneous width under certain circumstances, persistent spectral hole-burning.[28] This effect produced marks or dips in absorption, termed "holes", in an inhomogeneous line caused by photochemical changes[29] or photophysical changes[30] in the nearby environment. One of us (W. E. Moerner), was employed at IBM Research in San Jose, California, where hole-burning was being developed for an optical storage scheme called "frequency-domain optical storage".[31] Happily, it was important at this industrial research lab to not only develop new materials and methods for spectral hole-burning storage (see various chapters in ref. 28), but it was also a key goal to define and understand fundamental limits of signal-to-noise ratio, *etc.*[32] Due to unavoidable number fluctuations in the actual number of molecules present in any spectral interval, there should exist a "spectral roughness" on an inhomogeneous absorption profile scaling as the square root of the number of molecules in resonance (*i.e.*, the number of molecules per homogeneous width, N_H). We named this effect "statistical fine structure" (SFS), and the *relative* size of SFS should scale as $1/\sqrt{N_H}$, while the absolute root-mean-square (rms) size of the fine structure should grow as $\sqrt{N_H}$. Surprisingly, prior to the late 1980s, SFS had not been detected.

In 1987, SFS was observed for the first time by Moerner and Carter,[33,34] using a powerful zero-background optical absorption technique; laser frequency-modulation (FM) spectroscopy invented by Gary C. Bjorklund earlier.[35,36] FM spectroscopy (FMS) probes the sample with a phase-modulated laser beam which produces two out-of-phase sidebands on the laser carrier.

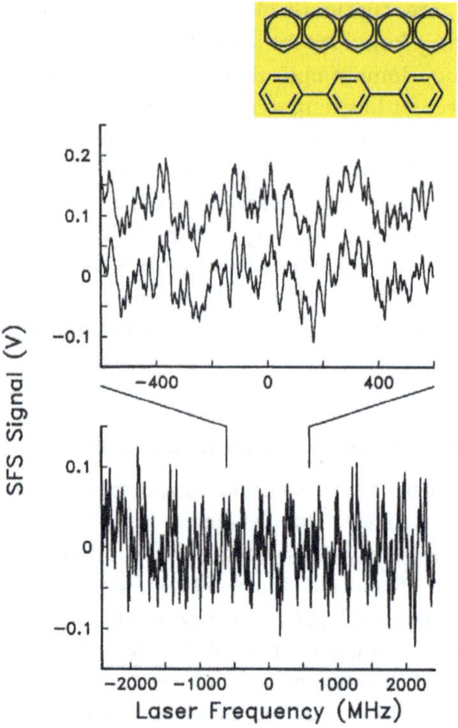

Fig. 2 Observation of Statistical Fine Structure (SFS) (right) for pentacene in *p*-terphenyl (schematic structure) using laser FM spectroscopy. Reprinted with permission from ref. 34. Copyright 1988, AIP Publishing LLC.

When a narrow spectral feature is present, the imbalance in the laser sidebands leads to amplitude modulation in the detected photocurrent at the modulation frequency. A key feature of the method is that it senses only the deviations of the absorption from the average value, so that detection of SFS could be easily accomplished, but only if a test sample was chosen with minimal spectral hole-burning so that the shape of the inhomogeneous profile could be measured without disturbance from the scanning. The choice of sample was critical: pentacene dopant molecules in a transparent *p*-terphenyl crystal (Fig. 2), in which spectral hole-burning was very weak. In the lower traces of Fig. 2, SFS is the repeatable spectral roughness seen in the lower trace, a small part of which is shown in the upper panel, with two repeated acquisitions – this is not time-dependent noise! SFS is clearly an unusual spectral feature, in that its (rms) size depends not upon the total number of resonant molecules, but rather upon the square root of the number.

With the detection of SFS in hand, it became possible to push on toward single-molecule detection. The first SMS experiments in 1989 by Moerner and Kador utilized either of two powerful double-modulation FMS absorption techniques; laser FMS with Stark secondary modulation (FM-Stark) or FMS with ultrasonic strain secondary modulation (FM-US).[1,37] Secondary modulation was required in order to remove the effects of residual amplitude modulation produced by the imperfect phase modulator.[38] The optical absorption experiments on pentacene in *p*-terphenyl indeed showed that this material has sufficiently inefficient spectral hole-burning to make it a useful model system for single-molecule studies. In 1990, Michel Orrit and Jacky Bernard produced a further critical advance: they demonstrated that sensing the optical absorption by detection of the emitted fluorescence produces superior signal-to-noise if the emission is collected efficiently and the scattering sources are minimized.[39] Due to its relative simplicity, subsequent experiments have almost exclusively used this method, which is also called "fluorescence excitation spectroscopy". In fluorescence excitation, a tunable narrowband single-frequency laser is scanned over the absorption profile of the single molecule, and the presence of absorption is detected by measuring the fluorescence emitted to long wavelengths, away from the laser wavelength itself. The method is often background-limited, and it requires the growth of ultrathin crystal-clear sublimed flakes to reduce the scattering signals that could arise from the *p*-terphenyl crystal, but it does not suffer from the difficult trade-off between SNR and optical broadening that occurred with FM spectroscopy.

With the ability to detect single molecules in crystals and polymers, in the early 1990s many investigators all over the world jumped into the field in order to take advantage of the extremely narrow optical absorption lines available at low temperatures and the removal of ensemble averaging, two of the greatest motivations for the study of single molecules.[13] Investigations were sometimes directed at specific observations of particular processes like the Stark effect,[40] two-level system dynamics,[41] or polarization effects,[42] to name a few. At other times experiments were performed simply to observe, because surprises would be expected when a new regime is first explored. The great body of work done is too large to review here, and the reader is referred to selected texts[13,28] and selected review articles[2–5,43–45] for more information. The Moerner group of postdoctoral researchers and collaborators completed a wide array of experiments, including measurements of the lifetime-limited width, temperature-dependent dephasing, and optical saturation effects,[46,47] photon antibunching correlations,[48] vibrational spectroscopy,[49–51] magnetic resonance of a single molecular spin,[52] and near-field spectroscopy.[53] Some experiments have particular relevance for super-resolution microscopy and will be discussed next.

1.2. Surprises from single-molecule spectroscopy: blinking and photocontrol

During the early SMS studies on pentacene in *p*-terphenyl, an unexpected phenomenon appeared: resonance frequency shifts of individual pentacene molecules in a crystal at 1.5 K,[46,54] mentioned briefly by Orrit.[39] We called this effect "spectral diffusion" due to its close relationship to similar spectral-shifting behavior long postulated for optical transitions of impurities in amorphous systems.[55] Here, spectral diffusion means changes in the center (resonance) frequency of a probe molecule due to configurational changes in the nearby host which affect the frequency of the electronic transition *via* guest–host coupling. Ambrose and Moerner observed this at the single-molecule level as shown in Fig. 3a, which displays a sequence of fluorescence excitation spectra of a single pentacene molecule in *p*-terphenyl taken as fast as allowed by the available SNR, every 3 s. The spectral shifting or hopping of this molecule from one resonance

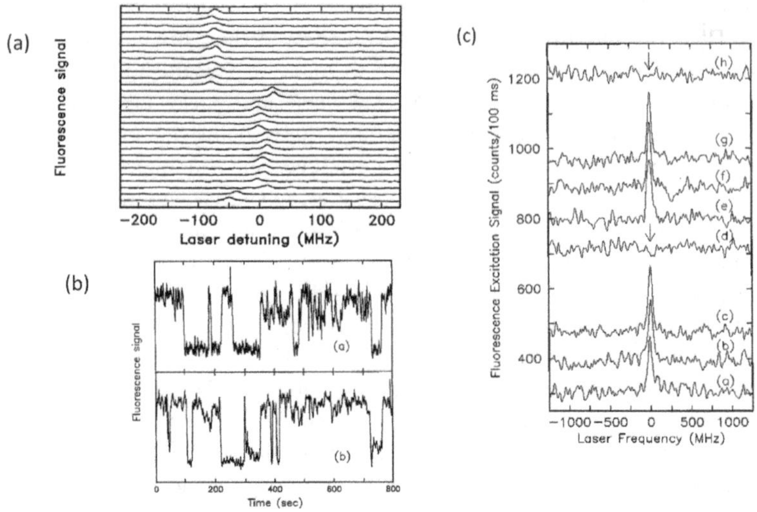

Fig. 3 (a and b) Spectral diffusion for pentacene in *p*-terphenyl and (c) light-induced spectral shifts for perylene in poly(ethylene). From ref. 46, 54 and 60 respectively, by permission. Panel b is reprinted from ref. 54, copyright (1991) by The American Physical Society.

frequency to another from scan to scan is clearly evident. Now if the laser frequency is held fixed near the molecular absorption, then the molecule appears to blink on and off as it jumps into and out of resonance (Fig. 3b, at two power levels). Due to the lack of power dependence on the rate, these spontaneous processes suggested that there are two-level systems available in the host matrix which can undergo thermally induced transitions even at these low temperatures. One possible source for the tunnelling states in this crystalline system could be discrete torsional librations of the central phenyl ring of the nearby *p*-terphenyl molecules about the molecular axis. The *p*-terphenyl molecules in a domain wall between two twins or near lattice defects may have lowered barriers to such central-ring tunnelling motions, a useful model proposed by Jim Skinner and co-workers.[56–58] These studies illustrate the power of SMS in probing details of the local nanoenvironment and the importance of theoretical insight to further understanding. Spectral shifts of single-molecule lineshapes were observed not only for certain crystalline hosts, but also for essentially all polymers studied, and even for polycrystalline Shpol'skii matrices.[59] This is a dramatic example of the heterogeneity that was uncovered by the single-molecule studies.

In addition, Basché and Moerner observed light-driven shifts in absorption frequency for perylene dopant molecules in poly(ethylene), in which the rate of the process clearly increased with increases in laser intensity (Fig. 3c).[60,61] This photoswitching effect may be called "single-molecule hole-burning" by analogy with the earlier hole-burning literature;[28] however, since only one molecule is in resonance with the laser, the absorption line simply disappears. Subtraces (a), (b), and (c) show three successive scans of one perylene molecule. After subtrace (c) the laser was tuned into resonance with the molecule, and at this higher irradiation fluence, eventually the fluorescence signal dropped, that is, the molecule apparently switched off. Subtrace (d) was then acquired, which showed that the resonance frequency of the molecule apparently shifted by more than ±1.25 GHz as a result of the light-induced change in the nearby environment. Surprisingly, this effect was reversible for a good fraction of the molecules: a further scan some minutes later (subtrace (e)) showed that the molecule returned to the original absorption frequency. After subtrace (g) the molecule was photoswitched again and the whole sequence could be repeated many times, enabling us to measure the Poisson kinetics of this process from the waiting time before a spectral shift.[61] Optical modification of single-molecule spectra not only provided unique insights into the photophysics and low-temperature dynamics of the amorphous state; this effect also presaged another area of current interest at room temperature: photoswitching of single molecules between emissive and dark forms, a powerful tool currently being used to achieve super-resolution imaging (*vide infra*).

2. Room temperature studies of single molecules

Soon after the first low-temperature experiments, studies of single molecules at room temperature began. A selection of crucial early milestones are described in Table 1.

Early steps arose out of the development of "fluorescence correlation spectroscopy" (FCS),[63,64] a large body of work which has been extensively reviewed in ref. 79–81. The method depends upon the fluctuations in emission from a tightly focused spot in solution arising from passage of molecules diffusing through a laser beam. Autocorrelation analysis of the fluorescence provides a window into a variety of dynamical effects on time scales less than the transit time on the order of 1–10 ms. The contrast ratio of the autocorrelation degrades at high concentrations but improves at low concentrations, and in 1990, correlation functions were recorded from concentrations so low that much less than one molecule was in the probe volume.[68] The passages of many single molecules must be averaged;

Table 1 Room temperature milestones of single-molecule detection and imaging

Solution: correlation functions	Fluorescence Correlation Spectroscopy (FCS)[62–67]	1972
	Autocorrelation detected from 1 fluorophore or less in the volume[68]	1990
Solution: single bursts	Multichromophore emitter bursts (phycoerythrin)[69]	1989
	Single bursts of fluorescence from 1 fluorophore[70,71]	1990
Solution and surface	Single antibody with multiple (∼80–100) labels[72]	1976
Near-field scanning optical microscopy	Imaging a single fluorophore[73–75]	1993
Confocal imaging	Single molecule in a polymer[76]	1996
Widefield, single fluorophore imaging	*In vitro*, single myosin on actin[77]	1995
	Cell membrane, single-lipid tracking with super-localization[78]	1996

it is impossible to study one and the same molecule for a long time with FCS – for a solution to this, see Section 5.

A crucial advance occurred in 1990, when the Keller lab at Los Alamos used a carefully designed hydrodynamic flow to reduce the volume producing interfering background signals and directly detected the individual fluorescence bursts as individual single rhodamine 6G molecules passed through the focus.[70] This was a key step in reducing backgrounds, but there is great value in being able to watch the same single molecule for extended periods, measuring signal strength, lifetime, polarization, fluctuations, and so on, all as a function of time and with the express purpose of directly detecting any heterogeneity from molecule to molecule. Hirschfeld reported detection of a single antibody with 80–100 fluorophores in a short report much earlier in 1976,[72] but severe photobleaching and the optical apparatus available at the time limited further work.

A further milestone was reached in 1993 when single-molecule imaging at room temperature was demonstrated using near-field scanning optical microscopy (NSOM).[73–75] It was subsequently demonstrated that with careful sample preparation and optimal detection, single molecules could be imaged with far field techniques such as confocal microscopy,[76] wide-field epifluorescence, and total internal reflection fluorescence microscopies.[77] Of particular importance for cell biology applications, in 1996 Schmidt *et al.* explored the diffusion of single labeled lipids on a cell surface.[78] The explosion of methods allowing single-molecule detection and imaging has led to a wealth of exciting research in this area, with advances far too numerous to review comprehensively,[7,82,83] and two sets of Nobel Conference Proceedings have appeared.[84,85]

During the years since the 1990s, a stunning array of single-molecule experiments have been performed in many environments to explore biophysics, polymer dynamics, cellular processes, and dynamics. Reviewing all these studies is

- Förster resonance energy transfer
- Polarization microscopy
- Photon antibunching
- Counting subunits (stoichiometry)
- Probing metallic nanoantenna structures for electromagnetic enhancements
- Vibrational mode spectra
- Stark effects
- Optically detected magnetic resonance
- Surface-enhanced Raman scattering
- Detection without fluorescence at room T
- Force spectroscopy (optical tweezers, magnetic tweezers, AFM...)
- Nanoscale materials studies: polymers, crystals, glasses
- Biophysics and Cell Biology – HUGE number of applications: enzymes, signaling, molecular motors, actin bands, DNA processing and dynamics, RNA, gene expression, chaperonins, viral entry, photosynthetic antenna proteins, ..., ...
- Tracking and trapping single molecules in solution and in cells
- New fluorophores and fluorophore dynamics
    - Blinking effects
    - Optical and chemical control of single molecule emission
    - Color centers in diamond
- Nanoparticles: metals, semiconductors, polymers

Fig. 4 A selection from the huge variety of single-molecule methods and applications over the decades – with apologies for possible omission.

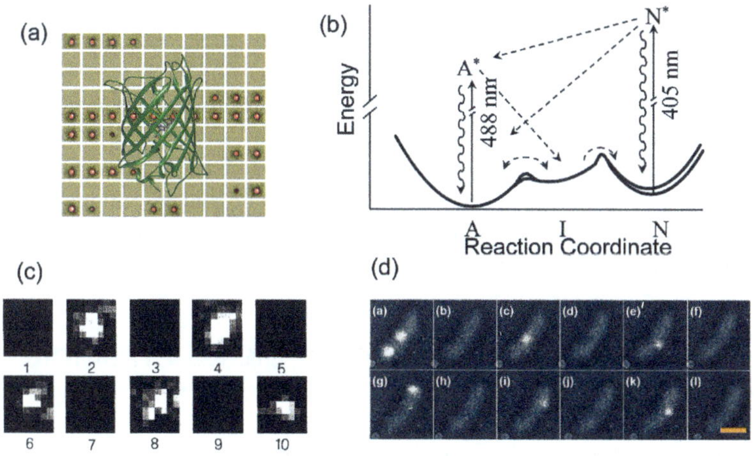

Fig. 5 (a) Structure of the green fluorescent protein[91] superimposed on a series of images of a single GFP trapped in a gel, 100 ms per image. (b) Schematic of the energy-level structure consistent with the blinking and photoswitching effects.[86] Reprinted with permission from ref. 86. (c) Images (600 nm × 600 nm) of 488 nm-pumped emission from the long-lived dark state (odd panels) with the photoreactivated state (even panels) produced by 405 nm irradiation for the same single molecule of the T203F yellow mutant of GFP. Similar results occurred for the T203Y mutant.[86] Reprinted with permission from ref. 86. (d) Reactivation of EYFP–MreB fusions in live *C. crescentus* cells. Fluorescence images show single EYFP–MreB molecules (white spots) overlaid on a reversed-contrast white-light image of the cell being examined. Only a few molecules are reactivated by 407 nm light in each image. Bar, 1 μm. For details, see ref. 92. Reprinted with permission from ref. 92.

impossible at this time, as is properly referencing all the work that has been done, so in Fig. 4, a partial list is provided. The textbooks and book chapters referenced at the start of this paper should be consulted for details. One particular imaging study in 1997 deserves a bit of additional mention: Dickson and Moerner used the water-filled pores of poly(acrylamide) gels to achieve optical imaging of single copies of a green fluorescent protein (GFP) mutant using total-internal-reflection fluorescence microscopy.[86] As was the case with the first low-temperature studies, an unexpected surprise occurred: the first example of blinking and optical switching of a single fluorescent protein. The experiments utilized two yellow-emitting GFP mutants (S65G/S72A/T203Y denoted "T203Y" and S65G/S72A/T203F denoted "T203F"), which differ only by the presence of a hydroxyl group near the chromophore, both of which are quite similar to the currently widely used enhanced yellow fluorescent protein EYFP (S65G/V68L/S72A/T203Y). Fig. 5a illustrates the blinking behavior: a single molecule emits frame after frame but then turns off, only to turn on again after a time in a stochastic fashion. This blinking behavior likely results from transformations between at least two states of the chromophore (A and I, Fig. 5b), only one of which (A) is capable of being excited by the 488 nm pumping laser and producing fluorescence. Additionally, a much longer-lived dark state N is also accessible. Thermally stable in the dark for many minutes, this long-lived dark state is not permanently photobleached, but can be excited at 405 nm to regenerate the original fluorescent state as shown in the sequence of images in Fig. 5c. This means that the protein can be used as an emitting label until it enters the long-lived dark state, and it can be photo-reactivated back to the emissive form with the 405 nm light, a reversal of the apparent photobleaching. This can occur many times for the same single molecule. Switching and blinking effects also occur for single fluorescent protein fusions expressed in bacteria as shown in Fig. 5d. These effects, and similar effects in a variety of other engineered photoswitchable fluorescent proteins (such as Kaede,[87] PA-GFP,[88] EosFP,[89] and DRONPA[90]), can be utilized for super-resolution imaging,[17,19] to be described next.

3. Super-resolution imaging based on single molecules

One continuing driving force in single-molecule fluorescence studies is the study of biomolecules, *in vitro* and *in vivo*.[82] As is well-known, biological fluorescence microscopy depends upon a variety of labeling techniques to light up different structures in cells, but the price often paid for using visible light is the relatively poor spatial resolution compared to X-ray crystallography or electron microscopy. When attempting to image nanoscale structures with visible light, a serious problem arises: fundamental diffraction effects first noted by Abbe in the late 1800s[93] limit the resolution to a dimension of roughly the optical wavelength λ divided by two times the numerical aperture (NA) of the imaging system, $\lambda/(2 \times \mathrm{NA})$. Here, resolution means the ability to distinguish two point objects which are closely positioned; when individual molecules are far apart as shown in Fig. 1a, there is no problem imaging them separately. Since the largest values of NA for state-of-the-art, highly corrected microscope objectives are in the range of about 1.3–1.6, the spatial resolution of optical imaging has been limited to about $\sim$200 nm for visible light of 500 nm wavelength.

In recent years, several optical methods have appeared which allow optical microscopy to surpass the diffraction limit to achieve "super-resolution", recognized by the Nobel Prize in Chemistry in 2014. All such methods rely on the use of fluorophores which can have two states: an emissive "on" state and a dark "off" state. One class of methods are based on optical patterning of a depletion beam to turn off molecules whose emission is not desired, as in Stimulated Emission Depletion (STED) microscopy of Stefan Hell and its variants.[94] A related method by Mats Gustafsson (Structured Illumination Microscopy or SIM) uses interfering beams to achieve resolution beyond the diffraction limit.[95] Neither of these methods requires single-molecule sensitivity, so subsequently in this paper we will focus on the methods which do require single-molecule imaging.

The basic approach is illustrated in a general way in Fig. 6, which illustrates the two critical ideas. In the first Key Idea, one must be able to super-localize single molecules in an image, which means finding the position of each molecule to a precision better than the diffraction limit. In the upper left of the figure, one of the original low temperature images of single molecules is shown from ref. 46. At room temperature, with a wide-field image of single molecules shown in the white-on-black panel, the diffraction-limited spots are evident. It is essential to spread out each detected spot on multiple pixels of the camera as shown in the next panel to the right. Then, illustrated by a 1D cross-section at the upper right,

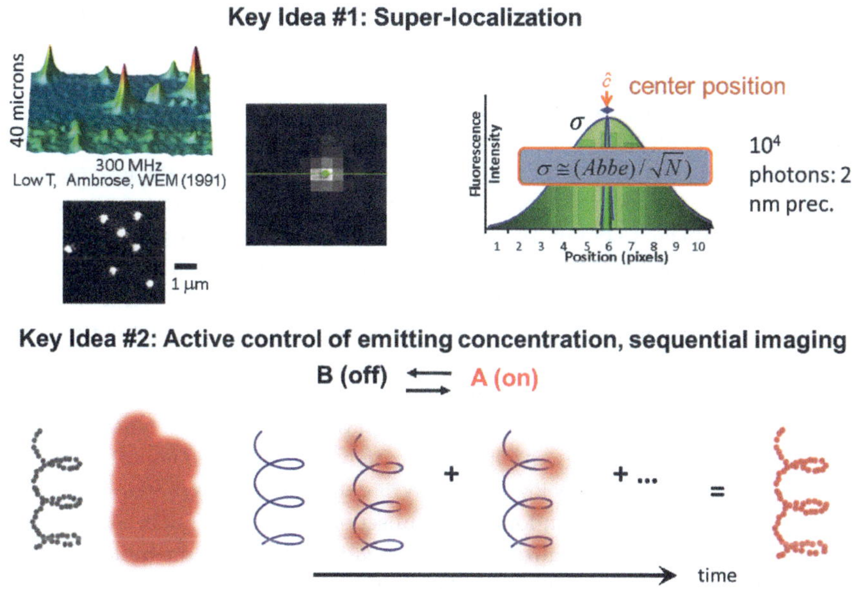

Fig. 6 Illustration of the key ideas underlying super-resolution optical microscopy based on single-molecule localization combined with active control of the emitting concentration and sequential imaging.

the various pixels detect different numbers of photons according to the shape of the point spread-function (PSF) of the microscope. Formally the PSF is an Airy function, but it may be approximated by a Gaussian function for simplicity, especially in the presence of background. The photon numbers detected in the various pixels provide samples of the function, which may be fit mathematically. While the width of this fit is still diffraction-limited with width $\hat{w}$, the estimate of the center position $\hat{c}$ from the fit follows a much narrower error distribution with standard deviation σ, which is generally called the "localization precision". The precision with which a single molecule can be located by digitizing the PSF depends fundamentally upon the Poisson process of photon detection, so the most important variable is the total number of photons N detected above background, with a weaker dependence on the size of the detector pixels and background.[96–98] The leading dependence of σ is just the Abbe diffraction limit divided by the square root of the number of photons detected in the image. This functional form makes sense, since each detected photon is an estimate of the molecular position, so for N measurements, the precision improves as expected. Super-localization means that if 100 photons are detected, then the precision can approach 20 nm, and so on. Clearly, then, emitters with the largest numbers of emitted photons before photobleaching are preferable. This is a commonly used method for determining the position of single objects from a diffraction limited image,[99] and early applications in biology include determinations of the centroid of large LDL particles[100] or fluorescent beads attached to single myosin motors.[101] Of course, this method also works with single molecules, for example, to track the motion of a single lipid molecule on a cell membrane[78] and many other examples.

But super-localization only works for a single molecule far away from its peers, so that the diffraction-limited images do not overlap.

True super-resolution requires the ability to localize single molecules when their PSFs do overlap, and this requires Key Idea #2. The early history of various fledgling approaches to solve this problem is described in Table 2 of ref. 102, so here, only the general form of the idea is described as shown in the lower part of Fig. 6. The key concept is to actively control the emitting concentration and then sequentially image to obtain the single-molecule positions. A structure has been labeled with many fluorescent labels as shown on the left, but when all are allowed to emit simultaneously, the blurry image results because the many PSFs overlap. The key idea is simply to not allow all the molecules to emit at the same time! Let us suppose that there is some mechanism which allows the emitters to be "On" part of the time and emitting photons (state A), and "Off", or dark, another part of the time (state B), even though the pumping light is still present. The experimenter uses this mechanism to actively control the concentration of emitting molecules to a very low level such that the PSFs do not overlap in any one image. Effectively then, a random small subset of the emitters are in the emissive state. Then using super-localization in one acquired image of the molecules, the positions of those are determined and recorded. Then these molecules are turned off or photobleached, and another subset is turned on, super-localized, *etc.* In the end, after a number of sequential imaging cycles, many locations on the structure have been sampled using the tiny single-molecule "beacons", and the underlying image is reconstructed in a pointillist fashion to show the detail previously hidden by the diffraction limit on the right. In this way, the locations of the single molecule labels are learned by sequential imaging, analogous to time-domain multiplexing.

This idea was first presented by Eric Betzig and his primary collaborator, Harald Hess, in April 2006 at the Frontiers in Live Cell Imaging Conference at the NIH main campus in Bethesda, Maryland. They used the PA-GFP photoactivatable fluorescent proteins of George Patterson and Jennifer Lippincott-Schwartz[88] and other photoswitchable fluorescent proteins as an active control mechanism, terming the method PALM (for PhotoActivated Localization Microscopy).[17] Light-induced photoactivation of GFP mutant fusions is used to randomly turn on only a few single molecules at a time in fixed cell sections or fixed cells. In their experiment, individual PSFs were recorded in detail to find their positions to ~20 nm, then the emitters were photobleached so that others could be turned on, and so on until many thousands of PSF positions were determined, and a super-resolution reconstruction was produced.

Very quickly after the NIH meeting, a flood of researchers demonstrated super-resolution imaging with single molecules and additional active control mechanisms and additional acronyms. The laboratory of Xiaowei Zhuang utilized controlled photoswitching of small molecule fluorophores for superresolution demonstrations[18] (STORM, for STochastic Optical Reconstruction Microscopy). Samuel Hess *et al.* published a nearly identical approach with an acronym termed F-PALM (Fluorescence PhotoActivation Localization Microscopy),[19] which also utilized a photoactivatable GFP with PSF localization to obtain superresolution. Also in 2006, an alternative approach was reported by the laboratory of Robin Hochstrasser based on accumulated binding of diffusible probes, which are quenched in solution yet de-quench in close proximity of the surface of the object to be imaged[103] (termed PAINT, for Points Accumulation for Imaging in Nanoscale Topography). The method relies upon the photophysical behavior of certain molecules that light up when bound or constrained, and they demonstrated the idea with the twisted intermolecular charge transfer (TICT) of Nile Red.[104] PAINT has advantages such that the object to be imaged need not be labeled and that many individual fluorophores are used for the imaging, thus relaxing the requirement on the total number of photons detected from each single molecule.

Other active control mechanisms quickly appeared such as dSTORM[105] (direct STORM), GSDIM (Ground-State Depletion with Intermittent Return),[106] blinking, as in BLINK-microscopy,[107] SPDM (Spectral Precision Determination Microscopy),[108] and the list goes on. In 2008, Julie Biteen in the Moerner laboratory used the EYFP photorecovery mechanism described above to perform super-resolution imaging in bacteria,[92] and an example image is presented below in Fig. 7. There are photochemical methods for single-fluorophore turn-on[109] and even enzymatic methods for turn-on which may be controlled by the concentration of substrate and the enzymatic rate.[110] The experimenter must actively choose some method to control the emitting concentration. Of course, the imaging is still time-sequential, thus this approach is best for quasi-static structures or fixed cells, but significant progress has been made in increasing the imaging speed.[111] Selected reviews may be consulted for additional detail of modern challenges and progress in super-resolution imaging,[112–120] and some current thrust areas are described in Section 4.

Fig. 7 shows a few examples of super-resolution images. In Fig. 7a, *Caulobacter crescentus* cell surfaces are imaged far beyond the diffraction limit using a small-

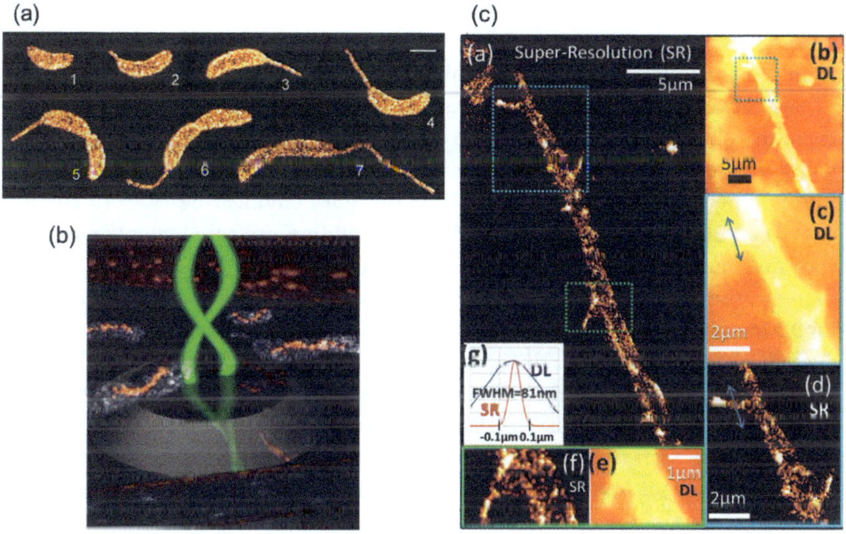

Fig. 7 Selected super-resolution images in cells. (a) The cell surface of *Caulobacter crescentus* bacteria. Scale bar = 1 μm. From ref. 121 by permission. (b) Two-color 3D imaging of the cell surface and a protein fiber-like structure in *Caulobacter*. Wikimedia file Bacteria-3D-Double-Helix.jpg Creative Commons Attribution-Share-Alike 4.0 International license. (c) Super-resolution and diffraction-limited (DL) images of the locations of voltage-gated sodium (NaV) channels in a differentiated PC12 cell. Reprinted from ref. 122 (A. E. Ondrus, H. D. Lee, S. Iwanaga, W. H. Parsons, B. M. Andresen, W. E. Moerner and J. Du Bois, *Chem. Biol.*, 2012, **19**, 902–912), with permission from Elsevier.

molecule photoactivatable rhodamine molecule, described in detail in ref. 121. The subdiffraction-sized cell stalks of various lengths are easily visualized. Fig. 7b also shows *Caulobacter* cells, but now the cell surface has been imaged using PAINT and the Nile Red molecule, shown in gray. At the same time, a fiber made of CreS proteins has been imaged by fusion to EYFP, with light-induced blinking of the fluorescent proteins used as the active control mechanism. These images are actually three-dimensional images acquired using the double-helix point-spread function microscope described in Section 4, and the "double-helix" is sketched in an artist's conception. The details of the imaging procedure are described in ref. 123. Finally in Fig. 7c, super-resolution and corresponding diffraction-limited images are shown for a differentiated PC12m neuronal model cell.[122] Here, a completely different active-control mechanism was used, where a fluorescent saxitoxin molecule was added to the buffer outside the cell. The diffusing neurotoxin binds to voltage-gated Na channels on the cell membrane, and the resulting flash of light from the bound fluorescent ligand in several imaging frames yields the images shown. From these few examples, it is hoped that the breadth and variety of super-resolution imaging as a powerful tool for cell biology can be appreciated.

4. Three-dimensional localization microscopy

Although super-resolution localization microscopy of thin ($\sim$two dimensional) samples has proven to be extremely informative, life happens in 3D: a cell (or even a nucleus) is typically several microns thick, which is larger than the <1 μm depth of focus of a high NA objective. As a result, with a standard microscope imaging a thick sample, a point source that is more than a few hundred nanometers away from the focal plane appears large, blurry ('defocused') and very dim.

Several schemes exist for adding axial (z) information to localization microscopy, thereby enabling 3D super-resolution imaging: multi-plane imaging,[124–126] using a tilted mirror to generate a side view of the sample,[127] single-photon interferometry,[128] and point-spread-function (PSF) engineering.[129–137] In the last case, the PSF of the microscope is altered by adding optical elements along the emission path, such that the shape of the acquired image of a single emitter (namely, the PSF) is indicative of the emitter's axial (z) position. PSF engineering for 3D imaging has the benefit of being a relatively simple method, requiring only a small number of optical elements that can be added to augment an existing standard microscope. Nevertheless the method exhibits extremely high precision, on the order of 10 nm in 3D for single-molecule localization for $\sim$1000 photons detected. It is scan-free, and has been recently extended to allow a very large, variable z-range, of up to 20 μm with a concomitant reduction in precision.[137]

4.1 PSF engineering

A general way to control the PSF of a microscope is to place an optical phase-changing element in a plane conjugate to the pupil plane, *i.e.* the Fourier plane of the microscope (Fig. 8).[129] Since there is a Fourier transform relation between the pupil plane and the image plane, multiplying the electromagnetic field (originating from the emitter) by the complex transmission function of the pupil-plane phase element is equivalent to a convolution operation in the image plane.[138] This corresponds to modifying the PSF.

Various PSF designs have been used in recent years to encode z information (Fig. 8). These include an elliptical (astigmatic) PSF,[130,131] the rotating double-helix,[132,139] the phase-ramp,[134] the accelerating-beam[135] and the saddle-point/tetrapods.[136,137] The different PSFs differ in the way they are produced, *e.g.* by using a cylindrical lens,[131] a phase-ramp,[134] or a specially designed phase mask in the Fourier plane.[132,135–137] However, their purpose is the same – to encode the axial position of the emitter by the shape of its image. Different PSFs have different characteristics, and their performance varies. For an example of the use of these ideas in cellular imaging, see Fig. 7b.

Two key parameters by which one can evaluate the applicability of a PSF to an experimental measurement are its precision and its applicable depth (z) range. The precision associated with a specific PSF determines how well (how *precisely*)

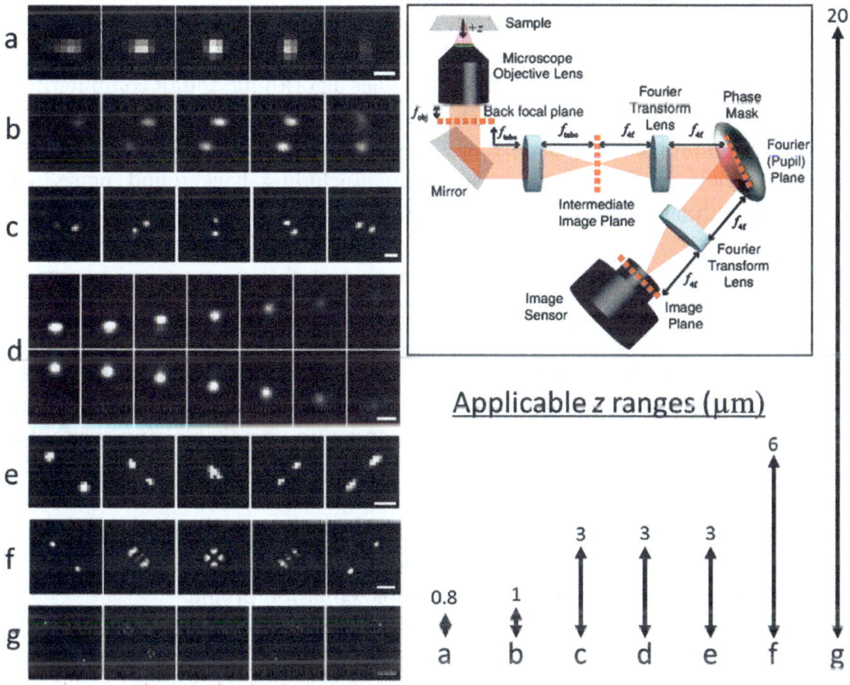

Fig. 8 Various PSFs for 3D localization microscopy shown as a function of z-position of the emitter (experimentally measured). (a) Astigmatic.[131] Scale bar = ~0.5 μm. From ref. 131. Reprinted with permission from AAAS. (b) Phase-ramp.[134] Reprinted with kind permission from Springer Science and Business Media (c) Double-helix.[132] Scale bar = 2 μm. Reprinted with permission from ref. 132. (d) Accelerating beam.[135] Scale bar = 1 μm. Reprinted by permission from Macmillan Publishers Ltd, copyright (2014) (e) saddle-point.[136] Scale bar = 1 μm. Reprinted from ref. 136, copyright (2014) by The American Physical Society. (f) and (g) Tetrapods,[137] scale bars = 2 μm and 5 μm, respectively. The arrows (right) represent the z-ranges over which the PSFs on the left were imaged, which correspond to their applicable depth ranges. Top right: experimental setup for pupil plane modulation-based PSF engineering.[136] Adapted with permission from ref. 136, copyright (2015) American Chemical Society.

one can localize, *i.e.* determine the position of the emitter, by measuring its image (the PSF). Emitter localization is a parameter estimation problem: given a noisy measurement (pixelated image of a PSF), and a known optical system and noise model, one must determine the 3D position of the emitter and possibly its brightness and background level (3–5 parameters).[136,140] As such, tools from information theory can be used in order to quantify the precision obtainable from a PSF.

A useful measure to evaluate the theoretical precision of a PSF, borrowed from information theory, is Fisher information.[141,142] Fisher information is a mathematical measure of the sensitivity of an observable quantity (the PSF) to changes in its underlying parameters (emitter position). From the Fisher information function, one derives the Cramér–Rao lower bound (CRLB), which is the optimal (best) x, y, z precision that can be attained with any unbiased estimator. It has been shown that the CRLB can be approached in practice.[98,143,144]

Fisher information analysis has been used to evaluate the effect of different imaging parameters on precision (*e.g.* pixel-size, multifocal *vs.* single plane imaging),[140,145] to show that the double-helix PSF is more precise and uniform in z than astigmatism and bi-plane imaging over a $\sim$3 μm z-range,[141] and to fine tune the double-helix design.[146] Importantly, treating Fisher information as a design variable (rather than an analysis metric) enables algorithmic and/or computational design of optimally precise PSFs. This has been recently achieved by solving an optimization problem of maximizing Fisher information for determination of x, y, and z subject to the system's constraints, including background noise. The resulting optimal PSFs are the saddle-point,[136] and more generally, tetrapod PSFs.[137]

4.2 Single-molecule orientation

One specific consideration for single-molecule microscopy stems from the fact that single molecules (with low rotational mobility) are not isotropic emitters, and the dipole nature of a single fluorophore can be non-negligible, and even informative.[147–152] A practical implication of the dipole nature of fluorophore emission is that the emitter's orientation may need to be taken into account when localizing single molecules from microscope images. Ignoring this effect can lead to localization errors: out-of-focus molecules can appear to be shifted in the lateral (x–y) plane, simply due to their non-isotropic emission profile.[133,153,154]

One approach to handling dipole-induced localization bias is to measure the orientation of a fluorophore, and later correct for the localization error by post processing. The orientation can be determined by a variety of methods, ranging from the use of polarization optics[148,149,155] to analysis of the defocused emitter's image.[150,156] An especially useful method to determine orientation is PSF engineering, which has the advantage of being simple, precise, and robust to minor defocus errors.[129] This can be achieved, for example, by using the double-helix PSF,[133] a quadrated-pupil,[157] or a bisected mask.[158]

Alternatively, it has been recently suggested that PSF polarization engineering can be used to avoid dipole-induced localization bias by removing the effect of dipole orientation from the measurement itself.[159] This approach utilizes an azimuthal polarizer in the Fourier plane to remove transmitted light arising from the z component of the dipole moment, resulting in unbiased lateral localization determination, and experimental demonstrations of this idea are to be expected in the near future.

5. Single-molecule photodynamics and transport properties in solution

Single molecules are frequently used as sensors to probe nanoscale dynamics.[160] To obtain the most information from each molecule, it is highly desirable to observe the molecule for as long as possible in a non-perturbative environment. Although free diffusion-based measurements are routinely implemented and provide invaluable insight into problems such as protein folding,[161] RNA polymerase–DNA interaction[162] and molecular aggregation,[163] such methods are limited by Brownian motion to a per-molecule-observation window of only ~1 ms, which is too short to observe dynamics of many biologically relevant processes.

The most commonly adopted remedy is to tether the molecule to a surface,[164] but the act of immobilization could perturb the system of interest.[165] It is also possible to encapsulate single molecules in nano-containers to spatially limit Brownian motion for extended measurements.[166–169] Another approach is to directly follow the motion of a single molecule over long distances as it moves around, using feedback control. Many clever instruments have been designed around this scheme (Table 2). Although feedback tracking is particularly powerful in a cellular context, revealing the heterogeneous environment around a single-molecule probe, the slow mechanical movement of a piezoelectric actuator currently limits the application of this approach to only slowly-diffusing molecules.

An alternative approach is to "trap" a single molecule in solution, so it does not escape the field-of-view of observation. Note that the forces provided by optical tweezers are generally too weak to capture single biomolecules at reasonable laser intensities, so other forces have to be considered. The anti-Brownian electrokinetic (ABEL) trap has been developed for this purpose.[182–184] Unlike an optical trap, where the trapping force is provided by a physical potential, the ABEL trap uses feedback control to counteract Brownian motion. Its basic operating principle is extremely simple (Fig. 9): a high-speed imaging module continuously monitors the position of a single molecule and depending on where the molecule is, appropriate feedback voltages are applied in a microfluidic environment to drive electrokinetic motion that restores the molecule's position. In the ABEL

Table 2 Feedback tracking of single molecules in solution and cells

Experiment	Ref.
Orbital tracking in cells	170 and 171
Confocal 3D tracking of single nanoparticles in solution	172 and 173
3D orbital tracking of nanometer-sized objects	174–176
Non-scanning, bi-plane tracking of single nanoparticles in solution	177
3D tracking in cells using a tetrahedral arrangement of detection/excitation spots	178–180
Cellular uptake of a single nanoparticle	181

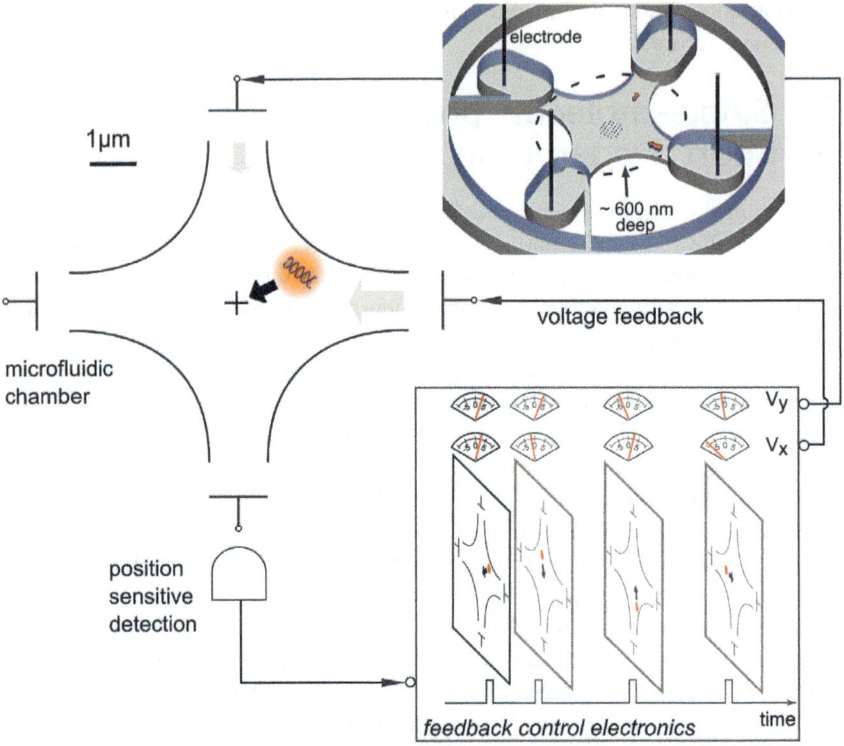

Fig. 9 The working principles of the ABEL trap: single-molecule motion in a microfluidic environment is tracked with high speed and Brownian motion is compensated for by applying appropriate electrokinetic forces (black arrow) in 2D in a feedback loop. The motion of the molecule along the third dimension (*z*) is confined within the ~600 nm depth of the microfluidic chip (inset).

trap, electrokinetic forces are utilized due to their favorable strength at the nanoscale and fast (~μs) response times.

Since its inception a decade ago,[182] the ABEL trap has evolved from a proof-of-principle device to a reliable and versatile platform for single-molecule biophysics research. First-generation traps used fast EMCCD cameras and centroid fitting algorithms to determine position.[185] Later it was realized that the camera-based approach was too slow for trapping biomolecules and laser-scanning methods were subsequently implemented[186] to detect position. In the most advanced version of the trap, a focused laser spot undergoes a 32-point "knight's tour" scanning pattern at the sample plane at a speed of 600 ns per point, and every photon-stamped beam position is taken as an estimation of molecule position.[187] This photon-by-photon position sensing scheme, together with a real-time Kalman filter[188,189] implemented on dedicated hardware to reduce measurement noise, greatly extended the trap's capability to capture molecules of the smallest scale. With the choice of a photostable label, single proteins and short strands of nucleic acids can be trapped for 10s of seconds, limited only by probe photobleaching. Even individual fluorophores, the smallest fluorescent objects, can be trapped for a couple of seconds.[190,191] More recently, other groups have

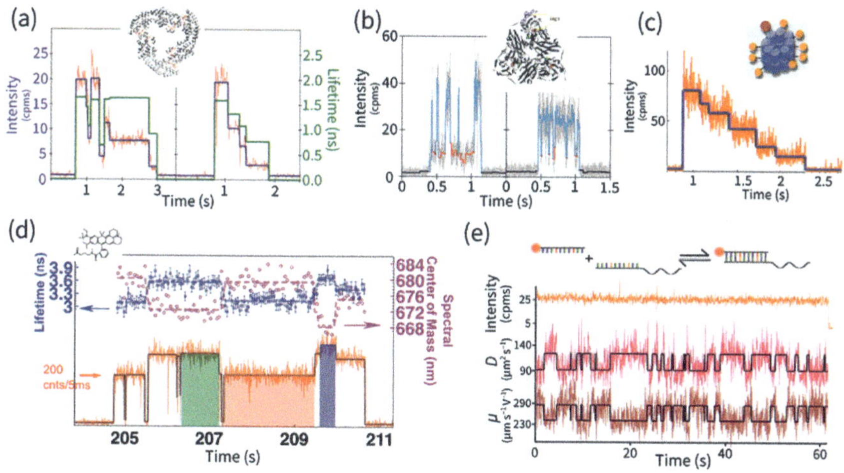

Fig. 10 Examples of single-molecule dynamics probed in the ABEL trap. (a) Photodynamics of allophycocyanin. Different intensity and lifetime states represent partially photobleached and quenched intermediates. Adapted with permission from ref. 194. (b) Redox cycling of nitrite reductase. Example molecules of nitrite reductase fluctuate between two digital intensity levels, corresponding to the oxidized (red) and reduced (blue) Cu redox states caused by single-electron transfer events. Adapted with permission from ref. 196. (c) Counting the number of ADP molecules on a multi-subunit chaperonin enzyme (TRiC). Digital steps represent photobleaching of the individual Cy3-ADP molecules bound to the protein. Adapted with permission from ref. 197. (d) Correlated intensity-lifetime-spectral photodynamics of a single Atto647N fluorophore in solution. Shaded areas indicate the three typical emissive states of this molecule. Adapted with permission from ref. 191, copyright (2012) American Chemical Society. (E) Visualization of single DNA binding and unbinding dynamics by ABEL-trap measurements of the diffusion coefficient (*D*) and electrokinetic mobility (μ). Reproduced with permission from ref. 200.

implemented ABEL traps with unique capabilities such as trapping in all three dimensions[192] and integrated RF circuits.[193]

By feedback suppression of Brownian motion, the ABEL trap extends the observation time of a single biomolecule in solution by more than three orders of magnitude, permitting direct observation of nanoscale dynamics without immobilization. We have used the device to study the behaviors of a variety of biomolecules. In one of the first applications, we examined the photodynamics of single allophycocyanin proteins[194] (Fig. 10a), an important antenna complex in cyanobacteria. Correlated dynamics of fluorescence intensity and excited-state lifetime report on photobleaching processes and subtle structure-function relations of pigment–protein interactions. A more elaborated detection scheme with additional fluorescence parameters has been subsequently used to study other antenna complexes in nature[195] and the photophysics of single fluorophores (Fig. 10d). As another example,[196] an engineered FRET redox sensor was used to directly visualize the transitions between discrete redox states of the enzyme nitrite reductase (Fig. 10b), providing new insights into the mechanism of catalysis. In yet another experiment,[197] the number of ADP molecules (each labeled with a fluorophore) bound to a chaperonin enzyme TRiC can be directly counted by enumerating the number of photobleaching steps for a single enzyme

Table 3 Application of the ABEL trap to study biomolecules, chemistry and physics

Experiment	Ref.
DNA conformational dynamics	201
Photodynamics, photoprotection, pigment organization principles of photosynthetic antenna complexes	194 and 202–205
Conformational dynamics of single G-protein coupled receptors in solution	206
ATP binding stoichiometry of single multi-subunit enzymes	197
Redox cycling of single nitrite reductase enzymes	196
Test of Landauer's principle	199
Direct visualization of biomolecular interactions	200
Electron spin resonance of nitrogen-vacancy centers in a single nanodiamond in solution	193

in the trap (Fig. 10c). Meanwhile, other groups have built their own ABEL traps to expand the application to broader areas, such as sensing the rotary motion of FoF1 ATPase,[198] probing electron spin resonance of NV centers,[193] and testing Landauer's principle in statistical physics,[199] as listed in Table 3.

Recently, we showed that it is possible to extract size and charge information from trapped single molecules.[200] This advance relies on statistical analysis of the molecule's residual motion in the trap to extract diffusive and electric-field-induced motion parameters. Knowledge of single-molecule transport properties opens up a variety of new sensing possibilities in fluorescence spectroscopy. For example, we demonstrated visualization of the size and charge fluctuations of a single DNA molecule as it binds and unbinds complementary strands in real time (Fig. 10e).

With the capability to observe single-molecule dynamics in solution for extended periods without perturbation and the unprecedented amount of both spectroscopic and transport information from each molecule, we envision the ABEL trap to become an essential tool in nanoscience and technology.

6. Conclusion and outlook

At the present time, single-molecule spectroscopy and microscopy continue to have a broad impact across the sciences. Fig. 11 illustrates some of the areas

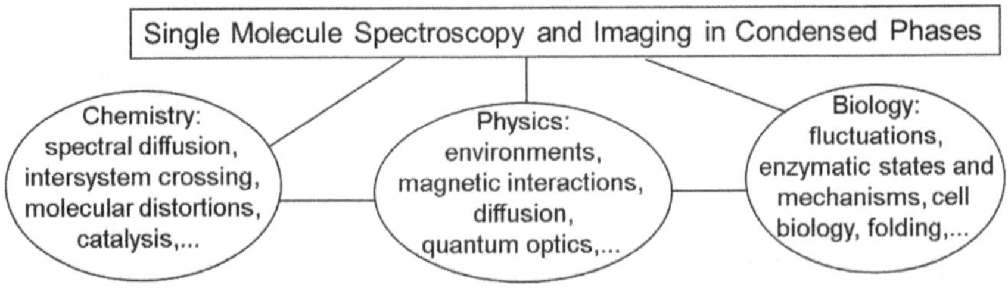

Fig. 11 Broad impact of single-molecule spectroscopy and imaging.

where these efforts have led to deeper insight. The advances in this area are the result of the dedicated work of many researchers all over the world, and the various papers presented during this Faraday Discussion provide a useful summary of many of the current areas of interest.

Acknowledgements

The authors thank Prof. Michel Orrit for organizing a stimulating and exciting Faraday Discussion, and we gratefully acknowledge the research and collaboration of all members of the Moerner Laboratory, past and present. This work was supported in part by the National Institutes of Health, National Institute of General Medical Sciences Grant No. R01GM085437 and R01GM086196, and by the Division of Chemical Sciences, Geosciences, and Biosciences, Office of Basic Energy Sciences of the U.S. Department of Energy through Grant DE-FG02-07ER15892.

References

1 W. E. Moerner and L. Kador, *Phys. Rev. Lett.*, 1989, **62**, 2535–2538.
2 W. E. Moerner and T. Basché, *Angew. Chem., Int. Ed.*, 1993, **105**, 537.
3 W. E. Moerner, *Science*, 1994, **265**, 46–53.
4 M. Orrit, J. Bernard, R. Brown and B. Lounis, *Prog. Opt.*, 1996, **35**, 61–144.
5 T. Plakhotnik, E. A. Donley and U. P. Wild, *Annu. Rev. Phys. Chem.*, 1997, **48**, 181–212.
6 S. Nie and R. N. Zare, *Annu. Rev. Biophys. Biomol. Struct.*, 1997, **26**, 567–596.
7 W. E. Moerner and M. Orrit, *Science*, 1999, **283**, 1670–1676.
8 W. E. Moerner, *J. Phys. Chem. B*, 2002, **106**, 910–927.
9 W. E. Moerner and D. P. Fromm, *Rev. Sci. Instrum.*, 2003, **74**, 3597–3619.
10 P. Tinnefeld and M. Sauer, *Angew. Chem., Int. Ed.*, 2005, **44**, 2642–2671, DOI: 10.1002/anie.200300647.
11 P. V. Cornish and T. Ha, *ACS Chem. Biol.*, 2007, **2**, 53.
12 W. E. Moerner, P. J. Schuck, D. P. Fromm, A. Kinkhabwala, S. J. Lord, S. Y. Nishimura, K. A. Willets, A. Sundaramurthy, G. S. Kino, M. He, Z. Lu and R. J. Twieg, in *Single Molecules and Nanotechnology*, ed. R. Rigler and H. Vogel, Springer-Verlag, Berlin, 2008, pp. 1–23.
13 Ed. T. Basché, W. E. Moerner, M. Orrit and U. P. Wild, *Single Molecule Optical Detection, Imaging, and Spectroscopy*, Verlag Chemie, Munich, 1997.
14 Ed. C. Zander, J. Enderlein and R. A. Keller, *Single-Molecule Detection in Solution. Methods and Applications*, Wiley-VCH, Berlin, 2002.
15 C. Gell, D. J. Brockwell and A. Smith, *Handbook of Single Molecule Fluorescence Spectroscopy*, Oxford Univ. Press, Oxford, 2006.
16 Ed. P. R. Selvin and T. Ha, *Single-Molecule Techniques. A Laboratory Manual*, Cold Spring Harbor Laboratory Press, Cold Spring Harbor, NY, 2008.
17 E. Betzig, G. H. Patterson, R. Sougrat, O. W. Lindwasser, S. Olenych, J. S. Bonifacino, M. W. Davidson, J. Lippincott-Schwartz and H. F. Hess, *Science*, 2006, **313**, 1642–1645.
18 M. J. Rust, M. Bates and X. Zhuang, *Nat. Methods*, 2006, **3**, 793–796.
19 S. T. Hess, T. P. K. Girirajan and M. D. Mason, *Biophys. J.*, 2006, **91**, 4258–4272.

20 A. M. Stoneham, *Rev. Mod. Phys.*, 1969, **41**, 82.
21 K. K. Rebane, *Impurity Spectra of Solids*, Plenum, New York, 1970.
22 K. K. Rebane, *Chem. Phys.*, 1994, **189**, 139–148.
23 R. I. Personov, E. I. Al'Shits and L. A. Bykovskaya, *Opt. Commun.*, 1972, **6**, 169–173, DOI: 10.1016/0030-4018(72)90220-9.
24 M. Orrit, J. Bernard and R. I. Personov, *J. Phys. Chem.*, 1993, **97**, 10256–10268.
25 H. de Vries and D. A. Wiersma, *J. Chem. Phys.*, 1979, **70**, 5807.
26 D. A. Wiersma, *Adv. Chem. Phys.*, 1981, **47**, 421.
27 M. Berg, C. A. Walsh, L. R. Narasimhan, K. A. Littau and M. D. Fayer, *J. Chem. Phys.*, 1988, **88**, 1564–1587, DOI: 10.1063/1.454136.
28 Ed. W. E. Moerner, *Topics in Current Physics 44; Persistent Spectral Hole-Burning: Science and Applications*, Springer, Berlin, 1988.
29 J. Friedrich and D. Haarer, *Angew. Chem., Int. Ed. Engl.*, 1984, **23**, 113–140, DOI: 10.1002/anie.198401131.
30 J. M. Hayes and G. J. Small, *Chem. Phys. Lett.*, 1978, **54**, 435–438, DOI: 10.1016/0009-2614(78)85255-5.
31 W. E. Moerner, *J. Mol. Electron.*, 1985, **1**, 55–71.
32 W. E. Moerner and M. D. Levenson, *J. Opt. Soc. Am. B*, 1985, **2**, 915–924, DOI: 10.1364/JOSAB.2.000915.
33 W. E. Moerner and T. P. Carter, *Phys. Rev. Lett.*, 1987, **59**, 2705.
34 T. P. Carter, M. Manavi and W. E. Moerner, *J. Chem. Phys.*, 1988, **89**, 1768.
35 G. C. Bjorklund, *Opt. Lett.*, 1980, **5**, 15.
36 T. P. Carter, D. E. Horne and W. E. Moerner, *Chem. Phys. Lett.*, 1988, **151**, 102.
37 L. Kador, D. E. Horne and W. E. Moerner, *J. Phys. Chem.*, 1990, **94**, 1237–1248.
38 E. A. Whittaker, M. Gehrtz and G. C. Bjorklund, *J. Opt. Soc. Am. B*, 1985, **2**, 1320.
39 M. Orrit and J. Bernard, *Phys. Rev. Lett.*, 1990, **65**, 2716–2719.
40 M. Orrit, J. Bernard, A. Zumbusch and R. I. Personov, *Chem. Phys. Lett.*, 1992, **196**, 595.
41 A. Zumbusch, L. Fleury, R. Brown, J. Bernard and M. Orrit, *Phys. Rev. Lett.*, 1993, **70**, 3584–3587.
42 F. Güttler, M. Croci, A. Renn and U. P. Wild, *Chem. Phys.*, 1996, **211**, 421–430.
43 W. E. Moerner, R. M. Dickson and D. J. Norris, *Adv. At. Mol. Opt. Phys.*, 1997, **38**, 193–236.
44 J. L. Skinner and W. E. Moerner, *J. Phys. Chem.*, 1996, **100**, 13251–13262.
45 W. E. Moerner, *Acc. Chem. Res.*, 1996, **29**, 563.
46 W. P. Ambrose and W. E. Moerner, *Nature*, 1991, **349**, 225–227.
47 W. P. Ambrose, T. Basché and W. E. Moerner, *J. Chem. Phys.*, 1991, **95**, 7150–7163.
48 T. Basché, W. E. Moerner, M. Orrit and H. Talon, *Phys. Rev. Lett.*, 1992, **69**, 1516–1519.
49 P. Tchénio, A. B. Myers and W. E. Moerner, *J. Phys. Chem.*, 1993, **97**, 2491.
50 P. Tchénio, A. B. Myers and W. E. Moerner, *Chem. Phys. Lett.*, 1993, **213**, 325.
51 A. B. Myers, P. Tchénio, M. Zgierski and W. E. Moerner, *J. Phys. Chem.*, 1994, **98**, 10377.
52 J. Köhler, J. A. J. M. Disselhorst, M. C. J. M. Donckers, E. J. J. Groenen, J. Schmidt and W. E. Moerner, *Nature*, 1993, **363**, 242–244.
53 W. E. Moerner, T. Plakhotnik, T. Irngartinger, U. P. Wild, D. Pohl and B. Hecht, *Phys. Rev. Lett.*, 1994, **73**, 2764.

54 W. E. Moerner and W. P. Ambrose, *Phys. Rev. Lett.*, 1991, **66**, 1376.
55 J. Friedrich and D. Haarer, in *Optical Spectroscopy of Glasses*, ed. I. Zschokke, Reidel, Dordrecht, 1986, p. 149.
56 P. D. Reilly and J. L. Skinner, *Phys. Rev. Lett.*, 1993, **71**, 4257–4260.
57 P. D. Reilly and J. L. Skinner, *J. Chem. Phys.*, 1995, **102**, 1540.
58 E. Geva and J. L. Skinner, *J. Phys. Chem. B*, 1997, **101**, 8920–8932.
59 T. Plakhotnik, W. E. Moerner, T. Irngartinger and U. P. Wild, *Chimia*, 1994, **48**, 31.
60 T. Basché and W. E. Moerner, *Nature*, 1992, **355**, 335–337.
61 T. Basché, W. P. Ambrose and W. E. Moerner, *J. Opt. Soc. Am. B*, 1992, **9**, 829.
62 D. Magde, E. Elson and W. W. Webb, *Phys. Rev. Lett.*, 1972, **29**, 705.
63 E. L. Elson and D. Magde, *Biopolymers*, 1974, **13**, 1–27.
64 D. L. Magde, E. L. Elson and W. W. Webb, *Biopolymers*, 1974, **13**, 29–61.
65 D. Magde, W. W. Webb and E. L. Elson, *Biopolymers*, 1978, **17**, 361–367.
66 M. Ehrenberg and R. Rigler, *Chem. Phys.*, 1974, **4**, 390–401, DOI: 10.1016/0301-0104(74)85005-6.
67 S. R. Aragón and R. Pecora, *J. Chem. Phys.*, 1976, **64**, 1791–1803, DOI: 10.1063/1.432357.
68 R. Rigler and J. Widengren, Proceedings of Bioscience, ed. B. Klinge and C. Owman, Lund, Lund University Press, 1990, **3**, pp. 180–183.
69 K. Peck, L. Stryer, A. N. Glazer and R. A. Mathies, *Proc. Natl. Acad. Sci. U. S. A.*, 1989, **86**, 4087–4091.
70 E. B. Shera, N. K. Seitzinger, L. M. Davis, R. A. Keller and S. A. Soper, *Chem. Phys. Lett.*, 1990, **174**, 553–557.
71 S. Nie, D. T. Chiu and R. N. Zare, *Science*, 1994, **266**, 1018–1021.
72 T. Hirschfeld, *Appl. Opt.*, 1976, **15**, 2965–2966, DOI: 10.1364/AO.15.002965.
73 E. Betzig and R. J. Chichester, *Science*, 1993, **262**, 1422–1425.
74 W. P. Ambrose, P. M. Goodwin, J. C. Martin and R. A. Keller, *Phys. Rev. Lett.*, 1994, **72**, 160–163.
75 X. S. Xie and R. C. Dunn, *Science*, 1994, **265**, 361–364.
76 J. J. Macklin, J. K. Trautman, T. D. Harris and L. E. Brus, *Science*, 1996, **272**, 255–258.
77 T. Funatsu, Y. Harada, M. Tokunaga, K. Saito and T. Yanagida, *Nature*, 1995, **374**, 555–559.
78 T. Schmidt, G. J. Schutz, W. Baumgartner, H. J. Gruber and H. Schindler, *Proc. Natl. Acad. Sci. U. S. A.*, 1996, **93**, 2926–2929.
79 Ed. R. Rigler and E. Elson, *Fluorescence Correlation Spectroscopy, Theory and Applications*, Springer, Berlin, 2001.
80 P. Schwille, *Cell Biochem. Biophys.*, 2001, **34**, 383–408.
81 S. T. Hess, S. Huang, A. A. Heikal and W. W. Webb, *Biochemistry*, 2002, **41**, 697–705.
82 S. Weiss, *Science*, 1999, **283**, 1676–1683.
83 W. E. Moerner, in *Single Molecule Spectroscopy in Chemistry, Physics and Biology: Nobel Symposium 138 Proceedings*, ed. A. Graslund, R. Rigler and J. Widengren, Springer-Verlag, Berlin, 2009, pp. 25–60.
84 Ed. R. Rigler, M. Orrit and T. Basché, *Single Molecule Spectroscopy, Nobel Conference Lectures*, Springer-Verlag, Berlin, 2001.

85 Ed. A. Gräslund, R. Rigler and J. Widengren, *Single Molecule Spectroscopy in Chemistry, Physics and Biology: Nobel Symposium 138 Proceedings*, Springer-Verlag, Berlin, 2009.
86 R. M. Dickson, A. B. Cubitt, R. Y. Tsien and W. E. Moerner, *Nature*, 1997, **388**, 355–358.
87 R. Ando, H. Hama, M. Yamamoto-Hino, H. Mizuno and A. Miyawaki, *Proc. Natl. Acad. Sci. U. S. A.*, 2002, **99**, 12651–12656.
88 G. H. Patterson and J. Lippincott-Schwartz, *Science*, 2002, **297**, 1873–1877.
89 J. Wiedenmann, S. Ivanchenko, F. Oswald, F. Schmitt, C. Röcker, A. Salih, K. Spindler and G. U. Nienhaus, *Proc. Natl. Acad. Sci. U. S. A.*, 2004, **101**, 15905–15910, DOI: 10.1073/pnas.0403668101.
90 R. Ando, H. Mizuno and A. Miyawaki, *Science*, 2004, **306**, 1370–1373.
91 M. Ormo, A. B. Cubitt, K. Kallio, L. A. Gross, R. Y. Tsien and S. J. Remington, *Science*, 1996, **273**, 1392–1395.
92 J. S. Biteen, M. A. Thompson, N. K. Tselentis, G. R. Bowman, L. Shapiro and W. E. Moerner, *Nat. Methods*, 2008, **5**, 947–949.
93 E. Abbe, *Arch. Mikrosk. Anat.*, 1873, **9**, 413–468.
94 S. W. Hell, *Science*, 2007, **316**, 1153–1158, DOI: 10.1126/science.1137395.
95 M. G. L. Gustafsson, *J. Microsc.*, 2000, **198**, 82–87.
96 R. E. Thompson, D. R. Larson and W. W. Webb, *Biophys. J.*, 2002, **82**, 2775–2783.
97 X. Michalet and S. Weiss, *Proc. Natl. Acad. Sci. U. S. A.*, 2006, **103**, 4797–4798.
98 K. I. Mortensen, L. S. Churchman, J. A. Spudich and H. Flyvbjerg, *Nat. Methods*, 2010, **7**, 377–381.
99 N. Bobroff, *Rev. Sci. Instrum.*, 1986, **57**, 1152–1157.
100 L. S. Barak and W. W. Webb, *J. Cell Biol.*, 1982, **95**, 846–852.
101 J. Gelles, B. J. Schnapp and M. P. Sheetz, *Nature*, 1988, **331**, 450–453.
102 W. E. Moerner, *Rev. Mod. Phys.*, 2015, **87**, 1183–1212, DOI: 10.1103/RevModPhys.87.1183.
103 A. Sharonov and R. M. Hochstrasser, *Proc. Natl. Acad. Sci. U. S. A.*, 2006, **103**, 18911–18916.
104 E. Mei, F. Gao and R. M. Hochstrasser, *Phys. Chem. Chem. Phys.*, 2006, **8**, 2077–2082.
105 M. Heilemann, S. van de Linde, M. Schüttpelz, R. Kasper, B. Seefeldt, A. Mukherjee, P. Tinnefeld and M. Sauer, *Angew. Chem., Int. Ed.*, 2008, **47**, 6172–6176, DOI: 10.1002/anie.200802376.
106 I. Testa, C. A. Wurm, R. Medda, E. Rothermel, C. von Middendorf, J. Foelling, S. Jakobs, A. Schoenle, S. W. Hell and C. Eggeling, *Biophys. J.*, 2010, **99**, 2686–2694.
107 T. Cordes, M. Strackharn, S. W. Stahl, W. Summerer, C. Steinhauer, C. Forthmann, E. M. Puchner, J. Vogelsang, H. E. Gaub and P. Tinnefeld, *Nano Lett.*, 2010, **10**, 645–651, DOI: 10.1021/nl903730r.
108 P. Lemmer, M. Gunkel, D. Baddeley, R. Kaufmann, A. Urich, Y. Weiland, J. Reymann, P. Mueller, M. Hausmann and C. Cremer, *Appl. Phys. B: Lasers Opt.*, 2008, **93**, 1–12, DOI: 10.1007/s00340-008-3152-x.
109 H. D. Lee, S. J. Lord, S. Iwanaga, K. Zhan, H. Xie, J. C. Williams, H. Wang, G. R. Bowman, E. D. Goley, L. Shapiro, R. J. Twieg, J. Rao and W. E. Moerner, *J. Am. Chem. Soc.*, 2010, **132**, 15099–15101, DOI: 10.1021/ja1044192.

110 M. K. Lee, J. Williams, R. J. Twieg, J. Rao and W. E. Moerner, *Chem. Sci.*, 2013, **4**, 220–225, DOI: 10.1039/c2sc21074f.
111 F. Huang, T. M. P. Hartwich, F. E. Rivera-Molina, Y. Lin, W. C. Duim, J. J. Long, P. D. Uchil, J. R. Myers, M. A. Baird, W. Mothes, M. W. Davidson, D. Toomre and J. Bewersdorf, *Nat. Methods*, 2013, **10**, 653–658, DOI: 10.1038/NMETH.2488.
112 B. Huang, H. Babcock and X. Zhuang, *Cell*, 2010, **143**, 1047–1058.
113 M. A. Thompson, M. D. Lew and W. E. Moerner, *Annu. Rev. Biophys.*, 2012, **41**, 321–342, DOI: 10.1146/annurev-biophys-050511-102250.
114 M. A. Thompson, J. S. Biteen, S. J. Lord, N. R. Conley and W. E. Moerner, *Methods Enzymol.*, 2010, **475**, 27–59.
115 J. S. Biteen and W. E. Moerner, *Cold Spring Harbor Perspect. Biol.*, 2010, **2**, a000448.
116 M. D. Lew, S. F. Lee, M. A. Thompson, H. D. Lee and W. E. Moerner, "Single-Molecule Photocontrol and Nanoscopy", in *Far-Field Optical Nanoscopy*, ed. P. Tinnefeld, C. Eggeling and S. W. Hell, Springer, Berlin, Heidelberg, 2012, pp. 1–24.
117 W. E. Moerner, *J. Microsc.*, 2012, **246**, 213–220.
118 A. Gahlmann, J. L. Ptacin, G. Grover, S. Quirin, A. R. S. von Diezmann, M. K. Lee, M. P. Backlund, L. Shapiro, R. Piestun and W. E. Moerner, *Nano Lett.*, 2013, **13**, 987–993, DOI: 10.1021/nl304071h.
119 S. J. Sahl and W. E. Moerner, *Curr. Opin. Struct. Biol.*, 2013, **23**, 778–787, DOI: 10.1016/j.sbi.2013.07.010.
120 A. Godin, B. Lounis and L. Cognet, *Biophys. J.*, 2014, **107**, 1777–1784, DOI: 10.1016/j.bpj.2014.08.028.
121 M. K. Lee, P. Rai, J. Williams, R. J. Twieg and W. E. Moerner, *J. Am. Chem. Soc.*, 2014, **136**, 14003–14006, DOI: 10.1021/ja508028h.
122 A. E. Ondrus, H. D. Lee, S. Iwanaga, W. H. Parsons, B. M. Andresen, W. E. Moerner and J. Du Bois, *Chem. Biol.*, 2012, **19**, 902–912, DOI: 10.1016/j.chembiol.2012.05.021.
123 M. D. Lew, S. F. Lee, J. L. Ptacin, M. K. Lee, R. J. Twieg, L. Shapiro and W. E. Moerner, *Proc. Natl. Acad. Sci. U. S. A.*, 2011, **108**, E1102–E1110, DOI: 10.1073/pnas.1114444108.
124 E. Toprak, H. Balci, B. H. Blehm and P. R. Selvin, *Nano Lett.*, 2007, **7**, 2043–2045.
125 M. F. Juette, T. J. Gould, M. D. Lessard, M. J. Mlodzianoski, B. S. Nagpure, B. T. Bennett, S. T. Hess and J. Bewersdorf, *Nat. Methods*, 2008, **5**, 527–529.
126 P. Prabhat, S. Ram, E. S. Ward and R. J. Ober, *IEEE Transactions on Nanobioscience*, 2004, **3**, 237–242.
127 J. Tang, J. Akerboom, A. Vaziri, L. L. Looger and C. V. Shank, *Proc. Natl. Acad. Sci. U. S. A.*, 2010, **107**, 10068–10073, DOI: 10.1073/pnas.1004899107.
128 G. Shtengel, J. A. Galbraith, C. G. Galbraith, J. Lippincott-Schwartz, J. M. Gillette, S. Manley, R. Sougrat, C. M. Waterman, P. Kanchanawong, M. W. Davidson, R. D. Fetter and H. F. Hess, *Proc. Natl. Acad. Sci. U. S. A.*, 2009, **106**, 3125–3130, DOI: 10.1073/pnas.0813131106.
129 A. S. Backer and W. E. Moerner, *J. Phys. Chem. B*, 2014, **118**, 8313–8329.
130 L. Holtzer, T. Meckel and T. Schmidt, *Appl. Phys. Lett.*, 2007, **90**, 053902.
131 B. Huang, W. Wang, M. Bates and X. Zhuang, *Science*, 2008, **319**, 810–813.

132 S. R. P. Pavani, M. A. Thompson, J. S. Biteen, S. J. Lord, N. Liu, R. J. Twieg, R. Piestun and W. E. Moerner, *Proc. Natl. Acad. Sci. U. S. A.*, 2009, **106**, 2995–2999, DOI: 10.1073/pnas.0900245106.

133 M. P. Backlund, M. D. Lew, A. S. Backer, S. J. Sahl, G. Grover, A. Agrawal, R. Piestun and W. E. Moerner, *Proc. Natl. Acad. Sci. U. S. A.*, 2012, **109**, 19087–19092, DOI: 10.1073/pnas.1216687109.

134 D. Baddeley, M. B. Cannell and C. Soeller, *Nano Res.*, 2011, **4**, 589–598.

135 S. Jia, J. C. Vaughan and X. Zhuang, *Nat. Photonics*, 2014, **8**, 302–306.

136 Y. Shechtman, S. J. Sahl, A. S. Backer and W. E. Moerner, *Phys. Rev. Lett.*, 2014, **113**, 133902.

137 Y. Shechtman, L. E. Weiss, A. S. Backer, S. J. Sahl and W. E. Moerner, *Nano Lett.*, 2015, **15**, 4194–4199, DOI: 10.1021/acs.nanolett.5b01396.

138 J. W. Goodman, *Introduction to Fourier Optics*, Roberts & Company Publishers, Greenwood Village, CO, 2005.

139 R. Piestun, Y. Y. Schechner and J. Shamir, *J. Opt. Soc. Am. A*, 2000, **17**, 294–303.

140 R. J. Ober, S. Ram and E. S. Ward, *Biophys. J.*, 2004, **86**, 1185–1200.

141 M. Badieirostami, M. D. Lew, M. A. Thompson and W. E. Moerner, *Appl. Phys. Lett.*, 2010, **97**, 161103, DOI: 10.1063/1.3499652.

142 S. M. Kay, *Fundamentals of Statistical Signal Processing: Estimation Theory*, Prentice-Hall PTR, Englewood Cliffs, NJ, 1993.

143 A. V. Abraham, S. Ram, J. Chao, E. S. Ward and R. J. Ober, *Opt. Express*, 2009, **17**, 23352–23373.

144 C. S. Smith, N. Joseph, B. Rieger and K. A. Lidke, *Nat. Methods*, 2010, **7**, 373–375.

145 S. Ram, P. Prabhat, J. Chao, E. S. Ward and R. J. Ober, *Biophys. J.*, 2008, **95**, 6025–6043.

146 G. Grover, K. deLuca, S. Quirin, J. deLuca and R. Piestun, *Opt. Express*, 2012, **20**, 26681.

147 T. Ha, T. Enderle, D. S. Chemla, P. R. Selvin and S. Weiss, *Phys. Rev. Lett.*, 1996, **77**, 3979–3982.

148 J. N. Forkey, M. E. Quinlan and Y. E. Goldman, *Biophys. J.*, 2005, **89**, 1261–1271, DOI: 10.1529/biophysj.104.053470.

149 S. A. Rosenberg, M. E. Quinlan, J. N. Forkey and Y. E. Goldman, *Acc. Chem. Res.*, 2005, **38**, 583–593.

150 E. Toprak, J. Enderlein, S. Syed, S. A. McKinney, R. G. Petschek, T. Ha, Y. E. Goldman and P. R. Selvin, *Proc. Natl. Acad. Sci. U. S. A.*, 2006, **103**, 6495–6499.

151 D. Axelrod, *Methods Cell Biol.*, 1989, **30**, 333–352.

152 E. J. Peterman, H. Sosa and W. E. Moerner, *Annu. Rev. Phys. Chem.*, 2004, **55**, 79–96.

153 J. Enderlein, E. Toprak and P. R. Selvin, *Opt. Express*, 2006, **14**, 8111–8120, DOI: 10.1364/OE.14.008111.

154 J. Engelhardt, J. Keller, P. Hoyer, M. Reuss, T. Staudt and S. W. Hell, *Nano Lett.*, 2011, **11**, 209–213, DOI: 10.1021/nl103472b.

155 J. T. Fourkas, *Opt. Lett.*, 2001, **26**, 211–213.

156 M. Böhmer and J. Enderlein, *J. Opt. Soc. Am. B*, 2003, **20**, 554–559.

157 A. S. Backer, M. P. Backlund, M. D. Lew and W. E. Moerner, *Opt. Lett.*, 2013, **38**, 1521–1523.

158 A. S. Backer, M. P. Backlund, A. R. Diezmann, S. J. Sahl and W. E. Moerner, *Appl. Phys. Lett.*, 2014, **104**, 193701–193705.
159 M. D. Lew and W. E. Moerner, *Nano Lett.*, 2014, **14**, 6407–6413.
160 C. Joo, H. Balci, Y. Ishitsuka, C. Buranachai and T. Ha, *Annu. Rev. Biochem.*, 2008, **77**, 51–76.
161 B. Schuler, E. A. Lipman and W. A. Eaton, *Nature*, 2002, **419**, 743–747.
162 A. N. Kapanidis, E. Margeat, S. O. Ho, E. Kortkhonjia, S. Weiss and R. H. Ebright, *Science*, 2006, **314**, 1144–1147, DOI: 10.1126/science.1131399.
163 A. Orte, N. R. Birkett, R. W. Clarke, G. L. Devlin, C. M. Dobson and D. Klenerman, *Proc. Natl. Acad. Sci. U. S. A.*, 2008, **105**, 14424–14429, DOI: 10.1073/pnas.0803086105.
164 S. D. Chandradoss, A. D. Haagsma, Y. K. Lee, J. Hwang, J. Nam and C. Joo, *J. Visualized Exp.*, 2014, **86**, e50549.
165 M. Friedel, A. Baumketner and J. E. Shea, *Proc. Natl. Acad. Sci. U. S. A.*, 2006, **103**, 8396–8401, DOI: 10.1073/pnas.0601210103 ER.
166 I. Cisse, B. Okumus, C. Joo and T. Ha, *Proc. Natl. Acad. Sci. U. S. A.*, 2007, **104**, 12646–12650.
167 M. J. Shon and A. E. Cohen, *J. Am. Chem. Soc.*, 2012, **134**, 14618–14623.
168 H. H. Gorris, D. M. Rissin and R. W. David, *Proc. Natl. Acad. Sci. U. S. A.*, 2007, **104**, 17680–17685.
169 Y. Rondelez, G. Tresset, T. Nakashima, Y. Kato-Yamada, H. Fujita, S. Takeuchi and H. Noji, *Nature*, 2005, **433**, 773–777.
170 V. Levi, Q. Q. Ruan and E. Gratton, *Biophys. J.*, 2005, **88**, 2919–2928.
171 Y. Katayama, O. Burkacky, M. Meyer, C. Bräuchle, E. Gratton and D. C. Lamb, *ChemPhysChem*, 2009, **10**, 2458–2464.
172 H. Cang, C. M. Wong, C. S. Xu, A. H. Rizvi and H. Yang, *Appl. Phys. Lett.*, 2006, **88**, 223901.
173 H. Cang, C. S. Xu, D. Montiel and H. Yang, *Opt. Lett.*, 2007, **32**, 2729–2731.
174 K. McHale, A. J. Berglund and H. Mabuchi, *Nano Lett.*, 2007, **7**, 3535–3539, DOI: 10.1021/nl0723376.
175 K. McHale and H. Mabuchi, *J. Am. Chem. Soc.*, 2009, **131**, 17901–17907.
176 D. Ernst, S. Hain and J. Koehler, *J. Opt. Soc. Am. A*, 2012, **29**, 1277–1287.
177 M. F. Juette and J. Bewersdorf, *Nano Lett.*, 2010, **10**, 4657–4663.
178 G. A. Lessard, P. M. Goodwin and J. H. Werner, *Appl. Phys. Lett.*, 2007, **91**, 224106–224113.
179 N. P. Wells, G. A. Lessard, P. M. Goodwin, M. E. Phipps, P. J. Cutler, D. S. Lidke, B. S. Wilson and J. H. Werner, *Nano Lett.*, 2010, **10**, 4732–4737.
180 E. P. Perillo, Y. Liu, K. Huynh, C. Liu, C. Chou, M. Hung, H. Yeh and A. K. Dunn, *Nat. Commun.*, 2015, **6**, 7874.
181 K. Welsher and H. Yang, *Nat. Nanotechnol.*, 2014, **9**, 198–203.
182 A. E. Cohen and W. E. Moerner, *Appl. Phys. Lett.*, 2005, **86**, 093109.
183 A. E. Cohen and W. E. Moerner, *Proc. Natl. Acad. Sci. U. S. A.*, 2006, **103**, 4362–4365.
184 Q. Wang, R. H. Goldsmith, Y. Jiang, S. D. Bockenhauer and W. E. Moerner, *Acc. Chem. Res.*, 2012, **45**, 1955–1964, DOI: 10.1021/ar200304t.
185 A. E. Cohen and W. E. Moerner, *Proc. SPIE*, 2005, **5699**, 296–305.
186 A. E. Cohen and W. E. Moerner, *Opt. Express*, 2008, **16**, 6941–6956.
187 Q. Wang and W. E. Moerner, *ACS Nano*, 2011, **5**, 5792–5799.
188 Q. Wang and W. E. Moerner, *Appl. Phys. B: Lasers Opt.*, 2010, **99**, 23–30.

189 A. P. Fields and A. E. Cohen, *Opt. Express*, 2012, **20**, 22589–22601.
190 A. P. Fields and A. E. Cohen, *Proc. Natl. Acad. Sci. U. S. A.*, 2011, **108**, 8937–8942.
191 Q. Wang and W. E. Moerner, *J. Phys. Chem. B*, 2012, **117**, 4641–4648.
192 J. K. King, B. K. Canfield and L. M. Davis, *Appl. Phys. Lett.*, 2013, **103**, 043102.
193 M. Kayci, H. Chang and A. Radenovic, *Nano Lett.*, 2014, **14**, 5335–5341.
194 R. H. Goldsmith and W. E. Moerner, *Nat. Chem.*, 2010, **2**, 179–186, DOI: 10.1038/NCHEM.545 ER.
195 G. S. Schlau-Cohen, S. Bockenhauer, Q. Wang and W. E. Moerner, *Chem. Sci.*, 2014, **5**, 2933–2939.
196 R. H. Goldsmith, L. C. Tabares, D. Kostrz, C. Dennison, T. J. Aartsma, G. W. Canters and W. E. Moerner, *Proc. Natl. Acad. Sci. U. S. A.*, 2011, **108**, 17269–17274.
197 Y. Jiang, N. R. Douglas, N. R. Conley, E. J. Miller, J. Frydman and W. E. Moerner, *Proc. Natl. Acad. Sci. U. S. A.*, 2011, **108**, 16962–16967.
198 B. Su, M. G. Düser, N. Zarrabi, T. Heitkamp, I. Ilka Starke and M. Börsch, *Proc. SPIE*, 2015, 9329.
199 Y. Jun, M. Gavrilov and J. Bechhoefer, *Phys. Rev. Lett.*, 2014, **113**, 190601.
200 Q. Wang and W. E. Moerner, *Nat. Methods*, 2014, **11**, 555–558.
201 A. E. Cohen and W. E. Moerner, *Proc. Natl. Acad. Sci. U. S. A.*, 2007, **104**, 12622–12627, DOI: 10.1073/pnas.0610396104 ER.
202 G. S. Schlau-Cohen, Q. Wang, J. Southall, R. J. Cogdell and W. E. Moerner, *Proc. Natl. Acad. Sci. U. S. A.*, 2013, **110**, 10899–10903.
203 G. Schlau-Cohen, H. Yang, T. P. J. Krueger, P. Xu, M. Gwizdala, R. van Grondelle, R. Croce and W. E. Moerner, *J. Phys. Chem. Lett.*, 2015, **6**, 860–867, DOI: 10.1021/acs.jpclett.5b00034.
204 Q. Wang and W. E. Moerner, *Proc. Natl. Acad. Sci. U. S. A.*, 2015, **112**, 13880–13885.
205 S. Bockenhauer and W. E. Moerner, *J. Phys. Chem. A*, 2013, **117**, 8399–8406.
206 S. Bockenhauer, A. Fuerstenberg, J. Y. Yao, B. K. Kobilka and W. E. Moerner, *J. Phys. Chem. B*, 2011, **115**, 13328–13338.

Faraday Discussions

PAPER

Polarization effects in lattice–STED microscopy

B. Yang,[ab] C.-Y. Fang,[c] H.-C. Chang,[c] F. Treussart,[d] J.-B. Trebbia[ab] and B. Lounis*[ab]

Received 22nd May 2015, Accepted 17th June 2015
DOI: 10.1039/c5fd00092k

Massive parallelization of STED-like nanoscopies is now achievable using well-designed optical lattices for state depletion. Yet, only the lattice intensity distribution was considered for the description of the super-resolved point spread function. This holds for fast-rotating fluorescent emitters. Here, we study the effects of electric field topography in lattice–STED microscopy. The dependence of the super-resolved point spread function on the number of dipoles and their orientation is investigated. Single fluorescent nano-diamonds are imaged using different optical lattice configurations and the measured resolutions are compared to theoretical simulations.

Introduction

Fluorescence images of emitters separated by less than half of the wavelength are not resolvable due to the diffraction limit. By surpassing this limit and providing resolution down to a few nanometers, super-resolution microscopy techniques[1,2] have revolutionized far-field optical microscopy in the last decade. These techniques exploit the photo-physical and photo-chemical properties of emitters to make them fluoresce sequentially and therefore to optically resolve them. The sequential selective imaging is achieved either in stochastic or deterministic ways. Stochastic super-resolution methods are based on the localization of isolated single emitters.[3–9] Deterministic methods,[10–15] such as stimulated emission depletion (STED)[10] are based on patterned laser illuminations, which restrict the fluorescence emission to pre-determined nanoscale regions. This sequential imaging modality implies that the resolution gain is at the expense of the imaging speed, which could be a severe limitation when studying dynamic structures in living cells, since many biological phenomena occur at timescales from milliseconds to seconds.[16,17]

[a]Univ Bordeaux, LP2N, F-33405 Talence, France. E-mail: brahim.lounis@u-bordeaux.fr
[b]Institut d'Optique & CNRS, LP2N, F-33405 Talence, France
[c]Institute of Atomic and Molecular Sciences, Academia Sinica, Taipei 106, Taiwan
[d]Laboratoire Aimé Cotton, UMR 9188 CNRS, ENS Cachan, Université Paris Sud, 91405 Orsay, France

Although localization-based methods are intrinsically parallelized, their imaging speeds are limited by the density of the fluorescent emitters per raw image and by the stochastic process of on/off switching of emitters. The acquisition time of a super-resolved image usually lies in the range of 1–100 s.[2,18–20] Standard STED microscopy is a point-scanning technique[21] which requires only a limited number of fluorescence photons integrated over dwell times down to several tens of μs,[22] to obtain the local density of emitters. The imaging speed of large fields of view is however strongly limited since dense pixelization is required.

An important step towards massive parallelization of STED-like nanoscopies has been achieved using well-designed optical lattices for state depletion.[23,24] These lattices contain an array of unit cells, each of which has a point-like intensity minimum surrounded by a nearly isotropic intensity distribution. Combined with a homogeneous wide-field illumination for fluorescence excitation, each unit cell plays the same role as a doughnut beam of confocal STED, which reduces the fluorescence emission volume to a sub-diffraction scale by stimulated emission depletion. To obtain a super-resolution image, one only needs to scan the sample over a unit cell of the optical lattice rather than the entire field of view. With this lattice–STED approach, imaging of a 3 μm × 3 μm field of view with a resolution of 70 nm and a speed of 80 ms per super-resolution image has been achieved.[23] The low pulse energy of the high repetition rate laser used for the depletion and the readout time of the camera set the limit of the field of view and the imaging speed, respectively. Using a laser with lower repetition rate and higher pulse energy, larger fields of view can be imaged.[25]

In the aforementioned experiments, the polarization effects in the depletion lattice have been neglected. Only the lattice intensity distribution was considered for the description of the super-resolved point spread function (s-PSF). This description holds for fast-rotating fluorescent emitters with rotation characteristic time much shorter than the depletion pulse for which the polarization effects are averaged. In this paper, we study the effects of electric field topography in lattice–STED microscopy. The dependence of the s-PSF on the number of dipoles and their orientations is investigated. Single nano-diamonds containing a few nitrogen-vacancy (NV) fluorescent defects are imaged using different optical lattice configurations and the measured s-PSF are compared to theoretical simulations.

Optical lattices

Optical lattices are periodic patterns generated by multi-beam interferences. They were first introduced in the trapping of cold atoms and their manipulation.[26–28] Their applications have been extended to other fields such as photonic crystal fabrication[29] and microfluidics.[30] The electric field topology (*i.e.* spatial distributions of its amplitude and its polarization) of an optical lattice depends on the number of interfering beams, their directions, polarizations and relative phases.[27,31]

The total electric field $\vec{E}(\vec{r}, t)$ created by the interference of a number p of plane waves having the same amplitude ε is written as:

$$\vec{E}\left(\vec{r},\ t\right)=\sum_{j=1}^{p}\varepsilon \mathrm{e}^{\mathrm{i}\left(\vec{k}_j\cdot\vec{r}-\omega t+\phi_j\right)}\vec{e}_j=\varepsilon \mathrm{e}^{\mathrm{i}\left(\vec{k}_p\cdot\vec{r}-\omega t+\phi_p\right)}\left(\vec{e}_p+\sum_{j=1}^{p-1}\mathrm{e}^{\mathrm{i}\left(\left(\vec{k}_j-\vec{k}_p\right)\cdot\vec{r}+\phi_j-\phi_p\right)}\vec{e}_j\right) \tag{1}$$

where $\vec{k}_j$, ω, ϕ_j, $\vec{e}_j$ are, respectively, the wave vector, the angular frequency, the phase and the polarization directions of each plane wave. The vector position $\vec{r}$ has q spatial coordinates corresponding to the space dimension q. Since the arbitrary choice of the time origin allows cancelling the phases ϕ_p, the number of independent relative phases in eqn (1) is $(p-1)$. Furthermore, the arbitrary choice of the space origin ($\vec{r}=\vec{r}'+\vec{r}_0$) allows rewriting the spatial dependence of the electric field as:

$$\vec{E}\left(\vec{r}\right)=\varepsilon\left(\vec{e}_p+\sum_{j=1}^{p-1}\mathrm{e}^{\mathrm{i}\left(\vec{k}_j-\vec{k}_p\right)\cdot\vec{r}'}\mathrm{e}^{\mathrm{i}\left(\left(\vec{k}_j-\vec{k}_p\right)\cdot\vec{r}_0+\phi_j-\phi_p\right)}\vec{e}_j\right) \tag{2}$$

This equation clearly shows that any relative phase variation between the beams results in a global spatial translation of the optical lattice ($\vec{r}_0$, without any change in the field topography) if the system of $(p-1)$ linear equations of q variables $(\phi_j-\phi_p+(\vec{k}_j-\vec{k}_p)\cdot\vec{r}_0=0,\ 1\leq j\leq p-1)$ has a solution.[31] For independent wave vectors, this requirement is fulfilled if the number of interfering beams p is equal to $q+1$.

In the case of 2D imaging ($q=2$), the interference of three beams will lead to an optical lattice which keeps both the topographies of the electric field and of the intensity independent of the beams phases. A four-beam optical lattice has an electric field topography that depends on the beams relative phases. However, in certain configurations, the intensity topography is insensitive to the beams phases and leads to better STED parallelization.

Three beam configuration

To build a three-beam optical lattice in the focal plane of a microscope objective suitable for STED parallelization, we consider three beams propagating in the direction of the optical axis of the objective ($\vec{e}_z$) with the same linear polarization ($\vec{e}_x$), as shown in Fig. 1a. The beams intersect the objective at the vertices of a centered equilateral triangle. After passing through the objective they are deviated by an angle θ toward the focal region where they interfere and form an optical lattice. The incident beams are slightly focused Gaussian beams with beam waists located at the back focal plane of the objective. The transmitted beams are collimated and deviated by the objective and have their polarizations rotated. The interference of the three beams leads to a lattice with hexagonal structures with a periodicity of $\frac{2\lambda}{3n\sin(\theta)}$, n being the refractive index of the medium and $\lambda=760$ nm the depletion wavelength. Optical lattices useful for STED nanoscopy (local zero-intensity minima, each of which is surrounded by a nearly isotropic intensity distribution) can be obtained when the three incident beams have their polarizations parallel to one of the three sides of the incidence triangle shown in Fig. 1a. Fig. 1b shows the intensity distribution of the optical lattices obtained for an incidence angle $\theta=72°$. At this angle, the hexagonal pattern displays a period of 355 nm and an optimal intensity contrast better than 99.9%. The intensity

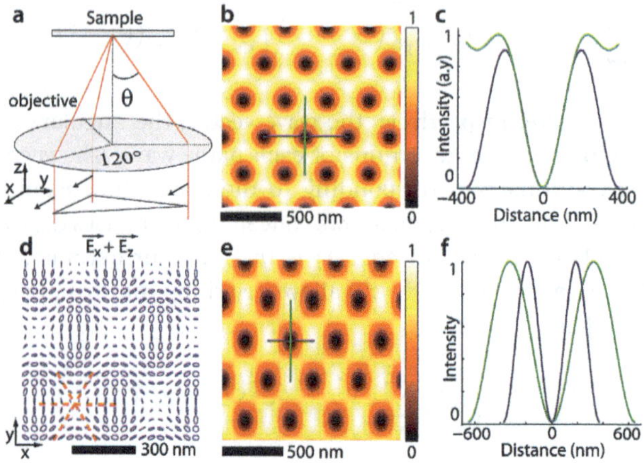

Fig. 1 (a) Three beam configuration. (b) Resulting intensity topography of the interference pattern for $\theta = 72°$. (c) Cross sectional intensity plots along the blue and green lines of b). (d) Electric field topography for $\theta = 72°$. Only the electric field polarization in the plane (x, z) is plotted. Straight lines, ellipses and circles indicate respectively linear, elliptical and circular polarizations. Red dashed lines correspond to the directions where the electric field is linear. (e–f) Resulting intensity topography and the associated cross sectional intensity plots for $\theta = 60°$.

profiles along two directions in the x–y plane (blue and green lines in Fig. 1c) are identical up to 150 nm away from the zero intensity position. Such depletion intensity profiles should lead to an almost isotropic s-PSF when imaging fast rotating emitters.

In the imaging plane (x, y), the amplitudes of the electric field projections along the x and z directions are significantly larger than that along the y direction since the incident beams have the same polarization $(\vec{e}_x)$. The polarization topography of the electric field projection in the (x, z) plane is shown in Fig. 1d. Around the intensity minima, the polarization is elliptical except along the three directions shown by the red dashed lines in Fig. 1d where it is linear. As discussed above, the field topography is insensitive to any phase variations of the incident beams.

In practice, the back aperture of the objective and the size of the beams at this aperture limit the maximum achievable incidence angle. In the experiment, we use an oil immersion objective (N.A. = 1.49, back aperture diameter ~10 mm) which allows us to achieve an angle $\theta = 60°$ with beam diameters of 1 mm. In this case, the optical pattern, shown in Fig. 1e, still displays "zero-intensity" regions with a contrast close to unity (98%). However, as shown in the intensity profiles of Fig. 1f, the intensity gradient around the locations of the zeros is steeper in the horizontal direction.

Four beam configuration

A key point for faster lattice-STED imaging is to generate lattices with the smallest possible unit cell. For this purpose, we use a four-beam interference configuration where a smaller unit cell can be obtained with the same angle θ. If one uses two orthogonally polarized pairs of beams (Fig. 2a), each pair will generate a one-dimensional interference pattern independently from the other pair of beams. The

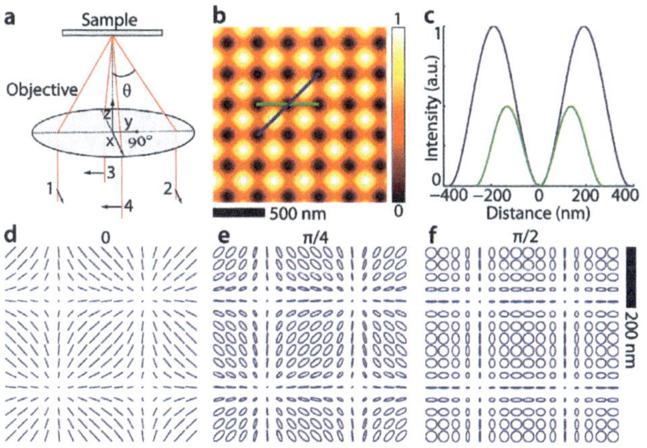

Fig. 2 (a) Geometrical four beam configuration. (b) Resulting intensity topography of the interference pattern for $\theta = 60°$. (c) Cross sectional intensity plots along the blue and green lines of b). (d–f) Plots of the electric field polarization in the plane (x, y) for $\Delta\phi = 0, \frac{\pi}{4}, \frac{\pi}{2}$. Straight lines, ellipses and circles indicate, respectively, linear, elliptical and circular polarizations.

intensity distribution of the final optical lattice is just the sum of the two one-dimensional interference patterns and a square optical lattice with a periodicity $\frac{\lambda}{2n\sin(\theta)} = 290$ nm can be obtained for $\theta = 60°$ and $\lambda = 760$ nm, as shown in Fig. 2b.

Fig. 2c shows the intensity plots along the x-axis (green) and the diagonal direction (blue). The maximal intensity along the diagonal direction is twice that along the x-axis. Nevertheless the intensity gradients around the intensity "zeros" are very similar. Therefore the resolution should be almost isotropic when imaging fast-rotating molecules. The intensity topography of the lattice (Fig. 2b) is insensitive to the relative phase between the beams. The electric field in the imaging plane is:

$$\vec{E}(x,\ y) = 2\varepsilon\left[\sin\left(\frac{2\pi n}{\lambda}(x - x_0)\sin(\theta)\right)\vec{e}_x + \sin\left(\frac{2\pi n}{\lambda}(y - y_0)\sin(\theta)\right)e^{i\Delta\phi}\vec{e}_y\right] \quad (3)$$

where x_0 and y_0 are the 2D spatial coordinates of $\vec{r}_0$ and $\Delta\phi$ is the relative phase between the two orthogonal standing waves. The electric field topography only depends on $\Delta\phi$ and it is depicted in Fig. 2d–f. For $\Delta\phi = 0$, the polarization of the field is linear everywhere in the imaging plane. If $\Delta\phi \neq 0$, the polarization around the zero intensity regions is linear only along the x and y axis, while it is elliptical anywhere else. It becomes circular along the directions $\pm 45°$ when $\Delta\phi = \pi/2$.

Influence of the polarization pattern on lattice–STED s-PSF

Fluorescent emitters commonly used in STED have a 1D transition dipole moment. Their excited state depletion efficiency depends on the squared projection of the electric field on the direction of their dipole moment. Rotating molecules have a time-dependent orientation of their dipole $\vec{d} = d\vec{e}_d(t)$ and a characteristic rotational time T_{rot}. The intensity relevant in the interaction of this dipole with the field $\vec{E}$ is the effective local intensity $I_{\text{eff}} = \langle|\vec{e}_{\text{d}}(t)\cdot\vec{E}(\vec{r})|^2\rangle$, averaged over the integration time.

To model the lattice–STED s-PSF, we construct a fluorescence emission image obtained when an ensemble of emitters localized in a nanoscale region (such as an ensemble of molecules in a nanosphere or a bunch of NV centers in a nanodiamond) are scanned in the focal plane of the objective in presence of a homogeneous wide-field excitation illumination and a depletion optical lattice. In this case, the fluorescence signal can usually be well-approximated by the expression $S_{\text{fluo}} \sim \text{e}^{-\left(\frac{\ln(2)I_{\text{eff}}}{I_{\text{sat}}}\right)}$, where I_{sat} is the saturation intensity which is defined as the intensity at which the probability of fluorescence intensity is reduced by half.[21,32] The image provides information about the depletion pattern (structure and periodicity). The spatial regions where the fluorescence is not depleted give direct access to the expected s-PSF.

Three different rotational mobility regimes should be considered: the first regime, introduced above, is the fast-rotation regime where T_{rot} is much shorter than the duration of the depletion pulse. The second one corresponds to the intermediate situation where the molecules display a slow rotational diffusion, characterized by a T_{rot} much longer than the pulse duration and much shorter than the integration time. The last one is the regime of "fixed" molecules, obtained when T_{rot} is much longer than the integration time.

In the fast-rotation regime, the transition dipole moment of the emitters can be considered as isotropic and the effective intensity profile is simply the intensity distribution of the depletion lattice. Time averaging over all possible orientations of the dipole during each depletion pulse results in a decrease of the intensity by an overall factor of 3. Therefore, the fluorescence signal can be depleted to zero in all directions around a zero intensity region. The images in Fig. 3a present the results of such simulations. The emitting dipoles are scanned over a region slightly larger than the unit cell of the optical lattice described above ($\theta = 60°$, intensity maxima of $6I_{\text{sat}}$). In the case of the three-beam hexagonal lattice (left image) the s-PSF is elongated reflecting directly the anisotropy of the intensity distribution of Fig. 1e and f. For the four-beam square lattice, the s-PSF is isotropic and does not depend on the relative phase difference $\Delta\phi$ between the two orthogonal standing waves (as shown in the images constructed for $\Delta\phi = 0, \frac{\pi}{4}$ and $\frac{\pi}{2}$).

In the two other regimes, the polarization pattern of the electric field has to be taken into account since dipoles perpendicular to the field polarization are not depleted, even at high STED intensities. Images of a small number of emitters exhibiting slow rotational diffusion or of a bead containing a large number of fixed emitters can be modeled in the same way. The fluorescence signal is calculated by averaging the fluorescence contributions of a large number of dipoles randomly oriented in a three-dimensional space:

$$S_{\text{fluo}} \sim \sum_{i=1}^{N} \text{e}^{-\left(\frac{\ln(2)I_{\text{eff}}^{i}}{I_{\text{sat}}}\right)} \tag{4}$$

where I_{eff}^{i} is the effective field intensity which depletes the excited dipole $\vec{d}_i$. The results of a calculation performed with $N = 500$ dipoles are shown in Fig. 3b. The

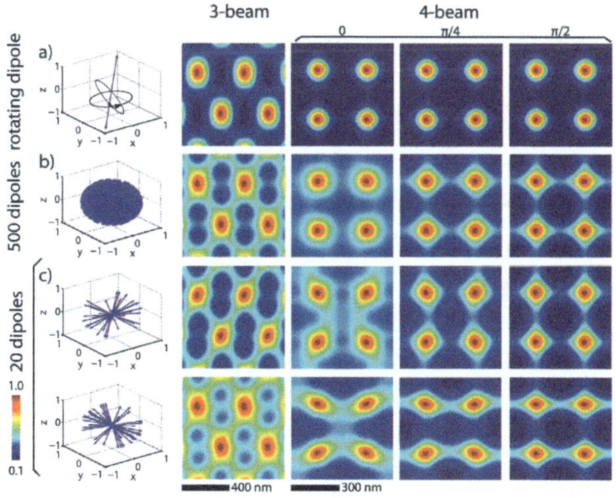

Fig. 3 Simulated fluorescence depletion patterns obtained for different dipole orientations in the three and the four beams optical lattices and for different $\Delta\phi$ $\left(0, \frac{\pi}{4} \text{and} \frac{\pi}{2}\right)$ in the cases of: a fast 3D rotating dipole (a), a large number of randomly oriented dipole directions (500) (b), and two different sets of 20 randomly oriented dipoles (3D) (c) at a depletion intensity of $I = 6I_{sat}$.

fluorescence images display strong correlations with the polarization patterns of Fig. 1d and 2d–f. For example, in the three beam configuration, the fluorescence image shows that from each zero-intensity region (corresponding to a fluorescence maximum in the image) there are three directions where the fluorescence is not efficiently depleted. They correspond to that presented in Fig. 1d giving the directions along which the lattice field is linearly polarized. An interesting situation is obtained in the four-beam configuration when $\Delta\phi = 0$. Since the electric field polarization is linear everywhere, excited emitters are not depleted where their dipole moments are perpendicular to the local field during the scan. Therefore the fluorescence signal cannot be decreased to zero even at higher STED intensities. For $\Delta\phi = \frac{\pi}{2}$, the emitters are efficiently depleted in the diagonal directions where the polarization of the field is close to circular. They cannot be completely depleted along the horizontal and vertical directions, where the polarization of the field is linear.

Polarization effects are more pronounced when the localized emission stems from a small number of fixed molecules. In this case, the s-PSF depends strongly on the orientation of dipoles as shown in Fig. 3c. The images are calculated for two configurations of 20 randomly oriented dipoles. In the four beam lattice, the shape of the s-PSF also varies strongly with the lattice relative phase $\Delta\psi$. The optimal configuration, which gives the most symmetric s-PSF is obtained for $\Delta\phi = \pi/2$.

Experimental setup

Our lattice-STED microscopy setup, sketched in Fig. 4a, is based on two synchronized laser sources delivering excitation and depletion pulses, akin to a

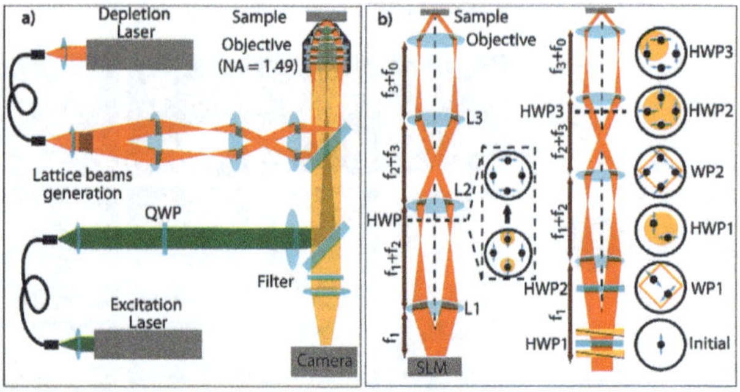

Fig. 4 (a) Optical setup of the lattice–STED microscope. The depletion beams are sent through an objective and interfere at the sample plane to form a 2D optical lattice. This optical pattern is superimposed on a wide-field excitation beam (~10 μm × 10 μm). A CMOS camera conjugated with the sample plane records the wide-field fluorescence images. The lattice generator is either a Spatial Light Modulator (SLM) or a set of two Wollaston prisms. It is conjugated with the sample plane by a set of 3 lenses (L1, L2 and L3) and the objective, with focal lengths f_1, f_2, f_3 and f_0, as shown in (b) for both configurations. Black spots illustrate the intersections of the beams with the different planes; blue double arrows show the polarizations. Orange circles and squares indicate, respectively, the half-wave plates (HWP) and the Wollaston prisms.

standard STED microscope.[32] The fluorescence excitation beam (wavelength 571 nm and pulse duration 2 ps) is delivered by a frequency-doubled optical parametric oscillator (Mira-OPO, Coherent) pumped by a picosecond Ti-sapphire laser (Mira 900 Coherent) at a repetition rate of 76 MHz. A second Ti-sapphire active mode-locked laser (Tsunami, Spectra Physics) emitting at 760 nm, provides the depletion beam. It is synchronized with the excitation laser and delivers Fourier transform-limited pulses of ~100 ps duration at an average power of 2.5 W. A Spatial Light Modulator or a set of two Wollaston prisms, conjugated with the sample plane, is used to generate multiple depletion beams (Fig. 4b). The laser beams are focused on the back focal plane of a high numerical aperture objective and illuminate a wide-field region of the sample (~7.5 μm × 7.5 μm). The optical lattice produced by interference of the depletion beams at the sample plane is superimposed on a uniform excitation beam. The fluorescence emission is collected with the same objective, filtered from excitation and depletion photons and sent to a fast CMOS camera (ORCA-Flash4.0, HAMAMATSU). A fast piezo-scanner is used for sample scanning.

The left scheme of Fig. 4b shows the generation of the multi-beam optical lattice with the SLM. The SLM surface is conjugated with the sample plane by three lenses and the objective. The focal lengths of the lenses are carefully chosen to set both the desired angle of incidence and size of the optical lattice at the sample plane. The size of the optical lattice is determined by the available STED laser power and the desired resolution. For the hexagonal lattice configuration, the SLM produces three beams with the same polarization. For the square lattice, the four beams diffracted by the SLM initially have the same linear polarization. Two half-wave plates are used to obtain two beam pairs of orthogonal polarization. With the diffraction efficiency of 4-beam generation with the SLM being only 40%, we can use two Wollaston prisms (WP) to generate the four beams so as to minimize power losses (right scheme of Fig. 4b). Starting from a linear vertical polarization, a first Wollaston prism splits the beam into two beams, with orthogonal polarizations $\pm 45°$. A half-wave plate (HWP1) rotates the polarization of the two beams into vertical and horizontal. A second Wollaston prism further splits the two beams into four. At last, two half-wave plates (HWP2 and HWP3) together set the polarizations of the four beams as in Fig. 2a. Depending on the chosen system, the phase difference $\Delta\phi$ could be adjusted either with the SLM hologram pattern or by the use of a tilted glass plate inserted in the beam paths.

The sample is composed of a thin layer of fluorescent nano-diamonds (size 35 nm) spin-coated on a glass coverslip. The number of NV defect centers per nano-diamond is ~15.[33] The emission from individual nano-diamonds shows an extreme photo-stability. The amplitude of the signal, which varies from one nano-diamond to another, can be used for a rough estimation of the number of emitting defects contained in the nanoparticle.

Experimental results

Fig. 5a and b represent a fluorescence image obtained by scanning a single nano-diamond with a piezo-stage over the field of illumination in the presence of both depletion and excitation beams, while recording its fluorescence with the CMOS camera. For each scanning step (15 nm), an image is acquired and the fluorescence intensity, integrated over the image spot of the tracked nano-diamond on the camera, is plotted. The signal maxima of the images correspond to the regions of minimal depletion, which occur at the zero-intensity positions of the optical lattice. As expected for the case of three interfering beams with an incidence angle $\theta = 60°$, we observed a hexagonal pattern with a periodicity of 390 nm. A field of view exceeding 6 μm × 6 μm can be achieved. The zoomed image shows that the s-PSF is elongated with an aspect ratio of ~2. Directions along which the depletion is not efficient are in agreement with the simulation of Fig. 3b for a large ensemble of emitting dipoles. Indeed, this nano-diamond was chosen among those presenting the highest fluorescence emission signal, *i.e.* those presenting a large number of NV defects. Using a total depletion average power of 850 mW an s-PSF of 80 nm could be achieved.

Lattice–STED super-resolved images have been acquired following the procedure described in ref. 23. The sample containing fluorescent nano-diamonds was scanned over a unit cell of the depletion lattice in the presence of the wide field excitation and the depletion pattern, while acquiring a fluorescence image for each scanning step with the s-CMOS camera. We then overlay a binary mask on these images which acts as an array of parallelized "point detectors". The complete lattice–STED image is then obtained by assembling the entire unit cell images together obtained for each "point detector".

Fig. 5c and d display 5.7 μm × 5.7 μm images of a sample containing fluorescent nano-diamonds spin-coated on a glass coverslip without and with the depletion pattern, respectively. Fig. 5c clearly shows that the PSF of the nano-diamonds are diffraction limited to ~290 nm, while the resolution of the lattice–STED image of Fig. 5d is well below this limit (typically ~80 nm).

The results with the four-beam square lattice are presented in Fig. 6. Fig. 6a–c were constructed using the same procedure as in Fig. 5b. Depletion patterns

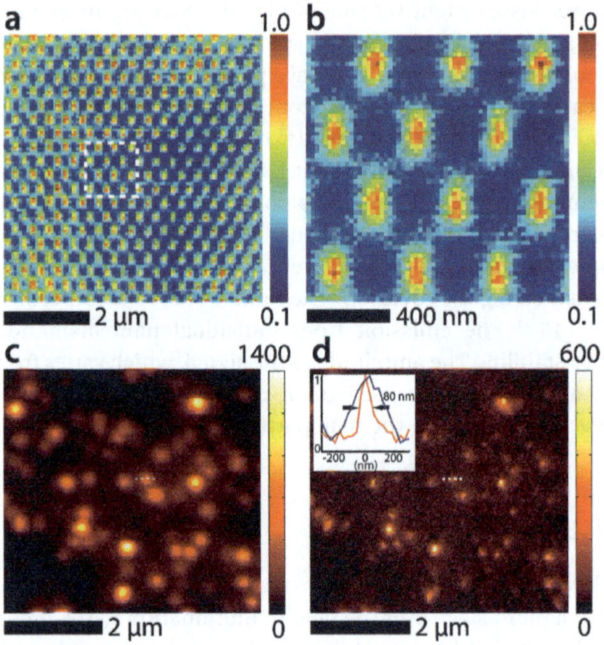

Fig. 5 (a) Experimental depletion patterns. Field of view: 6 µm × 6 µm recorded at an excitation power of 16 mW, and a depletion power of 850 mW (measured before the objective). Integration time per pixel: 20 ms. (b) Zoom of the central region (white dashed square). All images are normalized with respect to their maximal intensities. (c) and (d) Three-beam lattice–STED images of a sample containing 35 nm fluorescent nano-diamonds without and with the depletion optical lattice respectively. Field of view: 5.7 µm × 5.7 µm. Pixel size: 30 nm; pixel dwell time: 20 ms. Inset: intensity profiles of a single nano-diamond (white dashed line) which shows a FWHM of 80 nm. The images were obtained with the same excitation and depletion intensities. The scale bar corresponds to the counts per unit of pixel dwell time.

obtained at three different values of $\Delta\phi\left(0, \frac{\pi}{4} \text{and} \frac{\pi}{2}\right)$ were studied, and images recorded with three different weakly emitting nano-diamonds are shown. Using the total emission of the nanoparticle we can estimate the number of NV centers to a few tens. As predicted in Fig. 3, the shape of the s-PSF strongly changes from one nano-diamond to another. The s-PSF is not always symmetric but depends on the relative phase $\Delta\phi$ of the lattice, as expected.

For $\Delta\phi = 0$ (field pattern of Fig. 2d), the s-PSFs are elongated, indicating a preferential orientation of emitting dipoles. In each image, the variations of the s-PSF orientation from site to site are imposed by the local polarization distribution around the zero intensity regions: the s-PSFs are elongated along the direction where the electric field is perpendicular to the average orientation of the dipoles. For example, the nano-diamond used in Fig. 6c has dipoles preferentially oriented along a diagonal direction. The images could be reproduced (Fig. 6d) with a set of twenty random emitters (having a diagonal average orientation) using the depletion lattices of Fig. 2 with intensity maxima of $6I_{sat}$. As $\Delta\phi$ increases up to $\pi/2$, the anisotropy of the s-PSF becomes less pronounced. The configuration for

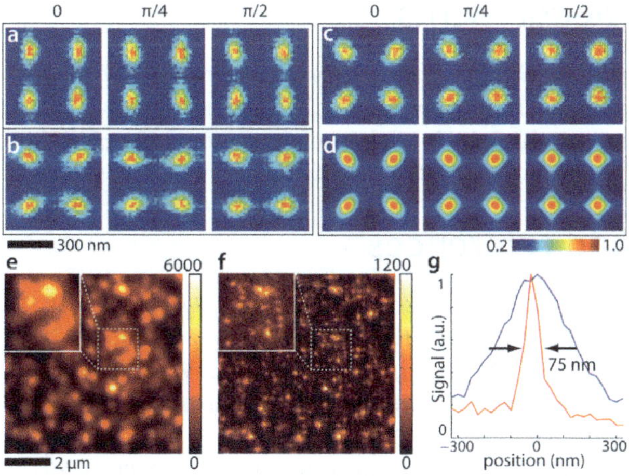

Fig. 6 (a–c) Experimental fluorescence depletion patterns measured for $\Delta\phi = 0$, $\pi/4$ and $\pi/2$. (d) Simulated image computed with a set of twenty randomly oriented emitters depleted at an intensity maxima of $6I_{sat}$. This simulation reproduces well the experimental s-PSF of image (c). Integration time per pixel: 20 ms. All images are normalized with respect to their maximal intensities. (e) and (f) Four-beam lattice–STED images of a sample containing 35 nm fluorescent nano-diamonds without and with the depletion optical lattice, respectively. Field of view: 7.5 μm × 7.5 μm. Pixel size: 30 nm, pixel dwell time: 20 ms. (g) Intensity profile of a single nano-diamond with a FWHM of 75 nm. The images were obtained with an excitation power of 16 mW and a depletion power of 850 mW measured before the objective.

which $\Delta\phi = \pi/2$ is therefore the most favorable to perform four-beam lattice-STED microscopy.

As for the three beam configuration, we acquired lattice–STED super-resolved images of a dense sample of nano-diamonds for $\Delta\phi = \pi/2$. Fig. 6e and f display 7.5 μm × 7.5 μm images without and with the four beam depletion pattern. A clear resolution improvement between these two images is visible. The s-PSF shown in Fig. 6g (~75 nm) is well below the diffraction limit. Compared to the three beam configuration, the lattice unit cell is smaller and the intensity gradients around the zero-intensity regions are steeper, enabling a larger field of view with the same averaged depletion power.

Conclusions

In this paper, we push further the investigations and the performance of lattice–STED which massively parallelizes STED microscopy. Super-resolution images of 75 nm resolution and a field view of 7.5 μm × 7.5 μm could be achieved with moderate powers of a high repetition rate laser. Images of fluorescent nano-diamonds acquired using different optical lattice configurations are compared to theoretical simulations. The results show that the s-PSF of lattice–STED varies with the electric field topography of the lattices and should depend on the rotational diffusion properties of the emitters. This allowed finding the optimum configuration of four-beam lattice–STED nanoscopy.

Acknowledgements

We warmly thank Philippe Tamarat for careful reading of the manuscript. We acknowledge financial support from the Agence Nationale de la Recherche, Région Aquitaine, the French Ministry of Education and Research, the European Research Council and FranceBioImaging (Grant no. ANR-10-INSB-04-01).

Notes and references

1 B. Huang, M. Bates and X. Zhuang, *Annu. Rev. Biochem.*, 2009, **78**, 993–1016.

2 A. G. Godin, B. Lounis and L. Cognet, *Biophys. J.*, 2014, **107**, 1777–1784.

3 E. Betzig, G. H. Patterson, R. Sougrat, O. W. Lindwasser, S. Olenych, J. S. Bonifacino, M. W. Davidson, J. Lippincott-Schwartz and H. F. Hess, *Science*, 2006, **313**, 1642–1645.

4 M. J. Rust, M. Bates and X. Zhuang, *Nat. Methods*, 2006, **3**, 793–795.

5 M. Heilemann, S. van de Linde, M. Schüttpelz, R. Kasper, B. Seefeldt, A. Mukherjee, P. Tinnefeld and M. Sauer, *Angew. Chem., Int. Ed.*, 2008, **47**, 6172–6176.

6 A. Sharonov and R. M. Hochstrasser, *Proc. Natl. Acad. Sci. U. S. A.*, 2006, **103**, 18911–18916.

7 G. Giannone, E. Hosy, F. Levet, A. Constals, K. Schulze, A. I. Sobolevsky, M. P. Rosconi, E. Gouaux, R. Tampé, D. Choquet and L. Cognet, *Biophys. J.*, 2010, **99**, 1303–1310.

8 B. Huang, W. Wang, M. Bates and X. Zhuang, *Science*, 2008, **319**, 810–813.

9 J. Folling, M. Bossi, H. Bock, R. Medda, C. A. Wurm, B. Hein, S. Jakobs, C. Eggeling and S. W. Hell, *Nat. Methods*, 2008, **5**, 943–945.

10 S. W. Hell and J. Wichmann, *Opt. Lett.*, 1994, **19**, 780–782.

11 S. W. Hell and M. Kroug, *Appl. Phys. B*, 1995, **60**, 495–497.

12 M. G. Gustafsson, *J. Microsc.*, 2000, **198**, 82–87.

13 R. Heintzmann, T. M. Jovin and C. Cremer, *J. Opt. Soc. Am. A*, 2002, **19**, 1599–1609.

14 M. G. L. Gustafsson, *Proc. Natl. Acad. Sci. U. S. A.*, 2005, **102**, 13081–13086.

15 E. H. Rego, L. Shao, J. J. Macklin, L. Winoto, G. A. Johansson, N. Kamps-Hughes, M. W. Davidson and M. G. L. Gustafsson, *Proc. Natl. Acad. Sci. U. S. A.*, 2012, **109**, E135–E143.

16 F. Huang, T. M. P. Hartwich, F. E. Rivera-Molina, Y. Lin, W. C. Duim, J. J. Long, P. D. Uchil, J. R. Myers, M. A. Baird, W. Mothes, M. W. Davidson, D. Toomre and J. Bewersdorf, *Nat. Methods*, 2013, **10**, 653–658.

17 P. Kner, B. B. Chhun, E. R. Griffis, L. Winoto and M. G. L. Gustafsson, *Nat. Methods*, 2009, **6**, 339–342.

18 S. A. Jones, S.-H. Shim, J. He and X. Zhuang, *Nat. Methods*, 2011, **8**, 499–505.

19 L. Zhu, W. Zhang, D. Elnatan and B. Huang, *Nat. Methods*, 2012, **9**, 721–723.

20 J. Huang, M. Sun, K. Gumpper, Y. Chi and J. Ma, *Biomed. Opt. Express*, 2015, **6**, 902–917.

21 S. W. Hell, *Nat. Biotechnol.*, 2003, **21**, 1347–1355.

22 V. Westphal, S. O. Rizzoli, M. A. Lauterbach, D. Kamin, R. Jahn and S. W. Hell, *Science*, 2008, **320**, 246–249.

23 B. Yang, F. Przybilla, M. Mestre, J.-B. Trebbia and B. Lounis, *Opt. Express*, 2014, **22**, 5581–5589.

24 A. Chmyrov, J. Keller, T. Grotjohann, M. Ratz, E. d'Este, S. Jakobs, C. Eggeling and S. W. Hell, *Nat. Methods*, 2013, **10**, 737–740.

25 P. Bingen, M. Reuss, J. Engelhardt and S. W. Hell, *Opt. Express*, 2011, **19**, 23716–23726.

26 A. Hemmerich and T. W. Hänsch, *Phys. Rev. Lett.*, 1993, **70**, 410–413.

27 G. Grynberg, B. Lounis, P. Verkerk, J. Y. Courtois and C. Salomon, *Phys. Rev. Lett.*, 1993, **70**, 2249–2252.

28 M. Greiner, O. Mandel, T. Esslinger, T. W. Hansch and I. Bloch, *Nature*, 2002, **415**, 39–44.

29 M. Campbell, D. N. Sharp, M. T. Harrison, R. G. Denning and A. J. Turberfield, *Nature*, 2000, **404**, 53–56.

30 M. P. MacDonald, G. C. Spalding and K. Dholakia, *Nature*, 2003, **426**, 421–424.

31 B. Lounis, PhD thesis, Université Paris Sud - Paris XI, 1993.

32 B. Harke, J. Keller, C. K. Ullal, V. Westphal, A. Schönle and S. W. Hell, *Opt. Express*, 2008, **16**, 4154–4162.

33 C.-C. Fu, H.-Y. Lee, K. Chen, T.-S. Lim, H.-Y. Wu, P.-K. Lin, P.-K. Wei, P.-H. Tsao, H.-C. Chang and W. Fann, *Proc. Natl. Acad. Sci. U. S. A.*, 2007, **104**, 727–732.

Faraday Discussions

PAPER

Single molecule fluorescence resonance energy transfer scanning near-field optical microscopy: potentials and challenges

S. K. Sekatskii,* K. Dukenbayev,† M. Mensi,‡ A. G. Mikhaylov, E. Rostova, A. Smirnov, N. Suriyamurthy§ and G. Dietler

Received 29th May 2015, Accepted 10th July 2015

DOI: 10.1039/c5fd00097a

A few years ago, single molecule Fluorescence Resonance Energy Transfer Scanning Near-Field Optical Microscope (FRET SNOM) images were demonstrated using CdSe semiconductor nanocrystal–dye molecules as donor–acceptor pairs. Corresponding experiments reveal the necessity to exploit much more photostable fluorescent centers for such an imaging technique to become a practically used tool. Here we report the results of our experiments attempting to use nitrogen vacancy (NV) color centers in nanodiamond (ND) crystals, which are claimed to be extremely photostable, for FRET SNOM. All attempts were unsuccessful, and as a plausible explanation we propose the absence (instability) of NV centers lying close enough to the ND border. We also report improvements in SNOM construction that are necessary for single molecule FRET SNOM imaging. In particular, we present the first topographical images of single strand DNA molecules obtained with fiber-based SNOM. The prospects of using rare earth ions in crystals, which are known to be extremely photostable, for single molecule FRET SNOM at room temperature and quantum informatics at liquid helium temperatures, where FRET is a coherent process, are also discussed.

1. Introduction

Fluorescence Resonance Energy Transfer Scanning Near-field Optical Microscopy (FRET SNOM) (the idea is illustrated in Fig. 1) consists of the use of a donor–acceptor FRET pair between a SNOM tip and a sample.[1,2] If the donor fluorescent centers of the imaging tip are excited and the fluorescence of the acceptor centers of the sample is monitored (or *vice versa*) when the tip is scanning along a sample

Laboratoire de Physique de la Matière Vivante, EPFL, CH1015 Lausanne, Switzerland. E-mail: serguei.sekatski@epfl.ch

† Present address: "National Laboratory Astana" PI, "Nazarbayev University" AEO, Kabanbay Batyr Avenue, 53, Astana, 010000 Kazakhstan.

‡ Present address: Optics and Photonics, KTH Isafjordsgatan 22, 16440 Stockholm, Sweden.

§ Permanent address: Radiological Safety Division, Indira Gandhi Centre for Atomic Research, Kalpakkam, India.

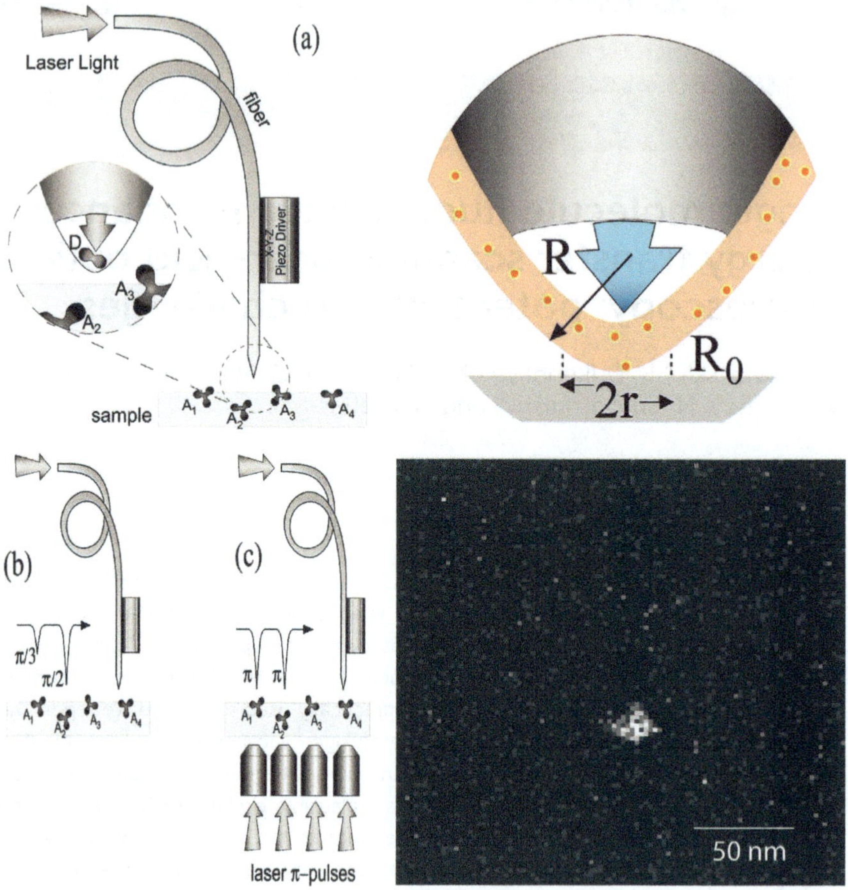

Fig. 1 Left (reproduced from ref. 22, with permission): (a) illustrating the FRET SNOM approach idea, with a single fluorescent centre case. The preparation of non-local entangled Greenberger–Horne–Zeilinger three-particle state (b; distances between the donor and acceptor centres are changed in a controllable fashion to ensure a desirable *partial* excitation transfer from the donor to acceptor, we name this $\pi/3$ and $\pi/2$-pulses) and Control Control Control NOT (CCCNOT) operator (c; donor is initially excited and remains excited if and only if the acceptors are initially excited by resonant light π-pulses) for the case of coherent FRET is shown. Right: illustrating the spatial resolution of FRET SNOM working at the many molecules level (upper), and a single molecule FRET SNOM image obtained for a CdSe nanocrystal–dye molecule donor–acceptor pair (below, reproduced from ref. 17, with permission).

surface, the spatial resolution is governed not by the aperture size of the microscope probe but by the value of a characteristic FRET (Förster) radius, R_0. The latter for typical donor–acceptor pairs ranges from 3–7 nm (ref. 3–5) and thus the spatial resolution of SNOM can be improved by ten times without a loss in the sensitivity: the resolution of a "standard" aperture SNOM is governed by the aperture size, which can not be made smaller than 50 nm due to the very rapid decrease in the light transmission through the tip when the aperture is decreasing.[6–8]

To achieve the subtip spatial resolution with this method, we do not need to work with its single molecule version: only a (partly filled) cylinder with a radius $r = \sqrt{2RR_0}$ contributes to the signal,[2] see Fig. 1, left. Here, R is the radius of the curvature of a SNOM probe apex, which is typically equal to 50–100 nm, and usually this exact radius determines the spatial resolution. It is easy to see that $r \ll R$.

Immediately after the idea of the method was proposed,[1] the corresponding experiments started and nanolocal FRET SNOM measurements and FRET SNOM images have been reported by a number of research groups using fluorescent centers, such as dye molecules,[9–12] semiconductor nanocrystals[13–15] and *F*-aggregated color centers in LiF crystals.[16] Finally, in 2008 these efforts led to the observation of single molecule FRET SNOM images in an experiment where 4.8–4.9 nm CdSe nanocrystals (this is the core diameter; the core was covered with a protective 1–2 nm thick ZnS layer) deposited onto a glass slide surface were used as donors and Alexa Fluor 594 dye molecules (Molecular Probes, Oregon, USA) contained inside a thin PMMA coating over the SNOM tip apex were used as acceptors.[17]

However, the results reported in ref. 17 remain isolated and did not lead to the common use of the single molecule FRET SNOM imaging method. We still prefer to characterize the current situation as "*towards single molecule FRET SNOM*", and the reason for this is the limiting photostability of the dye molecules and semiconductor nanocrystals used for imaging (our FRET SNOM research involving CdSe nanocrystals does *not* confirm that their photostability is much larger than that of rhodamine dyes, contrary to what is often stated). For a typical optical absorption cross section of a single fluorescent centre $\sigma \cong 10^{-16}$ cm^2, photon energy $h\nu \cong 3 \times 10^{-19}$ J and near-field illumination intensity $I \cong 10^3$ W cm^{-2}, the photostability (a good one) of $N_{\text{phst}} \cong 10^7$ cycles determines the life time of a center equal to $t = \frac{N_{\text{phst}} h\nu}{I\sigma} \cong 30$ s. In practice, this means that a special microscope construction enabling quite fast scanning should be exploited, and extreme precautions to avoid the premature killing of an acting single FRET centre by light (such as opening of the light source only after the scanning has already started, *etc.*) should be taken to observe single molecule FRET SNOM images. Despite all of this, such an observation still requires certain luck.

The situation would change dramatically if much (say, two orders of magnitude) more photostable fluorescent centres are used. Naturally, this suggests attempting to use the famous nitrogen vacancy (NV) defect centres in nanodiamond (ND) crystals, which are claimed to be extremely photostable,[18–21] for FRET SNOM imaging. Such research has not only been started in our group but also elsewhere. Still, to the best of our knowledge, this activity did not result in obtaining single molecule FRET SNOM images. The apparent reasons for this and possible further steps are discussed in our paper, together with necessary improvements to the scanning near-field microscope made specially to achieve this aim. In particular, we present what we believe is the first and unique observation of topographical images of *single strand* DNA molecules with an optical fibre-based SNOM device.

Other extremely photostable fluorescent centres are rare earth ions in crystals, and they naturally should be considered for FRET SNOM as well. Exploitation of rare earth ions in crystals looks interesting also from the viewpoint of quantum computing applications of FRET SNOM.[22] In the case of liquid helium temperatures, for numerous optical transitions of rare earth ions in crystals, decoherence (dephasing) time becomes quite comparable with the radiative decay time.[23–26] In some cases it can attain the value of a few milliseconds, while microsecond decoherence times are a widespread phenomenon. Under such conditions, FRET is a coherent process and the quantum dynamics of the system are approximated well by standard dynamics of a two-level quantum system: the excitation energy flows back and forth between the donor and acceptor, and for typical cases the period of this energy exchange is equal to the radiative decay time when the donor–acceptor distance is around 20 nm. Hence, all the parameters required (a few nanometers precision of the tip–sample motion on the time scale of a few microseconds, and liquid helium temperatures) are quite feasible with the modern Scanning Probe Microscopy (SPM) technique,[22] and different protocols for performing basic quantum computing operations (some of them are illustrated in Fig. 1) can be proposed.

The prospects of using rare earth ions in crystals for both incoherent and coherent FRET SNOM will be briefly discussed in the last section of the paper.

2. SNOM device: fast scanning and ultralow acting forces

The scanning near-field optical microscope used in our research has been discussed earlier,[27] and so here we only briefly characterize its peculiarities important for single molecule FRET SNOM imaging. This microscope is capable of fast scanning, necessary to (partly) overcome the severe photostability problem. The main tool to achieve this is the use of an original "double resonance principle": the working frequency of the tuning fork (which is close to 32 756 Hz) coincides with the second resonant frequency of the lateral dithering of a free-standing part of the fiber beam.[28] The use of proprietary low noise, precise and fast electronics measuring the resonance frequency f_{res} and Q-factor of a tuning fork[27] (previously designed to be exploited essentially with the sensor's "tuning fork and AFM cantilever glued upon it"[29]), is also important. The double resonant montage, which requires careful control of the length of the free standing part of the fiber beam and of the position of a thin (40 micron-thick) glass rod connecting the fiber probe with one of the tuning fork's prongs, as well as the careful assembly (gluing) of all parts, enables one to routinely achieve a quality factor of the sensor ranging from 3500–5500, and its exploitation results in a force sensitivity as small as 30 pN in a bandwidth of 300 Hz. This large force sensitivity enables imaging of double stranded DNA molecules using both glass and plastic[30] fiber-made SNOM probes.

An even better performance of the microscope was achieved using the electronics of Nanonis GmbH, Zürich, Switzerland, specially adapted by our laboratory and the company to work with SNOM, exploiting high Q-factor probes. The probes were oscillated at (or close to) the resonance frequency with a 1–5 mV_{rms} sinusoidal voltage applied to the tuning fork's electrodes, which caused a current on the order of 20 nA_{rms}. The current is amplified by a current–voltage preamplifier (Nanonis TFPA-3) with a gain of 1 V μA^{-1}, corresponding to a transimpedance of 1 MΩ. The bandpass ranges from DC to 150 kHz, and the noise level

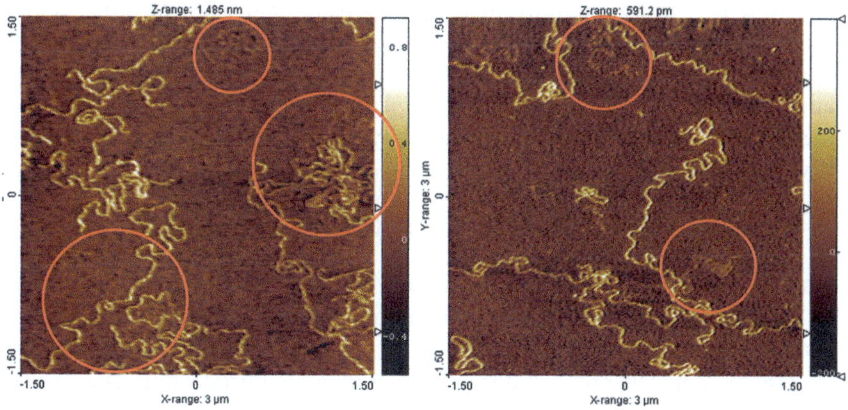

Fig. 2 Topographical images of DNA molecules deposited onto an APTES-modified mica surface (see text for details) obtained with SNOM. Single strand parts of the DNA are highlighted with red circles.

in the 10 Hz to 1 MHz window was smaller than 0.27 mV. Different glass fiber-made SNOM probes, including those commercially available from KDP company, Moscow, Russia, aperture probes modified in the scanning electron microscope vacuum chamber to grow ultrasharp whiskers onto their apex,[31] or home-etched probes made using the tube-etching approach in HF solution[32,33] were used.

In Fig. 2 we present what we believe are the first and unique topographical images of *single strand* DNA molecules obtained using the fiber-based SNOM technique. DNA sample preparation was the same as discussed earlier.[34] Very briefly, a double stranded linear DNA sample (Fermentas, Lambda Mix Marker 19) was diluted to a concentration of 1 µg ml^{-1} in 1 mM TRIS buffer solution at pH = 7.8. The single-strand DNA was prepared by heating up the linear DNA solution at 60 °C for 1 minute. After this, a 10 µl aliquot of the DNA solution was deposited onto mica modified by 3-aminopropyltriethoxysilane (APTES; AP-mica). This was incubated for 60 seconds, rinsed with 500 µl of ultrapure water and gently dried. AP-mica was prepared by treating a freshly cleaved mica surface with a solution of 0.05% APTES in water for 1 minute, rinsing thoroughly with ultrapure UHQ water (USF Elga, High Wycombe, England) and drying with dry air.

Another important characteristic for the application of the microscope is its capability for time-gated detection of the optical signal using an external TTL pulse triggering a light source (laser) and then, with an appropriate delay regulated by SNOM electronics, the photon detector. Alternatively, SNOM electronics themselves can generate a TTL pulse to trigger an external laser source and (again, with an appropriate delay) the photon detector. In both cases, the delay times can be regulated over a broad range with a quantum of 25 ns. This option was proved to be useful for imaging fluorescent nanospheres using 10 100 µs duration light pulses obtained by exploiting an acousto-optical modulator with a CW Nd:YAG laser at 532 nm,[27] and afterwards, also with a nanosecond (<2 ns duration) pulsed Nd:YAG laser, at 532 nm.

It is known and broadly accepted in the area of near-field optics (and fully corresponds with our own experience) that, when working with a CW laser, the light intensity at the subwavelength-size aperture of the SNOM probe cannot exceed the value of around $I \cong 10^3$ W cm^{-2}, and this corresponds to a laser power of *ca.* 0.5–3 mW at the entrance of the probe made of single-mode glass fiber. For larger values, the probe can be easily damaged and burnt by light. Correspondingly, when measuring the signal caused by a single fluorescent center, we can have around

$$n = \eta \frac{I\sigma}{h\nu} \cong 3 \times 10^3 \text{ to } 10^4 \tag{1}$$

photocounts per second. Here η is an overall efficiency coefficient, which includes all factors, such as the fluorescence quantum yield and photon collection- and detector-efficiencies, and it ranges from 0.01–0.03 (a good value indeed), while all the other parameters are the same as were used in the Introduction. This value is one to two orders of magnitude larger than the typical dark noise level of a single photon detector, but experimentally it is quite difficult to diminish the parasitic noise signal, caused essentially by unavoidable stray light, not to mention to sufficiently lower it.

Let us see what changes occur if pulsed laser excitation is used instead of a continuous one. For such a case, one still should use the same average value of the total laser energy passing through the fibre during 1 second, and consequently the same average value of intensity I, to avoid fibre damage: this damage is essentially a thermal effect (we have checked this circumstance with microsecond[27] and afterwards also with nanosecond laser pulses; the damage threshold does depend only on the total power introduced into the SNOM probe). However, now the gated detection drastically decreases both the dark detector- and (a part of) the stray light-related noise levels. Earlier[27] we considered the situation where the laser pulse duration t_p is much larger than the single fluorescent center radiative decay time t_{fl}. Then, for the laser pulse repetition rate F and gate duration T, this decrease is equal to $(FT)^{-1}$, while the overall signal level is unchanged, and this has been demonstrated experimentally. If, however, as it takes place now, $t_p < t_{fl}$, the total effect is smaller because the decrease in the noise level is partly "compensated" by the decrease of the signal. The number of optical absorption events per second is $N = 1/\tau = \frac{I\sigma}{h\nu}$, and it is clear that using a pulsed laser with the same average intensity I and pulse repetition rate F, if $F < N$, causes saturation of the optical absorption of the fluorescence center during a single pulse. However, for a single emitting centre, the signal level now is just ηF, hence the increase in the signal to noise ratio is given by τ/T. For $T = 25$ ns and the other parameters given above, we still have an improvement of two orders of magnitude.

Of course, this gating cannot diminish the noise coming from "not fully suppressed laser light itself" and caused by very fast parasitic fluorescence "of everything", but note that now we do not need any more strong light intensity: it is useless to oversaturate the transition during the single pulse, it just needs to be saturated. This decrease in the laser power leads to a corresponding decrease in the noise directly caused by excitation light. Certainly, a decrease in the signal level paid for the improvement in the signal to noise ratio necessitates slower scanning during imaging.

3. Single molecule FRET SNOM: room temperature case

3.1. From CdSe nanocrystals to color centres in nanodiamond crystals

In the Introduction section we briefly discussed the first single molecule FRET SNOM images, their limitations and the motivation to use ultraphotostable NV color centers in nanodiamond crystals. Indeed, these same color centers were considered from the very beginning of FRET SNOM research,[1] while their first use in the SNOM area (as a local fluorescence source) goes back to 2001.[35] The same motivation promotes the use (or attempts the use) of diamond NV fluorescent centers in many other areas. As an example, let us only underline their rather successful application in the Stimulation Emission Depletion Microscopy field,[36,37] where this same photostability problem defines the ultimate spatial resolution achievable.

The photophysics of NV color centers are still not fully understood and remain debated. This is especially true when details concerning the charge states NV^0 and NV^- of the nitrogen vacancy and their inter-transformations (either spontaneous or specially induced) are considered. Nowadays, it is almost exclusively the NV^- state that is exploited, but it is worthwhile to note that the NV^0 state was also discussed as a very prospective candidate for quantum informatics-related applications.[38] However, FRET SNOM imaging at room temperature does not involve any coherent and spin properties of the emitting center, hence, from the very beginning, we decided not to target a certain particular charge state but to search for imaging caused by either the NV^0 (luminescence band essentially in the 570–700 nm spectral range) or NV^- (620–750 nm) states. Indeed, detailed time-resolved experiments with single NV centers show the photo-switching from the neutral to negatively charged state of NV as well as reverse photochromic transformation with the time constant ranging from 0.3–3.6 microseconds (2.1 μs on average) under dark conditions.[39] Previously, photochromism has been shown for bulk samples.[40] There are not two different types of defects but "both fluorescence bands originate from the same defect" and "continuous switching between two charge states exists".[39] The laser ionization of nitrogen donors present in the vicinity of the NV defect is a plausible mechanism for the NV^0 to NV^- conversion, and this means that such a process should be strongly dependent on the concrete properties of the colored ND and the laser illumination conditions. In a sense, the NV^0 state should be considered as a "main" one, which is initially present (and consequently, its luminescence should "always" be observed), while the NV^- state should first be created by the irradiation itself so that its observation may be not be guaranteed at all.

Following this paradigm, different laser dye molecules (oxazine 1, oxazine 750, rhodamine 800)[41] were systematically tried in our experiments as potential acceptors for either the NV^0 or NV^- donors. This makes the much debated questions regarding to what extent the neutral charge state is preferable at small (a few nanometers) distances from the ND particle surface and to what extent (if any) the NV^-/NV^0 concentration ratio can be influenced by surface modification or other means mostly irrelevant.[42–46]

Two different sources of bright colored nanodiamond crystals were used in our experiments. The first was solutions (for simplicity, here and below we use the word "solution" to describe ND particle water-based suspensions) and powders of nanodiamond crystals purchased from Microdiamant AG, Lengwil, Switzerland, with an average size of 25 nm or 50 nm and communicated concentrations of the nitrogen vacancies around 25–100 ppm.

To color them, the powder of these diamonds (25 nm-size) was introduced into specially prepared "flat" envelopes made from thin 50 micron-thick Al foil and irradiated by a 2.4 MeV-energy electron beam in the accelerator ILU-6 (average current 2–3 mA) of the Institute of Nuclear Physics Siberian Branch of the Russian Academy of Sciences, Novosibirsk, Russia. The total dose of the irradiation was set to be equal to 1×10^{18} e^- cm^{-2} (for one half of the sample) and 2×10^{18} e^- cm^{-2} (for the other half). Afterwards, the irradiated diamond dust was thermally annealed in a technical vacuum (around 1 mTorr) at a temperature of 750–800 °C for two hours. No further treatment aiming to remove graphite layers possibly covering the as-irradiated nanodiamonds[37,47] was undertaken. This procedure is more or less similar to those suggested in many other papers, starting from the classic Davies and Hamer paper.[48] It has been reported that for a 3×10^{17} e^- cm^{-2} and 1.5 MeV dose and energy of electron beam, roughly 1 NV center in a 30 nm-diameter sphere is created,[49,50] and thus we anticipated somewhat similar results for our case.

The second sample of fluorescent nanodiamonds was commercially available colored nanodiamonds in solution obtained from L. M. van Moppes & Sons SA, Geneva, Switzerland (average size of 65 nm, concentration of NV color centres was not communicated), and Adamas Nanotechnologies, Inc., Raleigh, NC, USA. For the latter, an average of 1–4 NV fluorescent centres in one 40 nm-size nanodiamond particle was indicated.

In Fig. 3 we present the fluorescence spectra of a water solution of 40 nm-diameter nanodiamonds for different excitation wavelengths (results for other solutions of fluorescent NDs were similar with appropriate scaling). The first conclusion from these data is the very strong presence of all possible types of light

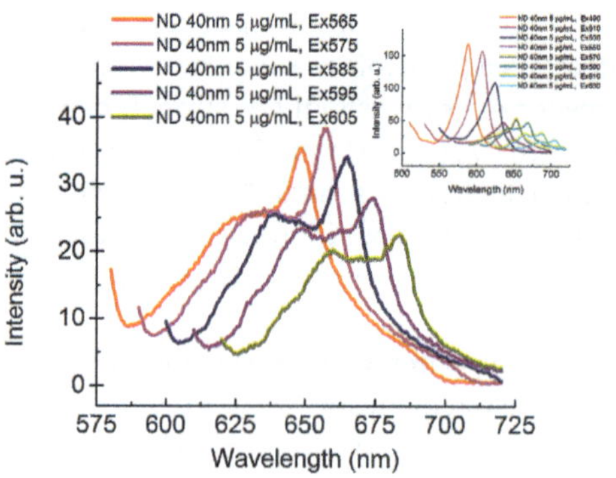

Fig. 3 Optical signal recorded at different excitation wavelengths for 5 μg ml^{-1} water-based suspensions of colored 40 nm-size nanodiamond particles. The known NV^0 luminescence band (590–700 nm) is seen when excited in the spectral range 540–630 nm and is not seen when excited at shorter wavelengths, see the inset. The spectra were recorded using a Jasco FP-8500 spectrofluorimeter (Easton, MD, USA) with a detector sensitive in the 200–750 nm spectral range, which may obscure observation of the red part of the NV center's fluorescence.

"scattering" in the recorded signal (we use this word as a generic term to describe any type of observed signal that is *not* NV fluorescence band(s). We did not perform detailed studies to trace out its origin, deeming this irrelevant for our purposes to construct FRET SNOM. Unfortunately, the existence and importance of this type of optical signal are almost never reported in the literature). The second conclusion is the strong dependence of the efficiency of the excitation of the NV fluorescence band, lying in the region 580–700 nm, on the excitation wavelength.

Direct comparison of the fluorescence signal obtained from a solution of colored 40 nm-diameter NDs with that from a sulphorhodamine B water solution[41] excited in the spectral range 530–570 nm, accepting the same quantum yield and an optical absorption cross-section of 3×10^{-17} cm^2 as a maximum for the NV centers[51] (the latter is equal to 6×10^{-17} cm^2 for dye molecules at the wavelength of 532 nm (ref. 41)), leads to an estimation of the average number of NV centers in one ND particle of 0.4–0.8, which is not too far from the data communicated by the producer. A similar estimation for colored 25 nm-diameter NDs in Novosibirsk gives a value around 0.2–0.4.

From the recorded graphs, the appearance of the NV fluorescence band (apparently, this is a NV^0 band) was unambiguously attested only in the excitation spectral range from roughly 540 to roughly 610 nm. For shorter wavelengths, the excitation is much less effective, and it is strongly obscured by a "light scattering signal". Still, the NV luminescence is definitely excited by a nanosecond pulsed 532 nm laser, as was attested by the observation of fluorescence kinetics with a life time equal to *ca.* 12 ns, and thus is quite comparable with the literature data[18–21] (a Time Correlated Single Photon Counting module PicoHarp 300, PicoQuant, Berlin, Germany, was used). Similar kinetics have been observed for ND samples in solution, in thin polymer films, and in thin layers of polymer deposited onto an aperture SNOM tip apex; see below. The efficiency of the excitation of the NV-center fluorescence by the ns pulsed 532 nm laser (as well as by a CW 514 nm laser) is difficult to estimate. Our data suggest that for the corresponding cross section, the value is something like one order of magnitude smaller than that at the maximum, *i.e.* around 3×10^{-18} cm^2 (note the value 1×10^{-17} cm^2 is reported for an optical excitation cross section at this wavelength in ref. 48).

For FRET SNOM experiments, the distance between the donor and acceptor should be smaller than the Förster radius, R_0. This adds one more constrain, which from first glance does not look too restrictive: for a spherical particle of radius R, only those color centers that lie at distances larger than $R-r$ with $r < R_0$ from the center, can contribute. For $R = 12.5$ nm and $r = 3$ nm, this signifies that more than one half of all the NV centers, *providing that they are evenly distributed inside a nanocrystal*, can deliver.

In brief, the numbers given above show that the task to realize single molecule FRET SNOM with NV color centers in nanodiamonds available to us is certainly demanding but could be realizable if one will work hard enough. During the last five years, we regularly prepared FRET-active SNOM probes containing colored nanodiamond particles by dipping a sharp aperture SNOM fiber probe into a solution (acetone, chloroform, ethanol, DMSO) containing a certain polymer (PMMA, chitosan, PDMS) at different concentrations and using fluorescent NDs with different sizes and NV center concentrations, and extracting them (in different ways) from these solutions. After solvent evaporation, one has an aperture SNOM probe that is covered with a thin polymer layer with ND particles; the presence of the latter has been confirmed by fluorescence and kinetic measurements as well as by scanning electron microscopy images. Examples of such FRET active probes are presented in Fig. 4. Previously, we prepared similar probes with dye molecules[12] and CdSe semiconductor nanocrystals[13] (we coined the name "self-sharpening pencil probe" for them), and their exploitation finally led to us obtaining single molecule FRET SNOM images,[17] while the recording of such images at a level of tens and hundreds of molecules involved was almost routine.

Nevertheless, despite all these efforts, we were unable to see FRET SNOM images with such probes and differently deposited layers of oxazine 1, oxazine 750 and rhodamine 800 dyes[41] (calculated Förster radii range 4–6 nm) even at the aforementioned "many molecules" level. These experiments were performed by exploiting the excitation from a nanosecond pulsed 532 nm laser and a CW argon ion laser line at 514 nm. In all cases, the necessary Kaiser Optical Systems (Ann Arbor, MI, USA) notch filter and an appropriate set of long pass optical filters were exploited. These probes were successfully used as local fluorescence probes, *cf.* ref. 52. What is the explanation for this failure? Of course, due to the quite demanding characteristics of the task, we cannot exclude the fact that we simply did not work hard enough or we were unlucky: it might be that no one probe prepared by us contains NV centers "correctly oriented" (close to the part of the ND that touches the sample rather than an opposite part), or such like. However, this does not seem very plausible, and, as of today, we are inclined to suggest the following explanation.

Recently, certain evidence has appeared that nitrogen vacancies may be *unstable* if they are too close to the nanodiamond crystal surface. In particular, there is a theoretical prediction that NV centers are thermodynamically unstable in nanodiamonds less than 2 nm in size.[53] Further, strong blinking was detected for ND particles 5 nm in diameter,[54] permanent bleaching of a material fraction of NV centers engineered 2–5 nm beneath the *F*-terminated diamond surface was reported,[44] and so forth. In our opinion, it might be possible that for distances of the order of 3 or so nanometers from the surface, the NV centers might be unstable, or, more probably, have a limited life time, possibly dependent on the irradiation conditions (*i.e.* they might be subject to strong enough photobleaching). This suggestion would explain all our experimental observations,

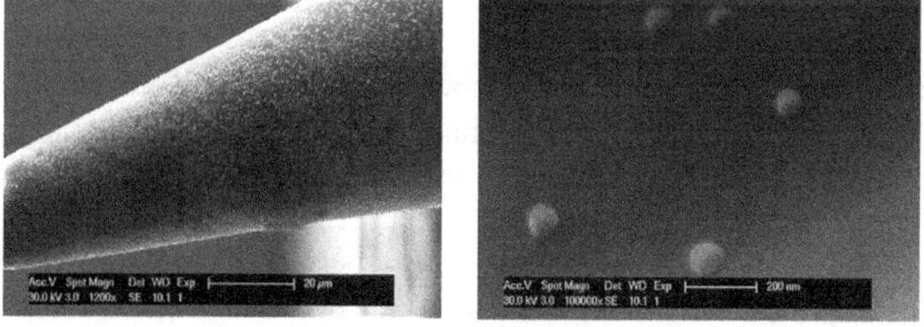

Fig. 4 Scanning electron microscopy images of the SNOM probe surface covered with a thin chitosan layer with fluorescent 40 nm-size nanodiamonds. In the right image, the isolated individual ND particles obtained with this approach are illustrated.

however, let us repeat again, at present we are not ready to claim that this is the only possible explanation as to why it is impossible to realize single FRET SNOM with NV centers in nanodiamonds.

To conclude this section of the paper, we would like to make the following remarks. FRET observations between a single ND (approximately 20 nm in size) and infrared IRdye-800CW (the calculated Förster radius is equal to 5.6 nm) have been reported.[55] An analysis of this paper shows that the fluorescent center is located somewhere close to the nanocrystal center, and a large number of dye molecules, attached to the surface of this particle, are responsible for the observed FRET. Thus, this is more of a confirmation rather than rebuttal of our suggestion.

In this research, the authors used NDs colored by 40 keV He^+ ion irradiation.[56] It is, of course, clear that much brighter particles can be obtained in such a way, and we tried a somewhat similar approach. For this, we started from a liquid solution of 50 nm diamond particles in water with an initial concentration of 7×10^{12} to 7×10^{10} particles per ml, prepared using initial water (with some additives to prevent coagulation) solutions obtained from Microdiamant AG, Lengwil, Switzerland. A 10 μl droplet of a solution was deposited onto a thin quartz plate, spread over the surface around 0.5 cm^2, and dried in air. Afterwards, these plates were immediately irradiated with a He^+ ion beam, either with an energy of 1 MeV and a dose of 3×10^{13} ions per cm^2 in a Van der Graaf generator at the Shevchenko Institute of Applied Physics Problems, Belarusian State University, Minsk, Belarus, or with an energy of 130 keV, 500 keV or 1.7 MeV and a typical dose of 10^{15} ions per cm^2 in a Tandetron accelerator at the Indira Gandhi Center for Atomic Research, Kalpakkam, India (proton irradiation with an energy of 1 MeV and similar doses were also attempted in India). After a few days, these quartz plates were annealed in a technical vacuum at 800 °C for two hours. Based on the data reported in ref. 56 and other available data concerning defect production in dielectrics by high energy ions, we estimated that under these conditions we should be close to the saturation, *i.e.* most of the nitrogen vacancies could be transformed into fluorescent centers. These samples were indeed very bright, and single fluorescent nanodiamond crystals and/or clusters of ND particles "glued" together were easily seen with our SNOM when excited by a 514 nm line of the CW argon ion laser in transmission mode, see Fig. 5. An appropriate Kaiser Optical Systems (Ann Arbor, MI, USA) notch filter and a set of color filters cutting the fluorescence shorter that 590 nm were used during signal detection. Unfortunately, all attempts to "lift off" the bright nanocrystal(s) from the quartz sample by an "approach–(gentle) pressure–contraction" cycle[52,57] using a bare or modified (for example, coated with a thin poly-L-lysine layer) SNOM tip to prepare an active FRET SNOM tip grafted with ND crystals, failed. Attempts to observe FRET SNOM images with these samples using "self-sharpening pencil" aperture SNOM probes coated with thin layers of the aforementioned appropriate dyes[12] were also unsuccessful.

Besides NV centers in NDs, other types of defects have been considered for use in quantum informatics and could be prospective candidates for FRET SNOM. Diamond contains more than 500 types of defect centers, and some of them look quite suitable, probably even more suitable than NV centres, for the task at hand. Analysis of the literature shows that, for example, chromium-related color centers, which possess a narrow fluorescence line around 750 nm and for which a

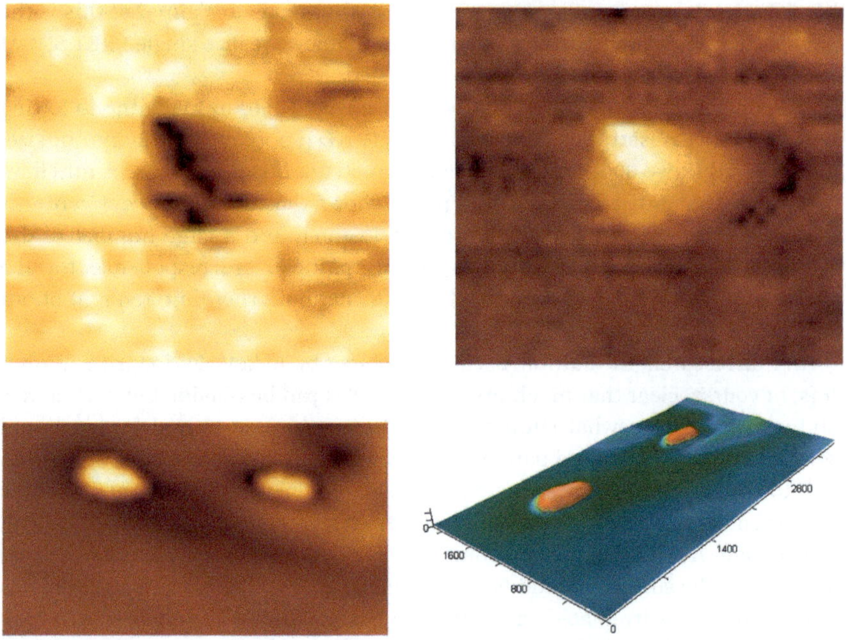

Fig. 5 Up: Topographical (left) and near-field fluorescence (right) images of 50 nm-size diamond nanoparticles colored by 1.7 MeV He^{+} ion irradiation in Kalpakkam, India; scan size: 150 × 150 nm. Down: Near-field optical fluorescence image (left) and its 3D reconstruction (right) of the same particles colored by 130 keV He^{+} ion irradiation in Kalpakkam, India; scan size 3.5 μm × 2 μm.

very large quantum efficiency was reported,[58] look quite interesting. Similarly, Si-related color centers (narrow fluorescence line around 740 nm) look interesting as well. Recently, the fact that SiV centers, contrary to the NV centers, are stable and bright emitters in particles as small as 1.6 nm in size, resulted in them becoming a main topic of research.[59] Note that these centers are also theoretically predicted to be stable in such small crystals.[60] Similarly, Xe-related vacancies[61] can be considered, and so on. However, questions regarding their thermodynamic stability when close to the nanoparticle surface, as well as detailed studies on the photostability of all these color centers (which is usually claimed to be quite large but we failed to find exact data) and their quantum efficiency must be investigated before further steps in this direction will become possible.

3.2. The possibilities of FRET SNOM with rare earth ions in (nano)crystals

The efficiency of FRET or, as is better to say, the value of the Förster radius, is independent of the donor dipole moment.[3–5] This is a prerequisite for the use of much weaker donors than dye molecules, semiconductor nanocrystals or color centres, for single molecule FRET SNOM. If one is willing to use rare earth ions in crystals for this purpose (for many of the optical transitions in them, the quantum efficiency is close to 100%, even at room temperature[23–26]), due to the fact that their optical absorption/emission cross sections (dipole forbidden intraband f–f transitions) are typically roughly three orders of magnitude smaller than those pertinent for dyes,[23–26] instead of thousands of counts per s, he or she will have a signal on the order of only 10 counts per s (see eqn (1)). Knowing that for relatively strong transitions of rare earth ions in crystals the fluorescence time ranges from tens–hundreds of microseconds, the exploitation of μs-duration laser pulses with a repetition rate around 100 Hz, *i.e.* something similar to what we actually used in our experiments with an acousto-optical modulator,[27] looks desirable and can lead to signal–noise ratio improvements overcoming the dark noise detector limit (see above; again, the total laser power at the SNOM fiber probe entrance should not exceed a few mW). But, of course, in any case, such a low level of the signal presupposes slow scanning, and it is hard to say whether such an approach is realizable and might be useful in practice.

Certainly, the optical *excitation* cross section of a rare earth ion can be a few orders of magnitude larger than the corresponding optical absorption/emission cross section if the energy transfer from the organic ligands or nanoparticles surrounding this ion is involved.[23] When can their use help? Due to the very nature of FRET, if the donor–acceptor distance is equal to the Förster distance, R_0, then the rate of energy transfer is equal to that of the radiative decay, and hence the process of imaging at such distances remains slow. Interesting possibilities might appear if we are able to work at donor–acceptor distances, say, twice as small as the R_0 value. This will lead to a roughly $2^6 = 64$ times increase in the FRET rate. Consequently, one may observe a sufficiently strong FRET signal when the active ion of the SNOM probe is very close to an acceptor, and see nothing when there is no appropriate acceptor in its vicinity. Besides, the scanning rate might also be essentially increased, and all this certainly looks promising for FRET SNOM.

Currently, we would not like to speculate more on this topic and will limit ourselves with just an indication of some optical transitions in rare earth ions in crystals, which seem interesting for future work in this direction. The list might be not exhaustive and we did not search for optimal host matrices for the rare earth ions and optical transitions in question (being fully aware of the large importance of this question). Analysis of the literature enables us to propose the following three transitions, each of which possesses an optical emission cross section of more than 10^{-19} cm^2 and a fluorescence time not exceeding 200 μs. First is the famous $^4F_{3/2}$–$^4I_{11/2}$ transition centred around 1.06 μm for the Nd^{3+} ion. In some cases, transitions to the level neighbouring $^4I_{11/2}$ might be used, which gives rise to a fluorescence wavelength around 880 nm. Then, there is the $^1D_2$–$^3H_4$ transition in the Pr^{3+} ion centred around 600–606 nm, and, finally, there is the $^3H_4$–$^3H_6$ transition in the trivalent thulium ion centred around 795 nm. Famous and extremely important for telecommunication, the *ca.* 1.5 microns $^4F_{13/2}$–$^4F_{9/2}$ transition in the erbium ion is strong but, besides probably being "too IR" for FRET SNOM purposes, it also possesses quite long fluorescence times, ≥2 ms.

In this section we would also like to indicate one more interesting possibility. This is not f–f intraband transitions, but electric dipole allowed f–d transitions in some rare earth ions in crystals. Indeed, a single (bivalent) samarium ion has been detected in a CaF_2 crystal already in 1988 (ref. 62) by observing the f–d transition at 708 nm, but this was done at 77 K, while at room temperature the quantum yield of the transition in question is prohibitively small. The situation is very different for three-valent Pr and Ce ions in different crystals, where, even at room temperature, the quantum efficiency of the f–d transition is close to unity,[63] which, in particular, enabled the observation of a single praseodymium ion in

YAG[64] and YSO[65] crystals recently. Different nanocrystals or nanoparticles containing these ions should be seriously considered as candidates for single molecule FRET SNOM in the UV and violet–blue spectral regions. Indeed, the reported parameters of *e.g.* $LiCaAlF_6:Ce^{3+}$ lasing medium, *viz.* the optical emission cross section $\sigma = 6.8 \times 10^{-18}$ cm^2 and the fluorescence time = 50 ns (band 280–315 nm),[58] show that the corresponding systems should be quite comparable with FRET SNOM probes exploiting dye molecules (but we hope much more photostable). Of course, a lot of things should still be clarified (including the influence of the nanoparticle surface on FRET and details of the UV SNOM construction), but actually the research aiming to use $LaF_3:Pr^{3+}$ and $LaF_3:Ce^{3+}$ nanoparticles (lasing in $LaF_3:Ce^{3+}$ crystals has also been reported[66]) made by microwave-hydrothermal techniques[67] as FRET SNOM probes has already started.

4. Single molecule coherent FRET SNOM at liquid helium temperatures

In the Introduction, we mentioned that for many rare earth ions in crystals at liquid helium temperatures, inter-ion FRET is a *coherent* process and electronic excitation energy flows back and forth between the donor and acceptor, having dipole moments of d_D and d_A, respectively, and located at a distance r from each other, with the characteristic time:[22]

$$T_{\mathrm{FRET}} = \frac{\pi\hbar r^3}{\alpha\kappa d_D d_A}. \tag{2}$$

Here α and κ are the geometrical and spectral factors (on the order of 1), respectively, which take into account the mutual orientation of two dipoles and their spectral overlap. Accepting $d_D = d_A$, we obtain the following estimation of the coherent FRET Förster radius, where the FRET rate is equal to that of the radiative decay:

$$r = \left(\frac{3\alpha\kappa\eta}{8\pi^5}\right)^{1/3}\lambda. \tag{3}$$

Here λ is the wavelength and η takes into account the ratio between the radiation decay time T_1 and the transverse relaxation time T_2, $T_2 = \eta T_1$. We argued that in typical cases this radius is around 20 nm (ref. 22) (afterwards this simple calculation has been confirmed by rigorous quantum electrodynamics treatment[68]), so that, given the characteristic microsecond-range decoherence times, the process can be controlled by a modern scanning probe microscope working at liquid helium temperatures. Using a pseudospin 1/2 model (the excited state of an ion corresponds to the state with the pseudospin z-projection equal to +1/2, and unexcited equal to −1/2) and its generalizations for systems involving many interacting dipoles,[69,70] protocols on how to create qubits in well-defined quantum states and how to realize CC···CNOT operators were proposed.[22] Later on, we reported the first experiments in the field (photon echo research on M-pair centers in Nd^{3+} ions in calcium fluoride; the interion interaction in this system is exactly the coherent interaction we are interested in)[69] and proposed their continuation.[70] Quite recently, the possibility of a *coherent cooperative* FRET process in rare earth ions has been predicted.[71]

Certainly, experiments in this field require a sophisticated SNOM arrangement. The appropriate device should not only simply work at liquid helium temperatures, but it should be capable of fast and precise scanning, which is necessary for *e.g.* addressing certain pre-selected (groups of) rare-earth ions or the like, *cf.* ref. 22, 69 and 70. Analysis of the literature shows that this problem, albeit difficult, can be solved and a few research groups have previously discussed how low temperature SNOM devices[72–74] look suitable for the task. In our case, we plan to use the recently constructed and available for joint work SNOM,[75,76] whose construction is rather similar to our SNOM.[27,28] It also exploits the proprietary double-resonance optical fiber probe montage onto a quartz tuning fork (thus proven to work at low helium temperature) so that this microscope has the same advantages, including the fast scanning, which makes it especially suitable for the realization of the aforementioned quantum informatics-related experiments.

As a first application, we are considering an experiment demonstrating the possibility of the entanglement of two spatially separated macroscopic bodies, *viz.* small microcrystals containing rare earth ions. The SNOM tip and sample that are presented in Fig. 1, left, can be used for illustration, but now we suppose that the sample is, say, a $LiYF_4$:Er single crystal with a concentration of Er ions *ca.* 0.1 at%, while other similar $LiYF_4$:Er single crystals of a few microns in size are attached onto the SNOM tip. The whole setup is at liquid helium temperatures (this particular crystal is selected because it is characterized by the especially small ratio of inhomogeneous to homogeneous spectral line broadening[77]). With a suitable laser excitation, numerous pairs of ions (donors and acceptors) divided between these two microcrystals are subject to coherent dipole–dipole interactions (FRET). It can be shown that starting from two microcrystals in contact and disengaging after the end of the laser pulse irradiation (by applying an appropriate HV electric pulse onto a *z*-piezo carrying the SNOM probe), one is able to entangle up to millions of rare earth ions belonging to two different macrobodies (microcrystals).

For the above estimation, we supposed an application of time-varying electric potential onto the SNOM tip to make different donor–acceptor pairs resonant with each other, by exploiting the Stark shift of the erbium ion optical transition frequency.[78] It is worthwhile to note that the possibility to exploit local Stark-induced resonance dipole–dipole interactions for coherent optical dipole coupling of two perylene molecules has been demonstrated earlier.[79]

At the current state of research, it would be premature to discuss in detail other planned combinations of FRET SNOM with quantum information processing approaches, such as the nanolocally controlled frequency shifts[80,24] and the coherent reversible inhomogeneous broadening of the dopant ion spectral line.[81,24] Still, we believe that this very short account demonstrates interesting and exciting perspectives of the coherent FRET SNOM approach for quantum computing and related topics.

5. Conclusion

First, we have discussed the current state of single molecule FRET SNOM at room temperature. Despite our hopes to profit from the ultrahigh photostability of NV color centers in nanodiamond crystals, this work did not result in the preparation of FRET-active SNOM probes containing single fluorescent centers at the apex suitable for everyday use, although there are many indications that this task can be solved with (alternative to NV) color centers in nanodiamonds and rare earth ions in crystals. The importance of the practical realization of the proposed method, in our opinion, is evident and does not need to be discussed at length.

We have also discussed novel possibilities concerning quantum informatics applications of coherent FRET SNOM at liquid helium temperatures. Here we profit a lot from the fact that, in comparison to 2003 when our proposal was put forward,[22] the number of papers discussing possible quantum computing applications of rare earth ions in crystals virtually exploded, and it has become much easier to search for appropriate crystals and optical transitions. In particular, this enabled us to design a sufficiently detailed scheme for an experiment aiming to demonstrate the entanglement of two macroscopic bodies (very briefly presented in the paper), as well as some other ideas.

Acknowledgements

The authors would like to thank M. Petrova and K. Keevend for help with the FRET SNOM experiments, and I. Sildos for useful discussions on the properties of color centers in NDs. We are grateful to the Swiss National Science Foundation for the financial support for this research over a number of years, including the latest grant N 200020-149424.

References

1 S. K. Sekatskii and V. S. Letokhov, *Appl. Phys. B: Lasers Opt.*, 1996, **63**, 525.
2 S. K. Sekatskii, Fluorescence Resonance Energy Transfer Scanning Near-Field Optical Microscopy, in *Nano-Optics and Near-Field optical Microscopy*, ed. A. Zayats and D. Richards, Artech, London, 2009.
3 V. M. Agranovich and M. D. Galanin, *Electron excitation energy transfer in condensed matter*, Amsterdam, North Holland, 1982.
4 R. M. Clegg, Fluorescence resonance energy transfer, in *Fluorescence imaging spectroscopy and microscopy*, ed. X. F. Wang and B. Herman, John Wiley and sons, New York, 1996.
5 E. A. Jares-Erijman and T. M. Jovin, *Nat. Biotechnol.*, 2003, **21**, 1387.
6 M. Ohtsu, *Near-field nano/atom optics and technology*, Springer, Tokyo, 1998.
7 R. C. Dunn, *Chem. Rev.*, 1999, **99**, 2891.
8 L. Novotny and B. Hecht, *Principles of nano-optics*, Cambridge Univ. Press, Cambridge, 2006.
9 S. A. Vickery and R. C. Dunn, *Biophys. J.*, 1999, **76**, 1812.
10 G. T. Shubeita, S. K. Sekatskii, M. Chergui, G. Dietler and V. S. Letokhov, *Appl. Phys. Lett.*, 1999, **74**, 3453.
11 S. A. Vickery and R. C. Dunn, *J. Microsc.*, 2001, **202**, 408.
12 G. T. Shubeita, S. K. Sekatskii, G. Dietler and V. S. Letokhov, *Appl. Phys. Lett.*, 2002, **80**, 2625.
13 G. T. Shubeita, S. K. Sekatskii, G. Dietler, I. Potapova, A. Mews and T. Basché, *J. Microsc.*, 2003, **210**, 274.
14 Y. Ebenstein, T. Mokari and U. Banin, *J. Phys. Chem. B*, 2004, **108**, 93.
15 F. Müller, S. Götzinger, N. Gaponik, H. Weller, J. Mlynek and O. Benson, *J. Phys. Chem. B*, 2004, **108**, 14527.

16 S. K. Sekatskii, G. Dietler, F. Bonfigli, S. Loreti, T. Marolo and R. M. Montereali, *J. Lumin.*, 2007, **122–123**, 362.
17 S. K. Sekatskii, G. Dietler and V. S. Letokhov, *Chem. Phys. Lett.*, 2008, **452**, 220.
18 R. Schirhagl, K. Chang, M. Loretz and C. L. Degen, *Annu. Rev. Phys. Chem.*, 2014, **65**, 83.
19 I. Aharonovich, S. Castelletto, D. A. Simpson, C.-H. Su, A.-D. Greentree and S. Prawer, *Rep. Prog. Phys.*, 2011, **74**, 076501.
20 F. Jelezko and J. Wrachtrup, *Phys. Status Solidi A*, 2006, **203**, 3207.
21 S. Pezzagna, D. Rogalla, D. Wildanger, J. Meijer and A. Zaitsev, *New J. Phys.*, 2011, **13**, 035024.
22 S. K. Sekatskii, M. Chergui and G. Dietler, *Europhys. Lett.*, 2003, **63**, 21.
23 G. Liu and B. Jacquier, *Spectroscopic properties of rare earths in optical materials*, Springer, Berlin, 2005.
24 P. Goldner, A. Ferrier and O. Guillot-Noel, *Handb. Phys. Chem. Rare Earths*, 2015, **46**, 1.
25 C. W. Thiel, T. Böttger and R. L. Cole, *J. Lumin.*, 2011, **131**, 353.
26 T. T. Basiev, I. T. Basieva, A. A. Kornienko, V. V. Osiko, K. K. Pukhov and S. K. Sekatskii, *J. Mod. Opt.*, 2012, **59**, 166.
27 D. V. Serebraykov, S. K. Sekatskii, A. P. Cherkun, K. Dukenbayev, I. V. Morozov, V. S. Letokhov and G. Dietler, *J. Microsc.*, 2008, **229**(2), 287.
28 A. P. Cherkun, D. V. Serebryakov, S. K. Sekatskii, I. V. Morozov and V. S. Letokhov, *Rev. Sci. Instrum.*, 2006, **77**, 033703.
29 D. V. Serebryakov, A. P. Cherkun, B. A. Loginov and V. S. Letokhov, *Rev. Sci. Instrum.*, 2002, **73**, 1795.
30 H. Chibani, K. Dukenbayev, M. Mensi, S. K. Sekatskii and G. Dietler, *Ultramicroscopy*, 2010, **110**, 211.
31 M. Mensi, G. Mikhailov, S. Pyatkin, J. Adamcik, S. Sekatskii and G. Dietler, *Proc. SPIE*, 2010, **7712**, 771222.
32 R. Stöckle, C. Fokas, V. Deckert, R. Zenobi, B. Sick, B. Hecht and U. P. Wild, *Appl. Phys. Lett.*, 1999, **75**, 160.
33 A. Sayah, C. Philipona, P. Lambelet, M. Pfeffer and F. Marquis-Weible, *Ultramicroscopy*, 1998, **71**, 59.
34 J. Adamcik, F. Valle, G. Witz, K. Rechendorff and G. Dietler, *Nanotechnology*, 2008, **19**, 384016.
35 S. Kühn, S. Hettich, S. Schmitt, J.-P. Poizat and V. Sandoghdar, *J. Microsc.*, 2001, **202**(1), 2.
36 E. Rittweger, K. Y. Han, S. E. I. Irvine, C. Eggeling and S. W. Hell, *Nat. Photonics*, 2009, **3**, 144.
37 S. Arroyo-Camejo, M. P. Adam, M. Besbes, J.-P. Hugonin, V. Jacques, J.-J. Greffet, J.-F. Roch, S.-W. Hell and F. Treussart, *ACS Nano*, 2013, **7**, 10912.
38 A. Gali, *Phys. Rev. B: Condens. Matter Mater. Phys.*, 2009, **79**, 235210.
39 T. Gaebel, M. Domhan, C. Wittmann, I. Popa, F. Jelezko, J. Rabeau, A. Greentree, S. Prawer, E. Trajkov, P. R. Hemmer and J. Wrachtrup, *Appl. Phys. B: Lasers Opt.*, 2006, **82**, 243.
40 K. Iakubovskii, G. J. Adriaenssens and M. Nesladek, *J. Phys.: Condens. Matter*, 2000, **12**, 189.
41 U. Brackmann, *Lambdachrome laser dyes*, Lambda Physik GmbH, Goettingen, 1997.

42 C. Santori, P. E. Barclay, K.-M. C. Fu and R. G. Beausoleil, *Phys. Rev. B: Condens. Matter Mater. Phys.*, 2009, **79**, 125313.
43 L. Rondlin, G. Dantelle, A. Slablab, F. Grosshans, F. Treussart, P. Bergonzo, S. Perruchas, T. Gacoin, M. Chaigneau, H.-C. Chang, V. Jacques and J.-F. Roch, *Phys. Rev. B: Condens. Matter Mater. Phys.*, 2010, **82**, 115449.
44 K. Ohashi, T. Rosskopf, H. Watanabe, M. Loretz, Y. Tao, R. Hauert, S. Tomizawa, T. Ishikawa, J. Ishi-Hayase, S. Shikata, C. L. Degen and K. M. Itoh, *Nano Lett.*, 2013, **13**, 4733.
45 M. Kaviani, P. Deak, B. Aradi, T. Frauenheim, J.-P. Chou and A. Gali, *Nano Lett.*, 2014, **14**, 4772.
46 I. I. Vlasov, O. Shenderova, S. Turner, O. I. Lebedev, A. A. Basov, I. Sildos, M. Rähn, A. A. Shiryaev and G. van Tendeloo, *Small*, 2010, **6**, 687.
47 G. Dantelle, A. Slablab, L. Rondin, F. Lainé, F. Carrel, P. Bergonzo, S. Perruchas, T. Gacoin, F. Treussart and J. E. Roch, *J. Lumin.*, 2010, **130**, 1655.
48 G. Davies and M. F. Hamer, *Proc. R. Soc. London, Ser. A*, 1976, **348**, 285.
49 A. Bevaratos, R. Brouri, T. Gacoin, J.-P. Poizat and P. Grangier, *Phys. Rev. A: At., Mol., Opt. Phys.*, 2001, **64**, 061802R.
50 A. Gruber, A. Dräbenstedt, C. Tietz, L. Fleury, J. Wrachtrup and C. Von Borczyskowki, *Science*, 1997, **276**, 2012.
51 T. L. Wee, Y. K. Tzeng, C. C. Han, H. C. Chang, W. Fann, J. H. Hsu, K. M. Chen and Y. C. Yu, *J. Phys. Chem. A*, 2007, **111**, 9379.
52 A. Drezet, Y. Sonnefraud, A. Cuche, O. Mollet, M. Berthel and S. Huant, *Micron*, 2015, **70**, 55.
53 A. S. Barnard and M. Sternberg, *J. Phys. Chem. B*, 2005, **109**, 17107.
54 C. Bradac, T. Gaebel, N. Naidoo, M. J. Sellars, J. Twamley, L. J. Brown, A. S. Barnard, T. Plakhotnik, A. V. Zvyagin and J. R. Rabeau, *Nat. Nanotechnol.*, 2010, **5**, 345.
55 Y.-Y. Chen, H. Shu, Y. Kuo, Y.-K. Tzeng and H.-H. Chang, *Diamond Relat. Mater.*, 2011, **20**, 803.
56 Y.-R. Chang, H.-Y. Lee, K. Chen, C.-C. Chang, D.-S. Tsai, C.-C. Fu, T.-S. Lim, Y.-K. Tzeng, C.-C. Fang, C.-C. Han, H.-H. Chang and W. Fan, *Nat. Nanotechnol.*, 2008, **3**, 284.
57 A. Cuche, A. Drezet, Y. Sonnefraud, O. Faklaris, F. Treussart, J. F. Roch and S. Huant, *Opt. Express*, 2009, **17**, 19969.
58 I. Aharonovich, S. Castelletto, D. A. Simpson, A. Stacey, J. McGallum, A. D. Greentree and S. Prawer, *Nano Lett.*, 2009, **9**, 3195.
59 I. I. Vlasov, A. A. Shiryaev, T. Rendler, S. Steinert, S.-Y. Lee, D. Antonov, M. Vörös, F. Jelezko, A. V. Fisenko, L. F. Semjonova, J. Biskupek, U. Kaiser, O. I. Lebedev, I. Sildos, P. R. Hemmer, V. I. Konov, A. Gali and J. Wrachtrup, *Nat. Nanotechnol.*, 2014, **9**, 54.
60 A. S. Barnard, I. I. Vlasov and V. G. Ralchenko, *J. Mater. Chem.*, 2009, **19**, 360.
61 Y. Deshko and A. Gorokhovsky, *Phys. Status Solidi B*, 2013, **250**, 278.
62 R. Lange, W. Grill and W. Martienssen, *Europhys. Lett.*, 1988, **6**, 499.
63 M. J. Weber, *Solid State Commun.*, 1973, **12**, 741.
64 R. Kolesov, K. Xia, R. Reuter, R. Stöhr, A. Zappe, P. R. Hemmer and J. Wrachtrup, *Nat. Commun.*, 2012, **3**, 1029.
65 T. Utikal, E. Eichhammer, L. Petersen, A. Renn, S. Götzinger and V. Sandoghdar, *Nat. Commun.*, 2014, **5**, 3627.
66 D. J. Ehrlich, P. F. Moulton and R. M. Osgood Jr, *Opt. Lett.*, 1979, **4**, 184.

67 A. S. Vanetsev, E. V. Samsonova, O. M. Gaitko, K. Keevend, A. V. Popov, U. Mäeorg, H. Mändar, I. Sildos and Y. V. Orlovskii, *J. Alloys Compd.*, 2015, **639**, 415.
68 V. V. Klimov, S. K. Sekatskii and G. Dietler, *J. Mod. Opt.*, 2004, **51**, 1919.
69 S. K. Sekatskii, T. T. Basiev, I. T. Basieva, G. Dietler, V. V. Fedorov, A. Y. Karasik, Y. V. Orlovskii and K. K. Pukhov, *Opt. Commun.*, 2006, **259**, 298.
70 I. T. Basieva, T. T. Basiev, G. Dietler, K. K. Pukhov and S. K. Sekatskii, *Phys. Rev. B: Condens. Matter Mater. Phys.*, 2006, **74**, 165329.
71 S. K. Sekatskii and K. K. Pukhov, *Opt. Spectrosc.*, 2014, **117**, 902.
72 J. Michaelis, C. Hettich, J. Mlynek and V. Sandoghdar, *Nature*, 2000, **405**, 325.
73 M. Gregor, R. Henze, T. Schroder and O. Benson, *Appl. Phys. Lett.*, 2009, **95**, 153110.
74 M. Skacel, M. Francardi, A. Gerardino, B. Alloing, L. Li and A. Fiore, *J. Phys.: Conf. Ser.*, 2010, **245**, 012040.
75 M. G. Petrova, G. V. Mishakov, E. I. Demikhov and A. I. Sharkov, *Bull. Lebedev Phys. Inst.*, 2010, **37**, 276.
76 V. V. Prokhorov, M. G. Petrova, S. I. Pozin, N. N. Kovaleva and E. I. Demikhov, *Curr. Nanosci.*, 2014, **10**, 700.
77 R. M. Macfarlane, A. Cassanho and R. S. Meltzer, *Phys. Rev. Lett.*, 1992, **69**, 542.
78 R. M. Macfarlane, *J. Lumin.*, 2007, **125**, 156.
79 C. Hettich, C. Scmitt, J. Zitzmann, S. Kühn, I. Gerhardt and V. Sandoghdar, *Science*, 2002, **298**, 385.
80 N. Ohlsson, R. Monah and S. Kroell, *Opt. Commun.*, 2002, **201**, 71.
81 A. L. Alexander, J. J. Longdell, M. J. Sellars and N. B. Manson, *Phys. Rev. Lett.*, 2006, **96**, 043602.

Faraday Discussions

PAPER

Mechanical matching between a ligand and receptor

Ana Peñaherrera

Received 10th June 2015, Accepted 7th July 2015
DOI: 10.1039/c5fd00106d

Interactions between ligands and receptors and subsequent "locking" must involve some resistance to unbinding, manifesting itself as an interaction force. At body temperature, spontaneous unbinding will occur, however, external forces are required to accelerate this process. Bearing in mind the potential forces that the receptor–ligand complex is likely to be subjected to in a biological environment, it might be hypothesised that there is some mechanical matching between the receptor and ligand. To test this hypothesis, various receptor and ligand pairs were unfolded in their entirety in order to determine their total unfolding force. In this way, the total force to unfold the protein could be determined, allowing a comparison between ligand and receptor pairs. The interest of this work is to examine the interaction between five proteins and a mica surface by AFM without any modification to preserve the natural elastic properties of the protein molecules during the force measurements. The results showed a mechanical matching between GP120 (ligand) and CD4 (receptor) when analysing the total force required to unfold the same number of domains or events shown by the force distance curves of these proteins.

Introduction

The first time AFM was used to mechanically unfold proteins was when the giant protein titin, found in muscle, was spanned between the AFM tip and the support. Separating these two from each other led to the unfolding of the titin domains and the corresponding forces were measured with the Bell's theory described by Evans and Ritchie in 1997.[1,2] Mechanical experiments performed on single biomolecules offer invaluable insights into their structure and function.[3] Importantly, the unfolding force could not be considered naively as a fixed characteristic value. As demonstrated by dynamic force spectroscopy, as the pulling rate increases, the unfolding force measured becomes higher.[2,4,5] However, the vertical tip jump during pull-off makes it possible to estimate the interaction forces, and the width and shape of the retraction curve are indicative of molecular elongation processes associated with entropically unfavourable

Department of Chemical Engineering and Advanced Materials, Newcastle University, Newcastle Upon Tyne, NE1 7RU, UK. E-mail: ab.penaherrera59@gmail.com

molecular unfolding and elongation.[6] AFM has been successfully utilized to measure intramolecular unfolding forces of individual proteins and intermolecular forces between various ligands and receptor pairs. Remarkably, the force needed to separate a ligand from its specific binding site is different from the force required to remove a non-specifically bound ligand.[6,7,8]

The elastic properties of stretching proteins can be described with the worm-like chain (WLC) model for entropic elasticity. Eqn (1) describes the approximate analytic formula.[3]

$$F(x) = \left(\frac{k\,T}{b}\right)\left[0.25\left(1-\frac{x}{L}\right)^{-2} - 0.25 + \frac{x}{L}\right] \tag{1}$$

This model dictates that the force needed to stretch a linear polymer in a solvent to a length x is given by Bustamante *et al.*, in which b is the persistence length and L is the contour length.[9,10]

When the adsorption properties of free floating molecules are desired, force spectroscopy of surface bound proteins with a naked tip can elucidate a wealth of information.[11] However, the surface reactivity of fibronectin (Fn) depends on how Fn exhibits its binding sites when adsorbed on a surface.[12] Moreover, several surface properties have been proposed to influence protein adsorption and cell behaviour, for instance: charge, topography, surface energy and surface chemistry.[13]

Many proteins with mechanical functions, such as the muscle protein titin, extracellular matrix (ECM) proteins fibronectin and tenascin, as well as other proteins that participate in cellular adhesion like cadherins, are composed of similarly structured globular domain connected series. Commonly, they consist of a single polypeptide chain that is able to locally fold into well-separated domains. These domains act like beads on a string and this design favours the entropic elasticity of the protein that can then be exploited for modulating the length and tension of various proteinaceous elastic cellular structures.[14]

Cellular adhesion is mainly mediated by integrins, cell-surface receptors that comprise an expanding family of transmembrane heterodimers of an α and β subunit. Integrins have been considered to play a key role in the malignant behaviour of neoplastic cells. Integrins recognize fibronectin through the RGD site that is comprised in the 10th type III repeat (10FNIII). Integrin binding to the Fn RGD-loop on 10FNIII results in the formation of a cytoskeletal apparatus that mechanically joins actin contractility to extracellular fibronectin fibers.[15]

Protein assembly at interfaces is essential for disease pathology and biological function. HIV-1 envelope glycoprotein mediates the fusion of viral and target cell membranes, the first critical step towards infection.[16] The precursor of the envelop protein, GP160, constitutes a trimer, which is then sliced by a furin-like protease into two noncovalently connected fragments, GP120 for receptor binding and GP41 for membrane fusion,[17] as schematically depicted in Fig. 1.

The self-assembly of the HIV envelop protein GP160 and its interaction with lipid bilayers have been studied.[18] Two types of assembly have been observed, pore and large aggregates formation. They also observed that GP160 sank in a one component lipid bilayer over time. Activation of GP160 demands cleavage into two fragments, GP41, which is the transmembrane section, and GP120, which is the protruding spike.[18] Spikes of GP120 protruding from the viral envelope bilayer

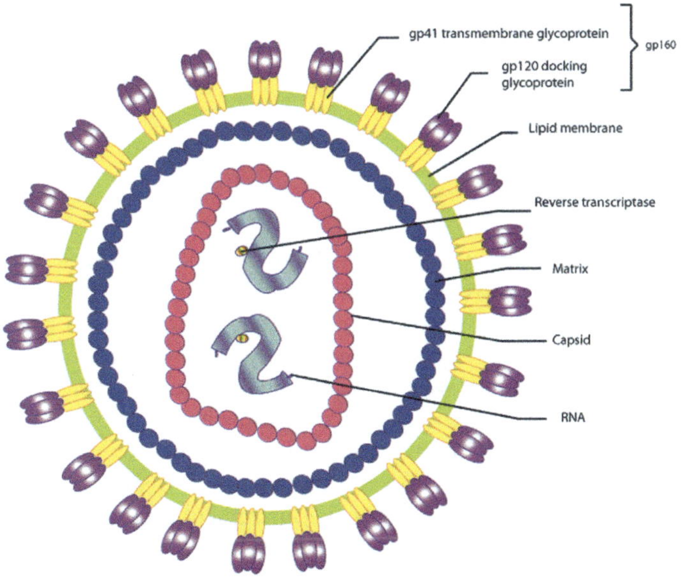

Fig. 1 Schematic diagram of the human immunodeficiency virion. A lipid bilayer delivered from the host contains the viral glycoproteins GP41 and GP120, which together form GP160.

interact with CD4 and co-receptors embedded in the target cell membrane, anchoring the virion to the cell,[19] as illustrated in Fig. 2.

CD4 belongs to the Type I membrane proteins that consist in four extracellular immunoglobulin-like domains, a transmembrane fragment and a cytoplasmatic tail. It is present on cell surfaces of T lymphocytes, monocytes, macrophages,

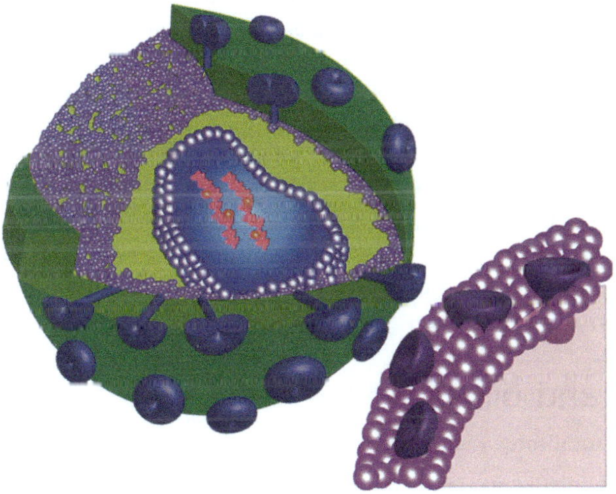

Fig. 2 Schematic illustration of GP120 spikes (blue) protruding from the viral envelope bilayer that interact with CD4 (purple).

Table 1 Receptor–ligand pairs

Receptor	Ligand
CD4	GP120
Integrin	Fibronectin

dendritic cells, and brain microglia, which are the main target cells for primate immunodeficiency viruses *in vivo* (Table 1).[15,19]

AFM is suitable for measuring the self-assembly properties of proteins and their interaction with lipids. Furthermore, AFM offers molecular level lateral resolution and subnanometer vertical resolution in a liquid environment. In force spectroscopy mode, AFM can unfold proteins of interest at defined locations, which facilitates nanomechanical identification of system components.[18]

Complications associated with ensuring proper molecular orientation, conformation and careful control over the instrument require careful interpretation of the force data.[5] With this in mind, there is a new way to consider protein domains, which may be interacting with the surface. Researchers suggest that by analysing all of the peaks in the fibronectin unfolding pattern, a parameter they refer to as the total unfolding force can be obtained. Any fibronectin domains that are interacting with a surface can be considered using the total unfolding force, even if the exact nature of each individual interaction is unknown.[18]

Interpreting sawtooth patterns relating to the unfolding of a protein from a surface is challenging. To this end, a number of factors need to be taken into consideration. Firstly, the conformation of the protein may vary according to the surface. Secondly, there is not a way to anticipate which orientation the tip will pick up the protein and whether it adsorbs to the tip *via* a single or multiple interactions. Finally, identifying which peak(s) in the sawtooth pattern relates to the protein–surface interactions is extremely difficult, and this may vary according to the characteristics of the underlying surface.[18]

Many reports on AFM studies have measured the force required to separate the ligand from the receptor. For instance, the streptavidin–biotin system with average rupture forces between 200 and 409 pN.[20,21] Moreover, Hussain *et al.* examined the interactions between integrin $\alpha_{IIb}\beta_3$ and fibrinogen.[22] However, the present work examines the interaction between various pairs of ligand and receptor molecules and a mica surface. By employing the strategy suggested by Donlon *et al.*,[18] all the peaks in the sawtooth patterns exhibited by each molecule were analysed. A mechanical matching has been observed between the ligand and receptor when comparing like for like force distance curves. This means that the curves that show the same number of events were compared for each ligand–receptor pair.

Results and discussion

Fibronectin total force measurement

By analysing all the peaks in the fibronectin unfolding pattern, a parameter, referred to as the total unfolding force,[18] considers domains that may be interacting with the mica. This analysis provides a valuable way to contrast ligand/

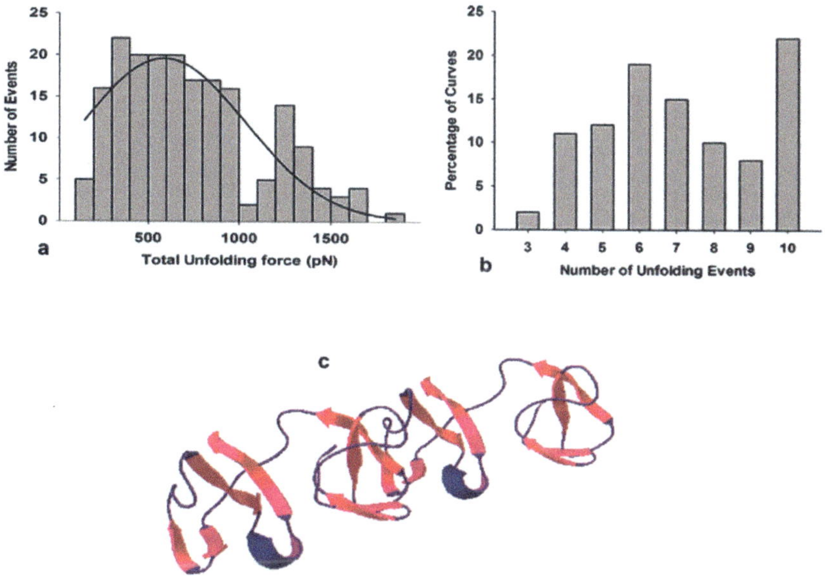

Fig. 3 (a) Histogram of the total unfolding force distribution for fibronectin on mica. The average of the total unfolding force distribution is 589.13 ± 64.58 pN (mean = 589.13, SD = 64.68, N = 195). (b) Distribution of the number of unfolding events in the sawtooth patterns for the force unfolding of fibronectin on mica. An unfolding event is recognized by a peak in the sawtooth pattern of the force distance curve. Final detachment of the protein from the tip (the last peak in the sawtooth pattern) is included in the count. The percentage of curves showing a given number of unfolding events is indicated. Only curves that could be fitted to the worm-like chain model were considered. (c) PDB (Protein Data Bank) image of fibronectin. Two fibronectin monomers join through disulphide bonds. β-strands from the upper and lower β-sheet are shown.

receptor behaviour on the same surface. The mean total unfolding force for fibronectin is 589.13 ± 64.58 pN, as can be seen in Fig. 3a.

Donlon *et al.* suggested a compact conformation of fibronectin on mica.[18] This conformation would minimize protein–surface interactions, and therefore the total unfolding force would decrease. As shown in Fig. 3b, a distribution of between three and ten events was observed. It is possible to compare like for like curves containing the same number of events,[18] which may suggest similar protein conformations between fibronectin (ligand) and the integrin (receptor).

As schematically depicted in Fig. 4, fibronectin is a multidomain protein where each domain can be classified into one of three distinct types (I, II, or III). The 10th III repeat (10FNIII) comprises the RGD site that is required for integrin recognition. This repeat has been suggested to unfold as a result of cell traction forces and this unfolding makes fibrillogenesis possible.[23] As such, 10FNIII has become a model to understand the unfolding pathway of Fn-III modules.[15] One of the most significant findings with AFM is that Fn-III modules vary significantly in mechanical stability, suggesting that the order of Fn-III unravelling is important to its function.[14,21]

The 10FNIII module was determined to be mechanically one of the weakest Fn-III modules and the β-strands progress from a twisted to an aligned state prior to

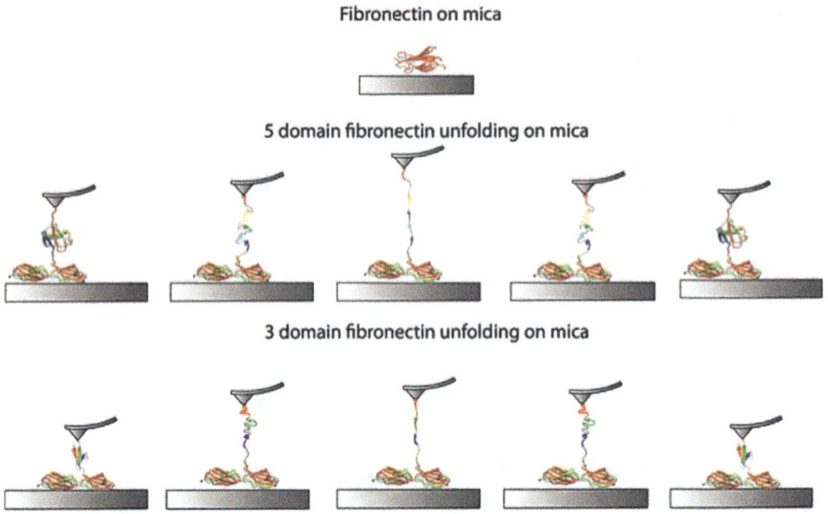

Fig. 4 Schematic representations of five and three domains of fibronectin being unfolded.

unravelling.[24,25] Gao *et al.* used multiple unfolding simulations of 10FNIII solvated in a water box to stretch it into its completely unfolded form.[15] It is worth mentioning that all –β domains from proteins of muscle or the extracellular matrix resist significantly higher forces than all –α domains, or mixed α/β proteins, even when they may be expected to experience stress *in vivo*.[24,26] In this report, fibronectin force distance curves exhibited 10 events, which relates to 10 domains, and the total force needed to unfold a certain number of domains was compared to its corresponding integrin receptor.

Integrin total force measurement

The mica surface is atomically flat over large terrace areas, and thus it is excellent for high resolution imaging of single molecules. Nordin *et al.* observed integrin aggregates on a mica surface, which may be necessary to stabilise the transmembrane protein because of the weak interaction with mica.[8] The mean total unfolding force identified for the integrin is 183.99 ± 21.36 pN (Fig. 5), which is lower than the total unfolding force determined for fibronectin. Besides, the force distance curves of the adsorbed integrin exhibit unfolding events, as well as detachment of the integrin from the tip. This result is consistent with the weak integrin–mica interactions observed by Nordin *et al.*

Force distance curves of the integrin on mica exhibit between two and ten events, which makes it possible to compare the ligand–receptor total unfolding force according to each number of events (Fig. 5b).

CD4 total force measurement

A histogram of the total unfolding force for CD4 on mica can be seen in Fig. 6. It shows an average value of 655.37 ± 33.17 pN.

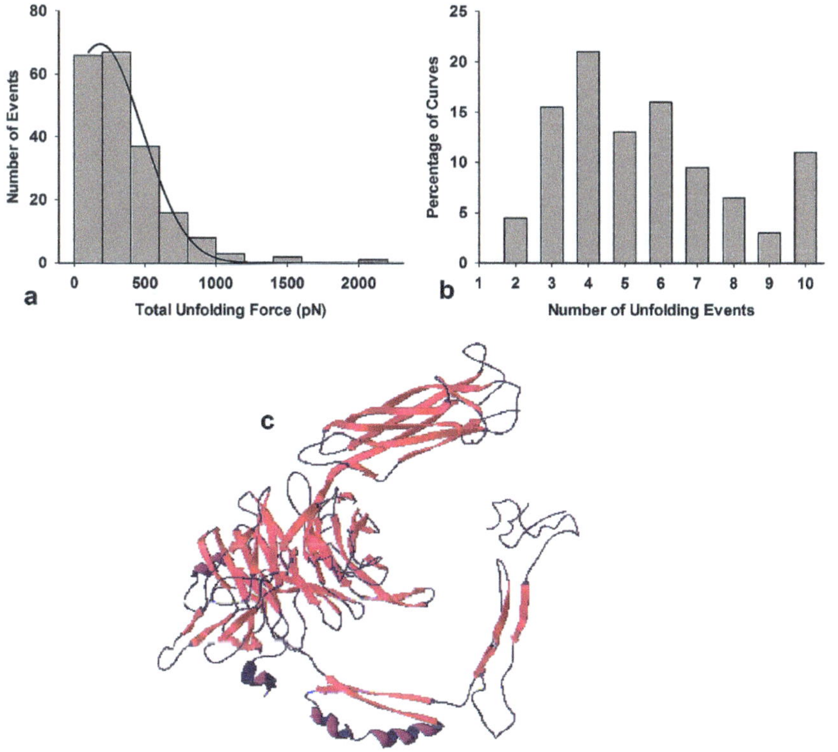

Fig. 5 (a) Histogram of the total unfolding force distribution for the integrin on mica. The average of the total unfolding force distribution is 183.99 ± 21.36 pN (mean = 183.99, SD = 21.36, N = 200). (b) Distribution of the number of unfolding events in the sawtooth patterns for the force unfolding of the integrin on mica. (c) PDB (Protein Data Bank) image of the integrin.

GP120 total force measurement

Force distance curves of GP120 on mica revealed well defined sawtooth patterns in the retraction curve, which are characteristic of a protein unfolding with individual peaks corresponding to the unfolding of individual protein domains.[18]

A histogram of the total unfolding force for GP120 on mica indicates an average value of 324.99 ± 35.52 pN (Fig. 7a).

GP160 total force measurement

Unfolding of single GP160 molecules on mica reveals characteristic, well defined sawtooth patterns in the retraction curve, which is indicative of protein unfolding with individual segments corresponding to the unfolding of single protein domains. Unfolding forces were extracted by fitting the peaks in the sawtooth pattern to the worm-like chain model. The final peak in the sawtooth pattern was discarded due to the fact that it corresponds to the detachment of the protein from the tip rather than the protein from the surface, and therefore it relates to the interaction between silicon nitride and GP160.[18]

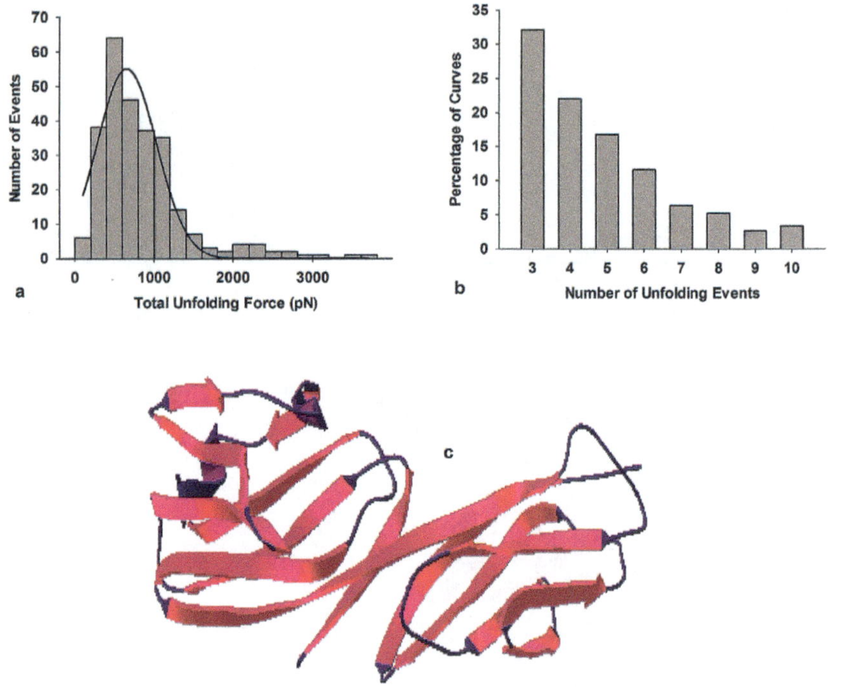

Fig. 6 (a) Histogram of the total unfolding force distribution for CD4 on mica. The average of the total unfolding force distribution is 655.37 ± 33.17 pN (mean = 655.37, SD = 33.17, *N* = 268). (b) Distribution of the number of unfolding events in the sawtooth patterns for the force unfolding of CD4 on mica. (c) PDB (Protein Data Bank) image of CD4.

A histogram of the total unfolding force for GP160 on mica can be observed in Fig. 8a. It shows an average value of 289.47 ± 11.07 pN. It is worthwhile noting that Donlon and Frankel found that GP160 is considerably easier to unfold when aggregated than isolated.[27]

Total force required to unfold each protein

For this study, the total unfolding forces of the ligand and its respective receptor were compared to look at their mechanical properties and to try to elucidate the mechanical relationship. For instance, the total force to unfold three domains of fibronectin was compared with the total force to unfold three domains of the integrin. The purpose was to compare the mean total force for all unfolding events because different proteins have different distributions of unfolding events, as can be observed in Fig. 9.

The results are presented as mean ± standard error. Statistical analysis was carried out using a Student's t test and two tailed ANOVA between each ligand and receptor pair according to the number of events.

In the case of GP120 and CD4, a mechanical matching was revealed due to the fact that the p values were higher than α ($\alpha = 0.05$) for each comparison (3–7 events). Therefore, there is not a significant difference between the total force required to unfold the same number of domains of GP120 and CD4. In contrast,

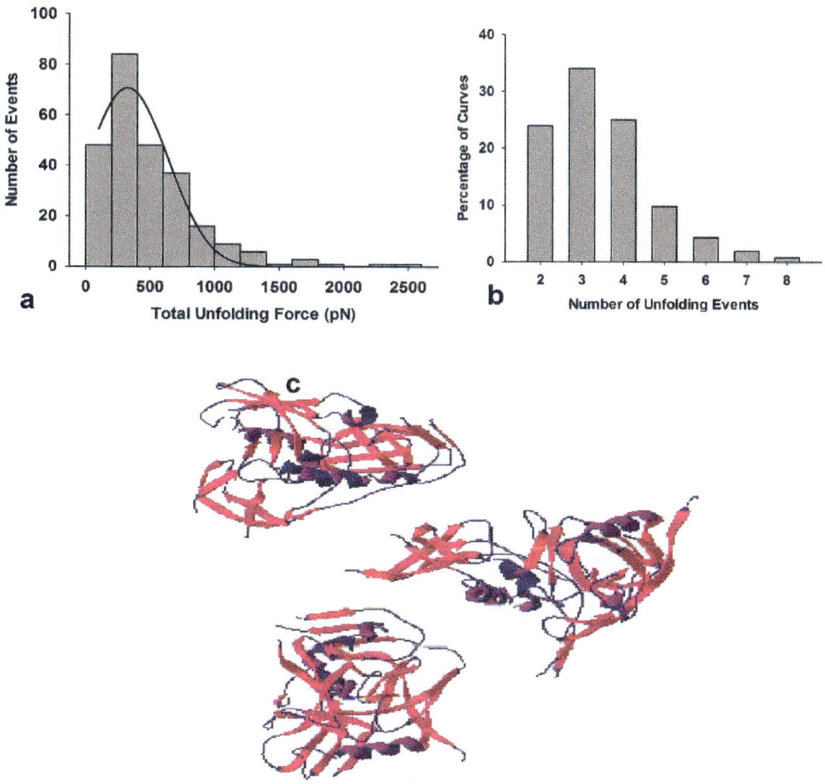

Fig. 7 (a) Histogram of the total unfolding force distribution for GP120 on mica. The average of the total unfolding force distribution is 324.99 ± 35.52 pN (mean = 324.99, SD = 35.52, N = 255). (b) Distribution of the number of unfolding events in the sawtooth patterns for the force unfolding of GP120 on mica. (c) PDB (Protein Data Bank) image of GP120.

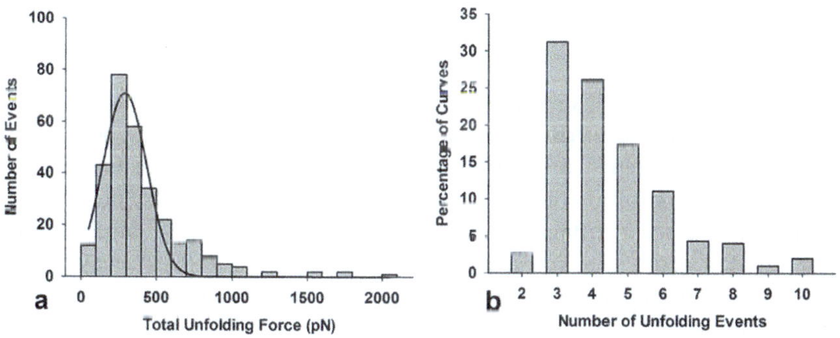

Fig. 8 (a) Histogram of the total unfolding force distribution for GP160 on mica. The average of the total unfolding force distribution is 289.47 ± 11.07 pN (mean = 239.47, SD = 11.07, N = 298). (b) Distribution of the number of unfolding events in the sawtooth patterns for the force unfolding of GP160 on mica.

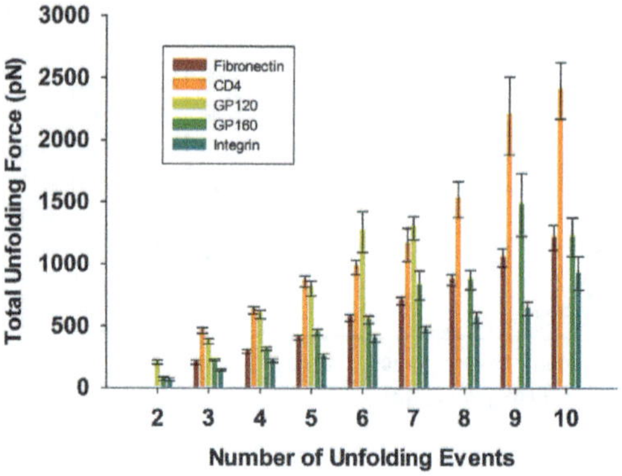

Fig. 9 The total force required to unfold each protein according to the number of events. All force distance curves displaying at least two unfolding events as well as the detachment of the protein from the tip were considered.

GP160 and CD4 do not show mechanical matching ($p < \alpha$) in any comparison as expected due to the fact that CD4 binds to GP120 and not to GP160.

The Fn and integrin total force values showed p values lower than α in almost all comparisons. The exception to this is when comparing the 10 events total unfolding force because the p value obtained for this test was $p = 0.11$. Then, there is not a significant difference in the total force when 10 Fn and integrin domains are unfolded.

As illustrated in Fig. 10, the force required to unfold the same number of domains between the ligand and receptor is compared.

Experimental

Fibronectin adsorption

Fibronectin from bovine plasma (Sigma-Aldrich, UK) was dissolved in phosphate buffered saline (PBS) (NaCl 0.138 M; KCl 0.0027 M; $Na_2HPO_4 \cdot 2H_2O$ 0.01 M; KH_2PO_4 0.002 M with pH 7.4 at 25 °C) to acquire a concentration of 50 μg ml^{-1} of the protein stock solution. Fibronectin was placed onto mica for 30 minutes (20 × 20 mm).

Integrin adsorption

The integrin used in this study was extracted from the Chinese Hamster Ovary (CHO) cell line, and derived as a Recombinant Human Integrin, $\alpha_5\beta_1$, also known as VLA5 (R&D Systems, Inc., Minneapolis, MN 55413 USA). The integrin was lyophilized from a 0.2 μm filtered solution in Tris-citrate and NaCl. Subsequently, the protein was reconstituted in sterile PBS to make a 100 μg ml^{-1} stock solution and stored at −20 °C. The integrin with a concentration of 1 μg ml^{-1} was used for adsorption onto a freshly cleaved mica surface, and left for a period of between 10 and 30 minutes.

HIV GP120 and CD4 adsorption

Aliquots of HIV GP120 (ab69717, Abcam, UK) and CD4 (ab39483, Abcam, UK) were diluted with ultrapure water to a working concentration of 20 μg ml^{-1}. Aliquots of

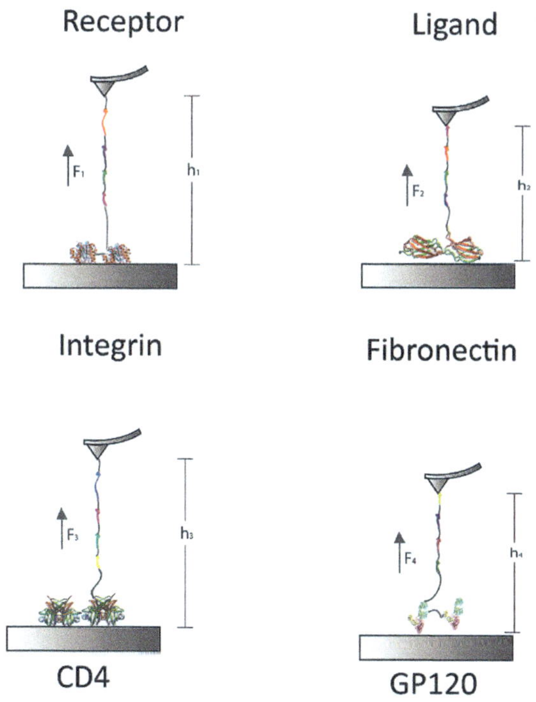

Fig. 10 Schematic representation of the ligand/receptor mechanical matching. Fn and the integrin showed a mechanical matching when 10 domains were unfolded, while GP120 and CD4 showed it in all comparisons (3–7 domains).

the stock solutions were stored at −20 °C with further dilutions carried out as necessary with ultrapure water. The proteins were adsorbed on mica surfaces (20 × 20 mm) for 30 minutes within an AFM liquid cell. Prior to imaging, the surfaces were gently washed 3 times with ultrapure water to remove residual, non-adsorbed protein.

HIV GP160 adsorption

Aliquots of HIV GP160 (ab73770, Abcam, UK) were diluted with ultrapure water to a final concentration of 20 μg ml^{-1}. Aliquots of the stock solution were stored at −20 °C. Further dilutions were carried out as necessary with ultrapure water. The protein was adsorbed for 30 minutes onto mica within an AFM liquid cell. The surfaces were then gently washed 3 times with ultrapure water to remove non-adsorbed protein prior to imaging.

Topographic AFM imaging

AFM experiments were conducted on an Agilent 5500 AFM/SPM microscope in a liquid environment (ultrapure water) at 20 °C. Images were obtained in contact mode utilizing standard nitrogen doped silicon tips with a nominal force constant of 0.02–0.77 N m^{-1}. Forces were minimized during the scans. Typical scan rates were in the range of 0.5–1 kHz at 512 pixel resolution. Imaging forces were kept below 1 nN.

Force spectroscopy

The force spectroscopy experiments are illustrated in Fig. 11 and they were performed using backside aluminium coated silicon cantilevers (PPP-CONTR, Nanosensors, Switzerland) with a nominal spring constant of 0.02–0.77 N m^{-1} and typical tip radius of less than 7 nm. The spring constants were calibrated using the equipartition theorem (Thermal K), as described by Hutter and Bechhoefer.[28] 1000 force distance curves were obtained per sample with an applied load in the range of 9–10 nN, tip velocities between 0.1 and 10 μm s^{-1} and a curve length of 1 μm. The force distance curves were analysed using Scanning Probe Image Processor (SPIP) software version 5.1.6 (Image Metrology, Lyngby, Denmark). Only force curves that fitted the worm-like chain model were accepted to represent single molecule pulling

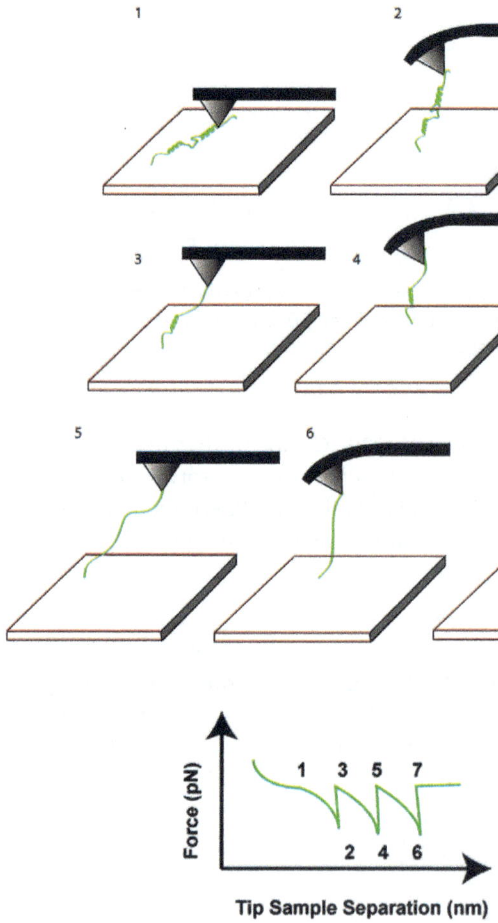

Fig. 11 Representation of two domains of protein unfolding on a mica surface (top) with a sawtooth pattern on the retraction curve that exemplifies the unfolding events and the subsequent detachment of the tip from the protein. The protein physisorbs on the tip (1) and becomes elongated as the tip retracts. When enough adhesion exists between the tip and the protein, domain denaturation events occur, corresponding to the rupture of more stable domains (2 and 4, respectively). Upon a critical elongation, the protein detaches from the tip, returns to its original conformation and remains on the surface (6–7).

events, from which unfolding forces were obtained. Determination of the rupture forces and loading rates from the force–distance spectra was carried out according to previously reported procedures (Faull *et al.*[29] and Slade *et al.*[30]).

Force spectra were taken on the bare mica before and after protein unfolding to rule out tip contamination. Drift was minimised by allowing the tip to equilibrate in the liquid cell prior to the unfolding measurements. Images before and after the spectra showed insignificant drift within the time frame of the experiments.

Conclusions

Mechanical matching between GP120 (ligand) and CD4 (receptor) has been demonstrated when comparing the unfolding total force for the same number of events. It is possible to compare like for like, which may suggest similar protein conformations and unfolding behaviour. These results provide a framework for understanding the complex entry of HIV into cells and should guide efforts to intervene.

The natural elastic properties of the protein molecules were maintained during the force measurements because there was not an AFM tip or substrate modification. This approach allows us to relate these findings to real biological systems due to the dependence of protein conformation, and thus the unfolding force, on the underlying surface characteristics, mainly features like surface chemistry and topography. It also allows us to focus our attention on the importance of measuring interactions under conditions that mimic those found *in vivo* as accurately as possible.

However, there is no way to know which orientation the tip will pick up the protein and if it adsorbs to the tip through a single or multiple interactions. Multiple orientation and attachment sites within the protein are possible although these may average out when enough curves are considered.

The total unfolding force is obtained from the sum of all individual unfolding events. The distribution of the number of unfolding events varies according to the protein. Fibronectin showed three to ten events while integrin and GP160 showed two to ten events. GP120 showed two to seven events. In contrast, CD4 exhibited three to ten events.

Acknowledgements

The author is grateful to Dr Daniel Frankel, Dr Darman Nordin and Dr Lynn Donlon for their assistance with the AFM analysis of the force distance curves.

References

1 G. Bell, *Science*, 1978, **200**, 618.
2 E. Evans and K. Ritchie, *Biophys. J.*, 1997, **72**, 1541.
3 D. Makarov, Z. Wang, J. Thompson and H. Hansma, *J. Chem. Phys.*, 2002, **116**, 7760.
4 M. Rief, M. Gautel, A. Schemmel and H. Gaub, *Biophys. J.*, 1998, **75**, 3008.
5 I. Casuso, F. Rico and S. Scheuring, *J. Mol. Recognit.*, 2011, **24**, 406.
6 C. Yip, *Curr. Opin. Struct. Biol.*, 2001, **11**, 567.

7 C. Lee, Y. Wang, L. Huang and S. Lin, *Micron*, 2007, **38**, 446.
8 D. Nordin, O. Yarkoni, N. Savinykh, L. Donlon and D. Frankel, *Soft Matter*, 2011, **7**, 10666.
9 C. Bustamante, J. Marko, E. Siggia and S. Smith, *Science*, 1994, **265**, 1599.
10 H. Mueller, H. Butt and E. Bamberg, *Biophys. J.*, 1999, **76**, 1072.
11 L. Donlon and D. Frankel, *Soft Matter*, 2013, **9**, 283.
12 J. Hull, G. Tamura and D. Castner, *Biophys. J.*, 2007, **93**, 2852.
13 E. Velzenberger, I. Pezron, G. Legeay, M. Nagel and K. Kirat, *Langmuir*, 2008, **24**, 11734.
14 P. Marszalek and Y. Dufrêne, *Chem. Soc. Rev.*, 2012, **41**, 3523.
15 M. Gao, D. Craig, V. Vogel and K. Schulten, *J. Mol. Biol.*, 2002, **323**, 939.
16 S. Harrison, *Nat. Struct. Mol. Biol.*, 2008, **15**, 690.
17 J. Kovacs, J. Nkolola, H. Peng, A. Cheung, J. Perry, J. C. Miller, M. Seaman, D. Barouch and B. Chen, *Proc. Natl. Acad. Sci. U. S. A.*, 2012, **109**, 12111.
18 L. Donlon, D. Nordin and D. Frankel, *Soft Matter*, 2012, **8**, 9933.
19 L. Donlon, A. Penaherrera and D. Frankel, *Soft Matter*, 2013, **9**, 2803.
20 S. Allen, J. Davies, A. Dawkes, M. Davies, J. Edwards, M. Parker, C. Roberts, J. Sefton, S. Tendler and P. Williams, *FEBS Lett.*, 1996, **390**, 161.
21 P. Williams, A. Moore, M. Stevens, S. Allen, M. Davies, C. Roberts and S. Tendler, *J. Chem. Soc., Perkin Trans. 2*, 2000, 5.
22 M. Hussain, A. Agnihotri and C. Siedlecki, *Langmuir*, 2005, **21**, 6979.
23 E. Gee, D. Ingber and C. Stultz, *PLoS One*, 2008, **3**, e2373.
24 A. Oberhauser, C. Badilla-Fernandez, M. Carrion-Vazquez and J. Fernandez, *J. Mol. Biol.*, 2002, **319**, 433.
25 D. Craig, A. Krammer, K. Schulten and V. Vogel, *Proc. Natl. Acad. Sci. U. S. A.*, 2001, **98**, 5590.
26 R. Best, S. Fowler, J. Toca Herrera, A. Steward, E. Paci and J. Clarke, *J. Mol. Biol.*, 2003, **330**, 867.
27 L. Donlon and D. Frankel, *Soft Matter*, 2012, **9**, 283.
28 J. Hutter and J. Bechhoefer, *Rev. Sci. Instrum.*, 1993, **64**, 1868.
29 R. Faull, N. Kovach, J. Harlan and M. Ginsberg, *J. Cell Biol.*, 1993, **121**, 155.
30 A. Slade, J. Luh, S. Ho and C. Yip, *J. Struct. Biol.*, 2002, **137**, 283.

Faraday Discussions

PAPER

An integrated system for optical and electrical detection of single molecules/particles inside a solid-state nanopore

Xin Shi,[a] Rui Gao,[a] Yi-Lun Ying,[a] Wei Si,[b] Yunfei Chen[b] and Yi-Tao Long*[a]

Received 4th May 2015, Accepted 10th June 2015
DOI: 10.1039/c5fd00060b

Nanopore techniques have proven to be useful tools for single-molecule detection. The combination of optical detection and ionic current measurements enables a new possibility for the parallel readout of multiple nanopores without complex nanofluidics and embedded electrodes. In this study, we developed a new integrated system for the label-free optical and electrical detection of single molecules based on a metal-coated nanopore. The entire system, containing a dark-field microscopy system and an ultralow current detection system with high temporal resolution, was designed and fabricated. An Au-coated nanopore was used to generate the optical signal. Light scattering from a single Au-coated nanopore was measured under a dark-field microscope. A lab-built ultralow current detection system was designed for the correlated optical and electrical readout. This integrated system might provide more direct and detailed information on single analytes inside the nanopore compared with classical ionic current measurements.

1 Introduction

Over the past two decades, nanopore-based techniques have proven to be useful tools for single-molecule detection. Nanopore techniques have become the most promising method for next-generation DNA sequencing.[1–3] In recent years, nanopores have been applied to many fields, such as the structural probing of peptides,[4] polymers,[5] protein-binding events,[6] the detection of small molecules and host–guest interactions.[7–9] Nanopore-based techniques can provide single-molecule properties during a dynamic process by driving an individual molecule that is traversing through a nanoscale channel. This characteristic enables the investigation of inter- and intramolecular interactions at the single-molecule level dynamically.[7,10]

[a]*Key Laboratory for Advanced Materials, Department of Chemistry, East China University of Science and Technology, Shanghai, 200237, P. R. China. E-mail: ytlong@ecust.edu.cn*
[b]*Jiangsu Key Laboratory for Design and Manufacture of Micro-Nano Biomedical Instruments, Southeast University, Nanjing 210096, P. R. China*

The core component of a typical nanopore setup is a nanoscale channel with a minimum diameter ranging from a few nanometers to several hundred nanometers. Conduction can only occur through a single nanochannel, which connects two chambers of an electrolyte separated by an insulating membrane. After a voltage bias is applied, a stable ionic current through the single channel (open-pore current) can be generated and recorded by a patch clamp.[11,12] If a charged analyte is electrophoretically driven through the nanopore, then a measurable current is induced.[13] Generally, the modulation of the ionic current results from the volume exclusion of the analyte, which is often defined as the "ionic-current blockade". By analyzing the duration and the amplitude of each blockade signal, physical information about an individual analyte can be determined.

The first nanopore-based sensor reported was composed of a single α-hemolysin (α-HL) with a constriction of 1.4 nm at the narrowest point.[14] To detect different analytes and to increase the sensitivity of the nanopore-based sensor, various membrane proteins have been studied, including areolysin,[15] MspA,[16,17] ClyA,[17] FhuA,[18] membrane-adapted phi29 motor protein,[19] and SP1.[20] In addition to biological nanopores, there is another category of nanopores known as solid-state nanopores, which are fabricated on thin solid-state membranes by industry standard semiconductor processing techniques, such as an ion/electron beam.[21] A dielectric material, such as low-stress silicon nitride (SiN_x), supported on silicon frames is probably the most widely used substrate for drilling the nanopore.[12] Solid-state nanopores can also be fabricated on other high-performance materials, such as SiO_2,[22] Al_2O_3[23] and graphene.[24–26] Modern industrial processing allows for the parallel production of nanopore chips and control of the nanochannel geometry.

Although an abundant amount of information can be deduced from the ionic current signal, ionic current-based nanopore methods still face challenges. For example, many signals from single-analyte events should be acquired for statistical analysis due to the randomness of nanopore-based single-molecule detection. To accelerate the measurements, a multichannel nanopore array could be introduced so that a parallel readout of multiple pores can be achieved by measuring the ionic current from each nanopore. Several studies on multiplexing detection have been performed on α-HL nanopores[27,28] and solid-state nanopores.[29] To simultaneously record an electrical signal from hundreds to thousands of channels, advanced microfluidics and complex circuits are required, thus enlarging and complicating the measurement system. Moreover, information about the species or chemical states of an analyte could not be obtained from the readout of the ionic current. Thus, it would be difficult to identify the species of analytes, which would hinder the application of nanopore-based methods in analyzing complex samples, such as clinical samples.

Optical approaches, in this case, are attractive methods to solve these problems. Detecting optical signals from multi-nanopore channels simultaneously provides a novel multiplexed method without the demand of an individual amplifier circuit for each nanopore,[30–33] which allows for a simple and compact instrument and a scaled-down microfluidic system. Optical signals could also reveal chemical states and properties of the specific analytes inside the nanopore. Recently, the translocation events of a single labeled analyte were detected by epi-fluorescence or total internal reflection fluorescence microscopy. The first proof-of-principle study of a single labeled DNA was accomplished in 2007 by integrating a nanopore with single-molecule fluorescence.[34] Since then, fluorophores on different subsets of ds-DNA have been detected by two-channel optical readouts.[31] In addition to fluorescence-based techniques, vibrational spectroscopy methods, such as surface enhanced Raman spectroscopy, have also been introduced for nanopore readouts.[35] Dye-coated gold nanoparticles were driven through a solid-state nanopore with a gold layered membrane, generating electromagnetic 'hot spots' between the particle and the wall of the metallic nanopore, which increased the surface-enhanced resonance Raman scattering (SERRS) signal from the analytes. As a result, the translocation events of a single nanoparticle could be detected by monitoring the characteristic band in the SERRS spectrum. These spectroscopy-based methods provide the possibility of identifying the molecular structures of different analytes by obtaining their corresponding vibrational spectra while recording their nanopore events simultaneously.

Although labeling approaches have increased the specificity of nanopore analysis, the bleaching of the fluorescent dyes and imperfect labeling can influence the detection. Moreover, the requirement for labeling places a restriction on the sample. Therefore, label-free methods for nanopore-based optical detection have been recently proposed. For example, "optical patch clamping",[36] a method based on calcium fluorescence, has been introduced to monitor the actions of a single-ion channel. Fluorescent dyes sensitive to Ca^{2+} are used in the electrolyte solution to analyze the concentration of Ca^{2+} in real time. Therefore, as Ca^{2+} ions flow from one side of the pore to the other side, the intensity of the ionic current can be converted into the intensity of fluorescence in a small region around the nanopore. Recently, this method has been successfully applied in solid-state nanopores detections.[37,38]

To integrate the advantages of optical and electrical readouts, the synchronized optical and electronic detection of nanopores not only reveals the volume and electrical information but also identifies the structures, conformation and compositions of the analytes. This is especially important for the application of a nanopore to complex samples. Synchronized optical and electrical detection of biomolecules was first achieved by an integrated system based on total internal reflection fluorescence excitation with a high temporal resolution of up to 1000 frames per second.[35] A correlated electrical and optical analysis of single nanoparticles and viruses has been demonstrated based on liquid-core anti-resonant reflecting optical waveguides (ARROWs).[39] To simultaneously acquire optical and electronic signals at a high bandwidth and high laser power, a low noise nanopore platform based on a Pyrex substrate and a silicon nitride membrane has been developed to reduce the background signal.[10] Meanwhile, "optical patch clamping" methods have also been applied to simultaneous optical and electronic detection on a solid-state nanopore without labeling of the analyte.[37,38]

Although fluorescence methods already represent a large portion of nanopore optical detection, the introduction of a fluorophore, either with the labeling procedure or without, would additionally influence the translocation process and the complexity of the experiment. Therefore, an alternative optical approach that does not introduce any additional molecules into the nanopore detection system would be preferable.

Subwavelength apertures have been studied for several years because of their fascinating properties,[41] such as their strongly enhanced transmission of light and wavelength filtering due to the interaction between light and the surface of the metal film. According to H. Bethe, an idealized subwavelength hole on a perfectly conducting metal screen of zero thickness would very weakly transmit light because light with a wavelength longer than the cutoff wavelength (λ) would not propagate through the hole.[42] Optical imaging methods beyond the diffraction limit have been developed based on his theory, such as near-field optical microscopy (NSOM).[43] At the end of the last century, the extraordinary transmission phenomenon through subwavelength holes was reported.[44] Orders of magnitude more light than that predicted by Bethe could be transmitted through an array of tiny holes or through a single subwavelength hole[44] because of surface plasmons, which are generated by the interaction between the free electrons of the metal and electromagnetic waves. Furthermore, light could be diffracted, and its direction of propagation could be regulated by sculpting the surrounding material at the wavelength scale.[45] The spectrum of transmitted light could also be regulated by changing the geometry of the nanoscale aperture.[44]

The nanoscale aperture has often been used in near-field optical microscopes to excite or collect optical signals from the subwavelength region beyond the diffraction limit. In the last decade, nanoholes in a metal film have also been used to detect single fluorescent biomolecules at high concentrations and have even achieved single-molecule sequencing, known as a zero mode waveguide.[46,47] Based on the surface plasmon generated around the nanopore, metal nanopores have also been used as biosensors on the bulk[48,49] and single-molecule scale.[50]

By exploiting subwavelength apertures with plasmonic effects, we have developed a new nanopore detection method for the simultaneous acquisition of optical and electrical readouts without the requirement of additional dyes. Our method is based on a single nanopore on a metal-coated silicon nitride (SiN_x) substrate, which can guide the incident light through the nanopore by generating surface plasmons at the edge of the nanopore. An integrated system was designed for simultaneous optical and electrical detection. As shown in Fig. 1, a dark-field microscopy system was used to increase the signal-to-noise ratio of diffracted light through the metal-coated solid-state nanopore. A flow cell acts as a port for connecting the two readouts. The detectors are designed to acquire the optical and electrical signal. A high-speed data acquisition system monitors the variations in the optical and ionic current signals simultaneously. We believe that such a strategy could provide more direct and detailed information from single analytes inside the nanopore compared with the classical ionic current measurements.

2 System composition

Our lab-built system was composed of three main parts: the dark-field microscopy system (Fig. 1a), the ultralow current acquisition system (Fig. 1b) and the metal-coated solid-state nanopore (Fig. 2).

2.1 Optical detection system

A standard dark-field microscopy setup was used for illuminating and monitoring the process on the nanopore chip. The dark-field illumination was constructed on an inverted microscope (Ti-U, Nikon, Japan). The unpolarized white light from a 100 W halogen lamp was focused on the nanopore chip by a dark-field condenser

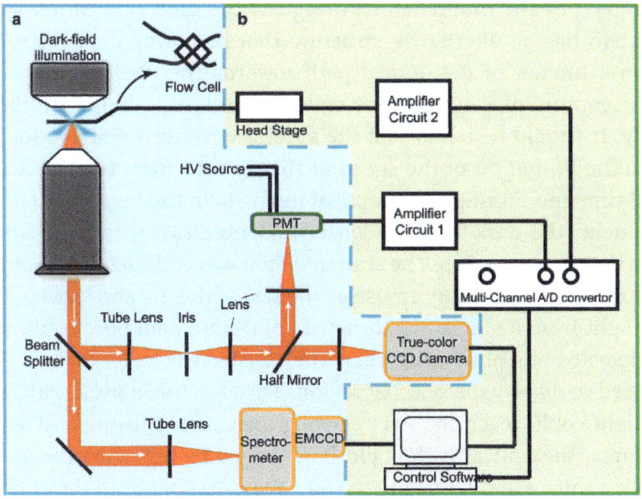

Fig. 1 Schematic of the integrated system. Experimental configuration of the dark-field microscopy system (a) and the home-built ultralow ionic current acquisition system (b).

(NA = 0.8–0.95). The scattered light signal was collected by a 40× objective (NA = 0.6) from the bottom of the flow cell. The light scattering spectra were dispersed by a Czerny–Turner spectrograph (Isoplane SCT320, Princeton Instrument, US) and recorded by an electron multiplying charge coupled devices array (EMCCD, ProEM, Princeton Instruments, US) camera. High-speed intensity monitoring was achieved by the integration of a lab-designed optical path equipped with a photomultiplier tube (PMT, R921, Hamamatsu, Japan), a lab-made current amplifier and a multichannel data acquisition system. The output image from the microscope was reconstructed on the photosensitive surface of the PMT and the assisted true-color digital CCD camera using lenses. The single-nanopore region was defined by an adjustable iris positioned on the output image plane in the optical path. All of these components were mounted along a common optical axis. The time resolution of the optical signal was determined by the bandwidth of the amplifier and the sampling rate of an analog–digital converter, which was 50 kHz in our experiments. The synchronized collection of electrical and optical signals

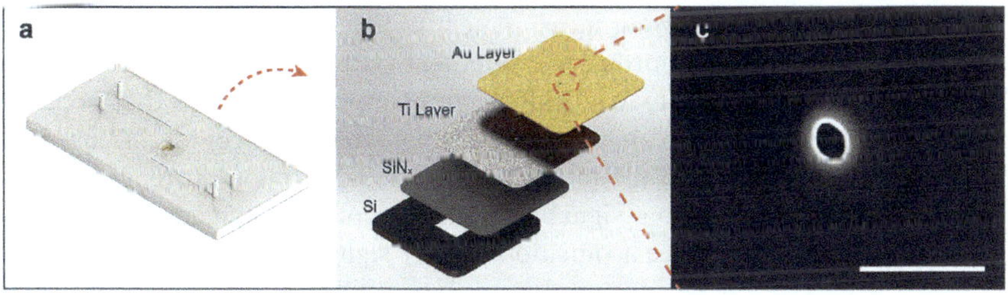

Fig. 2 (a) Schematic of the PDMS flow cell for a synchronized optical and electrical readout of a single nanopore. (b) Schematic of a metal-coated solid-state nanopore (not to scale). (c) Scanning electron micrograph of a typical metal-coated nanopore chip (scale bar is 500 nm).

was guaranteed by the multichannel design of the data acquisition system. Our optical system has an alternative structure that positions the spectrometer and EMCCD on a branch of the optical path towards the PMT. This configuration ensures the capture of spectra in the regions of interest along with the intensity monitoring. It should be noted that the spectra were recorded at a low sampling rate due to the limitation of the speed of the readout from the EMCCD.

For real-time monitoring of the optical intensity in the nanopore region during the experiment, the dark-field condenser and objective lens were adjusted and focused on the nanopore chip. The scattered light was collected by the objective and imaged on the CCD camera. By adjusting the size of the iris and moving the sample stage, the light from a single metal-coated solid-state nanopore could be selected and then detected by a photomultiplier. The assisted true-color digital CCD camera could be used to display the selected region. Through this optical path, 40% of the scattered light could reach the PMT sensing area. After moving and selecting the nanopore area, the optical path could be changed by replacing the 20 : 80 beam splitter with a mirror and removing the 50 : 50 beam splitter in the optical path so that all of the scattered light could be detected by the photomultiplier.

2.2 Ultralow current acquisition system

A lab-designed integrated and ultralow current measurement system was employed in the electrical detection. The entire measurement system contained four parts: the headstage, the current amplifier, the filter circuit and the analog–digital converter (ADC). For the headstage, a preamplifier was used to apply a constant potential on both sides of the nanopore and to amplify the ionic current at pico-ampere level with a 100 MΩ high-value feedback transimpedance amplifier circuit. The current amplifier could accomplish the subsequent signal conditioning with multistage amplifier circuits. A 3-pole Bessel low-pass filter was integrated in the current amplifier to eliminate noise. The ADC converted the analog signal into a digital signal so that the computer could acquire and manage the signals.

Our lab-built amplifier exhibited an appropriate current measuring range along with a good temporal resolution and current resolution in the unfiltered mode. The constant potential ranged from −1000 mV to 1000 mV, which was the typical range used in the nanopore experiment. The sampling rate of the ADC ranged from 10 kHz to 500 kHz, which satisfies the requirements for the different analytes. The low-pass Bessel filter, with a range from 1–10 kHz, is preferred for noise reduction in nanopore detection because of its outstanding whole-band frequency hysteresis feature. In addition, the shielding cage and grounding solutions were applied in this ultralow current measurement to reduce the various noises.

The lab-made ultralow current measurement system exhibited a performance comparable to that of commercial patch-clamp amplifiers, such as the Axopatch 200B and HEKA EPC 10 patch-clamp systems, which are commonly used in nanopore analyses. More importantly, the integrated ultralow current measurement system not only gave a signal amplification of the ionic current but also simultaneously achieved the acquisition of PMT signals.

2.3 Metal-coated nanopore and flow cell

Metal-coated solid-state nanopores were fabricated using standard semi-conductor processing of silicon chips. Low-stress silicon nitride membranes were covered by adhesion layers of titanium and gold to generate surface plasmons around the nanopore with thicknesses of 10 nm and 90 nm, respectively (Fig. 2b). The metal layer was deposited by electron beam evaporation after being cleaned with oxygen plasma to remove organic contamination. A focus ion beam was used to drill the nanopores with various diameters ranging from 150 nm to 300 nm. A scanning electron micrograph of a typical metal-coated solid-state nanopore is shown in Fig. 2c.

To achieve the acquisition of optical and ionic current signals, we designed and fabricated a flow cell containing an optical path for light illumination and acquisition, as well as a pair of chambers to hold the electrolyte solution. A diagram of the flow cell is shown in Fig. 2a. As a first step, two polydimethylsiloxane (PDMS, Sylgard 184, Dow Corning) layers were assembled to form the inlet and outlet channels. The thickness of each layer was less than 1 mm. There are two main reasons to maintain a low thickness of the PDMS layer. First, the edges of the upper chamber are a major source of stray light that significantly reduces the signal-to-noise ratio of the transmitted optical signal in a dark-field illumination. Second, the thickness of the flow cell is limited by the work distance from the dark-field condenser and the objective lens. The condenser and objective lens with a higher numerical aperture have a shorter work distance. High numerical aperture objective lenses provide a high-resolution in a dark-field imaging system while simultaneously offering fine focusing results for an objective-focused light beam. Thus, it is necessary to design an ultra-thin flow cell. In this study, the entire thickness of the designed flow cell was less than 4 mm. A metal-coated solid-state nanopore chip was sandwiched between the two layers of PDMS. The lower part of the PDMS was irreversibly bonded on a standard coverslip (ThermoFisher, US), which was thoroughly cleaned by an ultrasound bath in acetone and deionized water before assembly. The final assembly was achieved by directly positioning the PDMS layers on top of each other and then on the coverslip after exposure to the air plasma. Two Teflon tubes were used for injecting the solution into the chamber and holding the electrode.

3 Scattered light and ionic current from the single nanopore

To estimate the applicability for different sizes of nanopores, we fabricated an Au-coated solid-state nanopore array with diameters ranging from 150 nm to 300 nm. A 100 W halogen light source was used for the dark-field microscopy and exciting the surface plasmon. Scattered light from each single Au-coated nanopore could be collected by the inverted microscope and detected by the EMCCD. In this approach, the dark-field image and scattered light spectrum could be obtained.

In Fig. 3, we show the representative dark-field images and scattering spectra from single metal-coated nanopores. The nanopores were drilled on an Au layer-coated silicon nitride substrate, with diameters ranging from 150 nm to 300 nm (Fig. 3b). A bright reddish spot was observed for each nanopore, and the intensity of the diffracted light from the nanopore increased with the diameter, as shown in Fig. 3c. The scattered-light intensity from the nanopores with a diameter of 300 nm was approximately ten times higher than that of the nanopores with a diameter of 150 nm. The scattering spectrum of each nanopore exhibited a pronounced resonance peak with a maximum intensity at wavelengths varying from 650 nm to 750 nm, depending on the pore diameter. Such a resonance peak could be attributed to the generation of surface plasmons at the edge of the pore.[51]

According to the pioneering work, the resonance scattering spectra of nanopores in Au thin-films are qualitatively similar to those measured for Au nanodisks.[52] Using this knowledge, the plasmonic resonance position can be predicted for a single nanopore by calculating the scattering properties of a single nanodisk with the same diameter and height (depth). Fig. 4 shows a series of predicted surface plasmonic resonance spectra of isolated nanodisks, with a height of 90 nm and diameters of 150 nm, 200 nm, and 250 nm, using the discrete dipole approximation (DDA) simulation method. The simulated resonance peaks range from approximately 600 nm to 800 nm and are red-shifted with the increasing diameter of the nanodisk. The difference between the simulated spectra of the nanodisks and the experimental results of the Au-coated nanopores might be caused by the following factors. First, the geometry of the nanopores fabricated by

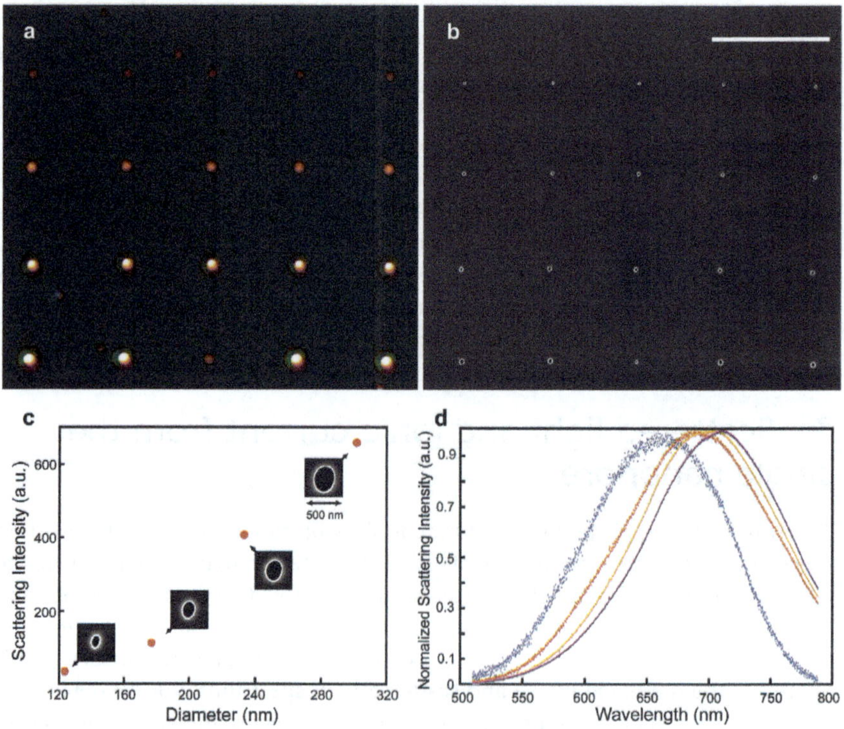

Fig. 3 Characterization of single metal-coated nanopores under a dark field. (a) True-color image of single metal-coated nanopores with diameters of ~150 nm, ~200 nm, ~250 nm and ~300 nm in ambient air. (b) Corresponding scanning electron micrograph for each nanopore array (scale bar is 10 μm). (c) Plot of the scattering intensity for each nanopore *vs.* diameter, which shows the intensity increasing along with an increase in the pore diameter. (d) Corresponding scattering spectra for each single metal-coated nanopore with diameters of 150 nm (blue), 200 nm (red), 250 nm (yellow) and 300 nm (purple) (an average of 4 nanopores).

a focused ion beam may not exactly match the shape we used for the simulation; second, the focused ion beam might induce the redistribution of Au atoms around the nanopore so that a thin layer of Au might form inside the pore and cover the wall of the silicon nitride. Hence, a volcano-like structure might be generated inside the pore that could change the dark-field scattering spectrum of the nanopore. Currently, we are working on exploring more suitable methods for predicting the scattering spectra of such metal-coated nanopores.

As previously demonstrated, the diffracted light from a single metal-coated nanopore covers a broad scattering spectrum because of dipole-localized surface plasmons at the edge of the nanopore. When a single analyte moves through the nanopore, the diffracted light behavior changes. To increase the sensitivity of an optical detection system and enlarge the signal variation caused by the translocation events, a reflected dark-field illumination method with a laser beam as the excitation light source could be used (Fig. 5). Briefly, we replaced the halogen light source and the dark-field condenser with a He–Ne laser beam. While the background light is reflected by the Au layer, a part of the monochromatic light can be diffracted through the metal-coated nanopore. Thus, a bright reddish spot could be observed on a CCD camera or detected by a photomultiplier. The dark-field image under laser illumination is shown in Fig. 5b. The intensity of each spot clearly depends on the size of each circular aperture.

To investigate the feasibility of this method for synchronized optical and electrical detection of the translocation events, we measured the baseline ionic current recorded by the current headstage with a pair of Pt electrodes and recorded the light scattering from a single nanopore with a photomultiplier. The ionic current and scattered light from a single nanopore were monitored simultaneously. The ionic current signal was low-pass filtered at 5 kHz before digitization at 50 kHz (Fig. 6a). Synchronized optical and electrical data acquisition was achieved using a multichannel-designed analog–digital converter (Axon Digidata 1550, Molecular Devices). The internal clock of the analog–digital converter ensures simultaneous data acquisition. Alternatively, we also examined the possibility of using a data acquisition board (PCI 6251, Nation Instruments) with

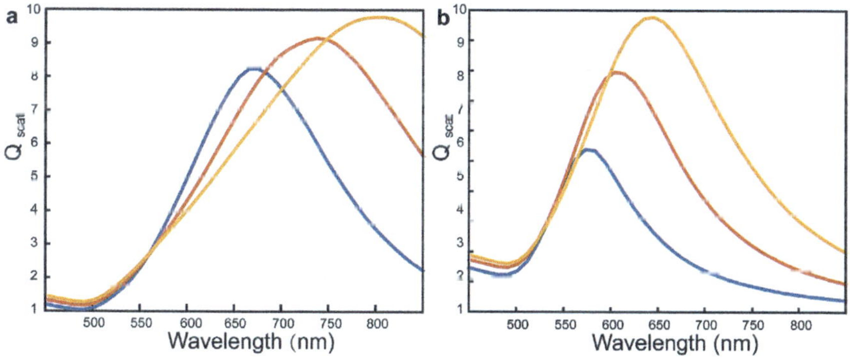

Fig. 4 Simulated scattering spectra of single gold nanodisks with a height of 90 nm and diameters of 150 nm (blue), 200 nm (red) and 250 nm (yellow) in ambient water (a) and air (b).

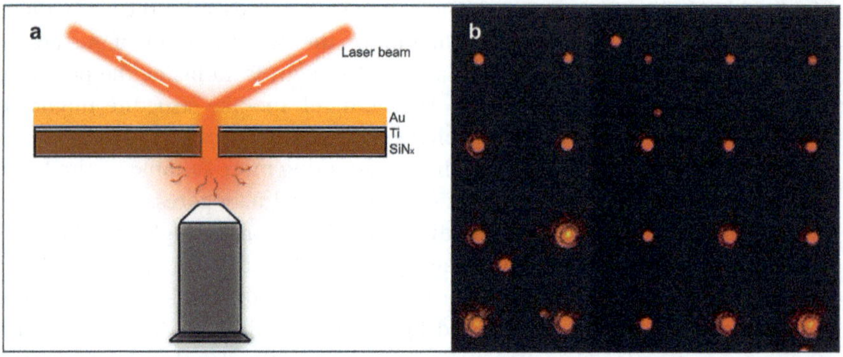

Fig. 5 (a) Schematic of the reflected dark-field illumination method for detecting a single nanopore. (b) True-color image of the same nanopore array corresponding to Fig. 3, under reflected dark-field illumination.

a custom LabVIEW program to achieve the synchronized optical and electrical measurements.

The dark-field imaging reported in this study was achieved using a dark-field condenser and 100 W halogen lamp. The scattered light transmitted through an Au-coated nanopore was collected by the objective lens, then the photocurrent was generated by the photomultiplier. The photocurrent was monitored by the data acquisition system with a sampling rate of 50 kHz, and was then sent to the computer and was simultaneously recorded with the ionic signal. In our experiment, the photocurrent generated by the scattered light from a single 150 nm Au-coated nanopore was typically 50 nA when illuminated by the halogen lamp (Fig. 6a).

The noise power spectrum density (PSD) of the nanopore measurements could help analyze noise conditions for the nanopore. Previous research showed that there is flicker noise ($1/f$ noise) that exists in the low frequency region (freq. < 1 kHz) because of the electrolyte solution fluctuations in the detection cell. Generally, for SiN_x solid-state nanopores, the flicker noise power density is less than 1 $pA^2\ Hz^{-1}$ at 1 Hz (the blue and red folding lines in Fig. 6a). During the synchronized optical and electrical measurements, the flicker noise level increased after turning on the illuminating white lamp. The reason for this might be the heat effect under light illumination, especially at the nanopore region because of the generation of surface plasmons.

Compared with the bare SiN_x solid-state nanopore, the Au coating might be able to suppress white noise with a frequency between 1 kHz and 10 kHz. Compared with the power spectral density of the bare SiN_x nanopore, which exhibits a white noise region with a similar power to the $1/f$ noise, the white noise of an Au-coated nanopore decreases over a long range of frequency.

Fig. 6b shows the current–voltage characteristics of the ionic current with and without illumination by a halogen lamp. According to previous work, the diameter of a solid-state nanopore can be estimated by the current–voltage curve using the following equation:[53]

$$G = \sigma \frac{\pi d^2}{4l}$$

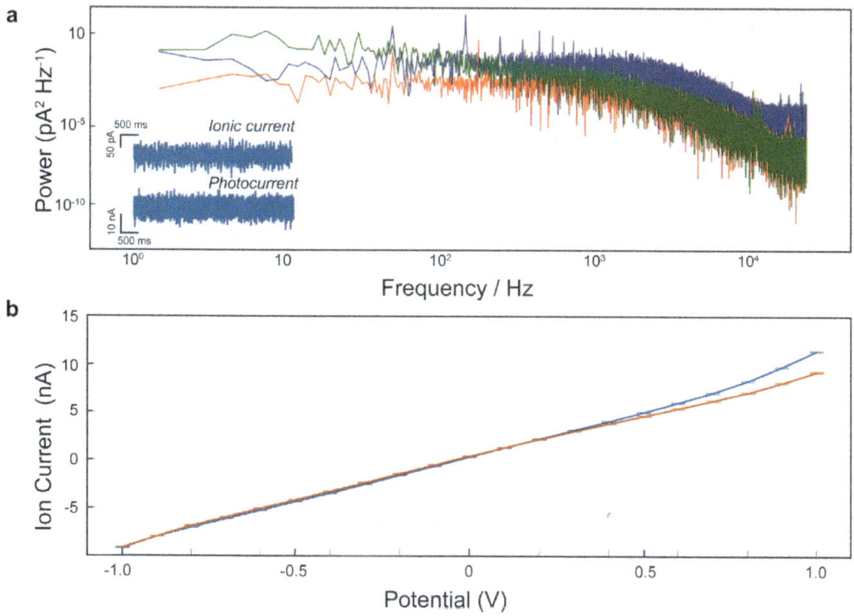

Fig. 6 (a) Power spectral densities of the ionic current detected from a single Au-coated nanopore, before (red) and after (green) turning on the halogen light, and from a single nanopore without the Au layer in a conventional flow cell (blue). Inset: baseline ionic current at 0 mV, under white light illumination for a single 150 nm Au-coated nanopore, and baseline photocurrent from PMT at 1100 V for the same nanopore at the same period time. (b) Ionic current–voltage characteristics of a single Au-coated nanopore with (blue) and without (red) white light illumination. The error bars are standard deviations of 1-s sampling traces under different bias. The light-induced effects of the ionic current through the nanopore might mainly originate from the plasmonic enhanced heat effects and plasmoelectric potential.

where G is the open pore conductance value, σ is the conductivity of the bulk solution (0.13 s m^{-1} for 10 mM KNO_3 solution, measured by a DDSJ-318 conductivity meter, INESA Instrument), l is the length of the nanopore, *i.e.* the thickness of the membrane, and d is the diameter of the circular pore. The estimated diameter of the Au-coated nanopore in Fig. 6b is 130 nm, which is smaller than the diameter measured using the SEM results. This is reasonable because the equation is more applicable to smaller solid-state nanopores. In addition, although there are more precise and complicated equations to model the relativity between conductance and nanopore diameter, the role of such an Au layer remains unclear. The current–voltage curve exhibits good linearity for the experimental voltage (−1.0 V to 1.0 V). During illumination by the halogen lamp, a small change in the ionic current could be observed, especially under the positive bias (0.5–1.0 V). This might be the result of the surface plasmon induced heat effect[57–59] and plasmoelectric potential induced by the incident light.[60] The results show that the noise introduced by illumination with white light has limited influence on the ionic current detection. Therefore, our integrated system that includes the metal-coated nanopore and the optical and electrical acquisition system, exhibits acceptable performance for both optical and electrical detection.

4 Conclusion

The present work demonstrates a novel integrated system for synchronized optical and electronic detection of nanopore translocation events. A metal-coated nanopore was fabricated on a silicon nitride substrate and assembled into a transparent flow cell. A dark-field microscopy system equipped with a high-speed optical intensity monitoring detector and an ultra-low current high-speed data acquisition system was designed for monitoring the scattering light and ionic current signal from the nanopore. Because of the plasmonic effect at the edge of the metal nanopore induced by the incident light, a single solid-state nanopore could be observed easily under the dark-field microscope, without introducing an additional fluorescent dye or a labeling procedure. With this system, the ionic current and elastic scattering signal could be detected simultaneously with a sampling rate of 10 to 100 kHz. The scattering spectra of a single metal-coated nanopore could be regulated by changing the diameter of nanopore, and the elastic scattering spectrum of a single pore could be predicted by the discrete dipole approximation simulation method. This system could provide a potential approach to simultaneously acquire electrical information, such as the charge, and optical information, including the refractive index, along with the single-analyte translocation process through the nanopore. At this stage, the system only monitors the ionic current through a single nanopore together with the recording of the light scattering. In the future, the integration of the multiplexing ionic current readouts with the optical detection will be of benefit to realize the parallel measurements of the current and light in multiple nanopores.

Using the interaction of surface plasmons between a single nanoparticle and a metal-coated nanopore, we are working on obtaining the single nanoparticles translocation events through the plasmonic nanopore by monitoring the variation of scattered light from the plasmonic nanopore. Furthermore, this novel integrated system based on a metal-coated nanopore also enables new possibilities for optical nanopore detection, including label-free detection of single-molecule events and the interaction process between biomolecules. Based on our plasmonic nanopore system, multiple types of plasmon-based optical detection methods, including not only Rayleigh scattering from a single nanopore but also surface-enhanced Raman spectroscopy[35] and plasmon-enhanced fluorescence,[54–56] can be performed.

Acknowledgements

This research was supported by the National Basic Research Program of China (2011CB707605), the National Natural Science Foundation of China (21327807, 21421004, 21125522), Shanghai Pujiang Program (12JC1403500), and Program for Professor of Special Appointment (Eastern Scholar) at Shanghai Institutions of Higher Learning (YJ0130504).

References

1 J. Shendure and H. Ji, *Nat. Biotechnol.*, 2008, **26**, 1135–1145.
2 D. Branton, D. W. Deamer, A. Marziali, H. Bayley, S. A. Benner, T. Butler, M. Di Ventra, S. Garaj, A. Hibbs and X. Huang, *Nat. Biotechnol.*, 2008, **26**, 1146–1153.

3 G. F. Schneider and C. Dekker, *Nat. Biotechnol.*, 2012, **30**, 326–328.
4 Y.-L. Ying, D.-W. Li, Y. Liu, S. K. Dey, H.-B. Kraatz and Y.-T. Long, *Chem. Commun.*, 2012, **48**, 8784–8786.
5 C. Cao, Y.-L. Ying, Z. Gu and Y.-T. Long, *Anal. Chem.*, 2014, **86**, 11947–11950.
6 H.-Y. Wang, Z. Gu, C. Cao, J. Wang and Y.-T. Long, *Anal. Chem.*, 2013, **85**, 8254–8261.
7 Y.-L. Ying, C. Cao and Y.-T. Long, *Analyst*, 2014, **139**, 3826–3835.
8 Y. Liu, Y.-L. Ying, H.-Y. Wang, C. Cao, D.-W. Li, W.-Q. Zhang and Y.-T. Long, *Chem. Commun.*, 2013, **49**, 6584–6586.
9 Y.-L. Ying, J. Zhang, F.-N. Meng, C. Cao, X. Yao, I. Willner, H. Tian and Y.-T. Long, *Sci. Rep.*, 2013, **3**, 1622.
10 Y. L. Ying, J. Zhang, R. Gao and Y. T. Long, *Angew. Chem., Int. Ed.*, 2013, **52**, 13154–13161.
11 S. Howorka and Z. Siwy, *Chem. Soc. Rev.*, 2009, **38**, 2360–2384.
12 B. N. Miles, A. P. Ivanov, K. A. Wilson, F. Doğan, D. Japrung and J. B. Edel, *Chem. Soc. Rev.*, 2013, **42**, 15–28.
13 M. Firnkes, D. Pedone, J. Knezevic, M. Döblinger and U. Rant, *Nano Lett.*, 2010, **10**, 2162–2167.
14 J. J. Kasianowicz, E. Brandin, D. Branton and D. W. Deamer, *Proc. Natl. Acad. Sci. U. S. A.*, 1996, **93**, 13770–13773.
15 M. Pastoriza-Gallego, L. Rabah, G. Gibrat, B. Thiebot, F. O. G. van der Goot, L. Auvray, J.-M. Betton and J. Pelta, *J. Am. Chem. Soc.*, 2011, **133**, 2923–2931.
16 E. A. Manrao, I. M. Derrington, A. H. Laszlo, K. W. Langford, M. K. Hopper, N. Gillgren, M. Pavlenok, M. Niederweis and J. H. Gundlach, *Nat. Biotechnol.*, 2012, **30**, 349–353.
17 T. Z. Butler, M. Pavlenok, I. M. Derrington, M. Niederweis and J. H. Gundlach, *Proc. Natl. Acad. Sci. U. S. A.*, 2008, **105**, 20647–20652.
18 M. M. Mohammad, R. Iyer, K. R. Howard, M. P. McPike, P. N. Borer and L. Movileanu, *J. Am. Chem. Soc.*, 2012, **134**, 9521–9531.
19 D. Wendell, P. Jing, J. Geng, V. Subramaniam, T. J. Lee, C. Montemagno and P. Guo, *Nat. Nanotechnol.*, 2009, **4**, 765–772.
20 H.-Y. Wang, Y. Li, L.-X. Qin, A. Heyman, O. Shoseyov, I. Willner, Y.-T. Long and H. Tian, *Chem. Commun.*, 2013, **49**, 1741–1743.
21 C. Dekker, *Nat. Nanotechnol.*, 2007, **2**, 209–215.
22 A. J. Storm, J. H. Chen, X. S. Ling, H. W. Zandbergen and C. Dekker, *Nat. Mater.*, 2003, **2**, 537–540.
23 B. M. Venkatesan, A. B. Shah, J. M. Zuo and R. Bashir, *Adv. Funct. Mater.*, 2010, **20**, 1266–1275.
24 S. Garaj, W. Hubbard, A. Reina, J. Kong, D. Branton and J. Golovchenko, *Nature*, 2010, **467**, 190–193.
25 G. F. Schneider, S. W. Kowalczyk, V. E. Calado, G. Pandraud, H. W. Zandbergen, L. M. Vandersypen and C. Dekker, *Nano Lett.*, 2010, **10**, 3163–3167.
26 C. A. Merchant, K. Healy, M. Wanunu, V. Ray, N. Peterman, J. Bartel, M. D. Fischbein, K. Venta, Z. Luo and A. C. Johnson, *Nano Lett.*, 2010, **10**, 2915–2921.
27 T. Osaki, H. Suzuki, B. Le Pioufle and S. Takeuchi, *Anal. Chem.*, 2009, **81**, 9866–9870.

28 G. Baaken, N. Ankri, A.-K. Schuler, J. R. Rühe and J. C. Behrends, *ACS Nano*, 2011, **5**, 8080–8088.

29 N. A. W. Bell, V. V. Thacker, S. Hernandez-Ainsa, M. E. Fuentes-Perez, F. Moreno-Herrero, T. Liedl and U. F. Keyser, *Lab Chip*, 2013, **13**, 1859–1862.

30 A. Kleefen, D. Pedone, C. Grunwald, R. Wei, M. Firnkes, G. Abstreiter, U. Rant and R. Tampé, *Nano Lett.*, 2010, **10**, 5080–5087.

31 B. McNally, A. Singer, Z. Yu, Y. Sun, Z. Weng and A. Meller, *Nano Lett.*, 2010, **10**, 2237–2244.

32 J. Larkin, M. Foquet, S. W. Turner, J. Korlach and M. Wanunu, *Nano Lett.*, 2014, **14**, 6023–6029.

33 M. Urban, A. Kleefen, N. Mukherjee, P. Seelheim, B. Windschiegl, M. Vor der Brüggen, A. Koçer and R. Tampé, *Nano Lett.*, 2014, **14**, 1674–1680.

34 G. A. Chansin, R. Mulero, J. Hong, M. J. Kim, A. J. Demello and J. B. Edel, *Nano Lett.*, 2007, **7**, 2901–2906.

35 M. P. Cecchini, A. Wiener, V. A. Turek, H. Chon, S. Lee, A. P. Ivanov, D. W. McComb, J. Choo, T. Albrecht and S. A. Maier, *Nano Lett.*, 2013, **13**, 4602–4609.

36 A. Demuro and I. Parker, *Cell Calcium*, 2003, **34**, 499–509.

37 A. Ivankin, R. Y. Henley, J. Larkin, S. Carson, M. L. Toscano and M. Wanunu, *ACS Nano*, 2014, **8**, 10774–10781.

38 B. N. Anderson, O. N. Assad, T. Gilboa, A. H. Squires, D. Bar and A. Meller, *ACS Nano*, 2014, **8**, 11836–11845.

39 S. Liu, Y. Zhao, J. W. Parks, D. W. Deamer, A. R. Hawkins and H. Schmidt, *Nano Lett.*, 2014, **14**, 4816–4820.

40 W. H. Pitchford, H.-J. Kim, A. P. Ivanov, H.-M. Kim, J.-S. Yu, R. J. Leatherbarrow, T. Albrecht, K.-B. Kim and J. B. Edel, *ACS Nano*, 2015, **9**, 1740–1748.

41 C. Genet and T. Ebbesen, *Nature*, 2007, **445**, 39–46.

42 H. Bethe, *Phys. Rev.*, 1944, **66**, 163–182.

43 E. Betzig and J. K. Trautman, *Science*, 1992, **257**, 189–195.

44 T. W. Ebbesen, H. Lezec, H. Ghaemi, T. Thio and P. Wolff, *Nature*, 1998, **391**, 667–669.

45 H. J. Lezec, A. Degiron, E. Devaux, R. Linke, L. Martin-Moreno, F. Garcia-Vidal and T. Ebbesen, *Science*, 2002, **297**, 820–822.

46 M. J. Levene, J. Korlach, S. W. Turner, M. Foquet, H. G. Craighead and W. W. Webb, *Science*, 2003, **299**, 682–686.

47 J. Eid, A. Fehr, J. Gray, K. Luong, J. Lyle, G. Otto, P. Peluso, D. Rank, P. Baybayan and B. Bettman, *Science*, 2009, **323**, 133–138.

48 A. G. Brolo, R. Gordon, B. Leathem and K. L. Kavanagh, *Langmuir*, 2004, **20**, 4813–4815.

49 T. Rindzevicius, Y. Alaverdyan, A. Dahlin, F. Höök, D. S. Sutherland and M. Käll, *Nano Lett.*, 2005, **5**, 2335–2339.

50 F. Nicoli, D. Verschueren, M. Klein, C. Dekker and M. P. Jonsson, *Nano Lett.*, 2014, **14**, 6917–6925.

51 E. Popov, N. Bonod, M. Nevière, H. Rigneault, P.-F. Lenne and P. Chaumet, *Appl. Opt.*, 2005, **44**, 2332–2337.

52 J. Prikulis, P. Hanarp, L. Olofsson, D. Sutherland and M. Käll, *Nano Lett.*, 2004, **4**, 1003–1007.

53 S. W. Kowalczyk, A. Y. Grosberg, Y. Rabin and C. Dekker, *Nanotechnology*, 2011, **22**, 315101.
54 A. G. Brolo, S. C. Kwok, M. G. Moffitt, R. Gordon, J. Riordon and K. L. Kavanagh, *J. Am. Chem. Soc.*, 2005, **127**, 14936–14941.
55 H. Aouani, O. Mahboub, E. Devaux, H. Rigneault, T. W. Ebbesen and J. Wenger, *Nano Lett.*, 2011, **11**, 2400–2406.
56 H. Aouani, O. Mahboub, N. Bonod, E. Devaux, E. Popov, H. Rigneault, T. W. Ebbesen and J. Wenger, *Nano Lett.*, 2011, **11**, 637–644.
57 U. F. Keyser, D. Krapf, B. N. Koeleman, R. M. Smeets, N. H. Dekker and C. Dekker, *Nano Lett.*, 2005, **5**, 2253–2256.
58 M. P. Jonsson and C. Dekker, *Nano Lett.*, 2013, **13**, 1029–1033.
59 C. R. Crick, P. Albella, B. Ng, T. Roschuk, A. P. Ivanov, M. Cecchini, F. Bresme, S. A. Maier and J. B. Edel, *Nano Lett.*, 2015, **15**, 553.
60 M. T. Sheldon, J. Van de Groep, A. M. Brown, A. Polman and H. A. Atwater, *Science*, 2014, **346**, 828–831.

Faraday Discussions

PAPER

Single occupancy spectroelectrochemistry of freely diffusing flavin mononucleotide in zero-dimensional nanophotonic structures

Lawrence P. Zaino, III†[a] Dane A. Grismer,†[b] Donghoon Han,[b] Garrison M. Crouch[b] and Paul W. Bohn*[ab]

Received 11th May 2015, Accepted 1st July 2015
DOI: 10.1039/c5fd00072f

Zero-mode waveguides (ZMW) have the potential to be powerful confinement tools for studying electron transfer dynamics at single molecule occupancy conditions. Flavin mononucleotide contains an isoalloxazine chromophore, which is fluorescent in the oxidized state (FMN) while the reduced state ($FMNH_2$) exhibits dramatically lower light emission, *i.e.* a dark-state. This allows fluorescence emission to report the redox state of single FMN molecules, an observation that has been used previously to study single electron transfer events in surface-immobilized flavins and flavoenzymes, *e.g.* sarcosine oxidase, by direct wide-field imaging of ZMW arrays. Single molecule electron transfer dynamics have now been extended to the study of freely diffusing molecules using fluorescence measurements of Au ZMWs under single occupancy conditions. The Au in the ZMW serves both as an optical cladding layer and as the working electrode for potential control, thereby accessing single molecule electron transfer dynamics at μM concentrations. Consistent with expectations, the probability of observing single reduced molecules increases as the potential is scanned negative, $E_{appl} < E_{eq}$, and the probability of observing emitting oxidized molecules increases at $E_{appl} > E_{eq}$. Different single molecules exhibit different electron transfer properties as reflected in the position of E_{eq} and the distribution of E_{eq} among a population of FMN molecules. Two types of actively-controlled electroluminescence experiments were used: chronofluorometry experiments, in which the potential is alternately stepped between oxidizing and reducing potentials, and cyclic potential sweep fluorescence experiments, analogous to cyclic voltammetry, these latter experiments exhibiting a dramatic scan rate dependence with the slowest scan rates showing distinct intermediate states that are stable over a range of potentials. These states are assigned to flavosemiquinone species that are stabilized in the special environment of the ZMW nanopore.

[a]*Department of Chemistry and Biochemistry, University of Notre Dame, Notre Dame, IN 46556, USA. E-mail: pbohn@nd.edu*
[b]*Department of Chemical and Biomolecular Engineering, University of Notre Dame, Notre Dame, IN 46556, USA*

† These authors contributed equally to this work.

Introduction

Since the first single molecule electrochemical experiments were performed with electrochemiluminescence[1] and redox cycling,[2,3] interest in investigating single molecule electrochemistry, particularly by means of coupling fluorescence and electrochemistry, has increased. In this context, the use of fluorescence microscopy, particularly total internal reflectance fluorescence, with electrochemistry gives researchers a powerful tool in investigating redox chemistry at high spatial resolution, while simultaneously minimizing background interference.[4] Early applications of coupled fluorescence microscopy and electrochemistry targeted absorption at chemical interfaces and the development of new biosensors.[5–8] In these experiments, redox-sensitive optical properties – either the absorption cross-section, or the fluorescence quantum efficiency, or both – of the molecule were exploited to observe electron transfer events spectroscopically. The utility of coupling electrochemical measurements and fluorescence is evidenced by recent studies of redox cycling, catalyst screening and diffusion zone imaging.[9–12] A common feature of coupled electrochemical–luminescence experiments is that a current measurement, with the attendant large background, is replaced by the detection of luminescent photons – in principle a zero background measurement.

Bard and Barbara first extended coupled luminescence and electrochemical measurements to single molecule detection.[13–16] Distinct differences in single molecule fluorescence dynamics were observed at E_{appl} values above and below the equilibrium potential, E_{eq}, thereby firmly establishing fluorescence spectroscopy as a tool to monitor the dynamics of single electron transfer events. The systems initially investigated by single molecule electrochemical fluorescence spectroscopy were confined to molecular solids and immobilized molecules, but these systems were extended to single molecule spectroelectrochemical behaviour under diffusive conditions with studies of cresyl violet[17,18] and other electrochemically active fluorophores.[19–26] Redox switching of single protein molecules has been investigated by observing the resonance energy transfer between the Cu redox center and covalently-linked Cy5 dye in the blue Cu protein azurin, the resulting studies yielding reaction kinetics and thermodynamic dispersion.[27–29] Previous work from this laboratory has exploited nanophotonic architectures, specifically zero-mode waveguides (ZMW) to investigate homogeneous electron transfer dynamics of sarcosine oxidase, both with its nominal substrate, *N*-methylglycine (sarcosine) as well as a non-canonical substrate, L-proline.[30] These experiments were extended to the use of the optical cladding layer of the ZMW as the working electrode in a spectroelectrochemical cell in studies of surface-functionalized flavin adenine dinucleotide (FAD).[31] Flavins are of particular interest, since they function as cofactors in many important enzymes, especially among the oxidoreductases.

Single molecule spectroscopy generally requires reducing the observation volume by means such as total internal reflection or tightly focused excitation in order to maximize the signal-to-background ratio and to produce reasonable signal magnitudes at single molecule levels.[32,33] More recently, single molecule experiments have been implemented by trapping molecules in nanophotonic architectures, such as ZMWs. These are usually realized physically as an array of metal-clad nanopores, which do not support any far-field propagating modes when the wavelength is greater than a critical wavelength $\lambda_c \sim 1.7d$, where d is the pore diameter.[34] Under these conditions an evanescent field is contained within the pore,[35,36] with a magnitude that decays exponentially along the longitudinal axis of the pore, of

$$I(z) = \frac{e^{-z}}{\Lambda} \tag{1}$$

where Λ is the decay constant and z is the distance in the pore. The decay constant is defined as

$$\frac{1}{\Lambda} = 2\left[\left(\frac{1}{\lambda_c}\right)^2 + \left(\frac{1}{\lambda_m}\right)^2\right]^{\frac{1}{2}} \tag{2}$$

where λ_m is the wavelength in the ZMW medium. The sharp decay of the evanescent field produces extremely small effective sampling volumes, V_{eff},[34,37] given by

$$V_{eff} = \frac{\pi d^2 \Lambda}{24} \tag{3}$$

Fabricating pores below 100 nm in diameter creates effective volumes $V_{eff} \sim$ 100 zL at 458 nm excitation in water. Utilizing pore volumes of this magnitude allows single molecule experiments to be conducted at relatively high concentrations. For example, at $V_{eff} \sim 100$ zL, single molecule occupancy is reached at a concentration $C \sim 17$ μM.

Zero-mode waveguides have been utilized for measurements of single enzyme kinetics, membrane diffusion, protein interactions, and a variety of other detailed studies at the single molecule level.[34,38–41] Generally, Al with its short skin depth is the metal of choice for the cladding layer in ZMW structures. Previously, our laboratory utilized Au-clad ZMWs due to the ability of Au to function as a working electrode (large ideally polarizable region), the wide range of thiol-based surface chemistries available to functionalize molecules to the Au surface, and the chemical orthogonality of Au and SiO_2, the common base material for ZMW supports.[42] Thus, Au can provide multi-functional utility, serving simultaneously as both the optical blocking layer of the ZMW and the working electrode of an electrochemical cell, a dual functionality that is at the heart of the work reported here.

Flavin mononucleotide, FMN, was chosen for detailed spectroelectrochemistry studies under single occupancy conditions (*i.e.* solution conditions where the expectation value for the ZMW nanopore occupancy ranges from ~0.13 to 1.3). FMN contains an isoalloxazine chromophore that is fluorescent in the oxidized state, while the reduced state, $FMNH_2$, exhibits dramatically lower light emission, characteristic of a dark-state. Intriguingly, FMN has a relatively constant fluorescence quantum efficiency over a large pH window, but the electrochemical behaviour varies significantly with pH.[43,44] These properties, coupled with the intrinsic biochemical relevance as important cofactors in flavoenzymes, make the flavins ideal candidates for monitoring single molecule fluorescence under active electrochemical potential control in a redox-active ZMW. FMN was chosen over flavin adenine dinucleotide, FAD, since the presence of the adenine base significantly complicates the radiative dynamics in free solution.[45] Au ZMWs were used to study freely diffusing, rather than immobilized, FMN/$FMNH_2$ using both statically and actively-controlled voltage programs. Constant potential experiments show strong correlation between E_{appl} and fluorescence intensity at the single molecule occupancy level, while potential step experiments (chronofluorometry) and cyclic potential sweep fluorescence (CPSF), are utilized to examine the coupling between single molecule transport and electron transfer in the ZMW nanopores.

Experimental

ZMW fabrication

ZMWs were fabricated as previously described.[30,31] Briefly, 5 nm of Ti and 200 nm of Au were deposited by electron beam evaporation on coverslips (no. 1.5H, Schott). Focused ion beam (FIB) milling was used to create the ZMW. The pores were milled at 9.7 pA with a dwell time of 0.1 ms. Typically, 11 × 11 arrays were fabricated with 4 large pores at the corners to assist in locating the ZMW array prior to single molecule experiments. The device was then cleaned with 1.0 M KOH at 80 °C for 45 min followed by 10 min in concentrated H_2SO_4 and 10 min in concentrated HNO_3. A PDMS well was affixed to the device though plasma bonding to hold solution over the area occupied by the ZMWs.

Finite element simulations

The numerical calculations are performed with a finite element method using COMSOL Multiphysics version 4.4. The simulation was performed over a two-dimensional domain representing the geometry and dimensions of the zero-mode waveguides employed in our experiments. The zero-mode waveguide is represented by a single pore consisting of a conical compartment. The geometry is adapted from scanning electron micrographs (SEMs) of FIB cross-sections. The domain above the single pore was drawn sufficiently large to avoid interference from boundaries, and the mesh was refined both within the single pore and in the region just above the single pore to provide sufficient resolution.

Electrochemical measurements

Electrochemical measurements were performed using a potentiostat (750E, CH Instruments). Both bulk and nanopore measurements made under single occupancy conditions were performed with a Pt counter electrode and a Ag/AgCl reference electrode. Bulk electrochemical measurements used a 3 mm Au disk working electrode (WE), and nanopore measurements used the Au ZMW array as the WE. Electrical contact to the ZMW was made at an exposed tab away from the ZMW area.

Fluorescence Measurements

A custom-built confocal microscope was used to collect all fluorescence data. 458 nm radiation from a continuous wave laser (Sapphire 458 LP, Coherent) was passed through a spatial filter to select the TEM_{00} mode and then a quarter wave plate to produce circularly polarized light. A dichroic filter (Z488RDC, Chroma Technology Corporation) reflected the collimated beam into the back aperture of a 40×, 1.30 NA oil immersion objective (420260-9900-000, Carl Zeiss, Inc.) to produce a diffraction-limited spot with 40 μW power at the focal plane. Fluorescence emission was collected by the same objective (epi-fluorescence) and passed through three emission filters (ET500lp, ET525/50m, ET500lp, Chroma Technology Corporation) before being focused through a 30 μm diameter confocal pinhole to define the radiation for a single-photon avalanche photodiode detector (SPCM-AQR-16, Perkin-Elmer, Inc.). A hardware correlator (TimeHarp260 PICO Single, PicoQuant) was used to collect single molecule fluorescence data. The fluorescence optical path was augmented with a trans-illumination path to assist in locating the ZMW pores. To access the trans-illumination path, collimated radiation from a 632.8 nm laser was used to illuminate the back surface of the ZMW array. Radiation emerging from the pores was collected by the objective, and a mirror was rotated into the detection path to deflect the transmitted radiation to a video-rate CCD camera (JE7462DC, Javelin Systems). Once located, the pore array was indexed to the motion of a piezoelectric stage (P-517.2CL, Physik Instrumente) to reliably center the x–y focal position of the objective on the ZMW pore of interest.

Results and discussion

Evanescent field attenuation and fabrication of ZMW working electrodes

11 × 11 square ZMW arrays of variable pitch (5 μm, typ.) were fabricated in Au to produce a working electrode array capable of supporting the optical and electrochemical requirements of spectroelectrochemistry under single molecule occupancy conditions. Despite having a smaller skin depth than Al, Au was chosen for the optical cladding layer due to its large ideally polarizable region, and ZMWs were designed and fabricated to effectively confine the optical field and reduce the active volume. As Fig. 1 illustrates, larger pores were FIB milled into the corners of the ZMW array as optical reference marks to assist in the location of the ZMW nanopores using trans-illumination. FIB direct writing of ZMWs produces pores with conical frustum shapes rather than the cylindrical shapes obtained using electron beam lithography to produce ZMW arrays. The SEMs and schematic diagram in Fig. 1 show a cross-section of a typical single ZMW nanopore, illustrating the conical pore cross-section as well as the overetched region, which is a common feature of pores produced by FIB milling.[30] The overetched region past the Au/glass interface is typically the same width as the pore bottom diameter (d_{bottom}) and well-controlled milling processes yield overetch depths less than 50 nm.

The evanescent field decays exponentially with distance in cylindrical pores at a rate that depends on the radius and diameter. To assess the behavior in the conical ZMW nanopores studied here, finite element simulations were performed at various conical frustum sizes and geometries. Fig. 2 shows the electric field in a series of conical frustum pores each having a ratio of top diameter, d_{top}, to bottom diameter, d_{bottom}, $d_{top}/d_{bottom} = 2$. As Fig. 2 illustrates, the field decays exponentially, and the nanostructures provide excellent attenuation, primarily determined by d_{bottom}. Structures fabricated for this study generally had $d_{top} \sim 80$–125 nm, and $d_{bottom} \sim 20$–60 nm with $d_{top} : d_{bottom}$ ratios between 1.5 and 4.0. At positions in the pore exhibiting attenuations in excess of $1/e^2$, the evanescent field strength is likely insufficient for excitation of fluorescence. The exact position in

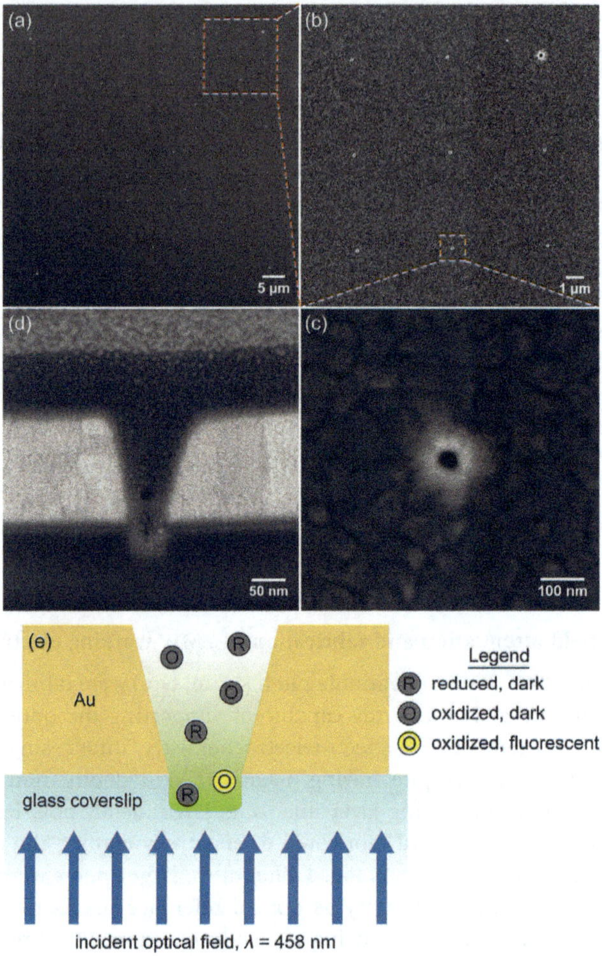

Fig. 1 Scanning electron micrographs of a ZMW array at different magnifications and perspectives: (a) top-down view of the entire 11 × 11 ZMW array, (b) top-down view of a 3 × 3 subset of the same array, (c) top-down view of a single nanopore with $d \sim 40$ nm, and (d) cross-section of a nanopore filled with Pt to assist imaging. (e) Schematic illustration of an over-etched ZMW with reduced and oxidized FMN molecules shown freely diffusing.

the pore at which the attenuation exceeds $1/e^2$ varies with pore diameter and shape, so the effective volume, where the majority of excitation/emission events occur, is typically constrained to the bottom 5–20% of the pore.

One of the advantages of using nanophotonic ZMWs is the extremely small effective sampling volume. Fig. 3 shows how the effective volume depends on d_{bottom} at $d_{\text{top}} : d_{\text{bottom}}$ values of 1.5, 2.0, and 3.0. The effective volumes for different pore shapes deviate from each other only at large d_{bottom} values. In addition, the concentration required to achieve a molecular occupancy expectation value of 1, *i.e.* $\langle N \rangle = 1$, is also shown as a function of d_{bottom}. Based on the simulations shown in Fig. 2 and taking into account the volume of the overetched region, the total effective volumes of the ZMWs used in this study are

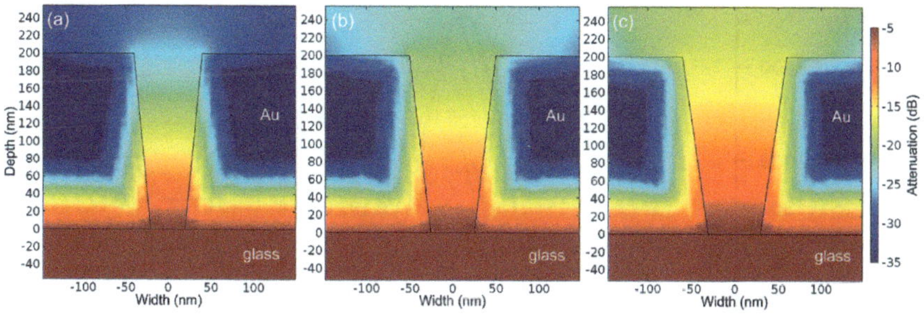

Fig. 2 Cross-sectional heat maps of evanescent field amplitudes in conical nanopores obtained by finite element simulations, showing attenuation as a function of nanopore geometry. All three structures exhibit $d_{top} : d_{bottom} = 2$, and $d_{bottom} =$ (a) 40 nm, (b) 50 nm, and (c) 60 nm.

approximately 200 zL, assuming that the volumes of the metal-enclosed pore and the overetched region are roughly equal. At this volume, molecular occupancies of $\langle N \rangle = 1.2$ and $\langle N \rangle = 0.12$ are expected at 10 and 1 μM, respectively. Additionally, Monte-Carlo simulations of Brownian motion were performed on the conical nanopore geometrics of relevance in these experiments. In these simulations, the mean residence time, defined as the time it takes for a randomly positioned particle to diffuse out of the effective volume of the pore, is calculated to be in the range ~0.5 to 3.5 μs depending on pore geometry.

Statically coupled fluorescence and electrochemistry

Initial experiments were implemented to characterize the emission rates of freely diffusing FMN molecules under static control of potential values in the range $E_{appl} = +0.06$ V to -0.44 V at FMN formal concentrations of 10 μM and 1 μM, where the average molecular occupancy is $\langle N \rangle = 1.2$ and 0.12 molecules, respectively. Solutions were held at constant potential for ~30 s in order to ensure that steady state was reached at the ZMW electrode, after which the fluorescence signal was recorded. Histograms of the signal intensities were then fit to Poisson functions

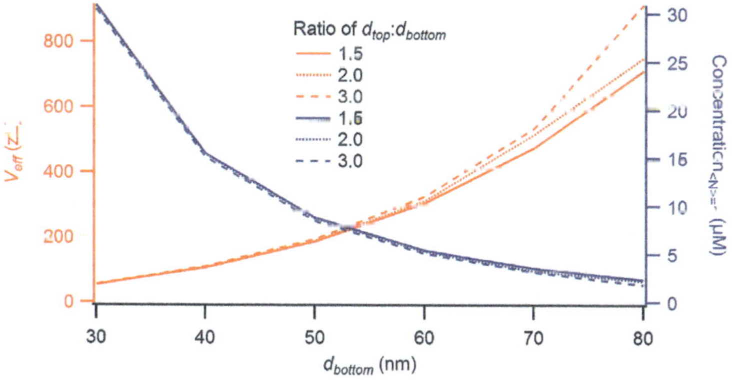

Fig. 3 Effective volumes (left ordinate) and concentrations for single molecule occupancy (right ordinate) for $d_{top} : d_{bottom}$ ratios of 1.5, 2.0, and 3.0 as a function of d_{bottom}.

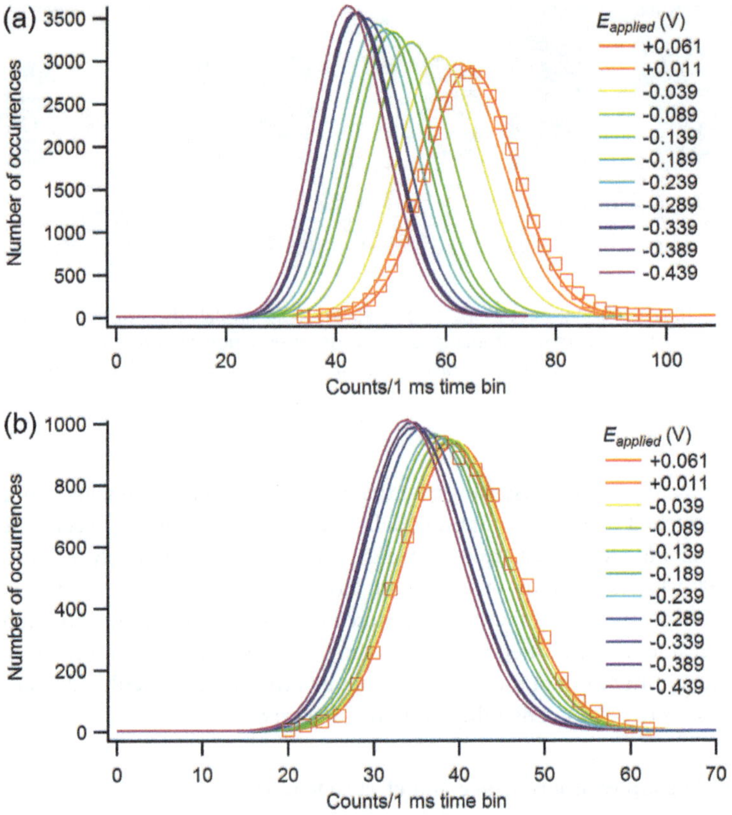

Fig. 4 Poisson distribution fits to histograms of fluorescence intensities at different values of E_{appl} for (a) 10 μM and (b) 1 μM FMN in 100 mM citrate buffer at pH 2.9. Experimental data are included with the fit for $E_{appl} = +0.061$ V to illustrate the quality of the fit.

to produce the distributions shown in Fig. 4. At both average occupancy levels, the distribution shifts to higher intensity values as the potential is made more positive, consistent with the molecules being more likely to be in the oxidized state; FMN is shown in Scheme 1.

With the potential poised at the most positive value, $E_{appl} = +0.06$ V, the peak of the fluorescence emission distribution occurs at $I_{ox} \sim 64$ ms^{-1}, while at the most negative value, $E_{appl} = -0.44$ V, the fluorescence signal decreases to $I_{red} \sim 41$ ms^{-1}. The intensity observed at oxidizing potentials is consistent with estimates of the emission rate for FMN excited at 458 nm based on single molecule occupancy, the measured irradiance at the ZMW array, and the estimated collection efficiency. However, the "dark state" emission observed at $E_{appl} = -0.44$ V is significantly higher than the apparent emission level, $I_{bkgnd} \sim 3$ ms^{-1}, observed in the absence of fluorophore in control experiments. The latter observation can be understood by considering three dynamic processes: (1) the intrinsic photopumping and luminescence dynamics of the fluorophore, (2) transport of both oxidized and reduced molecules into and out of the ZMW nanopore, and (3) heterogeneous electron transfer at the Au cladding/working electrode. The intrinsic photopumping and emission dynamics are

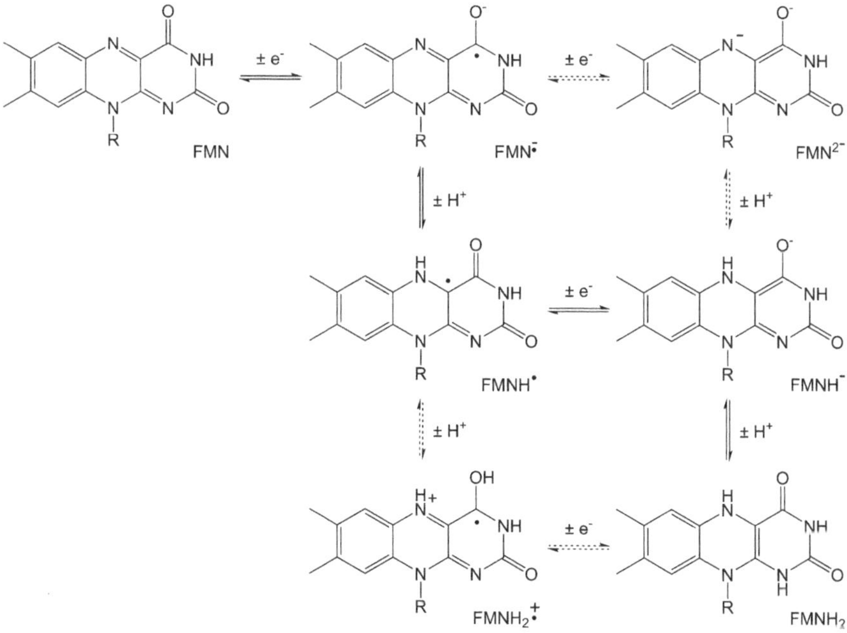

Scheme 1

overwhelmingly associated with the oxidized state, so the most likely explanation for the observation that $I_{red} > I_{bkgnd}$ involves occupancy of the ZMW nanopore by oxidized FMN under nominally reducing conditions, and the FMN can arise from two sources. Since the experiment begins with oxidized FMN, there is always a background population of fluorescent species that can diffuse into the excitation volume. In addition, $FMNH_2$ may react with dissolved O_2 in the nanopore to produce FMN, independent of the potential on the electrode. Some of these FMN molecules will diffuse to the Au surface and be reduced prior to diffusing out of the pore, so the level of emission observed at I_{red} likely reflects the balance between new oxidized species entering, or being generated in, the pore and oxidized species being reduced while in the ZMW. Since the time bin for data collection (1 ms) is $\geq 300\times$ the average residence time obtained from Monte-Carlo simulations, there is ample time to sample many individual molecules during the collection window.

The other interesting feature of these data concerns the narrowing of the intensity distribution at lower concentration. When the static potential experiment is repeated at [FMN] = 1 μM (Fig. 4(b)) the distribution shifts to higher intensity values, as the potential is made more positive, just as with the higher concentration, but the spread of fluorescence intensity distributions is considerably narrower. The negative end of the distribution decreases to $I_{red} \sim 34\ ms^{-1}$, but the positive end at $E_{appl} = +0.06$ V has a maximum at $I_{ox} \sim 40\ ms^{-1}$. The narrowed distribution is also consistent with the putative mechanism for I_{bkgnd} advanced above, as the lower concentration would be expected to yield a smaller departure from the background simply due to the smaller number of reduced species capable of diffusing into the ZMW.

Dynamics of coupled fluorescence and electrochemistry

Having established the conditions under which single molecule occupancy is achieved in ZMWs, the optimum geometries for coupling optical radiation into solutions supporting single redox-active molecules in close proximity to a working electrode, and the static potential dependence of FMN emission from ZMWs, the next fluorescence measurements were made under active potential control to explore how electrochemistry couples to a population of freely diffusing single molecules. Fig. 5 shows a comparison of the potential dependence of FMN luminescence from a single ZMW nanopore obtained at [FMN] = 10 μM in 100 mM pH 2.9 citrate buffer to a bulk measurement of the cyclic voltammetry (CV) for 5 mM FMN in the same buffer. At pH 2.9, the half-wave potential of FMN, determined from CV at [FMN] = 5 mM, is $E_{1/2} = -0.19$ V *vs.* Ag/AgCl. Furthermore, even though the luminescence data are acquired in a very different physical environment, the inflection in the potential-dependent luminescence curve is in reasonable agreement with the bulk electrochemical value.

A single molecule occupancy chronofluorometry experiment is shown in Fig. 6. Analogous to the more familiar chronoamperometry, this experiment monitors the fluorescence, as opposed to the current, response of a redox system to alternating potential steps. Given the $E_{1/2} = -0.19$ V determined for FMN in Fig. 5, the chronofluorometry experiment was designed to step between ±0.25 V overpotentials *vs.* Ag/AgCl (E_{appl} = +0.06 V and −0.44 V) in order to produce conditions in which FMN alternates between the fully oxidized and fully reduced states, represented in Scheme 1 as FMN and $FMNH_2$, the latter being the fully protonated form of reduced FMN, appropriate to experiments at pH 2.9. Although both longer and shorter time windows were investigated, the 5 s potential steps used in Fig. 6 are sufficiently long to allow the system to achieve steady state at both oxidizing and reducing conditions.

Clearly, the fluorescence response in Fig. 6 is strongly correlated to the applied potential steps. The experiment is initiated at an oxidizing potential, and stepping

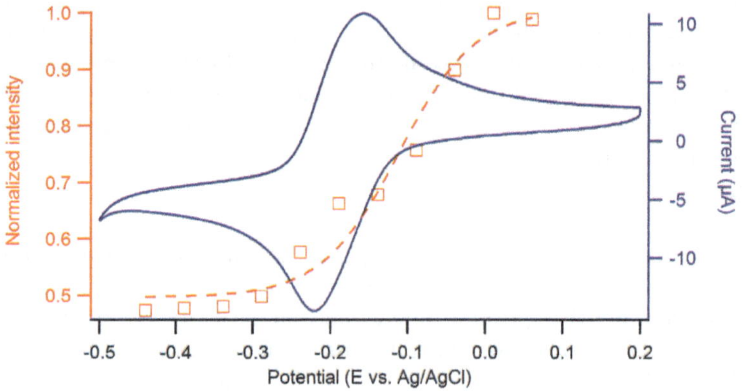

Fig. 5 Comparison of the potential dependence of FMN luminescence (discrete data points) from a single ZMW nanopore obtained at [FMN] = 10 μM in 100 mM pH 2.9 citrate buffer to a bulk measurement of the cyclic voltammetry for 5 mM FMN in the same buffer (solid curve). The dashed line gives the fit of the luminescence data to a sigmoidal optical response function.

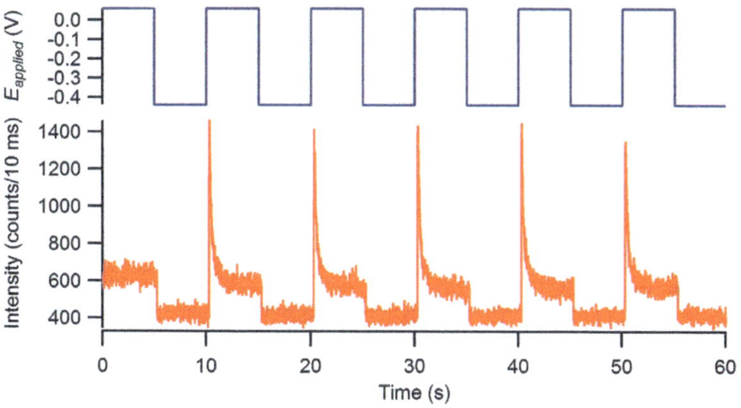

Fig. 6 Chronofluorometry of 10 μM FMN in 100 mM citrate buffer at pH 2.9. The potential (*vs.* Ag/AgCl) is modulated at 5 s intervals between 0.06 V and −0.44 V, and fluorescence emission is monitored from a single d_{top} = 80 nm, d_{bottom} = 40 nm ZMW nanopore using 10 ms time bins.

to a reducing potential causes an immediate decrease in emission intensity from ~64 ms^{-1} to ~41 ms^{-1}, consistent with the steady-state values shown in Fig. 4. Interestingly, when the potential is returned to an oxidizing value, the fluorescence signal exhibits a significant and reproducible positive transient before decaying to the steady state value $I_{ox} \sim 64$ ms^{-1}, a transient not observed in Fig. 4, because the static potential results were obtained at steady-state. Because the potential step occurs rapidly compared to molecular transport, any reduced molecules in the nanopore would be oxidized within the time required to diffuse to the pore wall; $\tau_{diff} \sim 1.5$ μs for a 60 nm diameter pore and $D \sim 10^{-6}$ cm^2 s^{-1}, thus establishing a momentarily larger population of oxidized FMN in and near the ZMW pores. This augmented population would be responsible for the initial increase in fluorescence observed on the positive potential step, after which a return to a steady-state population would be established by the balance between diffusion of FMN into the pore and oxidation of $FMNH_2$, against diffusion of FMN out of the pore and perhaps other processes, such as photobleaching, which would decrease the population of fluorescent species in and near the ZMW nanopore.

The single molecule occupancy fluorescence analog of the cyclic voltammetry experiments is shown in Fig. 7 at three distinct scan rates. At the fastest scan rate, 0.5 V s^{-1}, the luminescence behavior of 10 μM FMN is similar to that expected from a bulk measurement. As shown in Fig. 7(a), the fluorescence response is symmetric and reproducible, oscillating between extreme values of $I_{ox} \sim 70$ ms^{-1} and $I_{red} \sim 43$ ms^{-1}, similar to the results obtained at static potentials and with chronofluorometry. Decreasing the scan rate to 50 mV s^{-1}, the response develops an asymmetric shape, *cf.* Fig. 7(b), with a shoulder on the negative going scan and a constant intensity observed over an extended window of reducing potentials. Finally, at the slowest scan rate studied, 1 mV s^{-1}, *cf.* Fig. 7(c), the potential-dependent fluorescence develops distinct additional states with fluorescence emission intensities $I_{int} \sim 32$ ms^{-1} intermediate between the values $I_{ox} \sim 44$ ms^{-1} and $I_{red} \sim 25$ ms^{-1}.

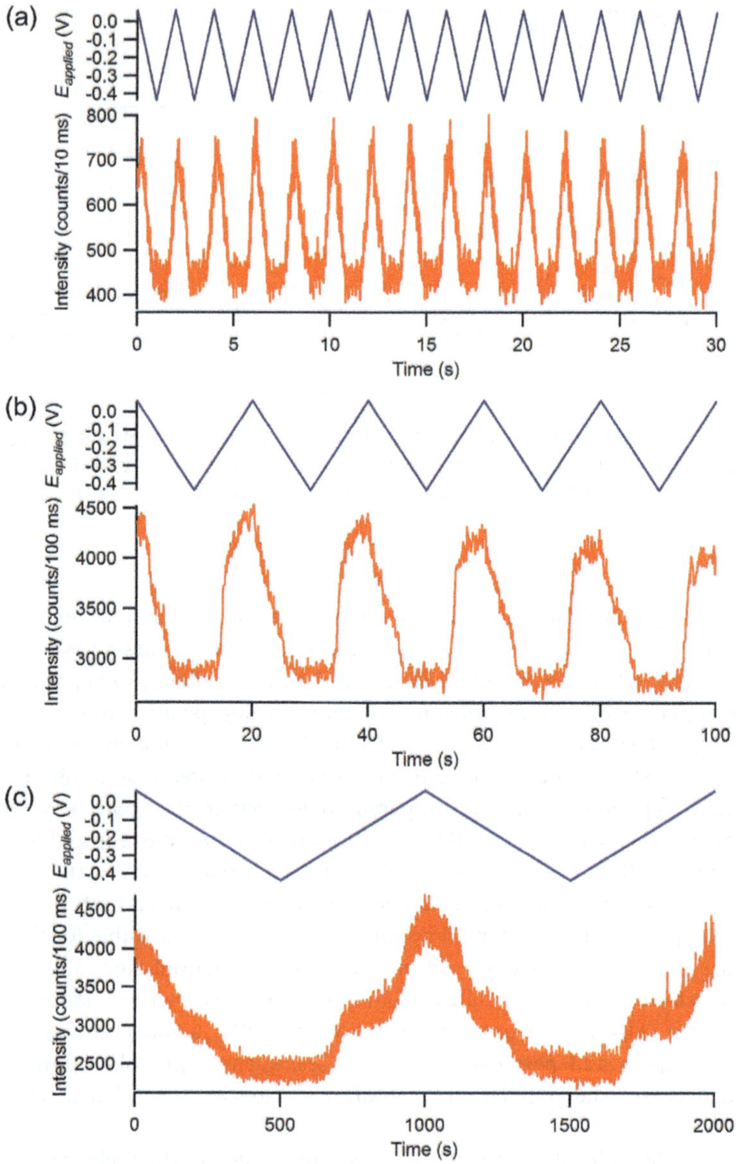

Fig. 7 Cyclic potential sweep fluorescence measurements of 10 μM FMN in 100 mM citrate buffer at pH 2.9 at scan rates of (a) 0.5 V s^{-1}, (b) 0.05 V s^{-1}, and (c) 0.001 V s^{-1}. In each panel, the potential program is shown at the top, and the fluorescence response at the bottom.

The asymmetric fluorescence potential response at 50 mV s^{-1} and the intermediate emission intensity states observed at 1 mV s^{-1} are likely due to intermediate reaction steps between FMN and $FMNH_2$. Scheme 1 shows all of the reaction pathways by which proton-coupled electron transfer can connect the fully oxidized FMN to the fully reduced (and protonated) $FMNH_2$, with the preferred pathway at pH 2.9 indicated by the solid equilibrium symbols. However, in the most general view at least three possible reaction pathways exist, with 5 possible intermediates including 3 semiquinones.[46] At pH 2.9 the reduction of FMN to $FMNH_2$ is thought to occur through two separate proton-coupled ($1e^-$/$1H^+$) steps through an intermediate flavosemiquinone ($FMNH^{\bullet}$ in Scheme 1) that is stabilized in the flavoprotein environment. In contrast, flavosemiquinone species are observed to undergo rapid dismutation in bulk solution. Thus, the appearance of a stable intermediate species in the slow cyclic potential sweep experiment suggests that the special environment of the ZMW nanopore facilitates the stabilization of the flavosemiquinone species, a behavior of radical redox species that is not observed in bulk solution.

Conclusions

ZMWs fabricated with Au cladding layers are attractive substrates for investigating electrofluorogenic behavior of freely diffusing single molecules. The combination of the small volume ($V_{eff} \sim 200$ zL) and the efficient confinement of optical radiation in the conical FIB-milled ZMW nanopores used here make it possible to study the electroluminescence behavior of single flavin (FMN/$FMNH_2$) molecules under both static and active potential control. Under static potential control fluorescence intensity histograms are consistent with expectations based on bulk studies, *i.e.* fluorescence increases as E_{appl} becomes more positive. However, a non-zero luminescence intensity is observed even at the most negative applied potentials, an observation that is ascribed to the interplay between the transport of oxidized and reduced molecules into and out of the ZMW nanopore and redox reactions at the Au cladding/working electrode.

Two types of actively-controlled electroluminescence experiments were performed at the single molecule occupancy level in these structures: chronofluorometry experiments, in which the potential is alternately stepped between oxidizing and reducing potentials, and fluorescence-reported cyclic potential sweep experiments, the fluorescence analog of cyclic voltammetry. Chronofluorometry measurements exhibit a fluorescence response that is strongly correlated to applied potential steps, but with an interesting asymmetry between the negative- and positive-going steps. The substantial fluorescence transient observed on the positive-going step is attributed to an augmented population of oxidized species that is depleted over time by diffusion of FMN out of the pore and possibly also by photobleaching. Fluorescence-reported cyclic potential sweep experiments under single occupancy conditions exhibit a dramatic scan rate dependence with the slowest scan rates showing distinct intermediate states that are stable over a range of potentials. These states are tentatively assigned to flavosemiquinone species that are stabilized in the special environment of the ZMW nanopore. The ultra small volume and efficient trapping of optical fields in the ZMW nanopores makes these systems attractive platforms for future single molecule and single enzyme spectroelectrochemical investigations.

Acknowledgements

This work was supported by the National Science Foundation under grant NSF1404744 (LPZ) and the Department of Energy Office of Science under grant DE FG02 07ER15851 (DAG) and a subcontract from Oak Ridge National Laboratory

(DH). Fabrication and structural characterization of the devices studied here were accomplished at the Notre Dame Nanofabrication Facility and the Notre Dame Integrated Imaging Facility, respectively, whose generous support is gratefully acknowledged. We thank David Heemstra of Notre Dame Nanofabrication Facility for useful advice on fabrication procedures.

References

1 M. M. Collinson and R. M. Wightman, *Science*, 1995, **268**, 1883–1885.

2 F. R. F. Fan and A. J. Bard, *Science*, 1995, **267**, 871–874.

3 F. R. F. Fan, J. Kwak and A. J. Bard, *J. Am. Chem. Soc.*, 1996, **118**, 9669–9675.

4 P. Audebert and F. Miomandre, *Chem. Sci.*, 2013, **4**, 575–584.

5 M. A. Jones and P. W. Bohn, *Anal. Chem.*, 2000, **72**, 3776–3783.

6 M. A. Jones and P. W. Bohn, *J. Phys. Chem. B*, 2001, **105**, 2197–2204.

7 F. Miomandre, E. Lepicier, S. Munteanu, O. Galangau, J. F. Audibert, R. Meallet-Renault, P. Audebert and R. B. Pansu, *ACS Appl. Mater. Interfaces*, 2011, **3**, 690–696.

8 F. Miomandre, R. Meallet-Renault, J. J. Vachon, R. B. Pansu and P. Audebert, *Chem. Commun.*, 2008, 1913–1915.

9 C. Ma, L. P. Zaino and P. W. Bohn, *Chem. Sci.*, 2015, **6**, 3484–3491.

10 Y. Fu, J. Zhang and J. R. Lakowicz, *Langmuir*, 2008, **24**, 3429–3433.

11 J. P. Guerrette, S. J. Percival and B. Zhang, *J. Am. Chem. Soc.*, 2013, **135**, 855–861.

12 S. M. Oja, J. P. Guerrette, M. R. David and B. Zhang, *Anal. Chem.*, 2014, **86**, 6040–6048.

13 R. E. Palacios, F. R. F. Fan, A. J. Bard and P. F. Barbara, *J. Am. Chem. Soc.*, 2006, **128**, 9028–9029.

14 Y. L. Chang, R. E. Palacios, F. R. F. Fan, A. J. Bard and P. F. Barbara, *J. Am. Chem. Soc.*, 2008, **130**, 8906–8907.

15 R. E. Palacios, F. R. F. Fan, J. K. Grey, J. Suk, A. J. Bard and P. F. Barbara, *Nat. Mater.*, 2007, **6**, 680–685.

16 R. E. Palacios, W. S. Chang, J. K. Grey, Y. L. Chang, W. L. Miller, C. Y. Lu, G. Henkelman, D. Zepeda, J. Ferraris and P. F. Barbara, *J. Phys. Chem. B*, 2009, **113**, 14619–14628.

17 C. H. Lei, D. H. Hu and E. J. Ackerman, *Chem. Commun.*, 2008, 5490–5492.

18 C. H. Lei, D. H. Hu and E. Ackerman, *Nano Lett.*, 2009, **9**, 655–658.

19 C. M. Hill, D. A. Clayton and S. Pan, *Phys. Chem. Chem. Phys.*, 2013, **15**, 20797–20807.

20 J. Liu, C. M. Hill, S. L. Pan and H. Y. Liu, *Phys. Chem. Chem. Phys.*, 2014, **16**, 23150–23156.

21 H. Shen, W. L. Xu and P. Chen, *Phys. Chem. Chem. Phys.*, 2010, **12**, 6555–6563.

22 G. Zhang, L. Xiao, R. Chen, Y. Gao, X. Wang and S. Jia, *Phys. Chem. Chem. Phys.*, 2011, **13**, 13815–13820.

23 A. J. Bard, *ACS Nano*, 2008, **2**, 2437–2440.

24 S. G. Lemay, S. Kang, K. Mathwig and P. S. Singh, *Acc. Chem. Res.*, 2013, **46**, 369–377.

25 K. Mathwig, T. J. Aartsma, G. W. Canters and S. G. Lemay, *Annu. Rev. Anal. Chem.*, 2014, **7**, 383–404.

26 J. M. Salverda, A. V. Patil, G. Mizzon, S. Kuznetsova, G. Zauner, N. Akkilic, G. W. Canters, J. J. Davis, H. A. Heering and T. J. Aartsma, *Angew. Chem., Int. Ed.*, 2010, **49**, 5776–5779.
27 N. Akkilic, F. v. d. Grient, M. Kamran and N. J. M. Sanghamitra, *Chem. Commun.*, 2014, **50**, 14523–14526.
28 A. V. Patil and J. J. Davis, *J. Am. Chem. Soc.*, 2010, **132**, 16938–16944.
29 N. Akkilic, M. Kamran, R. Stan and N. Sanghamitra, *Biosens. Bioelectron.*, 2015, **67**, 747–751.
30 J. Zhao, S. P. Branagan and P. W. Bohn, *Appl. Spectrosc.*, 2012, **66**, 163–169.
31 J. Zhao, L. P. Zaino and P. W. Bohn, *Faraday Discuss.*, 2013, **164**, 57–69.
32 E. Haustein and P. Schwille, *Annu. Rev. Biophys.*, 2007, **36**, 151–169.
33 W. E. Moerner and D. P. Fromm, *Rev. Sci. Instrum.*, 2003, **74**, 3597–3619.
34 M. J. Levene, J. Korlach, S. W. Turner, M. Foquet, H. G. Craighead and W. W. Webb, *Science*, 2003, **299**, 682–686.
35 H. A. Bethe, *Phys. Rev.*, 1944, **66**, 163–182.
36 C. J. Bouwkamp, *Philips Res. Rep.*, 1950, **5**, 321–332.
37 J. M. Moran-Mirabal and H. G. Craighead, *Methods*, 2008, **46**, 11–17.
38 J. Korlach, M. Levene, M. Foquet, S. W. Turner, H. G. Craighead and W. W. Webb, *Biophys. J.*, 2003, **84**, 141A.
39 J. Korlach, M. Levene, S. W. Turner, H. G. Craighead and W. W. Webb, *Biophys. J.*, 2002, **82**, 507A.
40 P. Zhu and H. G. Craighead, *Annu. Rev. Biophys.*, 2012, **41**, 269–293.
41 K. T. Samiee, J. M. Moran-Mirabal, Y. K. Cheung and H. G. Craighead, *Biophys. J.*, 2006, **90**, 3288–3299.
42 C. D. Kinz-Thompson, M. Palma, D. K. Pulukkunat, D. Chenet, J. Hone, S. J. Wind and R. L. Gonzalez, *ACS Nano*, 2013, **7**, 8158–8166.
43 S. D. M. Islam, T. Susdorf, A. Penzkofer and P. Hegemann, *Chem. Phys.*, 2003, **295**, 137–149.
44 A. Sengupta, R. V. Khade and P. Hazra, *J. Photochem. Photobiol., A*, 2011, **221**, 105–112.
45 Y. T. Kao, C. Saxena, T. F. He, L. J. Guo, L. J. Wang, A. Sancar and D. P. Zhong, *J. Am. Chem. Soc.*, 2008, **130**, 13132–13139.
46 S. L. J. Tan, J. M. Kan and R. D. Webster, *J. Phys. Chem. B*, 2013, **117**, 13755–13766.

Faraday Discussions

PAPER

Quantitative structural information from single-molecule FRET

M. Beckers, F. Drechsler, T. Eilert, J. Nagy and J. Michaelis*

Received 15th June 2015, Accepted 17th July 2015
DOI: 10.1039/c5fd00110b

Single-molecule studies can be used to study biological processes directly and in real-time. In particular, the fluorescence energy transfer between reporter dye molecules attached to specific sites on macromolecular complexes can be used to infer distance information. When several measurements are combined, the information can be used to determine the position and conformation of certain domains with respect to the complex. However, data analysis schemes that include all experimental uncertainties are highly complex, and the outcome depends on assumptions about the state of the dye molecules. Here, we present a new analysis algorithm using Bayesian parameter estimation based on Markov Chain Monte Carlo sampling and parallel tempering termed Fast-NPS that can analyse large smFRET networks in a relatively short time and yields the position of the dye molecules together with their respective uncertainties. Moreover, we show what effects different assumptions about the dye molecules have on the outcome. We discuss the possibilities and pitfalls in structure determination based on smFRET using experimental data for an archaeal transcription pre-initiation complex, whose architecture has recently been unravelled by smFRET measurements.

Introduction

Structural biology has, in the last few decades, revolutionised our understanding of many important cellular functions, such as transcription,[1] translation,[2] nucleic acid translocases,[3] and motor proteins, such as myosin,[4] among many others. While these studies provide (nearly) atomic models for these complex machineries, in order to mechanistically understand protein function, dynamic information is required as well. Here, single-molecule (sm) methods have, in particular, become the method of choice[5] since they provide a way to observe the action of a single enzyme over time. In particular, when attaching a pair of dye molecules at specific locations and measuring the Förster resonance energy transfer between them (smFRET), information about conformational changes becomes accessible. Since quite early on, FRET has been termed a *molecular*

Biophysics Institute, Ulm University, 89081 Ulm, Germany. E-mail: jens.michaelis@uni-ulm.de

ruler,[6] and obtaining quantitative information from FRET measurements has proven to be quite difficult.[7]

One idea used to obtain structural information from smFRET measurements was to determine an unknown position using a triangulation, or more correctly, a trilateration approach.[8–10] However, while measuring smFRET efficiencies can be done quite accurately if care is taken to perform the required corrections,[4,11] rather large uncertainties arise due to the relative orientation of the dye molecules, resulting in an uncertainty of κ^2 and thus an uncertainty in the Förster distance.[12] Moreover, the quantum yield of the donor could be sensitive to local changes and thus increase the experimental uncertainty in the distance determination.[13] Another important factor when using smFRET data to determine distances is the fact that dye molecules are attached *via* flexible linkers and can be located anywhere inside their so-called accessible volume (AV).[14]

In order to arrive at quantitative structural information, we have developed the Nano-Positioning System (NPS)[14,15] and have applied it to obtain insight into transcription elongation[14,16] and transcription initiation complexes.[17–19] NPS uses Bayesian parameter estimation together with a static model of the dye linker, assuming that the linker takes up a specific, however unknown, conformation. Quantum yield variations are accounted for by performing a molecule-wise correction of the quantum yield, and the relative orientation of the dye molecules is either modelled by a Förster radius prior computed using the measured anisotropies of the donor and acceptor,[14] or by determining the dye orientations explicitly, followed by a marginalisation to the pure position information.[15] NPS computes the three-dimensional probability distribution for each dye position, *i.e.* the marginalised posterior, which is best visualised by the smallest volume enclosing a certain probability, *i.e.* the credible volume. Depending on the number of measurements used, as well as on the geometry of the problem, the 1σ error can be as low as 3 Å,[14] but more typical values are on the order of 10–20 Å.[19] While at this resolution important structural insight into transcription elongation and transcription initiation was obtained,[14,16–20] in order to achieve such a resolution, large networks of FRET measurements were analysed globally, resulting in computation times of several weeks.

More recently, another related approach has been published, termed FRET-restrained positioning and screening (FPS).[21] FPS treats the dye molecules attached to the macromolecule differently. First, it is assumed that the dye molecules are free to re-orient on the time-scale of the fluorescence lifetime, implying $\kappa^2 = 2/3$. The second assumption is that, due to its flexible linker, each dye molecule is free to move within its AV. Furthermore, it is assumed that the dye occupies each position within its AV for an equally long time, such that its position can be represented by a mean position (MP). However, in order to correctly account for the R^6-dependence of the energy transfer, the distance of the mean positions R_{MP} is converted into the FRET-weighted donor–acceptor distance R_{DA} using a polynomial fit. The obtained quantitative smFRET distance constraints are then used for arranging several parts of a biological complex into a rigid-body docking scenario, thus allowing smFRET to be used for structure determination. In benchmark studies, the FPS-derived structure of the DNA primer-template bound to the HIV-1 reverse transcriptase yielded root mean square deviations as low as 0.5 Å compared to the X-ray structure, and the structure of an ssDNA overhang was determined with a precision of 2.9 Å.[21] While for several steps in the analysis a rigorous error estimate was performed,[22] the precision was not determined from the posterior shape and size but rather using a bootstrapping approach.

A related, less stringent approach for smFRET-restrained structure determination was published earlier.[23] Here, molecular dynamics (MD) simulations were used for determining the dye position attached *via* a flexible linker; however, no correction for the R^6-dependence of FRET was applied and no rigorous error estimate was performed.

Here, we present an improved NPS analysis method, termed Fast-NPS,† which uses Markov Chain Monte Carlo (MCMC) sampling, which has been sped up *via* parallel tempering, shortening computation times by two orders of magnitude. Moreover, we have added the capability to test different models for the respective dye molecules, thus enabling us to compare the results of different approaches. We test the new NPS scheme on recently published data of the archaeal transcription initiation complex.[19]

Results

Effective MCMC sampling of large smFRET networks

In NPS, the information from an arbitrarily large set of measurements between satellite dye molecules (SDM), attached to known positions of a macromolecule, and antenna dye molecules (ADM), attached to unknown positions, is analysed globally using Bayesian parameter estimation yielding the posterior for the set of all the dye molecules:[15]

$$p(\{\boldsymbol{x}_i,\boldsymbol{\Omega}_i\}|\{E_{ij}\},\{A_{ij}\},I) = p(\{\boldsymbol{x}_i,\boldsymbol{\Omega}_i\}|I) \times p(\{E_{ij}\},\{A_{ij}\}|\{\boldsymbol{x}_i,\boldsymbol{\Omega}_i\},I)/Z, \qquad (1)$$

where $p(\{\boldsymbol{x}_i,\boldsymbol{\Omega}_i\}|I)$ is the prior, which encodes the information before the measurement, and $p(\{E_{ij}\},\{A_{ij}\}|\{\boldsymbol{x}_i,\boldsymbol{\Omega}_i\},I)$ is the likelihood, which describes the expected data given the set of dye positions $\boldsymbol{x}_i$ and the average transition dipole orientation $\boldsymbol{\Omega}_i$. The evidence is denoted by Z, which, for a given set of data, is a constant and will in the following be treated as unity for simplicity. The measured FRET efficiency between molecules i and j is denoted by E_{ij}, and A_{ij} denotes the measured FRET anisotropy between the respective molecules.[15] The latter type of measurement is not required but improves the accuracy of the NPS analysis. Now, in order to obtain the probability distribution of the position $\boldsymbol{x}_i$ of one single ADM i, one marginalises the posterior by integration:[15]

$$p(\{\boldsymbol{x}_i\}|\{E_{ij}\},\{A_{ij}\},I) = \int \mathrm{d}\{\boldsymbol{x}_{j\neq i},\boldsymbol{\Omega}_j\}p(\{\boldsymbol{x}_j,\boldsymbol{\Omega}_j\}|\{E_{ij}\},\{A_{ij}\},I)$$

In our original publication, a nested sampling algorithm was used for computing the posterior. While reliable, we experienced excessively large computation times on the order of weeks for the large smFRET networks used to study transcription complexes.

† The developed Fast-NPS software package can be downloaded from http://www.uni-ulm.de/nawi/nawi-biophys/software.html.

We therefore decided to sample the posterior using an MCMC algorithm and employed Metropolis-within-Gibbs sampling, an efficient way for computing posteriors in Bayesian parameter estimation.[24]

In order to further speed up the analysis and to ensure that even posteriors consisting of several well-separated maxima are faithfully represented, we also employed parallel tempering, sometimes also referred to as replica exchange or Metropolis-coupled-MCMC.[25,26] This is a powerful technique to overcome several problems regarding simple MCMC algorithms, including multimodality and poor mixing (*i.e.* bad exploration of the state space). Parallel tempering is also often used as an enhanced sampling method for MD simulations; in contrast, here it is employed to achieve better mixing and to guarantee that all local maxima of the posterior are found. The idea behind parallel tempering is the simulation of the same system at different virtual temperatures – the higher the temperature, the lower the energy barriers in the respective potential energy landscape. Thus, a Markov chain is more likely to cross an energy barrier at higher temperatures and this way the trajectory can jump between different maxima. Replicas are run in parallel at different temperatures and after a certain number of steps of each, swaps of the complete states are proposed between different replicas, *i.e.* temperatures, and accepted or rejected using a Metropolis–Hastings criterion. These swaps force the lowest-temperature chain, *i.e.* the system of interest, to explore the energy landscape more widely. Thus, multimodalities are more likely to be found.

In our implementation, the temperature spacing of different replicas follows a geometric ladder,[27] *i.e.* the inverse temperatures of the replicas are given by $K_0 = 1$, and $K_i = K_{i-1} \times 0.4$ for $i = 1, \ldots, n$; the scale of 0.4 having been chosen experimentally for optimal performance.

In order to speed up the analysis even further, we used different adaptation algorithms of our MCMC sampler, details of which will be published elsewhere. The Fast-NPS implementation allowed us to analyse large smFRET networks using standard office computers on the order of hours, yielding one million samples to describe the posterior.

Archaeal transcription initiation complexes

One important question in the field of transcription is the architecture and dynamics of multi-component transcription initiation complexes. We have recently studied the architecture of the pre-initiation complex (PIC) of *M. jannaschii*, consisting of archaeal RNA polymerase, transcription factors TBP, TFB, TFE, and promoter DNA, using smFRET and the NPS system.[19] Using SDMs attached on various subunits of the polymerase and ADMs attached to the promoter DNA, TBP, TFE and TFB (Fig. 1), we assembled over 70 different complexes. Each of these complexes was labelled with a single smFRET pair attached to well defined positions. The information from all these measurements was used to build a model of the general architecture of the PIC. Here, we will use this smFRET network as a test-scenario for the new NPS algorithm. While NPS analysis using nested sampling took several weeks, using MCMC extended by parallel tempering we were able to compute the posterior on a time-scale of hours (Fig. 2B). Overall there is very good agreement in the computed credible volumes between NPS using nested sampling and MCMC. However, the faster computation times

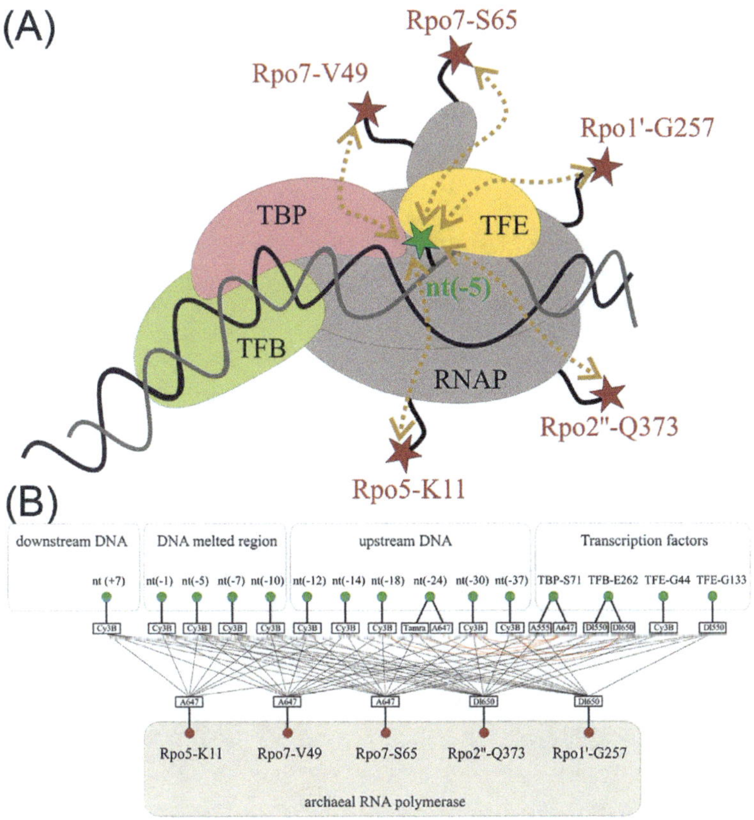

Fig. 1 Schematic of the NPS analysis of the archaeal PIC. (A) Cartoon depicting the archaeal PIC complex, including DNA (tDNA in dark grey, ntDNA in light grey) and transcription factors (TBP in pink, TFB in green and TFE in yellow). Labelling sites on the RNAP are marked with a red star (satellites) and an exemplary labelling site on the DNA is marked with a green star (antenna). SmFRET measurements between satellite and antenna dyes are indicated with dotted beige arrows. (B) Schematic representation of the global FRET network used to solve the complete architecture of the archaeal PIC. The unknown antenna positions (green circles) on the transcription factors as well as on the downstream, upstream and melted DNA region, and the five known satellite sites on the archaeal RNAP (red circles) are shown together with the corresponding attached dyes (A647 = Alexa647; A555 = Alexa555; Dl550 = DyLight550; Dl650 = DyLight650). FRET efficiencies were measured between pairs of satellite (acceptor) and antenna (donor) dyes (dotted black lines) and in between antennas (dotted red lines). In the case of the latter measurements, one of the antenna positions had to be labelled with an acceptor, as indicated.

allowed us to test the influence of several key assumptions in NPS on the overall precision and accuracy of the model.

Comparison of different dye models in NPS

While smFRET experiments allow one to obtain structural information about a particular macromolecule, the assumptions of how the dye molecule is modelled will impact the results (Fig. 3). In the original NPS analysis scheme, we chose a

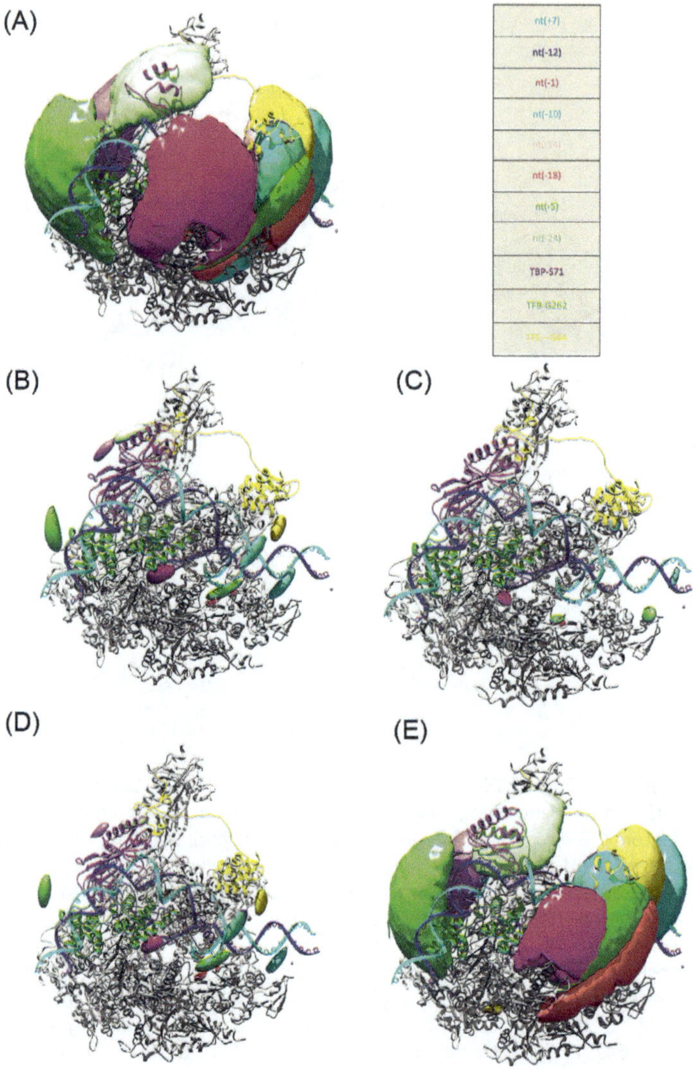

Fig. 2 Simulation results of the different model assumptions. All pictures show the front view of the archaeal RNA polymerase (pdb ID: 2WAQ) together with the model for promoter DNA (tDNA and ntDNA in blue and cyan, respectively), TBP (purple), TFB (green) and TFE (yellow) in the archaeal PIC. The credible volumes are superimposed for the NPS simulation results of (A) the *classic model*, (B) the *iso model*, (C) the *meanpos-iso model*, (D) the *var-meanpos-iso model* and (E) the *var-meanpos model*. All volumes are shown at 68% credibility. Comparing the credible volumes from the different models, one finds that the size of the volumes strongly depends on the model used. Moreover, one also observes a shift in the position of the volumes upon introduction of a fixed mean position (C). Comparison of (A) and (E) as well as (B) and (D) shows that the change from a free static position to a free mean position has only a minor effect upon the size of the volumes, whereas the introduction of isotropic averaging leads to a large reduction in the volume ((A) and (B), (D) and (E)).

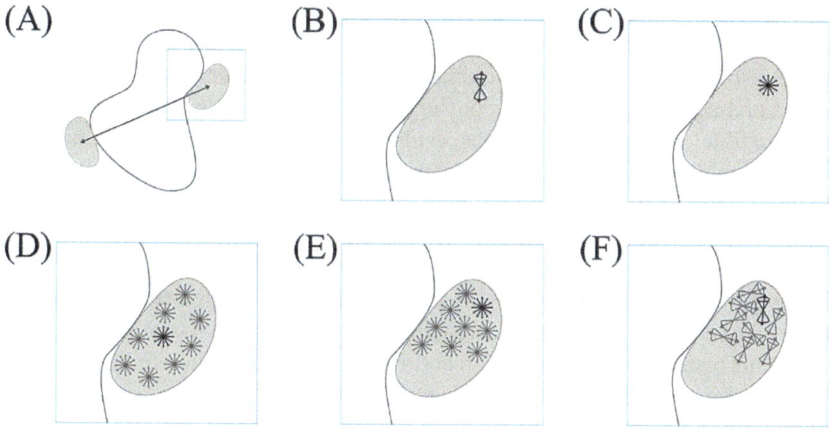

Fig. 3 Cartoon representation of different dye models used in NPS analysis. (A) NPS is based on data from different smFRET measurements between dye molecules attached to different positions on the protein. Due to the flexible linkers used, the dye molecules may reside anywhere within their accessible volumes (AVs). (B) *Classic model*. Within the AV, the dye molecule occupies one unknown position. At this position, the dye molecule can rotate, however, free rotation is only possible within a cone, whose size is determined from the respective fluorescence anisotropy. The orientation of the cone is not known, leading to large uncertainties in the conversion from the FRET efficiency to distance. (C) *Iso model*. The dye molecule resides at a fixed, yet unknown position within the AV. The fluorescence anisotropy is sufficiently small that dynamical averaging over all orientations is allowed. (D) *Meanpos-iso model*. The position of the dye molecule is not fixed, instead the molecule moves about the AV, however, all positions are occupied equally, so that the center of the AV volume, the so-called mean position, can be used in the analysis, together with a polynomial correction accounting for the R^6-law of energy transfer. At each of the positions, the dye molecule is free to rotate. (E) *Var-meanpos-iso model*. Like the *meanpos-iso model* with the only difference being that the mean position of the dye molecule is not known, and needs to be determined by the experiment. (F) *Var-meanpos model*. Like the *classic model*, with the difference being that the dye molecule dynamically switches between different positions, resulting in a dynamically averaged mean position.

fairly conservative model of the dye molecules: they were thought to reside in one particular location within their AV (static model), and the orientation of the transition dipole moment fluctuates within a cone about the mean orientation with the cone opening angle determined by the fluorescence anisotropy (*classic model*, Fig. 3B). Both these assumptions are conservative in the sense that they lead to large uncertainties and as a result the resulting credible volumes are fairly large. In the following, we want to relax these conditions and monitor the effect on the credible volume size and position. We expect that under less conservative assumptions, the determined credible volumes will shrink, however, we would like to develop criteria that help understand whether a particular assumption is justified, or if it may cause inconsistencies in the analysis.

First, we consider the situation where, for all measurements, both the donor and acceptor molecules are sufficiently free to rotate so that the orientation averages dynamically across the lifetime of the fluorescent state, and thus $\kappa^2 = 2/3$ can be used for all smFRET measurements (Fig. 3C). We termed this situation the *isotropic orientation NPS* (*iso model*). In the new Fast-NPS program, we have included the possibility to use an isotropic orientation for all dye molecules. In order to test the performance of the Fast-NPS and to compare the results of this more restrictive *iso model* to the *classic model*, we again use the smFRET scenario of the archaeal PIC and compute posteriors for all ADMs (Fig. 2B). By comparing the credible volumes, it becomes evident that the assumption of isotropic orientation drastically reduces the experimental uncertainty and as a result leads to much smaller credible volumes, with the average volume being about 40 times smaller than that for the *classic model* (Fig. 4).

Next, we wanted to test the effect of the static linker model used in NPS. While in the NPS system a static model is used, others have argued that the linker is better described by a dynamic averaging over all possible positions within the AV.[21] This results in a mean position at the center of the AV. Therefore, in addition to using the dynamically averaged orientation scheme, we also implemented the possibility of dynamic position averaging (*meanpos-iso model*, Fig. 2C). To account for the averaging, we use a polynomial conversion of R_{MP} to R_{DA}. While strictly

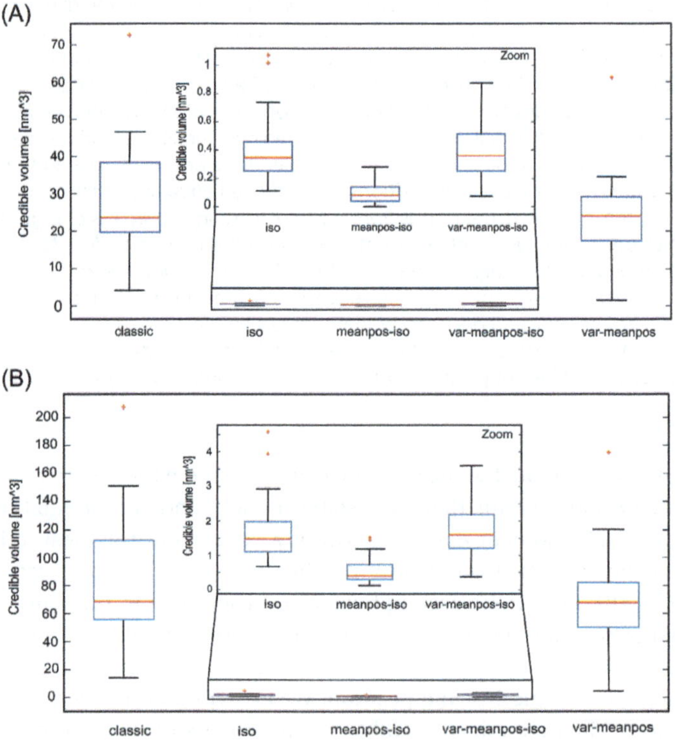

Fig. 4 Comparison of the size of the credible volumes for the different models. The computed credible volumes are compared enclosing 68% (A) and 95% (B) of the probability. Data is visualized by boxplots. The blue box contains the inner 50% of the data points, whereas the upper and lower bars, also called whiskers, represent the highest or lowest data points, respectively, still being in the 1.5 interquartile range, which is the length of the box in the *y*-direction. Red crosses represent outliers, defined by not being in the 1.5 interquartile range, and the red line inside the box depicts the median. The inner boxplot is zoomed in since the respective volumes are too small to be compared quantitatively on the same scale.

speaking the published values of the polynomial fit are only valid for the situation of dye molecules attached internally to DNA, we found that changing the shape of the AV had only small effects on the resulting polynomial. Considering the fact that the polynomial will result in changes only for very small distances, the use of an exact, specific polynomial is not likely to dramatically alter the results. We again analysed the archaeal PIC smFRET network with the *meanpos-iso model* (Fig. 2C). Comparison of the *iso model* and the *meanpos-iso model* shows that the additional assumption of a fixed mean position leads to a further reduction in the credible volumes by a factor of 3 (Fig. 4). When comparing to the *classic model*, one should note that the mean position is also a possibility within the static model (even though a static position of the linker pointing into free space seems artificial), however, in NPS the mean position would then lead to slightly altered smFRET values compared to the dynamically averaged position due to the R^6-dependence of FRET. While this difference is very small for low FRET efficiencies, it cannot be neglected for situations where the donor and acceptor molecules are quite close, resulting in FRET efficiencies of around 80% or higher.

While MD simulations have indicated that the mean position of the AV is a good model for non-interacting dye molecules attached to linear DNA molecules,[21] when dyes are attached to proteins, the situation might be more complicated because the dye molecule could interact with regions on the protein surface or simply stick to small cavities created by surface residues. In order to account for these effects, we have tested another model where we use the dynamical averaging of dipole orientations, and the dynamical averaging of positions within the AV, while allowing the freedom that the center of mass is no longer at the center of the AV but can take an arbitrary position within the AV (*var-meanpos-iso model*, Fig. 3E). To include this in our analysis scheme, we do not impose a fixed mean position but let the sampler find the most likely mean position. Again, we account for the dynamic position averaging by using the polynomial conversion of R_{MP} to R_{DA}. We again analysed the archaeal PIC smFRET network (Fig. 2D). The computed credible volumes have a size comparable to those of the *iso model* (Fig. 4). However, the positions of the credible volumes are shifted compared to those of the *iso model* and the *meanpos-iso model*.

In the last scenario, we wanted to investigate the effect of dynamic averaging of different positions for a situation where the dye molecule is not completely free to rotate (*var-meanpos model*, Fig. 3F). Such a situation could occur, for example, if the dye molecule binds reversibly to several different positions on the protein surface. When applying this model to the archaeal PIC smFRET network, we find credible volumes that are comparable in size to those of the *classic model* (Fig. 4), albeit at slightly shifted positions.

Influence of dye molecules with high fluorescence anisotropy on network analysis

From the comparison of the credible volumes of our five different dye models, we find that the largest effect on the size of the credible volume is the isotropic averaging over all possible dye orientations. The linker position and whether one considers one static or many dynamically averaged positions only have a minor effect on the size of the credible volumes. This is in accordance with our previous observation that the localisation accuracy of NPS is mainly dependent on our precise knowledge of the orientation of the dye molecules.[15] While the isotropic averaging of dye orientations is a good assumption when the FRET anisotropy is fairly low, high FRET anisotropies could lead to substantial errors in the analysis.[7] We therefore wanted to test whether in the analysed smFRET network, isotropic orientation averaging could lead not only to a reduction in uncertainty but also to a shifted position of the credible volume. For such a case, one could argue that the unjustified use of $\kappa^2 = 2/3$ could lead to wrong interpretations of the smFRET results.

In order to test this hypothesis, we investigated the overlap of the credible volumes of different models. To this end, we computed credible volumes at increasing probability to test at which confidence level the credible volumes of different models begin to overlap by at least one voxel (the voxel size in our analysis was 8 $Å^3$). This analysis was done for all ADMs. Since one can expect that the isotropic orientation averaging is a good hypothesis when the fluorescence anisotropy is low, but might fail for higher fluorescence anisotropies, we plotted

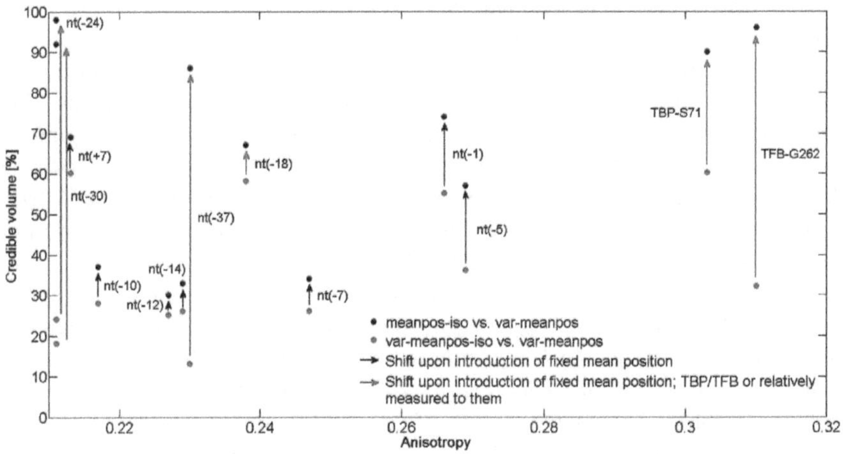

Fig. 5 Overlap of the credible volumes between different models. In order to investigate the overlap of the credible volumes of different models, we computed credible volumes at increasing probability to test when one observes the first voxel that is common to the credible volumes of different models. The data is plotted as a function of the fluorescence anisotropy of the respective ADMs in order to test the isotropic orientation hypothesis. The credibility volumes at the first voxel overlaps between the *var-meanpos model* and the *var-meanpos-iso model* (red dots) and between the *var-meanpos model* and the *meanpos-iso model* (black dots) are shown for the respective dyes. When comparing the *meanpos-iso model*, overlap occurs only at values as high as 98%, suggesting that the assumptions of the model are very unlikely for some dyes (*e.g.* nt(−24) or TBP-S71) because the assumed satellite state is also possible in the *var-meanpos model*. Additionally, we plotted arrows indicating the shift upon the introduction of a fixed mean position (switch from the *var-mean-iso model* to the *meanpos-iso model*). When looking at the data, we found that the anisotropies of the dye molecules attached to TFB and TBP were substantially higher than the other anisotropies. We used green arrows for describing localisations that used measurements to TBP and/or TFB and black arrows for all other localisations. The green arrows are generally much larger than the black arrows, suggesting that the assumption of a fixed mean position with respect to the accessible volume (AV) might not be valid here.

the described threshold confidence level as a function of fluorescence anisotropy (Fig. 5).

We compared the situation of a variable dynamic mean position (*var-meanpos-iso model*) as well as that of a fixed dynamic mean position (*meanpos-iso model*) to the *var-meanpos model*. For all the ADMs, the overlap for the *var-meanpos-iso model* occurred earlier than that for the *meanpos-iso model*. In fact, in this case, the volumes overlapped for all the ADMs before 68% credibility was reached, indicating that the two models are consistent with each other. When comparing the *meanpos-iso model* with the *var-meanpos model*, the first overlap occurs at values as high as 98%, suggesting that the assumptions of the *meanpos-iso model* are very unlikely for some dyes (*e.g.* nt(−24) or TBP) since the assumed ADM state in the *meanpos-iso model* is also possible in the *var-meanpos model*. At first sight, the level at which overlap occurs does not clearly correlate with the fluorescence anisotropy, *i.e.* small observed anisotropies do not necessarily result in overlaps at low confidence levels. However, one has to keep in mind that the analysis is done in a global scheme. Therefore, one ADM for which the $\kappa^2 = 2/3$ approximation or the fixed dynamically averaged mean position assumption is not valid could have detrimental effects on the localisation of another ADM, in particular if direct measurements between the two ADMs are included. In the smFRET scenario at hand, this becomes quite evident for the analysis of the positions −24, −30 and −37 on the non-template strand. All these ADMs have rather low fluorescence anisotropies, yet overlap occurs only at high confidence levels when comparing the *meanpos-iso model* with the *var-meanpos model*.

We plotted arrows indicating the shift in the credible level at which overlap occurs upon the introduction of a fixed mean position (switch from the *var-meanpos-iso model* to the *meanpos-iso model*). In view of the argument above, we used green arrows for describing localisations that used measurements to TBP and/or TFB, and black arrows for all other localisations. The green arrows are generally much longer than the black arrows, suggesting that the assumption of a fixed mean position with respect to the accessible volume (AV) might not be valid here. Therefore, when using both the $\kappa^2 = 2/3$ assumption and the fixed mean position, overlap occurs only at extremely high values, suggesting that the analysis model is not consistent with the experimentally observed data. In order to test whether the samples from ADMs used in smFRET measurements to TFB or TBP show a significant difference compared to the other ADMs, a two-sample Wilcoxon–Mann–Whitney test was performed at 95% confidence level upon the credible volumes, testing the percentage at which the first overlap occurred. This was done for the model comparison *meanpos-iso versus var-meanpos* (black dots in Fig. 5), as the difference between high and low anisotropies was especially significant. All in all, this led to the rejection of the null-hypothesis that the two sets of samples are from the same distribution, which can be taken as evidence for a significant difference in the medians of the two distributions. We interpret this as an inconsistency in the analysis of the *meanpos-iso model*.

Conclusions

We have exhibited a novel implementation of the NPS analysis scheme based on MCMC sampling and parallel tempering, decreasing the time needed for the analysis by ≈2 orders of magnitude. The new algorithm works very reliably and the resulting credible volumes are virtually identical to those published earlier. With this new analysis scheme, we also implemented the possibility to change the model of the conformational behaviour of dye molecules. In particular, we tested the effects of isotropic orientation averaging as well as dynamic position averaging about the center of the accessible volume or about another unknown position within the accessible volume.

We find that the isotropic orientation assumption reduces the size of the credible volumes dramatically. Moreover, when combined with a dynamic position averaging about the center of the accessible volume, the size of the credible volumes becomes so small that the precision of smFRET-based structure determination becomes comparable to what can be achieved by high resolution X-ray crystallography, as has previously been used in FRET-restrained position sensing.[21] However, careful analysis of the overlap of the credible volumes of different dye molecule priors showed that the isotropic orientation averaging can also lead to inconsistencies in the analysis. In particular, when combined with the stringent assumptions of the *meanpos-iso model*, this would result in false structural information. Thus, great caution has to be used when applying these assumptions. One can test the resulting models by bootstrapping approaches, given a sufficient amount of data. However, it might be beneficial to use other dye priors, such as the ones presented here instead.

Acknowledgements

We thank Adam Muschielok for insightful comments and discussion regarding NPS, as well as different sampling schemes. We also want to thank Timo Breßmer for help with the programming during the initial phases of this project.

Notes and references

1 P. Cramer, *Cell*, 2014, **159**, 985–994.
2 T. Schmeing and V. Ramakrishnan, *Nature*, 2009, **461**, 1234–1242.
3 K.-P. Hopfner and J. Michaelis, *Curr. Opin. Struct. Biol.*, 2007, **17**, 87–95.
4 J. Menetrey, A. Bahloul, A. L. Wells, C. M. Yengo, C. A. Morris, H. L. Sweeney and A. Houdusse, *Nature*, 2005, **435**, 779–785.
5 T. Ha, *Nat. Methods*, 2014, **11**, 1015–1018.
6 L. Stryer and R. P. Haugland, *Proc. Natl. Acad. Sci. U. S. A.*, 1967, **58**, 719–726.
7 J. Michaelis, in *Single Particle Tracking and Single Molecule Energy Transfer*, Wiley-VCH Verlag GmbH & Co. KGaA, Weinheim, Germany, 2009, pp. 191–214.
8 I. Rasnik, S. Myong, W. Cheng, T. M. Lohman and T. Ha, *J. Mol. Biol.*, 2004, **336**, 395–408.
9 M. Margittai, J. Widengren, E. Schweinberger, G. F. Schröder, S. Felekyan, E. Haustein, M. König, D. Fasshauer, H. Grubmüller and R. Jahn, *Proc. Natl. Acad. Sci. U. S. A.*, 2003, **100**, 15516–15521.
10 J. Andrecka, R. Lewis, F. Brückner, E. Lehmann, P. Cramer and J. Michaelis, *Proc. Natl. Acad. Sci. U. S. A.*, 2008, **105**, 135–140.
11 N. K. Lee, A. N. Kapanidis, Y. Wang, X. Michalet, J. Mukhopadhyay, R. H. Ebright and S. Weiss, *Biophys. J.*, 2005, **88**, 2939–2953.
12 R. E. Dale, J. Eisinger and W. E. Blumberg, *Biophys. J.*, 1979, 1–33.

13 J. J. Mccann, U. B. Choi, L. Zheng, K. Weninger and M. E. Bowen, *Biophys. J.*, 2010, **99**, 961–970.
14 A. Muschielok, J. Andrecka, A. Jawhari, F. Brückner, P. Cramer and J. Michaelis, *Nat. Methods*, 2008, **5**, 965–971.
15 A. Muschielok and J. Michaelis, *J. Phys. Chem. B*, 2011, **115**, 11927–11937.
16 J. Andrecka, B. Treutlein, M. A. I. Arcusa, A. Muschielok, R. Lewis, A. C. M. Cheung, P. Cramer and J. Michaelis, *Nucleic Acids Res.*, 2009, **37**, 5803–5809.
17 B. Treutlein, A. Muschielok, J. Andrecka, A. Jawhari, C. Buchen, D. Kostrewa, F. Hög, P. Cramer and J. Michaelis, *Mol. Cell*, 2012, **46**, 136–146.
18 D. Grohmann, J. Nagy, A. Chakraborty, D. Klose, D. Fielden, R. H. Ebright, J. Michaelis and F. Werner, *Mol. Cell*, 2011, **43**, 263–274.
19 J. Nagy, D. Grohmann, A. C. M. Cheung, S. Schulz, K. Smollett, F. Werner and J. Michaelis, *Nat. Commun.*, 2015, **6**, 6161.
20 J. Michaelis and B. Treutlein, *Chem. Rev.*, 2013, **113**, 8377–8399.
21 S. Kalinin, T. Peulen, S. Sindbert, P. J. Rothwell, S. Berger, T. Restle, R. S. Goody, H. Gohlke and C. A. M. Seidel, *Nat. Methods*, 2012, **9**, 1218–1225.
22 S. Sindbert, S. Kalinin, H. Nguyen, A. Kienzler, L. Clima, W. Bannwarth, B. Appel, S. Müller and C. A. M. Seidel, *J. Am. Chem. Soc.*, 2011, **133**, 2463–2480.
23 U. B. Choi, P. Strop, M. Vrljic, S. Chu, A. T. Brunger and K. R. Weninger, *Nat. Struct. Mol. Biol.*, 2010, **17**, 318–324.
24 R. V. Craiu and J. S. Rosenthal, *Annu. Rev. Stat. Appl.*, 2014, **1**, 179–201.
25 R. Swendsen and J. Wang, *Phys. Rev. Lett.*, 1986, **57**, 2607–2609.
26 C. J. Geyer, Markov Chain Monte Carlo Maximum Likelihood, *Computing Science and Statistics, Proceedings of the 23rd Symposium on the Interface*, New York, 1991, pp. 156–163.
27 A. Kone and D. A. Kofke, *J. Chem. Phys.*, 2005, **122**, 206101.

Faraday Discussions

PAPER

Quantitative single molecule FRET efficiencies using TIRF microscopy

Lasse L. Hildebrandt,[ab] Søren Preus[ab] and Victoria Birkedal*[ab]

Received 3rd June 2015, Accepted 24th July 2015
DOI: 10.1039/c5fd00100e

Förster resonance energy transfer (FRET) microscopy at the single molecule level has the potential to yield information on intra and intermolecular distances within the 2–10 nm range of molecules or molecular complexes that undergo frequent conformation changes. A pre-requirement for obtaining accurate distance information is to determine quantitative instrument independent FRET efficiency values. Here, we applied and evaluated a procedure to determine quantitative FRET efficiencies directly from individual fluorescence time traces of surface immobilized DNA molecules without the need for external calibrants. To probe the robustness of the approach over a wide range of FRET efficiencies we used a set of doubly labelled double stranded DNA samples, where the acceptor position was varied systematically. Interestingly, we found that fluorescence contributions arising from direct acceptor excitation following donor excitation are intrinsically taken into account in these conditions as other correction factors can compensate for inaccurate values of these parameters. We give here guidelines, that can be used through tools within the iSMS software (http://www.isms.au.dk), for determining quantitative FRET and assess uncertainties linked with the procedure. Our results provide insights into the experimental parameters governing quantitative FRET determination, which is essential for obtaining accurate structural information from a wide range of biomolecules.

Introduction

Single molecule Förster resonance energy transfer (smFRET) microscopy is a powerful technique to study conformational changes and dynamics of biomolecules with sub-nanometer resolution.[1] smFRET continues to provide new insights into conformational distributions, dynamic transitions and the kinetics governing diverse biological systems[2,3] as well as artificial molecular devices.[4,5] In addition, quantitative smFRET has the potential to report on accurate molecular distances and yield structural information on highly dynamic structures within single molecules, which is difficult to achieve with other techniques.[6,7]

[a]Interdisciplinary Nanoscience Center (iNANO), Aarhus University, Aarhus, Denmark. E-mail: vicb@inano.au.dk
[b]Centre for DNA Nanotechnology (CDNA), Aarhus University, Denmark

Quantitative Förster resonance energy transfer (FRET) values can indeed be linked to distances with adequate modelling of the donor and acceptor.[7–10] Some of the advantages of using FRET, compared to higher resolution techniques such as Nucleic Magnetic Resonance (NMR) is that very little material is needed and that highly polymorphic and dynamic molecules can be studied. The use of FRET has already successfully provided structural information in a number of systems, ranging from small DNA molecules[11,12] to proteins and nucleic acid protein complexes.[7,9,13]

Single molecule FRET efficiencies are determined from measured single molecule fluorescence intensities obtained with fluorescence microscopy. The measured values are instrument and fluorophore dependent and need to be corrected in order to determine the true FRET efficiency.[14,15] Instrument-independent FRET efficiencies, which are required in order to use FRET for distance measurements, are given by:

$$E = I_{\mathrm{FRET}}/(\gamma I_{\mathrm{DD}}+I_{\mathrm{FRET}}), \tag{1}$$

where I_{DD} is the measured fluorescence intensity of the donor dye following donor excitation and I_{FRET} is the fluorescence intensity of the acceptor dye resulting from energy transfer, which is given by the measured fluorescence intensity in the acceptor emission channel I_{DA} corrected for contributions from donor fluorescence bleed through into the acceptor emission channel (D-leak) and from direct excitation of acceptor fluorescence (A-Dir) at the donor excitation wavelength. Additionally, the γ factor corrects for differences in detection of the donor and acceptor fluorophores and is given by $\gamma = \eta_A\Phi_A/\eta_D\Phi_D$, where η and Φ are the detection efficiencies and quantum yields of the fluorophores respectively, with the subscript A denoting the acceptor and D the donor fluorophore.

There are a number of methods to determine the three correction factors D-leak, A-Dir and γ used in confocal[14,16,17] and widefield single molecule FRET microscopy.[18–20] Total internal reflection fluorescence (TIRF) widefield microscopy allows imaging of surface immobilized individual molecules over a long period of time (tens of minutes) with good temporal (~20 ms) and distance (<0.34 nm corresponding to one DNA base-pair separation) resolutions.[21,22] Procedures for obtaining quantitative FRET values from these experiments have been reported using the alternative laser excitation method (ALEX).[19] In this method, global values of the correction factors are determined and applied to all molecules. In some cases, correction factors can be determined directly from single molecule time traces of immobilized molecules by comparing fluorescence intensities before and after donor and/or acceptor photobleaching of individual molecules.[15,18,20,23] By observing each molecule individually, it becomes possible to determine and take into account correction factor distributions resulting from either molecular or experimental heterogeneities.

Here, we investigated the heterogeneity and accuracy of correction factors in smFRET TIRF microscopy when the factors are determined directly from fluorescence time traces of surface immobilized molecules. To assess the robustness of the approach and the importance of each correction factor for quantitative FRET measurements over a wide range of FRET efficiencies, we used a set of doubly labelled double stranded DNA samples. We found that the value of the direct acceptor correction factor A-Dir, which is often neglected, can affect both γ

factor values and the width of the FRET distribution, especially at low FRET values. However, peak FRET efficiency values were not affected by the A-Dir value when the intensity change during bleaching in single molecule time traces was used to determine the γ factor. Thus, this procedure can provide quantitative FRET, also when the A-Dir correction is neglected, as the γ factor can compensate for an inaccurate A-Dir correction. We find a good agreement between the measured and calculated FRET efficiencies. Our procedures for determining quantitative smFRET efficiencies using single molecule TIRF microscopy without external calibrants are implemented within the iSMS software platform and are easily applicable.[24]

Experimental

DNA constructs

PAGE grade purified single stranded DNA, labelled with either a donor (D, Cy3) or an acceptor (A, Alexa647), was purchased from IBA (IBA GmbH, Germany). DNA sequences were 5′-CGC GTC GGC AGC ATA CAA TAA CCT CAT CGA TAA GAA AGA AAT AAA GAA GAT CGC, where the underlined T base shows the donor position and a biotin was attached on the 5′ end. The complementary strand 5′-GCG ATC TTC TTT ATT TCT TTC TTA TCG ATG AGG TTA TTG TAT GCT GCC GAC GCG was synthesized with 5 different acceptor positions at T base numbers 16, 18, 20, 22 and 25, which are underlined. Fluorophores were attached using NHS-labeling through 5-C6-amino-2′-deoxythymidine modifications. Unlabeled strands were also purchased to produce control DNA constructs containing only a donor or acceptor label. Equimolar stoichiometries of donor and acceptor strands were annealed in TAE buffer (20 mM Tris/acetate/EDTA, pH 8.3) and 100 mM KCl (Sigma-Aldrich) for 5 minutes at 90 °C and left to cool down slowly over about one hour. This resulted in five different double stranded samples, which are named A39, A37 A35, A33 and A30, where the donor–acceptor dye pairs are separated by 20, 18, 16, 14 and 11 base pairs (bps), respectively.

Single-molecule FRET microscopy experiments

smFRET experiments were performed using alternating laser excitation, as described previously.[15] In short, DNA molecules with a concentration of 20 pM were immobilized via biotin–streptavidin linkage on a quartz coverglass for prism total internal reflection fluorescence microscopy. Fluorescence was measured using an inverted wide-field optical microscope. The sample was illuminated by alternate laser excitation at 514 and 630 nm of the donor and acceptor fluorophore, respectively, and fluorescence movies of several minutes were recorded with an EMCCD camera (Andor, iXon3 897) with a 200 ms integration time per frame. Typical excitation intensities were estimated to be ~0.3 and 0.1 kW cm^{-2} for the green and red laser, respectively.

The imaging buffer, used in smFRET experiments, was the same as the annealing buffer but contained in addition an enzymatic oxygen scavenging system consisting of glucose oxidase (Sigma-Aldrich, 16.67 units per mL), catalase (Sigma-Aldrich, 260 units per mL), B-D-(+)glucose (Sigma-Aldrich, 4.5 mg mL^{-1}), and (±)-6-hydroxy-2,5,7,8-tetramethylchroman-2-carboxylic acid (Trolox, Sigma-

Aldrich, 2 mM). Halfway through the acquisition the imaging buffer was substituted with the annealing buffer in order to facilitate fluorescence bleaching.

Data analysis

Data analysis was performed by the smFRET microscopy software package iSMS.[24] Fluorescence time traces of identified FRET-pairs were calculated by aperture photometry and only molecules ending with single-step donor and acceptor photobleaching were selected for further analysis, as this indicated that the source of the signal was a single doubly labelled molecule. Fluorescence intensities were corrected for background signal by subtracting the remaining signal after bleaching of both fluorophores.

Theoretical FRET efficiency determination

Ensemble absorption and fluorescence spectra were measured on control double stranded DNA constructs containing only a donor or an acceptor fluorophore using a Carry 60 absorption spectrometer (Agilent Technologies) and a FluoroMax 3 spectrofluorometer (HORIBA Scientific). Cy3 quantum yield was determined in annealing buffer using 5-carboxytetramethylrhodamine (Life Technologies) in methanol as a reference (with $\Phi_{ref} = 0.68$).[25] Based on the results of these measurements, the Förster radius was determined to be $R_0 = 5$ nm, assuming free fluorophore rotation ($\kappa^2 = 2/3$). Theoretical FRET efficiencies were calculated through the FPS software,[7] which takes into account the spatial distributions of donor and acceptor fluorophores using a geometrically accessible volume algorithm.

Results and discussion

We studied experimental parameters that influence the accuracy of FRET efficiencies determined using TIRF microscopy on surface-immobilized double-stranded DNA molecules. The DNA-duplexes were labelled with a donor–acceptor FRET pair, providing a series of samples with predefined donor–acceptor distances varying between ~3 and 7 nm as the acceptor position is systematically varied along the DNA (Fig. 1A). For each of the five DNA samples where donor–acceptor fluorophore pairs are positioned at different distances, single molecule fluorescence time traces of the donor and acceptor fluorophores were recorded (Fig. 2) using an alternating laser excitation scheme. The experimental conditions were adapted to have the majority of molecules show donor or acceptor photobleaching within the time frame of the measurements (minutes). Measured raw FRET values were different between the five samples and varied from low to high FRET according to the donor acceptor distance separation (Fig. 1B).

Quantitative FRET efficiency values were determined using eqn (1) with:

$$I_{\mathrm{FRET}} = I_{\mathrm{DA}} - \mathrm{Lk} - \mathrm{Dir}, \tag{2}$$

where Lk and Dir are the D-leak and A-Dir contributions. These fluorescence correction factors, as well as the γ factor, were determined directly from single molecule time traces (Fig. 2).

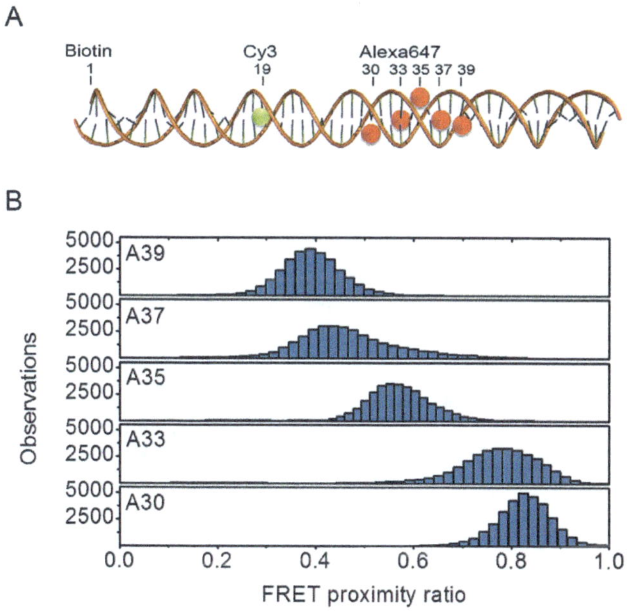

Fig. 1 DNA duplex used as a molecular ruler. (A) Schematic view of the double stranded DNA samples: one strand contains an internal Cy3, the second strand is labelled with Alexa 647 at one of the indicated positions. (B) Single molecule FRET efficiency histograms for each of the 5 DNA samples. The x-axis shows the proximity ratio, which is obtained using eqn (1), where the D-leak and A-Dir correction are set to zero and $\gamma = 1$.

Determination of fluorescence correction factors

The D-leak contribution is given by a fraction l of the donor fluorescence I_{DD} that leaks through to the acceptor channel with $l = I_{DA}/I_{DD}$ when only the donor fluorophore is active. l was determined from individual FRET-pairs, where the acceptor photobleached before the donor fluorophore (Fig. 2A). For the given measurement conditions, the acceptor fluorophore bleached first for ~90% of the molecules and the histogram over all individual l values showed a narrow peak (Fig. 2D). Thus l could be determined with good precision to a mean value of 0.105 ± 0.018. The obtained value matched theoretical expectations based on the donor fluorescence spectrum and specifications of the filters used in the single molecule experiments.

The direct acceptor excitation correction factor A-Dir is a fraction d of I_{AA}, the acceptor fluorescence following direct acceptor excitation.[14,19] The correction factor d is given by: $d = I_{Dex}/I_{Aex}\sigma^{A}_{Dex}/\sigma^{A}_{Aex}$, including the ratio of acceptor absorption cross sections at the donor and acceptor excitation wavelengths, and I_{Dex} and I_{Aex} are the excitation intensities of donor and direct acceptor fluorescence. d was determined from single molecule time traces showing donor bleaching as I_{DA}/I_{AA} using intensities after donor bleaching, which requires two color ALEX experiments (Fig. 2B). The distribution of d values over all measured DNA samples was broad (Fig. 2E), giving a mean value of $d = 0.06 \pm 0.03$. In comparison, theoretical predictions yielded $d \sim 0.06$. The large distribution of d values arose from the fact that the laser intensities used to excite the donor and

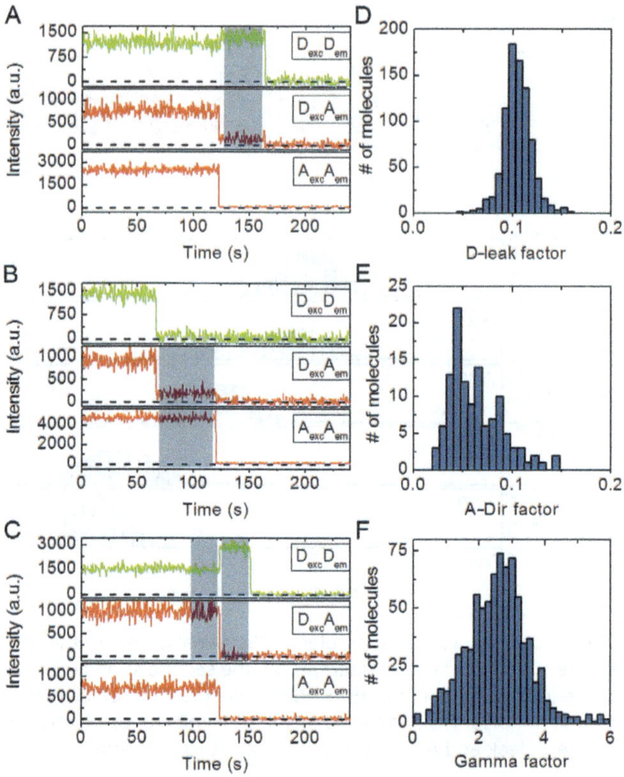

Fig. 2 Correction factors are determined from individual molecule fluorescence time traces. (A–C) Fluorescence time traces showing from top to bottom I_{DD} (green line), I_{DA} (red line) and I_{AA} (red line) as a function of time. (A) D-leak correction determination from molecules showing acceptor bleaching first. (B) A-Dir correction determination for molecules showing donor bleaching first. (C) γ factor determination from molecules showing acceptor bleaching first. The frames used when determining correction factors are highlighted in light blue. (D–F) Histograms of the correction factors obtained from individual molecule time traces on the whole Cy3/Alexa 647 data set (~1000 molecules) for l (D), d (E) and γ (F). The histograms contain data from 842 (D), 119 (E), and 816 (F) molecules.

acceptor molecules were not completely homogeneous throughout the CCD image field of view. It is to be noted that the A-Dir correction is independent of I_{Aex} and thus should have a similar distribution width as the D-leak correction factor. Statistics on this correction were however poor, as it could be determined for only ~10% of molecules and the A-Dir correction distribution was observed to be wider than the D-leak. Thus, using an average A-Dir value on all time traces may introduce an error in quantitative FRET determination.

γ correction factors were determined from single molecule fluorescence time traces as:

$$\gamma = -\Delta I_{FRET}/\Delta I_{DD}, \tag{3}$$

where ΔI denotes the difference in the average fluorescence intensities before and after acceptor photobleaching (Fig. 2C). Because E drops to 0 at the acceptor bleaching event, the post bleaching intensities can be used as a reference for the relative brightness of donors and acceptors.[18] The γ factor correction was here determined using average d and l values (Fig. 2F). The histogram of individual values shows a relatively broad distribution and yields an average value of $\gamma = 2.6 \pm 1.2$ with a large uncertainty.

Applying the D-leak and A-Dir corrections will decrease measured energy transfer values. These corrections have proportionally a larger effect for small I_{DA} values and thus will play a larger role for lower energy transfer efficiencies. The more different the γ factor is from 1 the more it will affect the FRET efficiency in a non-linear way. As I_{FRET} is used to determine the γ factor value, it will also depend on the value of the D-leak and A-Dir corrections. Fluctuations of the D-leak factor were small from molecule to molecule and using either the average value from single molecule measurements or the theoretical value yielded very similar γ factors and FRET efficiency values.

Effect of direct acceptor excitation

Since determination of the A-Dir correction factor was associated with a large uncertainty, we explored the effect of A-Dir on the value of the γ factor and FRET efficiency. Although direct excitation of the acceptor is a small effect (the ratio of the Alexa 647 fluorophore absorption cross sections at the donor and acceptor excitation wavelengths is 0.02), the A-Dir correction is of about the same order of magnitude as the D-leak correction under our experimental conditions. d factor values depend on excitation spot profiles and thus may well vary from molecule to molecule. Direct determination from single molecule data requires the donor fluorophore to photobleach before the acceptor bleaches. Thus, it is not possible to simultaneously extract A-Dir and γ factor corrections for individual molecules. To gain insight into how the uncertainties for the A-Dir correction factor influence the γ factor, we determined γ factor corrections assuming different d factor values.

The d factor value influenced the average values of γ factors determined using the photobleaching method (vertical comparison in Table 1). γ factor values were observed to decrease for higher d values. This observation can be understood by the fact that higher values of A-Dir will lead to smaller I_{FRET} signals through eqn (2), and thus result in smaller values of the γ factor through eqn (3).

Table 1 Average of individual correction factors for quantitative FRET determination, determined from single molecule time traces for samples containing Cy3 and Alexa647

	γ factor				
d	A39	A37	A35	A33	A30
0.0	3.0 ± 0.9	2.9 ± 0.9	2.9 ± 0.6	2.8 ± 0.7	3.3 ± 0.7
0.025	2.7 ± 0.8	2.6 ± 0.9	2.7 ± 0.6	2.7 ± 0.7	3.1 ± 0.7
0.058	2.2 ± 0.8	2.2 ± 0.8	2.5 ± 0.5	2.5 ± 0.7	3.0 ± 0.7
0.091	1.8 ± 0.8	1.8 ± 0.8	2.3 ± 0.5	2.4 ± 0.7	2.9 ± 0.7

Separate determination of the γ factor for each of the five DNA duplexes (horizontal comparison in Table 1) showed differences between the different constructs. For the larger A-Dir value (bottom row in Table 1), the average γ factor value increased as the separation between donor and acceptor fluorophores decreased. As the A-Dir correction is a larger fraction of the I_{FRET} signal at low FRET efficiencies, its effect on the γ factor will be larger for large donor acceptor separations and almost negligible at high FRET efficiencies for small donor acceptor separations. γ factor values obtained for the A30 sample were minimally influenced by the value of the A-Dir correction, which is in agreement with the above argument. As the A-Dir correction influences γ factor values differently as a function of donor–acceptor distance separation, the γ factor will not necessarily be constant for a given donor–acceptor pair. It is also to be noted that a larger uncertainty is associated with individual γ factor determination for larger donor–acceptor separations (Table 1). This effect is intrinsic to the use of eqn (3) for γ factor determination, as large donor–acceptor separations result in only small fluorescence changes in both the numerator and denominator of the equation. In order to determine which γ factor value to use for quantitative FRET determination, FRET efficiency histograms were determined for all conditions summarized in Table 1 (data not shown).

We illustrate the effect of d and γ on FRET efficiency values by showing FRET efficiency histograms for three individual molecules that have different donor–acceptor separations (Fig. 3). Using γ correction factor values obtained by the photobleach method for a given d value: $d = 0.058$ (Fig. 3A) and $d = 0$ (Fig. 3C) yielded the same FRET efficiency peak values, despite the use of different γ correction factor values. For each molecule, as illustrated by three examples in Fig. 3, we observed that the same peak FRET efficiencies are obtained throughout the whole FRET range when overestimating, underestimating or even ignoring the A-Dir correction through an adjustment of the γ factor value. However, changing

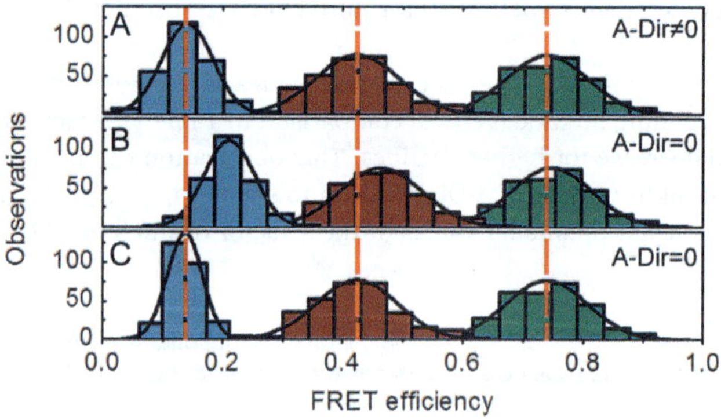

Fig. 3 Correction factors for quantitative FRET directly determined for individual molecules. Histograms showing FRET efficiency distributions for 3 representative molecules from samples A39, A35 and A30, respectively, when systematically varying the value of the A-Dir correction. (A) $d = 0.058$ and l and γ were individually determined for each molecule: blue curve: $l = 0.12$, $\gamma = 1.90$; red curve: $l = 0.09$, $\gamma = 3.10$; green curve: $l = 0.11$, $\gamma = 2.54$ (B) $d = 0$ and the same l and γ values were used as in (A). (C) $d = 0$ and l and γ were individually determined for each molecule. Blue curve: $l = 0.12$, $\gamma = 3.22$; red curve: $l = 0.09$, $\gamma = 3.64$; green curve: $l = 0.11$, $\gamma = 2.64$.

one parameter without adjusting the other resulted in different FRET peak values (Fig. 3B). The dependence of the γ correction factor values on A-Dir can be understood by fully expressing eqn (3) using eqn (2). As I_{FRET} after acceptor photobleaching is zero, the larger the A-Dir correction the smaller ΔI_{FRET} and thus smaller γ values are obtained. Thus γ correction factors obtained with the photobleach method can to a large extend compensate for errors in the determination of the A-Dir correction to yield quantitative FRET efficiencies. As such, these γ values include an A-Dir component in addition to the quantum yield and detection efficiency terms. This observation also extends to sample averaged results, as peak FRET efficiencies in FRET distribution histograms that include all molecules for a given DNA construct were also observed to be independent of the chosen A-Dir value when using a sample averaged γ factor.

A broadening of FRET efficiency distributions was observed for large d values (Fig. 3A). The effect was most visible for molecules showing small FRET efficiencies ($E < \sim 0.3$), for which the A-Dir correction represents a significant fraction of I_{FRET}. Setting the A-Dir correction to zero and determining γ factor values under these conditions resulted in the narrowest FRET efficiency peaks (Fig. 3C). We have thus used this approach for quantitative FRET determination.

Quantitative FRET efficiencies with TIRF microscopy

Quantitative FRET efficiency histograms were determined for each of the five samples (Fig. 4A). Histograms contain only molecules where the acceptor fluorophores bleached first to allow for γ factor determination for individual

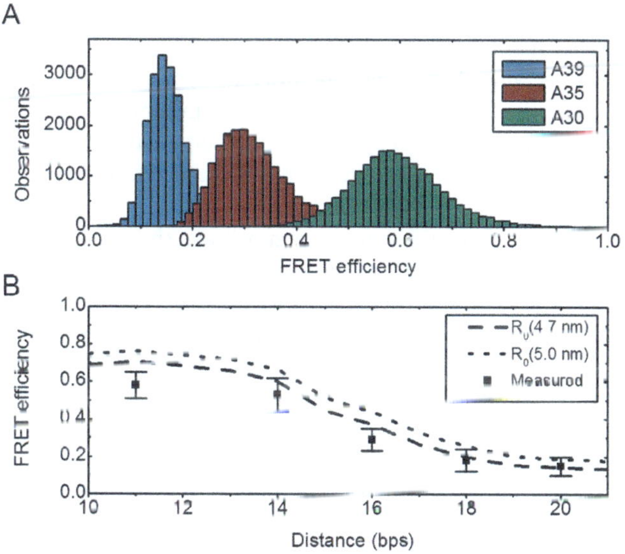

Fig. 4 Quantitative FRET efficiency. (A) Single molecule FRET histograms showing data for samples A39, A35, A30. Average D-leak and γ-factor corrections were used for each sample. The A-Dir correction was set to zero. (B) Comparison between experimentally determined FRET efficiencies and theoretical predictions following procedures from ref. 7 using $R_0 = 4.7$ nm and $R_0 = 5$ nm. The parameters used to describe the Alexa 647 dye were adapted from ref. 26. Average γ factors are given in Table 1.

molecules. FRET values were determined using both sample averaged (Fig. 4A) and individually determined γ factors (data not shown). Uncertainties on individual γ factors were calculated from eqn (3) using error propagation and uncertainties linked with the determination of the average fluorescence intensities. We observed that the FRET distribution was broader when using individual γ factors than when using sample averaged γ values. This effect arose due to a large uncertainty on individually determined γ factors for a subset of molecules. Removing these molecules from the FRET histogram yielded narrower FRET distributions with a width comparable to or slightly wider than those obtained using a sample averaged γ factor. Peak FRET values were the same in both cases.

The use of individually determined γ factor values has previously been reported to significantly decrease the width of FRET distribution peaks.[20] This is expected to be the case, provided that the uncertainty of each individually determined γ factor is not larger than the width of the γ factor distribution and that γ factors are not intrinsically heterogeneous. In the present experiments, the uncertainty linked to individual γ factors was in most cases larger than the uncertainty of the average γ factor. Furthermore, the observed γ factor distribution was broad (Fig. 2F). These two factors would contribute to broaden FRET efficiency distribution histograms, and increase the uncertainty of quantitative FRET determination. Broad γ factor distributions have previously been observed for Cy3,[23] and could be intrinsic to the used fluorophore.

FRET efficiency peak values increased with decreasing donor acceptor distance (Fig. 4). Peak FRET efficiencies were compared with theoretical predictions based on the known DNA and fluorophores' structures[7] and a Förster radius R_0 of 4.7 nm and 5 nm (Fig. 4B). Uncertainties in the experimentally determined FRET values were determined from uncertainties on the γ factor (Table 1). Experimental FRET values are in good agreement with theoretical predictions, especially when using the lower R_0 value. This R_0 value is slightly lower than our estimate, but is within experimental uncertainty. One source of uncertainty is the used value of κ^2, which may not be completely applicable for immobilized molecules.[22] We conclude that the procedure used here for quantitative FRET determination gives good results.

For the smallest donor acceptor separations, the larger observed FRET efficiency was $E \sim 0.6$ and the agreement between experimentally determined FRET values and theoretical predictions was poorer. Comparatively low FRET efficiencies compared to the expected donor–acceptor separation were previously reported for Cy3.[23] We observed a larger average γ factor value for the sample with the smallest donor–acceptor separation, which could indicate a change in the fluorescence properties of the dyes. Fluorescence quenching effects can influence FRET efficiencies and have been reported to occur for small donor–acceptor separations.[27] Although individual single molecule time traces did not show clear evidence of acceptor fluorescence quenching, donor or acceptor fluorescence quenching could well play a role in explaining our observations. This effect can also affect the value of the γ factor so that it depends on donor–acceptor separation. Thus, procedures for determining γ factors directly for individual samples will give more accurate results than those where a global correction factor is used for a given donor–acceptor fluorophore pair.

Conclusions

In conclusion, we evaluated the robustness of a procedure to determine quantitative FRET efficiencies directly from fluorescence time traces of surface immobilized molecules. We found that the FRET values are not influenced by contributions from direct acceptor excitation following donor excitation in these conditions. Indeed, the correction arising from direct acceptor excitation is intrinsically taken into account and contributes to the value of the γ factor in this procedure. As the correction from direct acceptor excitation is proportionally more important at low FRET values, the γ factor value is expected to vary depending on acceptor donor fluorophore separation if an inaccurate value for the direct acceptor correction is used. A global γ factor for a given donor–acceptor fluorophore pair will thus introduce an error for quantitative FRET determination when neglecting the direct acceptor correction. Our results provide insights into the parameters governing quantitative instrument independent FRET determination from TIRF based single molecule experiments. The used procedures are implemented for further use in the iSMS single molecule FRET data analysis software (http://www.isms.au.dk) that allows determination of correction factors from fluorescence time traces of individual molecules.

Acknowledgements

This work was supported by the Danish Council for Independent Research's carrier program Sapere Aude, the Danish National Research Foundation (Centre for DNA Nanotechnology, DNRF81) and the Aarhus University Research Foundation.

Notes and references

1 T. Ha, *Nat. Methods*, 2014, **11**, 1015–1018.

2 A. A. Deniz, S. Mukhopadhyay and E. A. Lemke, *J. R. Soc., Interface*, 2008, **5**, 15–45.

3 S. Myong, B. C. Stevens and T. Ha, *Structure*, 2006, **14**, 633–643.

4 S. L. Kragh and V. Birkedal, in *Nucleic Acid Nanotechnology*, ed. J. Kjems, E. Ferapontova and K. V. Gothelf, 2014, vol. 29, pp. 53–68.

5 L. L. Hildebrandt, S. Preus, Z. Zhang, N. V. Voigt, K. V. Gothelf and V. Birkedal, *J. Am. Chem. Soc.*, 2014, **136**, 8957–8962.

6 S. Preus and L. M. Wilhelmsson, *ChemBioChem*, 2012, **13**, 1990–2001.

7 S. Kalinin, T. Peulen, S. Sindbert, P. J. Rothwell, S. Berger, T. Restle, R. S. Goody, H. Gohlke and C. A. M. Seidel, *Nat. Methods*, 2012, **9**, 1218–1225.

8 A. Muschielok, J. Andrecka, A. Jawhari, F. Bruckner, P. Cramer and J. Michaelis, *Nat. Methods*, 2008, **5**, 965–971.

9 A. Muschielok and J. Michaelis, *J. Phys. Chem. B*, 2011, **115**, 11927–11937.

10 A. T. Brunger, P. Strop, M. Vrljic, S. Chu and K. R. Weninger, *J. Struct. Biol.*, 2011, **173**, 497–505.

11 T. Sabir, G. F. Schroeder, A. Toulmin, P. McGlynn and S. W. Magennis, *J. Am. Chem. Soc.*, 2011, **133**, 1188–1191.

12 A. K. Wozniak, G. F. Schroder, H. Grubmuller, C. A. M. Seidel and F. Oesterhelt, *Proc. Natl. Acad. Sci. U. S. A.*, 2008, **105**, 18337–18342.

13 U. B. Choi, P. Strop, M. Vrljic, S. Chu, A. T. Brunger and K. R. Weninger, *Nat. Struct. Mol. Biol.*, 2010, **17**, 318–324.
14 N. K. Lee, A. N. Kapanidis, Y. Wang, X. Michalet, J. Mukhopadhyay, R. H. Ebright and S. Weiss, *Biophys. J.*, 2005, **88**, 2939–2953.
15 A. C. Kruger and V. Birkedal, *Methods*, 2013, **64**, 36–42.
16 E. Sisamakis, A. Valeri, S. Kalinin, P. J. Rothwell and C. A. M. Seidel, in *Methods Enzymol*, ed. N. G. Walter, 2010, vol. 475, pp. 455–514.
17 V. Kudryavtsev, M. Sikor, S. Kalinin, D. Mokranjac, C. A. M. Seidel and D. C. Lamb, *ChemPhysChem*, 2012, **13**, 1060–1078.
18 T. J. Ha, A. Y. Ting, J. Liang, W. B. Caldwell, A. A. Deniz, D. S. Chemla, P. G. Schultz and S. Weiss, *Proc. Natl. Acad. Sci. U. S. A.*, 1999, **96**, 893–898.
19 E. Margeat, A. N. Kapanidis, P. Tinnefeld, Y. Wang, J. Mukhopadhyay, R. H. Ebright and S. Weiss, *Biophys. J.*, 2006, **90**, 1419–1431.
20 J. J. McCann, U. B. Choi, L. Q. Zheng, K. Weninger and M. E. Bowen, *Biophys. J.*, 2010, **99**, 961–970.
21 S. J. Holden, S. Uphoff, J. Hohlbein, D. Yadin, L. le Reste, O. J. Britton and A. N. Kapanidis, *Biophys. J.*, 2010, **99**, 3102–3111.
22 R. Roy, S. Hohng and T. Ha, *Nat. Methods*, 2008, **5**, 507–516.
23 C. R. Sabanayagam, J. S. Eid and A. Meller, *J. Chem. Phys.*, 2005, **122**, 061103.
24 S. Preus, S. L. Noer, L. L. Hildebrandt, D. Gudnason and V. Birkedal, *Nat. Methods*, 2015, **12**, 593–594.
25 D. Magde, J. H. Brannon, T. L. Cremers and J. Olmsted, *J. Phys. Chem.*, 1979, **83**, 696–699.
26 H. Hoefig, M. Gabba, S. Poblete, D. Kempe and J. Fitter, *Molecules*, 2014, **19**, 19269–19291.
27 N. Di Fiori and A. Meller, *Biophys. J.*, 2010, **98**, 2265–2272.

Faraday Discussions

Superresolution techniques, biophysics with nanostructures, and fluorescence energy transfer: general discussion

Thorben Cordes, William Moerner, Michel Orrit, Lisa-Maria Needham, Sergey Sekatskii, Sanli Faez, Himangshu Prabal Goswami, Alex Clark, Alfred J. Meixner, Lukasz Piatkowski, Vahid Sandoghdar, Gary M. Skinner, Francisco Balzarotti, Wolfgang Langbein, Lukas Novotny, Victoria Birkedal, Felix Koberling, Jens Michaelis, Phil Williams, Arindam Chowdhury, Niek van Hulst, Maxime Dahan, Ana Peñaherrera, Brahim Lounis, Paul W. Bohn and Xin Shi

DOI: 10.1039/c5fd90094h

Maxime Dahan opened the discussion of the paper by William Moerner: You present a method for 3D localization over more than 10 microns. What is the trade-off between axial depth and localization accuracy?

William Moerner replied: It is important to note that since the new PSFs are calculated to maximize Fisher information over a specified range in the presence of background, then the localization precision is (almost) the best that it can be over this range. I say "almost", because the optimization problem is not convex, but the optimization does find a maximum in Fisher information. In our paper[1] we show precisions for specified numbers of photons for a 3 micron PSF. For longer ranges, the localization precision certainly degrades, and that is one of the tradeoffs if longer axial ranges are needed. In our tetrapod paper,[2] a 6 micron PSF and a 20 micron PSF are compared in Fig. 2, and this increase in range was accompanied by a worsening of localization precision by a factor of roughly 3, depending upon the z position.

1 Y. Shechtman, S. J. Sahl, A. S. Backer and W. E. Moerner, Optimal Point Spread Function Design for 3D Imaging, *Phys. Rev. Lett.*, 2014, **113**, 133902.
2 Y. Shechtman, L. E. Weiss, A. S. Backer, S. J. Sahl and W. E. Moerner, Precise Three-Dimensional Scan-Free Multiple-Particle Tracking over Large Axial Ranges with Tetrapod Point Spread Functions, *Nano Lett.*, 2015, **15**, 4194–4199.

Jens Michaelis commented: You described various tailored PSFs for improving the resolution and in particular the sensitivity also in the axial direction. Some of those PSFs allowed accurate axial localisation also in quite deep sample depths. Here, aberrations are known to be problematic. Could you please elaborate whether these engineered PSFs are more or less sensitive to aberrations than standard PSFs?

William Moerner responded: An excellent question. In one specific case, we have a clear answer: in our study of the use of the DH-PSF to measure the orientation of single molecules to remove dipole-induced shifts,[1] we found the following. Measurements of the same molecules with defocused clear-aperture imaging to separately determine orientation absolutely required correction of aberrations, while the DH-PSF did not require removal of the same optical aberrations. The DH-PSF phase pattern apparently imparts a large enough (controlled) aberration itself so that this dominates over the smaller optical aberrations of the microscope. We would expect this to be the case for a number of the engineered PSFs, but we do not have specific data on this aspect so far other than for the DH-PSF.

1 M. P. Backlund, M. D. Lew, A. S. Backer, S. J. Sahl, G. Grover, A. Agrawal, R. Piestun and W. E. Moerner, Simultaneous, accurate measurement of the 3D position and orientation of single molecules, *Proc. Nat. Acad. Sci.*, 2012, **109**, 19087–19092.

Lukas Novotny opened the discussion of the paper by Brahim Lounis: You are interfering two or four beams. That gives you a cubic lattice. Would there be any advantage of using three or six beams to generate a hexagonal lattice? The hexagonal lattice has a higher packing density.

Related to this, any periodic lattice generates cross-talk between unit cells. How about extending your concept to a non-commensurate lattice, for example by interfering five or seven beams? Would you still be able to implement the technique? What would the advantages/disadvantages be?

Brahim Lounis answered: Using three-beam interference, we can generate 2D hexagonal lattices. In this case, any relative phase variation between the beams results in a global spatial translation of the optical lattice and does not change its topography of the electric field. When using more than three beams to generate 2D lattices, the topography of the electric field of the lattice will depend on the relative phases of the beams. This is often undesirable since our approach of data analysis is based on *a priori* knowledge of the minimal intensity positions. Nevertheless, it is possible to generate desirable 2D lattices with four-beam interference under certain conditions. If the four beams are composed of two orthogonally polarized pairs of beams, the topography of intensity of the lattice stays unchanged with relative phase variations. The period of the lattices is given by the incident angle, which is limited by the back aperture of the objective. With the same incidence angle, the three-beam hexagonal lattice has a larger period compared to the four-beam squared lattice, thus does not have higher packing density.

In a lattice-STED microscope, the camera together with a digital mask acts as an array of "point detectors", each of which records the fluorescence signal originating from corresponding intensity "zero".[1] Since the distance between the "point detectors" is ~300 nm, comparable to the diffraction limit, one possible issue is that a "detector" might detect fluorescence from a fluorophore located in a region of neighbor intensity minimum. Therefore we apply a linear unmixing process to the signals delivered by the "detectors" to eliminate possible artifacts due to this cross-talk.

1 B. Yang, F. Przybilla, M. Mestre, J.-B. Trebbia and B. Lounis, *Opt. Express*, 2014, 22, 5581–5589.

William Moerner remarked: You presented a beautiful analysis of the complex polarization behaviors which occur in lattice-STED microscopy. The variety of PSFs which occur are fascinating, and should be taken into account in the image creation process. Furthermore, it should be useful to analyze the detected photons as a function of time delay during the STED depletion pulse, because this might allow more spatial discrimination.

Francisco Balzarotti asked: From the paper, it looks like uniform excitation light was used. Is it possible to gain speed (or alternatively use less depletion power) by using a patterned excitation? How much is there to win?

Brahim Lounis responded: We do not expect a gain in speed by using patterned excitation. With structured exception we reduce the photobleaching.

Michel Orrit questioned: Can you summarize the current limitations in the field of view and rate of acquisition of your parallelized STED technique?

Brahim Lounis replied: In lattice-STED, the field of view is a compromise between the desired resolution and the available STED laser system (average power and repetition rate). Recently, we could achieve a field of view of ~10 μm diameter with a resolution down to 70 nm using pulsed laser with a repetition rate of 80 MHz and an average power of about 850 mW. The field of view area scales linearly with STED laser average power. For the same average power, one can also significantly increase the field of view using a laser having lower repetition rate. In a recent paper,[1] a field of view of ~20 μm diameter and a resolution of ~55 nm were obtained with a laser of 1 MHz repetition rate at comparable average powers (~1W).

The acquisition rate in lattice STED is limited mainly by the readout time of the camera. With our lattice-STED microscope, we record typically 100 camera frames to obtain a super-resolved image. The imaging speed is then limited to 80 ms per super-resolved image. The technical feasibility of massively parallelizing STED without degradation of the resolution heralds video-rate STED nanoscopy of large fields of view, when suitable high-speed sensitive cameras become available.

1 F. Bergermann, L. Alber, S. J. Sahl, J. Engelhardt and S. W. Hell, *Opt. Express*, 2015, **23**, 211.

Niek van Hulst said: I do like your parallelization approach to overcome the speed limits of STED. Yet the results are reconstructed from wide field images, and one still needs to scan and read-out each image. The ultimate frame rate will be limited by the camera, rather than the signal level, right? What frame rate can one expect at best?

Brahim Lounis answered: Currently, the frame rate of lattice-STED is limited by the readout time of the camera. For a region of 128 × 128 pixels, the readout time is 800 μs per frame of a fast CMOS camera (Hamamatsu, Orca Flash4) . The sample is scanned within a unit cell by typically 10 × 10 steps, leading to a lattice-STED frame rate of 12.5 Hz. With the advances in camera technology and the emergence of faster cameras, the ultimate frame rate will be limited by the signal level. Consider that the integration time per scan step in confocal STED can be as short as ~10 μs, the frame rate of lattice-STED can reach 1 kHz.

Alex Clark queried: Listening to Prof. Moerner's talk where depth can be scanned in a sample, and your talk on having many illuminated regions on a sample for parallel measurements, is there a chance of combining the lattice effects with depth effects to get image across a broad plane quickly and in three dimensions?

Brahim Lounis responded: Yes, it is possible to combine lattice-STED with PSF-engineered detections such as applying astigmatism or double helix function in the detection path if the emitters are resolved in a lattice-STED image. The emission PSF of single emitters can then be analyzed to obtain their axial positions with sub-diffraction precision. If more than one molecule emit fluorescence within one same zero intensity, their emission PSFs overlap and it becomes difficult to localize their axial positions. Therefore the combination of lattice-STED with PSF-engineered detections can be interesting to image samples with low molecular density to achieve fast three-dimensional super-resolution imaging.

William Moerner opened the discussion of the paper by Sergey Sekatskii: Were you able to find and test your method on a nanodiamond with only one NV-center?

Sergey Sekatskii answered: Yes, in many cases nanodiamond crystals which were fixed at the SNOM tip apex contained around one NV-fluorescent center. However, as has been already stated, we were unable to see FRET SNOM images with these probes.

Himangshu Prabal Goswami asked: Why do you not distinguish between dephasing and decoherence? Do they both represent the same physics in FRET SNOM? How do you compare the time scales of decoherence and radiative decay time and establish that the time scales are comparable?

Sergey Sekatskii replied: There is no difference between dephasing and decoherence in the context of low temperature FRET SNOM. As for the comparison of decoherence time T2 and radiative decay time T1, it is known that for numerous optical transitions of different rare-earth ions in different crystals, they do not differ too much from each other lying in micro- or even millisecond range. In our estimations we take the possible difference into account introducing a coefficient η which is the T2 : T1 ratio.

Gary M. Skinner enquired: You present some nice images of molecules, some of which you suggest were single-stranded DNA molecules, with the rest presumably being double-stranded. How do you know that these are single stranded DNA molecules?

Sergey Sekatskii answered: In our laboratory we have a large experience of imaging of DNA molecules with a "standard" (based on the reflection of light from the cantilever's backside) AFM, and single strand DNA molecules (ssDNA) are also quite regularly observed.[1] It is instructive to compare images presented in ref. 1 with those which are presented now and obtained with a SNOM glass fiber probe exploiting shear force mode: for the same preparation conditions they are quite similar. Note also that for ssDNA, in the context of scanning probe imaging the conception of the persistence length is not very sound one: contrary to the double strand DNA, the character of an image very strongly depends on the surface conditions, *i.e.* on the details of ssDNA–surface interaction.[1]

1 J. Adamcik, D. V. Klinov, G. Witz, S. K. Sekatskii and G. Dietler, *FEBS Letters*, 2006, **580**, 5671.

Alfred J. Meixner remarked: I can confirm the very high sensitivity and resolution of your shear-force images. We can resolve single molecule layers in thin molecular crystal films.

Michel Orrit questioned: How are the tuning fork and stretched fiber mechanically linked and which vibration modes are excited by the tuning fork?

Sergey Sekatskii replied: Briefly, first of all the length of the free standing part of the glass fiber beam should be equal to 4.6 ± 0.2 mm exactly; in this case the working frequency of the quartz tuning fork coincides with the second resonance frequency of the dithering of the fiber probe. The connector, which links the probe and one prong of the tuning fork, is a thin, 20–40 micron in diameter, glass rod. This connector needs to be quite accurately glued to both parts it connects with a small amount (tiny droplet) of hard glue. The position where this connector is placed is also not arbitrary but should be controlled with roughly ± 0.5 mm precision.

Alfred J. Meixner asked: How long did it take to record your shear-force images? Did your feed-back loop control the phase?

Sergey Sekatskii replied: The image acquisition time can be very short for "flat" samples and relatively small scan sizes. This was necessary to record true single molecule FRET SNOM images: the acquisition time was less than one minute for this case. For other cases, when such time is not of an uttermost importance, *e.g.* for single strand DNA (ssDNA) imaging, we recorded images during tens of minutes. Certainly, only rather low-rate scanning with such high Q (quality factor) probes is possible if one measures dithering amplitude, but in our case the feedback is based on the measurements of the resonant frequency and Q-factor values. This can be seen as a kind of phase measurements, and they can be done quite rapidly.

Phil Williams opened the discussion of the paper by Ana Peñaherrera† by communicating: In the paper you say that "...hypothesised that there is some mechanical matching between the receptor and ligand." My question is why? Why

† The paper by Ana Peñaherrera was not presented at the meeting. Subsequently, any questions regarding this paper were submitted after the meeting.

should the interaction between receptor **R** and ligand **L** be related to the mechanical properties of the individual partners, and why should these be 'matched'?

Ana Peñaherrera communicated in response: Mechanical matching is not seen between all individual partners. The hypothesis came out when it was observed that GP120 and CD4 required similar forces to be unfolded. Then, other pairs were tested and it was found that the force to unfold fibronectin and integrin was different according to the domains. However, the force required to unfold the same number of domains when comparing GP120 and CD4 was not statistically different.

Phil Williams communicated: The interactions between the "five proteins and a mica surface" were measured. If **R** has a strong interaction with the mica, then isn't its partner **L** more likely to have a weaker one? (consider charge, ion coordination, complementarity *etc.*)

Ana Peñaherrera replied by communicating: The answer is yes. When the receptor is adsorbed on mica and the ligand is added, the ligand shows a weaker interaction with mica as it identifies binding regions of its receptor. Nordin *et al.*[1] analyzed the behavior of fibronectin and integrin. In the case of integrin adsorbed on mica, the analysis of the dimensions of its features suggested that the transmembrane proteins are lying flat on the mica surface. This allows the exposure of the binding regions. In this case the mica surface does not contain the receptor when the ligand is added or *vice versa*, that is why the protein shows a different interaction with mica.

1 D. Nordin, L. Donlon and D. Frankel, *Soft Matter*, 2012, **8**, 6151.

Phil Williams communicated: The paper says that "...tip velocities between 0.1 and 10 μm s^{-1} ... loading rates according to previously reported procedure (Faull *et al.* and Slade *et al.*)". The references for Faull and Slade are not given in the bibliography and so I can't check and therefore have the following questions:

a) Is each force curve of a single protein, or of many? Did you compare the contour lengths of the WLC fitted to the curves to the unfolded length of the proteins?

b) How did you account for the difference in force with different loading rates between events and between recordings?

If you used the "loading rate at rupture" to determine the loading rate of the experiment as indicated in the 2012 and 2013 Soft Matter papers,[1,2] then such an analysis is incorrect: the probability of measuring an unbinding/unfolding event at a given force p(*f*) under a ramp of force is equal to the probability of unbinding at this force multiplied by the probability of the bond surviving this force (1 – integral of unbinding probability from 0 to force *f*). Since "... the elastic properties of stretching proteins can be described with the worm-like chain (WLC) model ..." then the loading rate is not constant as it depends on force and so the survival probability integral differs from one of a constant loading rate. The most probable rupture force, the mode of p(*f*), is not linear with the log of the loading rate at rupture.[3,4]

1 L. Donlon, D. Nordin and D. Frankel, *Soft Matter*, 2012, **8**, 9933–9940.
2 L. Donlon, A. Peñaherrera and D. Frankel, *Soft Matter*, 2013, **9**, 2803–2809.
3 E. Evans and P. Williams, *Dynamic Force Spectroscopy*, ed. Julicher, Ormos, David, Physics of Bio-Molecules and Cells, Flyvberg, Springer, 2002.
4 P. Williams, *Dynamic Force Spectroscopy with the AFM*, ed. Noy, Handbook of Molecular Force Spectroscopy, Springer, 2008.

Ana Peñaherrera communicated in response: a) Yes, each force curve is of a single protein. No, I did not compare the contour lengths.

b) Measurements at 1 $\mu m\ s^{-1}$ were reported. The rupture force was determined at the point of maximum extension of the cantilever.

The final rupture force which corresponds to detachment of the protein from the tip rather than the protein from the surface, was discounted as it relates to the interaction between silicon nitride and the protein. Please find below references to the work of Faull[1] and Slade.[2]

1 R. J. Faull, N. L. Kovach, J. M. Harlan, and M. H. Ginsberg, *J. Cell Biol.*, 1993, **121**(1), 155–162.
2 A. Slade, J. Luh, S. Ho, and C. M. Yip, *J. Struct. Biol.*, 2002, **137**(3), 283–291.

Sanli Faez opened the discussion of the paper by Xin Shi: It is envisioned that by measuring the variations of the optical scattering signal from the nanopore, one can detect bio-molecules without labeling. The scattering signal from the nanopore, however, can be several orders of magnitude larger than the variation induced by a single molecule. Can you please quantitatively estimate the size of the smallest particles that can be detected given a certain (realistic) detection sensitivity and dynamic-range?

It is suggested that a change in scattering can be used to detect the passage of analytes through the pore. Considering the large scattering cross section of the pore and the limited dynamic range of the detection apparatus, what would the size of smallest particle that can be detected with your method?

Xin Shi responded: Thank you very much for your suggestions. Our idea of detecting single particle/molecules by the single nanopore on a metal film is based on the fact that the extraordinary optical transmission through sub-wavelength nanoholes is highly depending on the material filled in the nanohole. Based on previous work (Ref. 48–50 cited in our manuscript (DOI: 10.1039/C5FD00060B)), the scattering light from the nanopores, both their intensities and spectra, will be affected by the analytes which are inside, or absorbed on these pores. We are still working on the numerical calculation to estimate the sensitivity that our method could achieved. Based on our recent results, our system currently could detect single dielectric nanoparticle with diameter between 50-100 nm and measure the scattering light from around 50 nm nanopore. Therefore, we think it would be very possible for our system to detect analytes less than 50 nm. Our study on label-free optical nanopore detection is at an early stage, the experiments for detection of bio-molecules such as DNA, peptides or proteins is under processing.

Michel Orrit commented: You showed images of double holes, as used by Reuven Gordon,[1] for enhancement and trapping of dielectric nanoparticles. Did you try to use those double holes in your experiments, and more generally, what prospect do you see in using those instead of simple holes?

1 S. Wheaton, R. M. Gelfrand and R. Gordon, *Nat. Photonics*, 2015, **9**, 68–72.

Xin Shi answered: Thank you so much for your comments. The double nano-hole structure, introduced by Prof. Reuven Gordon,[1,2] demonstrated an excellent application on single dielectric particle and single protein detection by sub-wavelength aperture. We think the scattering light through the double-hole structure could be observed under the dark-field microscope. The double nano-hole structure generated a hot-spot region between the tips formed by the overlap of the two holes, which make this nano region very sensitive to the dielectric index variations. As Prof. Gordon demonstrated in his previous paper, the double-hole structure could be used for plasmonic trapping the single dielectric particle[1] and even single proteins.[2] An individual analyte captured in the gap could be detected by the transmitted light. Therefore, scattering light through the double hole structure might be more sensitive to the analytes in the gap of the tips compared to the simple holes. Therefore, double-hole might be a potential structure for our further study on single bio-molecule detection. However, the electric field in the nanopores, which was used as the driven force for typical nanopore experiments, might be more complex in double nano-hole structures than the simple holes. More work needs to be done on modulating the driven force for traversing a single particle/molecule through the gap region electrically. Since our study on the metal-hole based optical nanopore detection is at an early stage, we are now focusing on the single hole on a metal film to prove the feasibility of our method.

1 Y. Pang and R. Gordon, *Nano Lett.*, 2011, **11**, 3763.
2 Y. Pang and R. Gordon, *Nano Lett.*, 2011, **12**, 402.

Wolfgang Langbein remarked: You have shown experiments on larger polystyrene beads of about 100 nm size. Can you estimate the sensitivity of your detection system in terms of bead diameter and measurement bandwidth? Which improvements from the present sensitivity can you realistically anticipate?

Xin Shi responded: Thank you very much for your questions. We are still working on the numerical calculation which might provide us with information on the sensitivity and dynamic range of our method. The first step of our presented method is to achieve the simultaneous electrical and optical detection of nano-scaled analytes such as nanoparticles, viruses and some bio-molecules. Based on our current experimental results shown in my presentation, we could currently detect single dielectric nanoparticle with diameter around 50–100 nm, and measure the scattering light from around the 50 nm nanopore. Therefore, we think it would be very possible for our system to detect analytes less than 50 nm. To achieve label-free electrical and optical nanopore readout of single bio-molecules, several strategies are being studied to enhance the performance of our methods. First of all, we are trying to fabricate a nanopore on a gold film with a diameter of 30–50 nm, as well as to design structures to enhance the localized electromagnetic field. This might be a promising approach to increase the detection sensitivity and signal to noise ratio. Secondly, we are improving the illumination approach. At present, we use a halogen lamp for the dark-field illumination, which is standardly equipped for single particle scattering experiments. We are now working on the reflected dark-field illumination approach to achieve a high sensitivity for both intensity and spectra of the scattering light variations as we demonstrated in our manuscript.

Niek van Hulst said: The size of DNA is 1.4 nm while your 100 nm holes are much larger. Other groups are currently making nanoholes in the 5 nm range, competitive with typical protein size. I think you will need much smaller holes. What is your strategy to go smaller?

Xin Shi replied: Thank you for your question. Yes, we do need much smaller holes for detection of single DNA or proteins. Typically, it is better to reduce the diameter of the nanopore to at least 20 nm for ds-DNA or ss-DNA sensing. There is a series work for us to get smaller. Firstly, smaller nanoholes should only be fabricated on thin membranes, usually with a thickness of 10-20 nm. Therefore, we need to fabricate a thin Au membrane on thin Si_3N_4, with a good optic performance, *i.e.* flat, clean and tough. We are working on fabricating 20 nm Au layer on 10 nm Si_3N_4 membrane for small nanoholes. Secondly, the drilling method needs to be improved. We used a focused ion beam to drill nanopores on our membrane. For small holes (<20 nm), we need to try to use a focused electron beam and dielectric breakdown for the fabrication. A focused electron beam is one of the typical approaches to fabricate small nanopores with a diameter less than 10 nm. Recently, fabrication of nanopores with a sub-10 nm diameter by controlled dielectric breakdown has been demonstrated by Tabard-Cossa's group.[1,2] We are also trying to use controlled dielectric breakdown to fabricate smaller holes with thinner membranes.

1 H. Kwok, K. Briggs and V. Tabard-Cossa, *PLoS ONE*, 2014, **9**(3): e92880.
2 K. Briggs, H. Kwok, and V. Tabard-Cossa, *Small*, 2014, **10**, 2077.

Maxime Dahan queried: What is the benefit of optically measuring the presence of the molecules compared to the conventional electrical measurements in nanopores?

Xin Shi answered: Thank you very much, Prof. Dahan. The optical detection of analytes translocation events could bring multiple advantages. Firstly, detecting optical signals from multi-nanopore channels simultaneously provides a novel multiplexed method without the demand of an individual amplifier circuit for each nanopore, which allows for a simple and compact instrument and a scaled-down microfluidic system. Many signals from single-analyte events, especially for plenty of biological applications, should be acquired for statistical analysis due to the randomness of nanopore-based single-molecule detection. The high throughput readout without losing the detection bandwidth and sensitivity would greatly decrease the time consummation. Secondly, the optical nanopore signal is also expected to reveal the chemical states and molecular fingerprints of the analytes inside the nanopore. Although a large amount of information can be deduced from the ionic current signal, the species and chemical states of analytes could not be directly obtained. Optical methods, such as the combination of nanopore technique and molecular spectroscopy, might provide the ability to obtain the molecular fingerprints during the nanopore detection. Besides, optical methods provide possible approaches to solve the flicker noise problem. The low frequency noise components involved in the ionic current are generally believed to be the key limiting factor to resolve fine molecular features in many sensing applications.

Vahid Sandoghdar stated: It is true that a liquid environment introduces an intrinsic background fluctuation. But unless the concentration of the background is really high, the particles/molecules that swim by do not stay long enough at one place to contribute to a signal. On the other hand, the signal of a protein that binds can be integrated for many 10–100 ms to generate a sufficient signal-to-noise ratio.

Wolfgang Langbein commented: I agree that intrinsic liquid fluctuations have a power spectrum which have a large bandwidth, effectively limited by the transit time of sound through the region measured. For water the sound speed is 1.5 km s^{-1}, and for a 1 μm focus size I find a microsecond timescale of the fluctuations. Averaging the fluctuations leads to a 1 per sqrt(time) factor, a supression of 100 for 10ms in the present example. However, when estimating the ultimate detection limit of reflection techniques, this needs to be considered.

Himangshu Prabal Goswami opened the discussion of the paper by Paul Bohn: What is the exact role of the evanescent field in single molecule spectroscopy? How do I include it in the Hamiltonian of the single molecule system so as to capture the essential physics?

Paul W. Bohn answered: The evanesecent field decays in amplitude over a distance away from the silica-solution interface at the bottom of the over-etched pore. The exact decay parameters are a function of the detailed geometry which varies from pore-to-pore, but we are certain that in all cases examined the excitation volume is significantly smaller than the geometric pore volume, *i.e.* the excitation field is confined to a volume near the bottom of the nanopore. Because the redox-active molecules are free to diffuse, the signal is integrated over the product of the spatial distribution of the field and the exact molecular location at the point of emission. Another way of looking at this is that the probability of excitation/emission is a function of position within the nanopore. The definitive early experiments on excitation in nanopores were done by Webb and Craighead – see, for example, references 34, 38, and 39 of our paper (DOI: 10.1039/C5FD00072F).

Alfred J. Meixner remarked: Concerning the protonation and associated spectral shifts of single molecules. We have observed spectral shifts associated with spontaneous conformation changes of single molecules embedded in polymers switching back and forth depending on the hybridization (sp2 *vs.* sp3) of the nitrogen of an amino group bound to the chromophore.[1] Did you see something similar?

1 F. Stracke, C. Blum, S. Becker, K. Müllen and A. J. Meixner, *Chem. Phys. Lett.*, 2000, **325**, 196–202.

Paul W. Bohn responded: Our experiments were implemented with filters to separate emission from scattering, so they were not very sensitive to spectral shifts. Acquisition of spectral data, as you suggest, is an ideal way to examine the detailed chemical changes occurring in these experiments. The spectral behavior of different flavosemiquinone species in enzymes varies depending on chemical (protonation) state. For example, there is a ~200 nm shift between FAD radical anion (in cryptochrome) and FMNH radical (in flavodoxin), and smaller, but observable, differences in solution between $FADH^-$ and $FADH_2$. Thus, there is a good deal to be gained from spectral measurements, especially in sorting out the detailed pathway between the fully oxidized and fully reduced states and the intriguing observation of a stable intermediate at slow potential scan rates. We plan to make these spectral measurements as a function of potential and protonation in the near future.

Sanli Faez commented: By changing the external electric potential, the counter-ions inside the pore are displaced and the measured molecule may experience a different electrostatic potential. This may in turn change the probability distribution of the position of the molecule inside the pore and hence result in a variation of the average intensity. How is it possible to distinguish between such external variations and intensity variations caused by internal electrochemical processes of the molecule?

Paul W. Bohn replied: Changing the applied potential on the cladding/working electrode (WE) will certainly change the ion distribution in the nanopore, as you suggest. This will have a number of consequences. Firstly, at pH 2.9 of the reported experiments, the FMN is net negatively charged due to the pK_a of the phosphate, so there is a permselectivity factor biasing against transport of FMN into the nanopore in the first place. We know, however, from the fact that we see signal, that the exclusion is not perfect. This effect will be somewhat stronger at more reducing potentials. Secondly, although these pores are very small, the Debye length at 100 mM citrate is ~1–2 nm, or about 10× smaller than the pore diameter. Thus most of the emission will be derived from molecules that are outside of the compact double layer, even accounting for the annular geometry of the nanopore WE. A small fraction of the emission will arise from molecules in the double layer where electrostatic effects may affect the emission behavior, but we believe that this will be no more than 5–10% of the total emitting species. Furthermore, while the potential distribution within the double layer changes with applied potential as the electrolyte ions reorganize, the physical size of the double layer is relatively insensitive (independent to zeroth order) to the applied potential, so the fraction of the population experiencing electrostatic effects is relatively small and potentially independent. Finally, note that electrochemistry can occur anywhere along the surface of the WE, whereas emission arises only where the optical field is strong near the bottom of the nanopore. Thus, a large fraction of the emission we observe arises from molecules that have reacted then diffused into the excitation volume.

Thorben Cordes asked: Do you find different types of behaviours of individual molecules due to varying locations within the zero-mode waveguide?

You mentioned that at low scanning rates a new redox-state could be seen. Is this real or an experimental artifact – what could that chemically be?

Paul W. Bohn responded: In our previous experiments with immobilized single redox fluorophores, we observed very different behavior for different individual molecules. For example, single FAD molecules covalently immobilized on the Au cladding exhibit different apparent values for the equilibrium potential, as evidenced by the different emission intensity histograms. Although we cannot definitively say that location is the reason, our working hypothesis is that the varying behavior results principally from two factors: orientation of the FAD isoalloxazine chromophore and location relative to the trapped optical field. In the present experiments, the molecules are freely diffusing, so we cannot assess where they are specifically when they emit other than to say they have to be in the bottom portion of the pore in order to have significant probability of being excited. The new redox state that we see in the intermediate potential range is reproducible in potential (between multiple cycles of potential), is observed at the same potential on both the positive-going and negative-going portion of the voltage ramp, and has been observed in multiple different device structures and experiments. We do know that the flavosemiquinone state is stable in some proteins (*e.g.* FAD radical anion in cryptochrome and FMNH radical in flavodoxin), but it is not generally observable in fluid solution.[1] It is an open question whether the same type of confinement effects apply to the ~20–40 zL active volume of the ZMW, thereby stabilizing the radical semiquinone. This is a question that we are currently addressing through experiments aimed at changing the volume (surface grafting) and the local molecular crowding (viscogens).

1 Zhong *et al.*, *J. Am. Chem. Soc.*, 2008, **130**, 13132.

William Moerner questioned: You reported the observation of single molecules in various redox states in your zero-mode waveguides based on the expected occupancy of the probe volume. Have you observed individual bursts of fluorescence at lower and lower concentrations, with no signal at even lower concentrations?

Paul W. Bohn replied: Two additional concentration conditions were examined. At 1 μM (corresponding to ~0.02–0.1 depending on the exact volume of the overetched region) we were able to observe a reduced (roughly 5–10×) number of bursts. In a solution containing only buffer, we were unable to see individual photon bursts and observed a background count level of the order 3–4 per ms. These latter experiments are hampered because it is difficult to find the focal volume accurately when switching from the transillumination geometry, used to find the pore array, to the fluorescence geometry. Thus, we took repeated measurements through a series of focal planes above and below the plane found by transmission (in order to find the plane where fluorescence is maximized). In measurements of the buffer only the background count was ~ 3 ms^{-1} compared to 40 ms^{-1} with $FMNH_2$ at 10 μM (~ 0.2-1). We did not examine concentrations below 1 μM because it became difficult to focus on the small pores by fluorescence after we had located them by transmission. These experiments would certainly benefit from examining a higher QE fluorogenic redox substrate, as FMN is not as bright, nor as differentiated between oxidized and reduced states, as would be ideal.

Alfred J. Meixner asked: Did you see any SERS?

Paul W. Bohn answered: As detailed in the answer regarding spectral shifts we are not currently equipped to make those measurements. As your question implies, the acquisition of redox-sensitive single molecule SERS in these geometries is certainly intriguing.

Himangshu Prabal Goswami opened the discussion of the paper by Jens Michaelis: How do you connect the Markov chain to the energy barrier by introducing the pseudo temperature? What are the physics behind it?

Jens Michaelis replied: I'd like to emphasize that we merely consider a virtual energy landscape that has nothing to do with the force potentials acting on the dyes. Our notion of temperature is merely a scaling of this virtual energy landscape. The "virtual particle" within this landscape corresponds to a configuration within the configuration space of the whole system.

Vahid Sandoghdar queried: I am not sure if I understand the method. How do you find the absolute x, y, z location of several molecules with respect to one satellite?

Jens Michaelis responded: We term satellites as molecules that are attached to a known position of a macromolecule whose structure at least in part is known. Thus the pdb file of the macromolecule defines our coordinate system. We then perform smFRET measurements to an unknown position termed antenna which gives us information about the distance between satellite and antenna. In case one makes only one such distance measurement of cause the resulting posterior will be not very informative, thus in order to increase our accuracy we typically measure to 5 or more satellites, include more than one antenna and also measure between antennas. Everything is then analysed in a global scheme giving us 3D probability densities for each antenna (so called posteriors) which are then typically displayed by credible volumes.

William Moerner commented: This is beautiful work, congratulations. Would you please contrast your accessible region calculations to those of C. Seidel?[1]

S. Sindbert, S. Kalinin, H. Nguyen, A. Kienzler, L. Clima, W. Bannwarth, B. Appel, S. Müller, and C. A. M. Seidel, *J. Am. Chem. Soc.*, 2011, 133(8), 2463–2480.

Jens Michaelis answered: We were the first to present a calculation of an accessible volume in the context of FRET measurements.[1] The Seidel laboratory later on used the same approach that we did. However, they found that sometimes a more elaborate description of the dye geometry gives improved results (elliptical instead of spherical shape assumed). It is clear that the more realistic you treat the shape of the dye molecule (as well as that of the linker) in your accessible volume calculations the better. However, I would like to point out that accessible volumes have been used quite differently for structural analysis. In our original NPS work we used a static model, meaning that the dye molecule could sit at any positions within the accessible volume and we gave every position a uniform probability in the prior. The Seidel lab[2] has used dynamic averaging over all positions within the prior. As a result, even though the prior volume is large, they assume they know precisely the averaged position of the dye molecule if they know the attachment point and accessible volume. We have used such an approach in one of the models in the paper presented here (DOI: 10.1039/C5FD00110B) (meanpos-iso model). However, we now also offer a different model termed var-meanpos model (or var-meanpos iso model if one additional assumes free rotation). In this model we assume dynamically averaged positions due to linker dynamics, but allow for a non-uniform sampling of the accessible volume. In our new software one can choose for every dye molecule which model one wants to use, making it easy to compare different approaches.

1 A. Muschielok *et al.*, *Nat. Methods*, 2008, **5**, 965–971.
2 S. Kalinin *et al.*, *Nat. Methods*, 2012, **9**, 1218–1225.

Alfred J. Meixner questioned: What is your essential goal? Is it to measure the distance between the chromophores? Have you tried plunge freezing of the proteins?

Jens Michaelis replied: The essential goal is to get structural information from a number of different measurements between chromophores attached to specific sites on macromolecular complexes. No, we haven't yet tried to freeze proteins in order to increase the photon budget. When working with transient complexes, where the goal is to obtain dynamic structural information including kinetic data, obtaining information from frozen samples yields different problems.

Thorben Cordes asked: Regarding the conversion of accurate FRET values into distances, how accurate can you be in general? The steady-state anisotropy values for different complexes are rather high (>0.2) – can they be used reliably? What about the works of Levitus[1] and others that suggest that steady-state anisotropy is not necessarily a reliable measure for free dye rotation. Is it better when determining distances *via* FRET to really measure all parameters – see Seidel and co-workers?

1 M. E. Sanborn, B. K. Connolly, K. Gurunathan and M. Levitus, *J. Phys. Chem. B*, 2007, **111**, 1064.

Jens Michaelis responded: The accuracy that one can achieve depends on several things, the freedom of the dye to rotate, the flexibility of the protein, the dynamics of the linker, the geometry of the situation that you are studying. The better your information about for example the rotation of the dye molecule or the linker dynamics, the more accurate you can get, in ideal cases down to the resolution of a couple of Angstroems. High anisotropies can be problematic but they don't have to be. Time resolved methods of measuring anisotropy generally give more information and thus should be preferred over steady state methods as pointed out by Marcia Levitus.[1] In our classic NPS model we make a conservative estimate, *i.e.* we assume that no more re-orientation occurs on slower time scale than can be probed using time-resolved methods. This limits our accuracy, meaning the computed posterior will be larger, but it will also contain the result for a more freely rotating molecule. In our own hands we get more accurate results when measuring immobilised molecules on surfaces and performing corrections at the single molecule level. We then do not get the same amount of information about polarisation then in a MFD setup, but we believe that determining the correction factors properly is more important at this point. Ideally one would have dye molecules that are either truly free to rotate, or a really at a fixed angle. Both scenarios would improve the accuracy.

1 M. E. Sanborn *et al.*, *J. Phys. Chem. B*, 2007, **111**, 11064–11074.

Victoria Birkedal enquired: Have you modeled the expected width of the FRET distribution for the different dye models? Do you expect that this observable can help deciding which dye model is most appropriate for a given system?

Jens Michaelis answered: The FRET histograms we are typically recording use integration times on the order of tens of ms. Dynamics such as re-orientation of dye molecules or linker dynamics on time scales shorter than the integration time are thus averaged out. Dynamics on longer time scales would lead to different states in the histogram (also visible in the time trajectories) which then can be analysed independently. Thus only dynamics on the time scale of the integration time are problematic. One can therefore perform measurements with different time resolution and compare the resulting histograms. For very fast dynamics on the order of nanoseconds also experiments on freely diffusing molecules can be helpful to decipher the dynamics.

Niek van Hulst said: I like your ambitious approach to map out protein structure using multi-FRET. Yet the work is tedious, and one needs to prepare a separate sample for each FRET-pair combination, hoping that the labelling does not affect the structure. If one prepares many protein-FRET systems, are all measurements consistent? Would it not be wise to first try the approach on a simple system for which the crystalline structure is known?

Jens Michaelis replied: In our initial publication,[1] we performed the benchmark test you refer to. We were investigating RNA polymerase II transcription elongation complexes and used NPS to infer a known position, namely the position of the first nucleotide of RNA in the active center cleft. We could show that our experiments and analysis are in good agreement with the position reported from X-ray crystallography. We have experienced situations with very hydrophobic dyes, were the dye molecule does change the protein structure locally. Thus, this is a potential problem and sufficient testing should be performed to rule out such effects. With the presented network analysis and the ability provided by the software to change the dye model for each dye used and get the analysis result quickly, these important tests are now feasible and are routinely done.

1 A. Muschielok *et al.*, *Nat. Methods*, 2008, **5**, 965–971.

Felix Koberling queried: How precise do I have to get with the triangulation to learn something about the 3D structure of the protein? Could you give an example for a structural question where the ultimate precision is necessary and an example where this is not so relevant?

Jens Michaelis responded: In the past we have experienced different situations where high-resolution information or medium resolution information are required. For example when trying to obtain structural information in order to build structural models of large macromolecular complexes such for transcription initiation complexes[1,2] or transcription elongation complexes,[3] the higher the resolution the better. By measuring not just one position but many positions on the same macromolecule we were able to build structural models at quite high resolution. At other times one would like to know whether a certain part of complex A can interact with another part of complex B in a certain scenario. Here, often the required resolution is not that high. For example we have shown that interaction of the nascent RNA of a Pol II transcription elongation complex with the so called stalk region of the polymerase requires the presence of transcription factor TFIIB.[4] As such we could learn something about the function of this transcription factor and its role in the transition from initiation to elongation.

1 J. Nagy *et al., Nat. Commun.*, 2015, DOI:10.1038/ncomms7161.
2 B. Treutlein *et al., Mol. Cell*, 2012, **46**, 136–146.
3 J. Andrecka *et al., Nucleic Acids Res.*, 2009, **37**, 5803–5809.
4 A. Muschielok *et al., Nat. Methods*, 2008, **5**, 965–971.

Vahid Sandoghdar asked: Could you please discuss some of the problems and limitations of the method? For example, how difficult is it to label the right residue/location with a specific dye?

Jens Michaelis answered: There are several different labeling approaches. When labeling nucleic acids, typically one uses bases with dye molecules directly attached during or after synthesis of the oligomer. When labeling proteins, the simplest way to achieve site specific labelling is to use single cysteine mutants. However, oftentimes this approach is not possible and instead one can use small peptide tags which are then modified using specific enzymes or one can incorporate unnatural amino acids with orthogonal chemical groups. In general all these methods are quite time consuming and need to be accompanied with functional assays, ensuring the proper functioning of the system under study.

Thorben Cordes commented: Structural biology requires high resolution preferably down to a (few) Angstroms. How good is the resolution that you can get regarding single distances with your experimental setup and your correction procedures? Is TIRF-based imaging or solution-based smFRET preferable?

The NPS modelling approach: how many distances do you need for validation of structural models? Would it help (and how much) to take into account distances that are known and static and by this verify the reliability of dye positions (*i.e.*, seeing that certain positions give reliable measures) to use them afterwards for measuring unknown distances? How does your model deal with this?

Jens Michaelis replied: In our experiments we typically include measurements between several unknown and several known positions. One should note, that while measuring a structurally determined distance gives additional information, there is still some uncertainty involved, due to the linker conformations and dynamics as well as due to the rotation of the dye molecules (or lack thereof). The resolution we can get when analysing networks of measurements can be as good as a couple of Angströms for ideal cases. When measuring a single distance, the uncertainty would be substantially higher. Typically one measures large networks (on the order of 50–100 distances) and analyses everything globally. However, as we show in the paper (DOI: 10.1039/C5FD00110B), it is important that the model of the dye prior is correct or at least conservative. If one makes strong (unjustified) assumptions, even a single problematic dye can push the network to a different (wrong) minimum. We therefore believe that it is best to either be conservative with the assumptions or to have redundant measurements that allow to identify problematic dyes in bootstrapping approaches. However, in the latter approach, determining the uncertainty of the method becomes very problematic.

Maxime Dahan noted: A lot of progress has been recently made in low temperature single molecule measurements. Do you think people will incorporate FRET measurements or is it incompatible with biological systems of interest?

Jens Michaelis responded: The advantage of going to a low temperature is that one is able to collect many more photons from single molecules, and with that, localisations down to the Angstrom level have been shown by the Sandoghdar and Enderlein labs. In our FRET measurements, we are however not limited by the number of emitted photons, since the $1/r^6$ dependence of the energy transfer provides a very sensitive means. Still, due to uncertainties in orientation, obtaining absolute distances from measured FRET signals is tricky. In that sense, low temperature measurements might become useful, since certainly the orientation of the dye molecules would be fixed during the single molecule measurement. However, it would be necessary to combine them with temperature jumps, so that different conformational changes can be probed. All in all while being potentially interesting doing such experiments on large macromolecular complexes remains extremely challenging.

Lisa-Maria Needham opened the discussion of the paper by Victoria Birkedal: Did you observe any change in the photo-stability of the donor fluorophore between high and low FRET states and, more generally, in the presence and absence of an acceptor?

Victoria Birkedal answered: We have not quantified fluorophore photostability as a function of donor acceptor separation in this study. In all samples with different donor acceptor separations, we observed individual molecules showing time traces with a long enough time before photobleaching to allow correction factor determination.

Lukasz Piatkowski remarked: In Fig. 4 of your manuscript in panel B you present FRET efficiency as a function of the donor–acceptor distance. Both the data points and the theoretical curves offset and 'flatten off' at very short distance (<12 bp) and very long (>19 bp) distances. What is the reason that FRET efficiency doesn't reach unity at short distances and also, in particular, it doesn't go to zero at long distances?

Victoria Birkedal replied: The FRET efficiency E is a function of donor–acceptor distance separation, R, to the power 6. Thus the FRET efficiency is expected to be less sensitive on donor acceptor distances for $R/R_0<0.5$ and $R/R_0>1.5$, with R_0 being the Förster radius. That said, we observe a decrease of calculated FRET efficiency as a function of distance above 19 base pairs (bps) and an increase below 10 bps. It is to be noted that the x-axis in this graph (Fig. 4b in the manuscript (DOI: 10.1039/C5FD001100E)) is shown as base pair separation while the FRET efficiency depends on donor acceptor distance.

Arindam Chowdhury asked: In Fig. 4A, why are the widths of the FRET efficiency distributions different for A39, A35 and A30, and why are these widths not reflected in the error bars of measured fret efficiency in Fig. 4B?

Victoria Birkedal answered: An important parameter is that the FRET process is a function of the distance between donor and acceptor fluorophores to the power six and thus FRET distribution widths will intrinsically vary with peak FRET values. The uncertainties in Fig. 4b were obtained from uncertainties on γ factors.

Sergey Sekatskii remarked: We completely agree that taking into account a direct excitation of acceptor molecules by laser light is extremely important for the interpretation of different types FRET experiments. Indeed, in our researches in the field of FRET SNOM we have seen that in many cases such direct excitation simply prevents the observation of FRET SNOM images. We also would like to note that one should be very careful when estimating the probability of the direct excitation of acceptors: optical absorption cross sections of dye molecules at the edges of the absorption line, which is important here, are strongly dependent on the detailed experimental conditions and vary from one sample of dye molecules to another. This parameter is much less stable than optical absorption cross sections pertinent to the center of absorption line.

Victoria Birkedal responded: Thank you for the comment. The nice thing about the procedure reported in our manuscript is that a precise determination of the direct acceptor correction is not needed (as another correction factor compensates for this value) to obtain quantitative FRET values from single molecule fluorescence time traces.

Thorben Cordes asked: Can you elaborate more on the benefits of your approach: how much more accurate will the E-values be after correction with individual gammas/d-exc values for each traces? If you evaluate your data in both ways – what is the difference? What is the trade-off between having to select traces due to need for a specific bleaching pattern (one of the two dyes needs to bleach first for your analysis) and the gain you get from the correction for each single-molecule trace?

Victoria Birkedal replied: The advantage of using single molecule time traces and photobleaching events is that one can obtain quantitative FRET efficiencies without the need of external calibrants. In our approach we evaluate uncertainties on extracted individual correction factors, which then can give information on the accuracy of the *E*-values in each particular situation. Different bleaching patterns are needed to extract all correction factors and thus it is not possible to extract all correction factors for individual molecules. Our work in this article shows that it is possible to obtain quantitative FRET efficiencies for single molecules when setting the direct acceptor correction to zero, see Fig. 3 in the manuscript (DOI: 10.1039/C5FD001100E). The experimental requirement to obtain individually corrected FRET efficiencies is to have the acceptor dye bleach first. Determining correction factors for individual molecules also allows determining the distribution of these correction factors (Fig. 2), which in our case was broad. Thus using average values of these correction factors may not always be the best approach.

Wolfgang Langbein commented: 1) Your single molecule emission traces have different noise floors after bleaching. Can you explain this?

2) In you analysis you assume that when the acceptor doesn't emit any more (photobleaching), it also does not absorb any more, *i.e.* does not lead to FRET. However, it is also possible that only the quantum efficiency is reduced. Has this been verified (for example using donor lifetime measurements)? What is known of different acceptors in different media in this respect ?

Victoria Birkedal answered: 1) All measured fluorescence time traces were composed of fluorophore fluorescence intensity and background. The background varied from molecule to molecule and also depended on excitation/detection wavelengths and thus different noise floors were observed in single molecule time traces.

2) Depending on the fluorophore, it may be able to absorb light even after photobleaching. I believe this effect is small for the Alexa 647 fluorophore used in this study as it has been used in numerous FRET studies where the molecular ruler properties of FRET were investigated (see for example ref. 14 and 26 in our manuscript – DOI: 10.1039/C5FD001100E).

Jens Michaelis asked: In your work you describe that when looking at single molecules, they show different gamma factors. However, at the end of the paper you come to the conclusion that the distributions of *E* you get become narrower when using an average gamma factor. You also give two explanations in your paper (broadening due to noise and broadening due to heterogeneity). Could you please elaborate on the second reason (heterogeneity)? It is not clear which heterogeneity would lead to a narrower FRET distribution when using the average rather than the molecule specific gamma factor.

Victoria Birkedal responded: By heterogeneity, we mean the intrinsic presence of several states, which can be merged together by using average correction factor values thus yielding a narrower FRET distribution.

Alfred J. Meixner addressed Jens Michaelis and Victoria Birkedal: Have you also looked at donor-lifetimes with and without an acceptor?

Jens Michaelis answered: We have a PIE-MFD setup, which allows us to look at lifetimes of dye molecules. With such a setup it is possible to measure lifetimes of molecules freely diffusing through the excitation volume. With this setup we have also studied donor fluorescence lifetimes and used them to determine FRET values. However, one should note, that more photons are required to accurately determine a lifetime (in particular if the decay is not perfectly mono-exponential) then to measure an intensity value. Thus smFRET values computed *via* lifetimes tend to show more noise than those computed using intensity ratios.

Victoria Birkedal replied: Measuring donor lifetimes in the presence and absence of an acceptor can provide FRET values and in some cases quantitative FRET efficiencies. An example of this approach can be found in ref. 26 of our manuscript (DOI: 10.1039/C5FD001100E). We have not done systematic lifetime measurements for the presented study.

Faraday Discussions

PAPER

Indirect quantum sensors: improving the sensitivity in characterizing very weakly coupled spins

Johannes N. Greiner, D. D. Bhaktavatsala Rao,* Philipp Neumann and Jörg Wrachtrup

Received 13th July 2015, Accepted 14th July 2015
DOI: 10.1039/c5fd00116a

We propose a scheme to increase the sensitivity and thus the detection volume of nanoscale single molecule magnetic resonance imaging. The proposal aims to surpass the T_1 limited detection of the sensor by taking advantage of a long-lived ancillary nuclear spin to which the sensor is coupled. We show how this nuclear spin takes over the role of the sensor spin, keeping the characteristic time-scales of detection on the same order but with a longer life-time allowing it to detect a larger volume of the sample, which is not possible by the sensor alone.

1 Introduction

Nuclear magnetic resonance (NMR) spectroscopy[1,2] provides a label-free method for chemical analysis, which allows to gain a detailed structural analysis as well as dynamical information from the specimen under study. Indeed, *e.g.* 80% of all structure information on proteins stem from the method. However, NMR is rather insensitive and requires chemical purification of the structure under study. Much effort has been directed to applying NMR to nanoscale samples. Indeed, NMR detection of a 4 nm^3 voxel of protons has been achieved with magnetic resonance force microscopy, a challenging experimental technique operating at an ultralow temperature under vacuum.[3,4] At ambient conditions, microcoil detectors have enabled the detection of liquid samples of 3000 nm^3 in volume, but further miniaturization of this technique is not straightforward.[5,6] Magnetic resonance imaging with a few or single proton sensitivity would have a significant impact on a number of research fields, including life sciences and solid state physics. In addition, if the detected spins do have good quantum properties they might become a valuable resource for quantum information protocols.[7]

Over the past decade single nitrogen-vacancy (NV) centers in diamond[8–11] have been proposed and used as a novel atomic-size magnetic field sensor for detecting

3. Physikalisches Institut, University of Stuttgart, Pfaffenwaldring 57, 70569 Stuttgart, Germany. E-mail: d.dasari@physik.uni-stuttgart.de

nanoscale nuclear spin ensembles or even single nuclear spins.[12] This center is a joint defect in the carbon lattice of diamond, consisting of a substitutional nitrogen atom and an adjacent vacancy. Its spin triplet ($S = 1$) groundstate can be polarized and read out optically, so that electron spin resonance experiments can be performed on a single spin. A single center can be used as a nanoscale magnetic field sensor, able to detect a magnetic field in the nanotesla range in an integration time of 1 s.[13] This corresponds to the field of a single nuclear spin at a distance of a few nanometers. Indeed, the detection of single ^{13}C nuclear spins and NMR signals from a 5 nm^3 voxel of various fluid and solid organic samples under ambient conditions has already been demonstrated.[14–18] In these experiments the detection volume of the sample is shown to be limited by the life time of the sensor spin.[19,20] In this paper we propose a detection scheme that is not limited by the T_1 of the sensor spin, allowing for larger detection volumes on a given sample, and improving the sensitivity of resolving the distances between adjacent spins in the sample. In our prototype system, the NV center in diamond, the electronic spin interacts with two types of nuclear spins. One type are the nuclear spins within the sample we want to measure. For those spins the electron spin acts as the quantum sensor sensing the fluctuations of the magnetic field caused by nearby spins in a given sample. Due to the distance dependent (dipolar) coupling between the sensor and these spins, the fluctuations of the latter are reflected in observable changes of the sensor spin state. The other type of nuclear spins are those in the diamond material itself (*e.g.* ^{13}C or ^{14}N), which are much stronger coupled to the NV spin, show long coherence times and thus can be used as ancillary quantum bits. While the strongly coupled spins, close to the sensor inside the diamond, have dominant contributions to these changes, the weakly coupled sample ones (that are farther away) are mostly masked by the stronger ones. To resolve these weakly coupled spins in the sample, the strongly interacting spins typically are dynamically decoupled (DD)[21,22] so that the weak effects caused by farther spins are slowly accumulated in the sensor which is then finally readout.[16] These weakly coupled sample spins can also become undetectable when their inverse coupling strength becomes comparable to the relaxation time of the sensor. While the masking of strongly interacting spins could be removed to a large extent by DD sequences, there is no way around to escape the relaxation time of the sensor placing an upper bound on the detectable spin volumes in the sample. To surpass this problem we propose an indirect or inductive method for sensing where a sensor spin is coupled to a long-lived ancillary spin (a strongly interacting nuclear spin), which now acts as a sensor for times longer than T_1 allowing one to detect larger parts of the sample.

The physical setup for our proposed scheme consists of a sensor described by a three-level (spin-1) system that is interacting with the spins (that are typically spin-1/2) of the sample. In the reference frame of the sensor one can think of the spins in the sample that are all distributed in a three-dimensional plane comprising of the NV axis, (the direction along which the sensor is aligned) and the plane perpendicular to it. For simplicity we consider the two-dimensional configuration where the position of each spin in the sample is described by polar coordinates r and θ. We assume that the sensor is placed along the z-axis and is parallel to the applied magnetic field. In addition to the spins of the sample we also have an ancillary spin which is strongly interacting with the sensor along the

z-axis. The relevant part of the Hamiltonian that determines the dynamics is simply given by

$$H = \omega_L \sum_i I_i^z + gS^z F^z + S^z \sum_{i=1}^{N} J_i(\cos\theta_i I_i^z + \sin\theta_i I_i^x), \tag{1}$$

where the sensor, ancillary and sample spins are respectively denoted by S, F and I. The zero-field splitting of the sensor and the magnetic field strength B_0 can be chosen so that the sensor can act as a two-level system formed by the magnetic levels $m_S = 0, -1$ (see Fig. 1(b)). In the presence of field B_0 the sample spins precess at a Larmor frequency ω_L.

2 Indirect sensing by an ancillary spin

In a standard detection scheme, using the sequence shown in Fig. 1(c),[14] the sensor is repetitively flipped (dynamically modulated) about the plane perpendicular to the z-axis at a rate that matches the energy of a particular sample spin I_i, which is $\omega_i = 2\omega_L + J_i\cos\theta_i$.[14] Under such matching conditions there is a

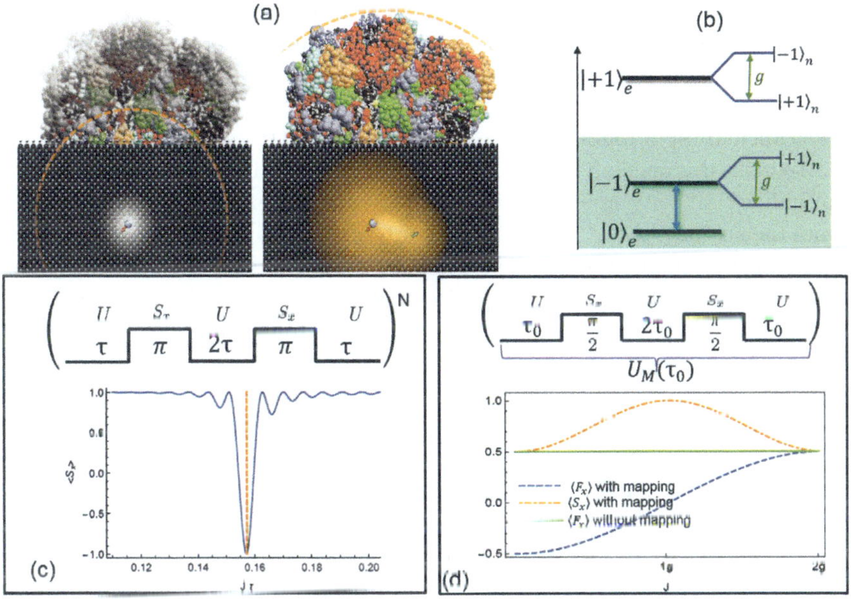

Fig. 1 (a) Schematic representation of the central idea, depicting the increase in detection volume of a sample on top of the diamond both through direct and indirect sensing. (b) The energy-levels of the sensor–ancillary system where only two levels of the sensor (within the triplet subspace) are resonantly coupled to a two-level ancillary (nuclear) spin. (c) We show the pulse sequence for a direct detection of a spin by the sensor. In the same figure we show the sharp transition of a sensor observable, $\langle S_x\rangle$, at $\tau = \pi/(2\omega_L + J)$. The parameters chosen for the simulation are $\omega_L/J = 10$, $\theta = \pi/4$ and the optimal number of repetitions $N = 23$. (d) Mapping pulse sequence $U_M(\tau_0)$ through which the sensor–sample coupling is directly reflected on the dynamics of the ancillary spin F. Also shown in the figure is the variation of the observables $\langle S_x\rangle$ and $\langle F_x\rangle$ both in the presence and absence of mapping.

transition in the sensor spin state which leads to a large observable effect in its optical or electrical readout (see Fig. 1(c)). The peak positions $\tau = \pi/\omega_i$ depend on the parallel component (z–z) of the dipolar coupling between the sensor and the sample spins, while the width depends on the perpendicular component (z–x) of the interaction, *i.e.* $\Delta \approx 1/(J_i \sin\theta_i)$. The optimal number of repetitions, $N_{\rm opt}$, required to obtain maximum contrast (height of the peak value) again depends on the values of J_i and θ_i. We will show later that in the presence of dissipation the peak width increases and the height decreases (thereby increasing the uncertainty of the detected position) but the peak position remains unchanged. Hence, by switching on a radio frequency (RF) field along the x-direction one can exchange the roles of the parallel and perpendicular components of dipolar coupling so that the peak now appears at $\tilde{\omega}_i = 2\omega_{\rm L} + J_i \sin\theta_i$. Thus both the variables can be readout through the peak-positions only. This detection method can also be understood as the well known spectral overlap of the applied control (the filter function) on the sensor with the spectral distribution of the sample spins (Fig. 1(c)), a widely used technique in magnetic resonance imaging.

As can be seen, the above detection scheme can only resolve spins that are spectrally separated by distances larger than the width of the detection peak. Over the past years there have been quite some advances in improving this resolution by using more complex pulse sequences so as to have better resolution by avoiding cross-talk between nearby spins, and also a high degree decoupling to the unwanted dephasing noise spectrum. But, the unavoidable relaxation of the sensor still remains a barrier, placing an upper bound on the detectable volumes of the sample. In the current proposal we overcome this effect by using a long-lived ancillary spin that is strongly coupled $g \gg J_i$ to the sensor.

To make this ancillary spin that is not directly coupled to the sample spins also an indirect sensor, the first step would be to map the sensor–sample dynamics onto the ancillary spin. Though the below discussion is valid for any number of spins in the sample, for simplicity we shall concentrate on the case where the sample is described by a single spin coupled to the sensor with strength J. The dynamics of this three-spin system, in which the central spin (sensor) is coupled to the sample spin I and to the ancillary spin F, is analytically solvable. Before we go ahead to design a mapping pulse sequence, we would like to first evaluate the behavior of the ancillary under free evolution generated by the Hamiltonian given in eqn (1) as a function of J. One finds that the time-dependent evolution of the coherence for the ancillary spin, given by

$$C_F(\tau) \equiv \langle F_x \rangle + \mathrm{i}\langle F_y \rangle = \frac{1}{2}[1 + \cos(g\tau)]C_F(0), \tag{2}$$

has no dependence on the sample parameters J and ω for any time τ. This can be seen from the structure of the Hamiltonian (eqn (1)), where the ancillary–sensor coupling term commutes with the remainder and hence the sensor–sample coupling (S–I) only leads to an overall phase shift and does not contribute to the dynamics of the ancillary spin. The situation becomes different if a microwave field that couples the energy states of the sensor is switched on. The total dynamics cannot be separated into the dynamics on the S–F and S–I subspaces as these terms no longer commute resulting in direct correlations between the ancillary and the sample. Instead of switching a continuous microwave field, we would rather do it stroboscopically using the pulse sequence shown in Fig. 1(d).

Under the action of this pulse sequence, the coherence of the ancillary spin now oscillates as a function of J as shown in Fig. 1(d). For a fixed time $\tau_0 = \pi/g$ and $B_0 = 0$ the analytical expression for C_F takes the simple form

$$C_F(\tau = \tau_0) = \cos(J\tau_0) - \mathrm{i}\cos(\theta)\cos(J\tau_0/2)\sin(J\tau_0). \tag{3}$$

Now one can clearly see that on time-scales of τ_0, the coherence of the ancillary spin purely oscillates on the frequency of the sample coupling parameters J and θ. For $J \ll g$ the purity of the ancillary state is close to one during the evolution, and changes as if it were directly coupled to the sample spin I. Equally the coherence of the sensor spin on this time scale only depends on J, given by

$$C_S(\tau_0) \equiv \langle S_x \rangle + \mathrm{i}\langle S_y \rangle = \frac{1}{2}\left(1 + \cos(J\tau_0)^2\right) - \mathrm{i}\cos(\theta)\cos(J\tau_0/2)^2\sin(J\tau_0). \tag{4}$$

We would like to note that by using the above mapping sequence the time units are now renormalized in multiples of τ_0. For example, with a Larmor frequency ω_L and interaction J, the detection sequence shown in Fig. 1(c) would give a sharp transition in the sensor coherence at $\tau = \pi/2\omega_L + J\cos\theta$. But with the mapping sequence, τ should now be varied in steps of τ_0, implying that spins whose total Zeeman energy is smaller than the S–F coupling strength g are detectable and the parameters should satisfy $g > \omega_L > J$, a commonly available situation in real physical systems. Having mapped the parameters of the sample onto the ancillary spin, the next step is to detect the sample spin by the ancillary. One should also remember

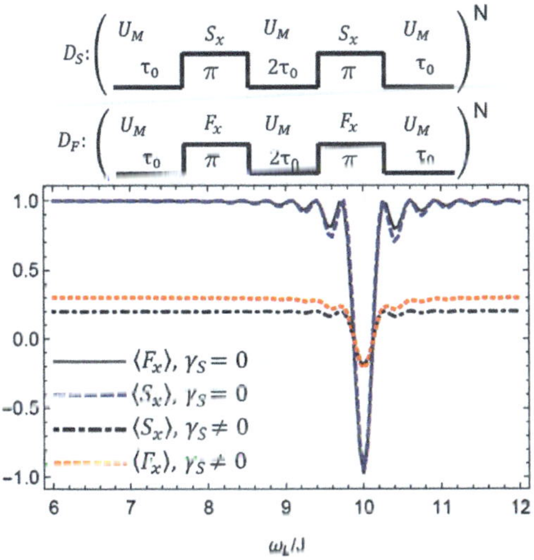

Fig. 2 Using the detection sequences D_S and D_F shown on the top of the figure we plot the expectation values of the sensor and ancillary spins $\langle S_x \rangle$ and $\langle F_x \rangle$ as a function of the Larmor frequency ω_L both in the presence and absence of relaxation. The sensor–ancillary coupling is much stronger than the sensor sample spin coupling and is taken to be $g = 80J$, and $\theta = \pi/4$. Relaxation is introduced through the Krauss form given in eqn (5) with $\gamma_S/g = 10^{-3}$. For consistency we have considered the relaxation of the sensor during the ancillary control pulses in the D_F sequence as they are assumed to be considerably slower than the manipulations on the sensor spin.

that the sample spin could also be detected by the sensor itself as its coherence is also a function of J and θ during every mapping sequence. One can see from Fig. 2 that we now have the possibility to sense the sample spin either by running the detection sequence on the sensor D_S or a detection sequence on the ancillary spin D_F, and see an almost identical transition in their observables $\langle S_x \rangle$ and $\langle F_x \rangle$, respectively. Though the difference is quite small one can find that sensing on the ancillary has a better contrast than on the sensor, and this amplifies in the presence of relaxation which we shall discuss later. The advantage of the mapping sequence becomes more evident when the sensor needs to be reinitialized quite frequently, making it ineligible to sense the weak sample spin, but the ancillary spin can still reflect the same transition independent of the sensor's reinitialization rate in the absence of any relaxation. If we now squeeze both the mapping and detection into one sequence, allowing for control only on the sensor (as they could be quite faster than controlling the ancillary) but still achieving the goal of reflecting the information of the sample spin on both the sensor and ancillary could be quite useful. For this we use a single asymmetric sequence (shown in Fig. 3) where a $\pi/2$-pulse (for mapping) and π (for detecting) are performed on the sensor during the first and second half of the sequence. Surprisingly, repeating this single sequence leads to identical dynamics on both the sensor and ancillary in the absence of relaxation. This can be seen from Fig. 3 where we have plotted the observables $\langle S_x \rangle$ and $\langle F_x \rangle$ that overlap perfectly.

We now turn to a realistic situation where the sensor is relaxing to its thermal equilibrium at a rate γ_S. Assuming a pure Markovian decay, we introduce relaxation as a depolarizing channel[23] through the Krauss form such that the sensor spin state at any time t is simply given by

$$\mathscr{E}(\rho(t)) = \frac{1}{2}(1 - e^{-\gamma_S t})\mathbb{1} \otimes \mathrm{Tr}_S(\rho(t)) + e^{-\gamma_S t}\rho(t), \tag{5}$$

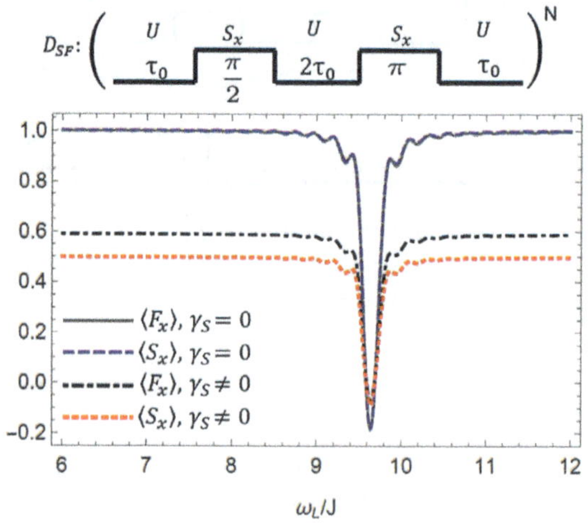

Fig. 3 Using the detection sequence D_{SF} shown on the top of the figure we plot the expectation values of the sensor and ancillary spins $\langle S_x \rangle$ and $\langle F_x \rangle$ as a function of the Larmor frequency ω_L both in the presence and absence of relaxation. The sensor–ancillary coupling is much stronger and is taken to be $g = 80J$, and $\theta = \pi/4$. Relaxation is introduced through the Krauss form given in eqn (5) with $e^{-\gamma_S \tau_0} = 10^{-3}$.

where $\rho(t)$ is the density matrix of the total system at any time t. Though the above way of introducing relaxation into the problem is simpler than solving the master equation for the total system exactly, it still could capture most of the physics relevant to the current discussion.

From the above analysis we have already seen that indirect sensing through the ancillary spin shows better contrast in the presence of relaxation even though they have equal contrast in the absence of relaxation. We now want to know if the detection through ancillary spin could be made independent of the sensor relaxation γ_S. We have already seen from Fig. 2 that by using the mapping sequence, an independent detection of the sample spin can be made through the ancillary using D_F. While running the detection sequence on the ancillary even if the sensor is reinitialized though it destroys the sensor–ancillary (S–F) correlations, the ancillary–sample (F–I) ones remain intact. Hence the availability of a fully coherent sensor spin state at the end of each repetition would map the sample spin parameters onto the ancillary with almost unit efficiency (as in the ideal case of $\gamma_S = 0$). Instead, if the reinitializations are performed at a slower rate, the relaxation effects of the sensor enter the dynamics of the ancillary, thereby reducing its contrast from the ideal value. This is shown in Fig. 4(a), where we plot the contrast observed on the ancillary spin readout as a function of the relaxation time $T_1 = 1/\gamma_S$, for different initializations of the sensor. Under very frequent

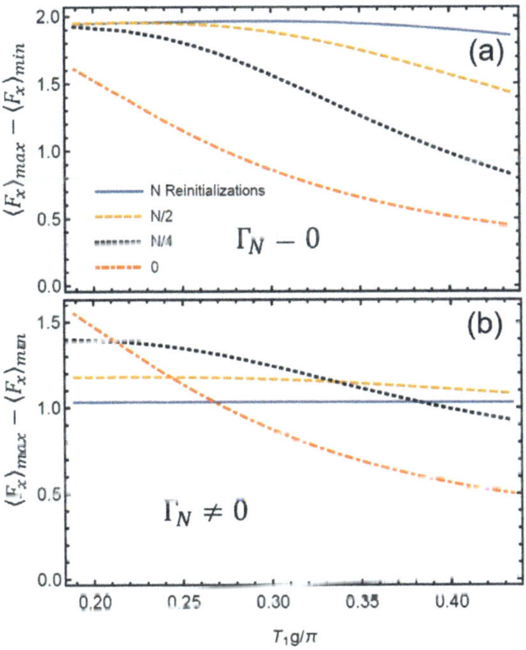

Fig. 4 The maximum contrast of the ancillary spin observable is plotted as function of the relaxation time for different reinitialization rates for two cases: (a) when the sensor reinitialization does not dephase the ancillary spin and (b) when the sensor reinitialization dephases the ancillary spin at a rate Γ_N. The dephasing Γ_N is chosen so that $e^{-\Gamma_N \tau_0} = 10^{-2}$, where the sensor–ancillary coupling and other parameters are similar to those given in earlier plots.

initializations the contrast remains independent of T_1, but for slower rates the loss of purity occurred until the reinitialization decreases the overall contrast. In a realistic situation one should also consider an additional dephasing incurred by the ancillary during the reinitialization of the sensor. This happens for example due to the difference in the hyperfine coupling between the sensor and the ancillary spin in the ground and excited states. Taking into account this additional dephasing, one immediately finds that frequent reinitialization may not be the ideal situation, which can be seen from Fig. 4(b).

We now make some numerical estimates on the new range of distant spins that the ancillary spin could sense. For NV centers in diamond one can find an ancillary spin such that the sensor–ancillary coupling is $g \approx 1$ MHz, and the sensor has a typical relaxation rate $\gamma_S = 1$ kHz. The sensitivity with which a sample spin whose coupling strength $J = 10$ kHz could be detected reduces to 17% of its maximum achievable case when $\gamma_S = 0$. Now running the detection sequence on the ancillary spin, with the possibility to make a π flip at rate g, we find that the sample spin could now be detected with an increased sensitivity of 25%. Now if we reinitialize the senor spin at a rate of $g/10$, *i.e.* the sensor is reinitialized three times in the entire sequence at every 10 μs, this increases the sensitivity to 33%. The dephasing incurred by the ancillary (nuclear) spin during the reinitialization process is proportional to the reinitialization time of the sensor (electron) spin τ_r and the hyperfine coupling between the ancillary and the sensor in the excited state g_e. While choosing strongly interacting nuclei as the ancillary spin will improve the spectral resolution of the sample spins, it equally will dephase quickly enough during frequent reinitializations of the sensor and *vice versa*. Thus finding an optimally coupled ancillary spin would become a key requirement for the indirect sensing proposed in this paper.

3 Conclusion

In conclusion we have proposed a new method to detect distant spins that are very weakly coupled to the sensor. When this coupling becomes much smaller than the relaxation rate of the sensor, direct sensing yields almost no contrast, while an indirect sensing through a long-lived ancillary spin coupled strongly to the sensor can be used to detect the same spins with a high contrast. We have shown that this kind of dual sensing could be achieved by a mapping sequence between the sensor and the ancillary so that the information related to the sample spin is stored in both the spins, and becomes detectable by either of them. Though the analysis is made on a simple model system having a single sample spin, the analysis could be straightforwardly generalized to highly dense samples with few hundreds of sample spins.

Acknowledgements

We thank Ingmar Jakobi for providing illustrative graphics, Sebastian Zaiser and Amit Finkler for useful comments and discussions. We acknowledge the financial support by the ERC project SQUTEC, DFG (FOR1693), DFG SFB/TR21, EU DIADEMS, SIQS, Max Planck Society and the Volkswagenstiftung.

References

1 F. Bloch, *Phys. Rev.*, 1946, **70**, 460–474.

2 E. M. Purcell, H. C. Torrey and R. V. Pound, *Phys. Rev.*, 1946, **69**, 37–38.

3 D. Rugar, R. Budakian, H. J. Mamin and B. W. Chui, *Nature*, 2004, **430**, 329–332.

4 C. L. Degen, M. Poggio, H. J. Mamin, C. T. Rettner and D. Rugar, *Proc. Natl. Acad. Sci. U. S. A.*, 2009, **106**, 1313–1317.

5 L. Ciobanu, D. A. Seeber and C. H. Pennington, *J. Magn. Reson.*, 2002, **158**, 178–182.

6 A. Blank and J. H. Freed, *Isr. J. Chem.*, 2006, **46**, 423–438.

7 J. Cai, A. Retzker, F. Jelezko and M. B. Plenio, *Nat. Phys.*, 2013, **9**, 168–173.

8 A. Gruber, A. Dräbenstedt, C. Tietz, L. Fleury, J. Wrachtrup and C. V. Borczyskowski, *Science*, 1997, **276**, 2012–2014.

9 F. Jelezko, T. Gaebel, I. Popa, A. Gruber and J. Wrachtrup, *Phys. Rev. Lett.*, 2004, **92**, 076401.

10 G. Balasubramanian, I. Y. Chan, R. Kolesov, M. Al-Hmoud, J. Tisler, C. Shin, C. Kim, A. Wojcik, P. R. Hemmer, A. Krueger, T. Hanke, A. Leitenstorfer, R. Bratschitsch, F. Jelezko and J. Wrachtrup, *Nature*, 2008, **455**, 648–651.

11 P. Maletinsky, S. Hong, M. S. Grinolds, B. Hausmann, M. D. Lukin, R. L. Walsworth, M. Loncar and A. Yacoby, *Nat. Nanotechnol.*, 2012, **7**, 320–324.

12 J. M. Taylor, P. Cappellaro, L. Childress, L. Jiang, D. Budker, P. R. Hemmer, A. Yacoby, R. Walsworth and M. D. Lukin, *Nat. Phys.*, 2008, **4**, 810–816.

13 G. Balasubramanian, P. Neumann, D. Twitchen, M. Markham, R. Kolesov, N. Mizuochi, J. Isoya, J. Achard, J. Beck, J. Tissler, V. Jacques, P. R. Hemmer, F. Jelezko and J. Wrachtrup, *Nat. Mater.*, 2009, **8**, 383–387.

14 T. H. Taminiau, J. J. T. Wagenaar, T. van der Sar, F. Jelezko, V. V. Dobrovitski and R. Hanson, *Phys. Rev. Lett.*, 2012, **109**, 137602.

15 S. Kolkowitz, Q. P. Unterreithmeier, S. D. Bennett and M. D. Lukin, *Phys. Rev. Lett.*, 2012, **109**, 137601.

16 N. Zhao, J. Honert, B. Schmid, M. Klas, J. Isoya, M. Markham, D. Twitchen, F. Jelezko, R.-B. Liu, H. Fedder and J. Wrachtrup, *Nat. Nanotechnol.*, 2012, **7**, 657–662.

17 T. Staudacher, F. Shi, S. Pezzagna, J. Meijer, J. Du, C. A. Meriles, F. Reinhard and J. Wrachtrup, *Science*, 2013, **339**, 561–563.

18 C. Müller, X. Kong, J.-M. Cai, K. Melentijević, A. Stacey, M. Markham, D. Twitchen, J. Isoya, S. Pezzagna, J. Meijer, J. F. Du, M. B. Plenio, B. Naydenov, L. P. McGuinness and F. Jelezko, *Nat. Commun.*, 2014, **5**, 4703.

19 A. Laraoui, F. Dolde, C. Burk, F. Reinhard, J. Wrachtrup and C. A. Meriles, *Nat. Commun.*, 2013, **4**, 1651.

20 H. J. Mamin, M. Kim, M. H. Sherwood, C. T. Rettner, K. Ohno, D. D. Awschalom and D. Rugar, *Science*, 2013, **339**, 557–560.

21 N. Bar-Gill, L. M. Pham, C. Belthangady, D. Le Sage, P. Cappellaro, J. R. Maze, M. D. Lukin, A. Yacoby and R. Walsworth, *Nat. Commun.*, 2012, **3**, 858.

22 G. S. Uhrig, *Phys. Rev. Lett.*, 2007, **98**, 100504.

23 M. A. Nielsen and I. L. Chuang, *Quantum computation and information*, Cambridge university press, 2000.

Faraday Discussions

PAPER

Towards quantum networks of single spins: analysis of a quantum memory with an optical interface in diamond

M. S. Blok,† N. Kalb,† A. Reiserer, T. H. Taminiau and R. Hanson*

Received 23rd June 2015, Accepted 3rd July 2015
DOI: 10.1039/c5fd00113g

Single defect centers in diamond have emerged as a powerful platform for quantum optics experiments and quantum information processing tasks. Connecting spatially separated nodes *via* optical photons into a quantum network will enable distributed quantum computing and long-range quantum communication. Initial experiments on trapped atoms and ions as well as defects in diamond have demonstrated entanglement between two nodes over several meters. To realize multi-node networks, additional quantum bit systems that store quantum states while new entanglement links are established are highly desirable. Such memories allow for entanglement distillation, purification and quantum repeater protocols that extend the size, speed and distance of the network. However, to be effective, the memory must be robust against the entanglement generation protocol, which typically must be repeated many times. Here we evaluate the prospects of using carbon nuclear spins in diamond as quantum memories that are compatible with quantum networks based on single nitrogen vacancy (NV) defects in diamond. We present a theoretical framework to describe the dephasing of the nuclear spins under repeated generation of NV spin-photon entanglement and show that quantum states can be stored during hundreds of repetitions using typical experimental coupling parameters. This result demonstrates that nuclear spins with weak hyperfine couplings are promising quantum memories for quantum networks.

1 Introduction

1.1 The NV-center as a quantum node

Spins associated with the nitrogen-vacancy (NV) center, an atomic defect in diamond, have recently emerged as a promising platform for quantum networks.[1–11] The NV-center's long-lived electron spin state ($S = 1$) can be controlled by magnetic resonance and can be initialized and read out optically. At cryogenic

QuTech and Kavli Institute of Nanoscience, Delft University of Technology, PO Box 5046, 2600 GA Delft, The Netherlands. E-mail: r.hanson@tudelft.nl

† These authors contributed equally to this work.

temperatures (<10 K), coherent optical transitions allow for the generation of spin-photon entanglement[12] and entanglement between spatially separated NV center electron spins.[6,13]

In addition, the electron spin couples to nuclear spins in the environment through the hyperfine interaction. Control over the host nitrogen spin and over multiple nearby ^{13}C spins has been demonstrated.[14–20] As these nuclear spins can be isolated from their environments, coherence times of more than one second have been demonstrated,[21] making them interesting candidates for quantum network memories.

1.2 Combining a quantum memory with an optical interface

A major challenge for realizing quantum memories based on nuclear spins is to overcome the dephasing that is introduced while using the electron spin as an optical interface to generate spin-photon entanglement. Consider the general case of an entanglement protocol that is inherently probabilistic due to lossy optical channels. The protocol therefore must be repeated many times in order to establish an entanglement link between adjacent network nodes. Whenever the entanglement attempt fails, the electron spin is projected in a random state. A fast and practical solution is to re-initialize the spin by optical pumping after each repetition. Because the exact time at which the electron spin is reset is uncertain (optical pumping is a stochastic process) and the electron–nuclear interaction is always present, the electron reset can cause nuclear spin dephasing (Fig. 1). A promising route to overcome this dephasing of the quantum memory is to use relatively distant ^{13}C spins with weak hyperfine couplings, which are less sensitive to fluctuations of the electron spin.

In this manuscript we explore the storage of quantum states in ^{13}C spins during the repeated generation of NV spin-photon entanglement. We first demonstrate a method to directly measure the frequency difference df for the electron-state-dependent nuclear spin precessions, which governs the nuclear dephasing (Fig. 1).[22] We then analyze the spin-photon entanglement protocol, develop a model to describe the dephasing of nuclear spins, and calculate the fidelity of the nuclear spin quantum memories with realistic coupling parameters under many repetitions of the entanglement protocol.

2 Control and characterization of nuclear spins in diamond

We start by discussing the experimental methods to control the NV center and nearby ^{13}C nuclear spins. We then introduce a Ramsey-spectroscopy method to directly determine the frequency df and characterize four candidate ^{13}C spins near a single NV-center. These experimental results highlight the universal presence of controllable nuclear spin memories and provide a realistic set of input parameters for the theoretical calculations.

2.1 Setup and electron spin control

At the heart of the experiment is a single nitrogen-vacancy (NV) center in high purity (Type IIa) single-crystal diamond grown by chemical-vapor-deposition. The diamond is held at a temperature of $T = 4.2$ K in a helium bath cryostat. The

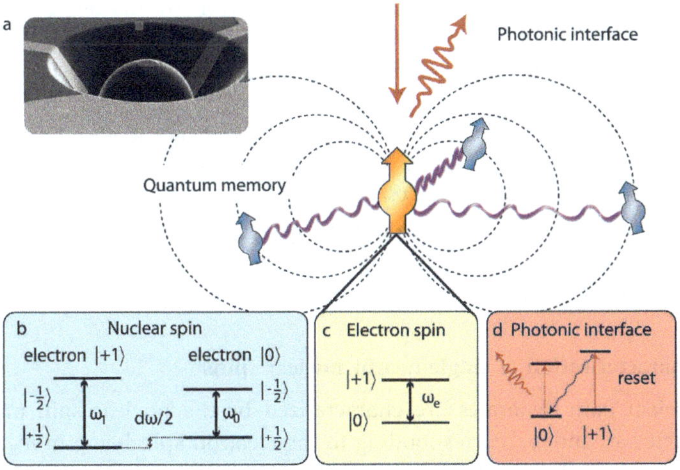

Fig. 1 The NV-center as a network node including a quantum memory. A single electron spin (orange) is coupled (purple curly arrows) to individual carbon spins (blue) *via* the magnetic dipole field (black dashed lines) associated with the electron spin. A laser beam (red straight arrow) is used to prepare and read-out the spin state by collecting the fluorescence (red curly arrow). (a) Scanning electron microscope image of the sample. A solid-immersion lens is fabricated with a focused ion beam in single-crystalline diamond for high collection efficiency. An on-chip gold stripline (bottom) enables microwave-control. (b) Level scheme for the quantum memory (^{13}C-spin, $I = 1/2$). The hyperfine coupling introduces energy level splittings (ω_0 and ω_1) that depend on the state of the electron spin, where $\omega_0 = (2\pi)\gamma_c B$, with γ_c the gyromagnetic ratio of the carbon spin and B the magnetic field. ω_1 depends on the hyperfine coupling, which is set by the distance to the electron spin. (c) Relevant ground-state energy levels of the electron spin ($S = 1$). The degeneracy of the $m_S = \pm 1$ states is lifted by applying a magnetic field along the quantization axis of the NV-center. We define the electron spin qubit in the $|0\rangle = |m_S = 0\rangle$, $|1\rangle = |m_S = +1\rangle$ manifold. Here, $B_z = (303 \pm 1)$G, leading to an energy level splitting of $\omega_e \sim (2\pi)$ 3.73 GHz. (d) Diagram of the electron spin including the relevant ground-, and excited-state levels. At low temperature the zero phonon line (~637 nm) exhibits spin-preserving optical transitions that can be addressed selectively. In the experiment, two lasers with different frequencies are used to address the E' transition for electron spin initialization (solid red) and the E_y transition for readout (fidelity = 0.93(5)) and generating spin-photon entanglement (dashed red).

diamond has a natural abundance (1.1%) of ^{13}C spins ($I = 1/2$) in an otherwise ^{12}C spin-free lattice. The NV electronic spin is polarized and measured optically by spin-selective resonant excitation.[12,23,24] To obtain high single-shot readout fidelities, a solid-immersion lens was fabricated on top of the NV center and a single-layer aluminum-oxide anti-reflective coating was deposited[13] (Fig. 1). The electronic spin is controlled by microwaves applied through an on-chip line (Rabi frequency: 3.3 MHz).

2.2 Carbon spin control by dynamical decoupling

We detect and control multiple ^{13}C nuclear spins in the spin bath surrounding the NV center using recently developed methods that coherently exploit the electron–nuclear interaction by periodically switching the electron state at well-defined times.[20,25–27] We apply sequences of the form $(\tau - \pi - 2\tau - \pi - \tau)^{\frac{M}{2}}$, where π

denotes a microwave pulse that rotates the electron by 180 degrees, 2τ is the interpulse delay and M is the total number of π-pulses. For τ precisely resonant with the electron–nuclear dynamics, this sequence imprints a phase on the electron spin conditional on the nuclear spin state.

Because the hyperfine interaction is determined by the specific position of each nuclear spin relative to the NV center, the resonance condition for τ is different for each nuclear spin. We can thus characterize the nuclear spin environment[20] by preparing the electron in a superposition state and measuring the phase that is acquired when sweeping τ. Here we select four individual ^{13}C spins to study in more detail, and design controlled gates following Taminiau *et al.*[27]

2.3 Characterization of single nearby nuclear spins

The nuclear spin dynamics are characterized by the nuclear spin precession frequencies ω_0 and ω_1, corresponding to the electron spin being in $m_S = 0$ and $m_S = 1$, respectively (see Methods section). To directly determine the frequencies ω_0, ω_1 and $df = (\omega_1 - \omega_0)/2\pi$ for each of the four nuclear spins, we perform the experimental sequence[22] shown in Fig. 2a. The electron is prepared in the state $\rho_{0,e} = |0\rangle\langle 0|$, whereas the nuclear spin state is un-polarized (mixed state

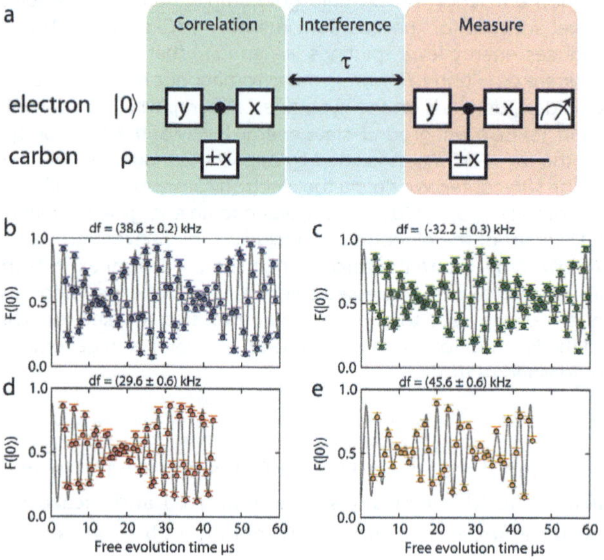

Fig. 2 Characterization of single ^{13}C spins. (a) Circuit diagram to determine ω_0 and ω_1 of individual ^{13}C spins *via* the electron spin. The conditional gates on the carbon spin are implemented using resonant dynamical decoupling techniques, as explained in the main text. Because the evolution of the carbon spin is correlated with an eigenstate of the electron spin during the interference, a coherent signal can be observed even for $\tau \gg T_{2,\text{electron}}^* = (4.18 \pm 0.01)$ μs. (b–e) The resulting interference signals measured for four individual ^{13}C spins near a single NV-center. Grey lines are fits to the data with function $F = (A/2)\cos(\omega_0\tau + \phi_0) + (B/2)\cos(\omega_1\tau + \phi_1) + c$. We find $\omega_0/2\pi = (326.0 \pm 0.2)$, (325.9 ± 0.2), (325.1 ± 0.5), and (325.9 ± 0.4) kHz (b–e), consistent with the gyromagnetic ratio of a ^{13}C spin in a field of (303 ± 1)G. For the second frequency component, we find $\omega_1/2\pi = (364.6 \pm 0.1)$, (293.7 ± 0.2), (354.7 ± 0.5), and (371.5 ± 0.4) kHz. The data is taken with 500 repetitions per data point and the error bars correspond to one standard deviation.

$\rho_{m,c} = (|0\rangle\langle 0| + |1\rangle\langle 1|)/2)$. The first set of gates correlates the electron state with the X-projection of the nuclear spin state, so that the state is $\rho_{0,e} \otimes \rho_{X,c} + \rho_{1,e} \otimes \rho_{-X,c}$, with $\rho_{\pm X,c} = |\pm X\rangle\langle \pm X|$ and $|\pm X\rangle = (|0\rangle_c \pm |1\rangle_c)/\sqrt{2}$. The controlled nuclear spin rotations are realized by the pulse sequences described above, with τ resonant for that specific spin. Second, the nuclear spin evolves freely, either with ω_0 or with ω_1, depending on the electron state. Finally, the phase accumulation of the nuclear spin is measured by correlating it to the electron spin before reading out the electron spin.

The beating observed in the signal directly yields the frequency difference df and therefore the additional phase picked up due to the time the electron spent in $m_S = +1$. For the four spins, we find d$f = (29.6 \pm 0.6)$, (-32.2 ± 0.3) (38.6 ± 0.2) and (45.6 ± 0.6) kHz, respectively. These values show that several nuclear spins with coupling strengths between approximately 20–50 kHz are readily available in diamond samples with a natural abundance of ^{13}C.

3 Modelling the dephasing of a carbon spin during entanglement generation

3.1 Sequence for generating heralded entanglement

We analyze the performance of ^{13}C spins as quantum memory in the context of the heralded entanglement protocol proposed by Barrett and Kok,[28] which was implemented by Bernien *et al.*[6] The protocol is based on the creation of spin-photon entanglement at both nodes, followed by two-photon interference and measurement of these photons. The protocol is probabilistic since it is susceptible to photon loss. Importantly, successful generation of entanglement is heralded by the detection of the two photons and thus the sequence can be repeated until successful.

Spin-photon entanglement is created using the following sequence. The electron spin is prepared in the state $|0\rangle_e$ by optical pumping (Fig. 3). Using microwaves, the electron spin is then brought in a coherent superposition. Next, the NV center is optically excited with a short laser pulse that is only resonant if the spin is in state $|0\rangle_e$. Spontaneous emission generates a photon that is entangled with the state of the spin: $|\Psi\rangle = 1/\sqrt{2}(|0\rangle_e|1\rangle_P + |1\rangle_e|0\rangle_P)$, where $|1\rangle_P$ ($|0\rangle_P$) denotes the presence (absence) of a photon. The goal is that the nuclear spin memory reliably stores quantum states during many repetitions of this sequence.

3.2 Evolution of carbon spin while generating entanglement

The performance of a quantum memory can be characterized by its ability to store an unknown quantum state $|\psi\rangle = \alpha|0\rangle_c + \beta e^{i\varphi}|1\rangle_c$. During the spin-photon entanglement sequences, the phase φ of the nuclear spin state is affected in two ways. First, when the emitted photon is lost (heralding fails), the electron spin state is randomly projected into either $|0\rangle_e$ or $|1\rangle_e$, resulting in nuclear spin evolutions with ω_0 or ω_1, respectively. Second, before the next repetition, the electron spin is reset by optical pumping, a stochastic process that introduces a distribution of spin flip times. These two effects are now analyzed separately.

After a single round of optical excitation that generates spin-photon entanglement, the electron spin is projected into an unknown eigenstate if the photon is lost. The carbon spin acquires a phase dωt if the electron is projected in $|1\rangle_e$,

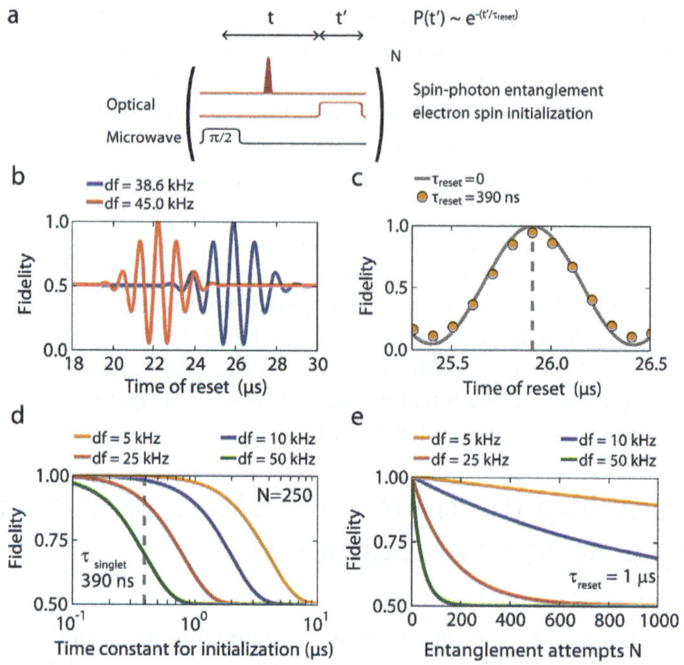

Fig. 3 Simulations of the dephasing of a ^{13}C spin quantum memory while generating entanglement. (a) Diagram for the protocol to create spin-photon entanglement. (b) Simulations of the fidelity for different ^{13}C spins after $N = 50$ repetitions of the protocol, assuming that the reset is instantaneous ($t' = 0$, formula (3.1) in the main text). The initial state of a carbon spin can be perfectly preserved by choosing the time between the $\pi/2$-pulse and the reset to $t = 2\pi/d\omega$. (c) The effect of the spin pumping process on the fidelity of the memory after $N = 50$ repetitions. The orange dots are a Monte-Carlo simulation where, for every electron spin reset, a time t' is drawn from an exponential probability distribution with $\tau_{reset} = 390$ ns. The grey line is a comparison with an ideal reset. $d\omega = (2\pi)$ 38.6 kHz. (d) Dependence of the memory fidelity on the characteristic reset time τ_{reset} using formula (3.2). (e) Dephasing of the memory as a function of entanglement attempts for different coupling strengths, fixing $\tau_{reset} = 1$ μs.

where $d\omega = 2\pi df$ and t is the time at which the reset is applied. When this process is repeated N times, the number of times k that the electron is projected in $|1\rangle_e$ is given by a binomial distribution and the final state fidelity F of the carbon spin state is given by:

$$F = \frac{1}{2} + \frac{1}{2^{N+1}} \sum_{k=0}^{N} \binom{N}{k} \cos[k d\omega t] \tag{3.1}$$

where the electron spin reset is taken to be instantaneous and we only consider that the initial memory state is $|\psi\rangle = 1/\sqrt{2}(|0_c\rangle + |1_c\rangle)$, which is most sensitive to dephasing. Fig. 3b shows the calculated fidelity as a function of the time t, for two carbon spins that were identified in Fig. 2. For each carbon spin there is a unique condition at $t = 2\pi/d\omega$, for which the phase is independent on the electron state, resulting in $F = 1$. Note that in the full entanglement protocol[6,28] an electron π-pulse is applied between rounds of excitation, so that this phase difference can be cancelled for all t.

In reality, the reset of the electron spin by spin pumping is a stochastic process involving multiple transitions to the optically excited state as well as mixing between multiple excited states. Here we model the dynamics of this process as an exponential distribution $e^{-t'/\tau_{reset}}$ with a characteristic time τ_{reset}. In Fig. 3c we compare the results of a Monte-Carlo simulation that includes the probabilistic reset time t' for $\tau_{reset} = 390$ ns with the curve of eqn (3.1). As expected, the same behavior is observed, but the maximum fidelity is reduced since the stochastic reset leads to dephasing of the carbon spin.

3.3 Dephasing of the memory including spin pumping

To analyze the effect of the electron spin reset on the nuclear spin in detail, we now assume that $t = 2\pi/d\omega$, which allows us to derive an analytical expression for the fidelity of the carbon spin (see Methods section for the derivation of this result):

$$F = \frac{1}{2} + \frac{1}{2^{N+1}}\left(1 + e^{-\frac{1}{2}\tau_{reset}{}^2 d\omega^2}\right)^N. \quad (3.2)$$

In Fig. 3d, we plot the resulting memory fidelity after 250 entanglement repetitions *versus* the reset constant τ_{reset}, for different values of d*f*. Although for an instantaneous reset ($\tau_{reset} \rightarrow 0$) the state can be perfectly preserved, a finite uncertainty in the reset time constant reduces the fidelity, with the effect being stronger for higher coupling strengths (d*f*). A natural lower limit to the reset time τ_{reset} is the slowest decay rate involved in the spin pumping process. For the NV center, this is expected to be the singlet lifetime $\tau_{singlet} \approx 390$ ns.[29] For this value, Fig. 3d predicts that for coupling strengths of d*f* < 10 kHz the state can be preserved with a fidelity of >98%, even after 250 entanglement attempts.

The reset constants currently reported in the literature are approximately 1 µs.[6] In Fig. 3e, we plot the fidelity as a function of the number of entanglement attempts for $\tau_{reset} = 1$ µs. These calculations predict that 25 repetitions of the entanglement protocol will yield a fidelity of 90.3% for the lowest coupling strength found in Fig. 2, which would already provide significant speed advantages in establishing entanglement links.[8] For coupling strengths of d*f* < 10 kHz, hundreds of repetitions become feasible. Such lower coupling strengths are available in isotopically purified diamonds.[21]

We emphasize that the model presented here does not include the detailed excited state dynamics of the spin pumping process. We expect that averaging over rapid spin flips and time spent in states with zero spin projection during these dynamics will further reduce actual dephasing. We therefore expect that our analysis sets a lower bound for the number of possible repetitions.

4. Conclusions

We have modelled the dephasing of nuclear spins quantum memories coupled to an NV electron spin that is repeatedly used to establish spin-photon entanglement. We find that nuclear spins with weak hyperfine couplings (20–50 kHz) are readily available in natural abundance diamonds. Our analysis shows that these spins can be used to store quantum states during 25 entanglement attempts with a fidelity of 90.3%, while nuclear spins in isotopically purified samples with coupling strengths below 10 kHz can even enable hundreds of repetitions. These results demonstrate that nuclear spins with weak hyperfine coupling strengths are promising quantum memories for quantum networks providing a route towards entanglement distillation and quantum repeaters.

5. Methods

5.1 Analytical derivation including the stochastic repumping process

We take the limit of $\gamma_c B \gg A_\perp$ (where $A_\perp$ is the hyperfine component perpendicular to the static magnetic field) such that the eigenstates of the ^{13}C-spin are independent of the electron and the only net effect of the electron–carbon coupling is that the carbon acquires a phase depending on the state of the electron. Choosing the rotating frame of the carbon resonant with the energy splitting for the electron in $|0\rangle_e$, the carbon state will acquire a phase $e^{id\omega t}$ for the electron in $|1\rangle_e$ (with $d\omega = 2\pi df$) and does not evolve otherwise.

We derive an expression for the maximally achievable memory fidelity. The scheme in Fig. 3a is repeated N times. Phase errors occur if the electron spin has to be reinitialized by pumping it to another spin state. During every execution of the protocol, the electron spin is projected into $|0\rangle_e$ or $|1\rangle_e$ with equal probability. The probability for k repumping events is then given by a binomial distribution:

$$P_{ek} = \frac{1}{2^N}\binom{N}{k}. \tag{5.1}$$

Every time the electron is reset from $|1\rangle_e$ into $|0\rangle_e$, the memory spin will pick up a random phase $\delta\theta = d\omega\,(t' - \tau_{\text{reset}})$, which is given by the difference between the energy levels of the carbon spin conditional on the electron spin $d\omega$ and the deviation $(t' - \tau_{\text{reset}})$ from the mean repumping time. The overall acquired phase for k repumping events is then the sum of the individual random phases. The fidelity with the initial memory state after N repetitions is thus given by:

$$F_k = \frac{1}{2}\left(1 + \cos\left[\sum_{k=0}^{N}\delta\theta\right]\right). \tag{5.2}$$

Under the assumption that the distributions for all repumping events are independent, the problem can be seen as a random walk in accumulated repumping time. Each step of this random walk is then exponentially distributed around the mean repumping time τ_{reset}. The probability distribution of the summed repumping time is given by:[30]

$$P(t') = \frac{t'^{k-1}}{\tau_{\text{reset}}{}^{k}(k-1)!}e^{-t'/\tau_{\text{reset}}} \approx \frac{1}{\sigma\sqrt{2\pi}}e^{-\frac{(t'-\mu)^2}{2\sigma^2}} \tag{5.3}$$

where we use the central limit theorem to approximate this distribution by a normal distribution with width $\sigma = \tau_{\text{reset}}\sqrt{k}$ and mean $\mu = \tau_{\text{reset}}k$, as we are interested in solutions for a large number of repumping events. The expected fidelity after N experimental runs is calculated by summing over the probability distributions for the electronic state and the corresponding accumulated repumping time:

$$\begin{aligned} F_N &= \sum_{k=0}^{N} P_{ek} \int F_k P(t')\mathrm{d}t' \\ &= \frac{1}{2} + \frac{1}{2}\sum_{k=0}^{N} P_{ek} \mathrm{e}^{-\frac{1}{2}k\tau_{\text{reset}}{}^2 \mathrm{d}\omega^2} \\ &= \frac{1}{2} + \frac{1}{2^{N+1}}\left(1 + \mathrm{e}^{-\frac{1}{2}\tau_{\text{reset}}{}^2\mathrm{d}\omega^2}\right)^N \end{aligned}$$

Acknowledgements

We thank C. Degen and M. Loretz for useful discussions. We acknowledge support from the Dutch Organization for Fundamental Research on Matter (FOM), the Netherlands Organization for Scientific Research (NWO), the Defense Advanced Research Projects Agency QuASAR program, and the European Research Council through a Starting Grant. THT is supported by a NWO VENI grant.

References

1 W. B. Gao, A. Imamoglu, H. Bernien and R. Hanson, Coherent manipulation, measurement and entanglement of individual solid-state spins using optical fields, *Nat. Photonics*, 2015, **9**, 363.
2 H. J. Kimble, The quantum internet, *Nature*, 2008, **453**, 1023.
3 D. L. Moehring, *et al.*, Entanglement of single-atom quantum bits at a distance, *Nature*, 2007, **449**, 68.
4 S. Ritter, *et al.*, An elementary quantum network of single atoms in optical cavities, *Nature*, 2012, **484**, 195.
5 J. Hofmann, *et al.*, Heralded entanglement between widely separated atoms, *Science*, 2012, **337**, 72.
6 H. Bernien, *et al.*, Heralded entanglement between solid-state qubits separated by three metres, *Nature*, 2013, **497**, 86.
7 C. H. Bennett, *et al.*, Purification of noisy entanglement and faithful teleportation *via* noisy channels, *Phys. Rev. Lett.*, 1996, **76**, 722.
8 E. T. Campbell and S. C. Benjamin, Measurement-based entanglement under conditions of extreme photon loss, *Phys. Rev. Lett.*, 2008, **101**, 130502.
9 H.-J. Briegel, W. Dür, J. I. Cirac and P. Zoller, Quantum repeaters: the role of imperfect local operations in quantum communication, *Phys. Rev. Lett.*, 1998, **81**, 5932.
10 L. Childress, J. M. Taylor, A. S. Sørensen and M. D. Lukin, Fault-tolerant quantum communication based on solid-state photon emitters, *Phys. Rev. Lett.*, 2006, **96**, 070504.
11 L. Childress and R. Hanson, Diamond NV centers for quantum computing and quantum networks, *MRS Bull.*, 2013, **38**, 134.
12 E. Togan, *et al.*, Quantum entanglement between an optical photon and a solid-state spin qubit, *Nature*, 2010, **466**, 730.
13 W. Pfaff, *et al.*, Unconditional quantum teleportation between distant solid-state quantum bits, *Science*, 2014, **345**, 532.

14 F. Jelezko, *et al.*, Observation of coherent oscillation of a single nuclear spin and realization of a two-qubit conditional quantum gate, *Phys. Rev. Lett.*, 2004, **93**, 130501.

15 M. V. G. Dutt, *et al.*, Quantum register based on individual electronic and nuclear spin qubits in diamond, *Science*, 2007, **316**, 1312.

16 P. Neumann, *et al.*, Multipartite entanglement among single spins in diamond, *Science*, 2008, **320**, 1326.

17 B. Smeltzer, J. McIntyre and L. Childress, Robust control of individual nuclear spins in diamond, *Phys. Rev. A: At., Mol., Opt. Phys.*, 2009, **80**, 050302.

18 G. D. Fuchs, G. Burkard, P. V. Klimov and D. D. Awschalom, A quantum memory intrinsic to single nitrogen-vacancy centres in diamond, *Nat. Phys.*, 2011, **7**, 789.

19 T. van der Sar, *et al.*, Decoherence-protected quantum gates for a hybrid solid-state spin register, *Nature*, 2012, **484**, 82.

20 T. H. Taminiau, *et al.*, Detection and control of individual nuclear spins using a weakly coupled electron spin, *Phys. Rev. Lett.*, 2012, **109**, 137602.

21 P. C. Maurer, *et al.*, Room-temperature quantum bit memory exceeding one second, *Science*, 2012, **336**, 1283.

22 A. Laraoui, *et al.*, High-resolution correlation spectroscopy of ^{13}C spins near a nitrogen-vacancy centre in diamond, *Nat. Commun.*, 2013, **4**, 1651.

23 P. Tamarat, *et al.*, Spin-flip and spin-conserving optical transitions of the nitrogen-vacancy centre in diamond, *New J. Phys.*, 2008, **10**, 045004.

24 L. Robledo, *et al.*, High-fidelity projective read-out of a solid-state spin quantum register, *Nature*, 2011, **477**, 574.

25 S. Kolkowitz, Q. P. Unterreithmeier, S. D. Bennett and M. D. Lukin, Sensing distant nuclear spins with a single electron spin, *Phys. Rev. Lett.*, 2012, **109**, 137601.

26 N. Zhao, *et al.*, Sensing single remote nuclear spins, *Nat. Nanotechnol.*, 2012, **7**, 657.

27 T. H. Taminiau, J. Cramer, T. v. d. Sar, V. V. Dobrovitski and R. Hanson, Universal control and error correction in multi-qubit spin registers in diamond, *Nat. Nanotechnol.*, 2014, **9**, 171.

28 S. D. Barrett and P. Kok, Efficient high-fidelity quantum computation using matter qubits and linear optics, *Phys. Rev. A: At., Mol., Opt. Phys.*, 2005, **71**, 060310.

29 M. Doherty, *et al.*, The nitrogen-vacancy colour centre in diamond, *Phys. Rep.*, 2013, **528**, 1–45.

30 M. Akkouchi, On the convolution of exponential distributions, *Journal of the Chungcheong Mathematical Society*, 2008, **21**, 4.

Faraday Discussions

PAPER

Towards witnessing quantum effects in complex molecules

T. Farrow,*[abc] R. A. Taylor[ab] and V. Vedral[abc]

Received 4th June 2015, Accepted 24th July 2015
DOI: 10.1039/c5fd00101c

Whether many-body objects like organic molecules can exhibit full quantum behaviour, including entanglement, is an open fundamental question. We present a generic theoretical protocol for entangling two organic molecules, such as dibenzoterrylene in anthracene. The availability of organic dye molecules with two-level energy structures characterised by sharp and intense emission lines are characteristics that position them favourably as candidates for quantum information processing technologies involving single-photons. Quantum entanglement can in principle be generated between several organic molecules by carefully interfering their photoluminescence spectra. Major milestones have been achieved in the last 10 years showcasing entanglement in diverse systems including ions, cold atoms, superconductors, photons, quantum dots and NV-centres in diamond, but not yet in molecules.

1 Introduction

"Quantum mechanics has only one mystery, and all of it is contained in the double-slit experiment," said Richard Feynman.[1] The image of a single particle going through two slits simultaneously and interfering with itself jars with our perception of reality, yet countless experiments have shown the effect to be as real as anything our eyes see.

Although interference has long been known classically, it is also a key signature of quantum phenomena, because it requires the quantum phase of superposed states to stay in step over the timescale of the experiment. In quantum mechanics we are of course talking about interference at the level of a single particle like a photon or an atom, or even a molecule undergoing a double-slit diffraction. Two closely related quantum phenomena are superposition and entanglement. The paradigm bench-top demonstration of quantum superposition is again the diffraction of photons or electrons that pass one-at-a-time through a double-slit and gradually build-up interference fringes. The

[a]Atomic & Laser Physics, Clarendon Laboratory, University of Oxford, OX1 3PU, UK. E-mail: tristan.farrow@physics.ox.ac.uk
[b]Oxford Martin School, Old Indian Institute, University of Oxford, OX1 3BD, UK
[c]Centre for Quantum Technologies, National University of Singapore, Singapore 117543

superposition here refers to the spatial superposition of different paths that the system can take when going through the slits to reach a particular spot at the detection screen.

One of the questions this throws up is the relationship between the notions of superposition and entanglement. To be able to talk about entanglement we need first to be able to identify two subsystems. In that experiment we can think of the paths that the electron takes as interfering, resulting from the superposition of the two paths. This can also be viewed as entanglement in the picture where the paths are seen as the entangled sub-systems (this is also known as mode-entangled). In this picture, a path containing zero electrons is still counted as a state, $|0\rangle$, and the path with 1 electron is counted as state $|1\rangle$. Hence the entangled state that arises where the electron is present equally in both paths is $|01\rangle + |10\rangle$. But this is not sufficient for entanglement. We also need to be able to measure two different complementary properties of each subsystem. In this case, this implies not only being able to measure the presence or absence of the electron in each of the paths, but also being able to measure in the complementary basis containing superpositions of zero and one electrons. This in itself is the main challenge. Looking ahead, it would be fascinating to conduct experiments that study and compare the fundamental relationship between spatial and temporal coherence, superposition, entanglement, and non-locality.[2]

Whether complex organic objects can maintain full quantum behaviour, including entanglement, is an open fundamental question. Until recently, the detection and characterisation of quantum mechanical effects has been confined to the scale of simple objects like atoms or particles (mostly over microscopic distances). That was before a series of experiments[3] in the last few years showed that quantum effects could also be observed in relatively large objects, such as proteins.[4]

The strongest manifestation of quantum coherence is usually considered to be quantum entanglement. There is a particularly strong interest in entanglement within many-body physical systems, rather than at the level of single-atoms or particles. This shift, from the small to the large, is remarkable considering that in less than one hundred years, we have gone from disbelieving entanglement, or "spooky action at a distance" as Einstein poked fun at it, to regarding it as being ubiquitous in the macroscopic world.[5–7] The overarching question, then, is in which kind of molecular structures are we most likely to find quantum coherence, and which can show interference and entanglement? And how will we know it when we see it? In other words, what are the practical ways in which we can witness entanglement? Here we suggest one.

2 Methods

2.1 Witnessing quantum properties

Quantum coherence, discord, and entanglement are quantum effects with an increasing degree of complexity. Quantum coherence, technically defined as the existence of off-diagonal elements in some pre-assigned measurement basis, is necessary in order to have entanglement.

Quantifying first-order derivatives of free energy (such as magnetization) as well as second and higher order quantities[8] (such as magnetic susceptibility and heat capacity) allow bounds to be obtained for different quantum and classically correlated states. The general logic is that we use correlation functions of the type

$$\langle \sigma_1 \otimes \sigma_2 \rangle$$

whose averages are sensitive to the nature of correlations and return different values depending on whether the state is fully separable (*i.e.* without any quantum entanglement)

$$\rho = \sum_i p_i {\rho_1}^i \otimes {\rho_2}^i$$

or if it contains quantum correlations. Note that the above state, which is considered a general disentangled state, can still contain quantum coherence within each subsystem 1 and 2. If none of the subsystems have any coherence (as defined with respect to their particular basis) then the state is merely classically correlated.

After this very brief tour of quantum entanglement we will look at one possible protocol with single organic molecules (used as qubits) to explore the kind of quantum effects that can be generated and observed.

2.2 Entangling molecules

Researchers still argue about whether quantum physics fully applies to more complex systems and where the classical-to-quantum transition occurs. Entangling two organic molecules would offer direct proof of (non-trivial) quantumness.[10]

An added motivation for testing the ability of single molecules to maintain entanglement comes from quantum information technologies, such as cluster state quantum computation[9] and quantum key distribution, where entanglement can be used to establish a communication channel.

Many of these technologies encode information into single photons and rely on traditional solid-state photon sources. Often, these have limitations in terms of reproducibility, photon-emission efficiency, indistinguishability, and high cost. All these reasons make it urgent to fully assess the technological potential, such as in entanglement of readily-available single organic molecules.

Fig. 1 illustrates a generic protocol for entangling two molecules, where they are projected into the state $|01\rangle + |10\rangle$ (designating entanglement between the internal energy levels of the two molecules) by impinging a single-photon on a beam-splitter such that it coherently interacts with both molecules.

The scheme uses path degrees of freedom (or modes) to construct a two-particle measurement where certain measurement outcomes reveal information about the combined state, but not about the individual particles.

The experiment must be constructed in such a way that no decoherence or information loss occurs that could provide information on the state of the individual molecules. The photons emitted from the two physical qubits must also be *indistinguishable*.

The ability to control emission into optical cavity modes is highly desirable as it offers a means of achieving indistinguishability without the application of molecular control fields (*e.g.* electric fields, strain).[12]

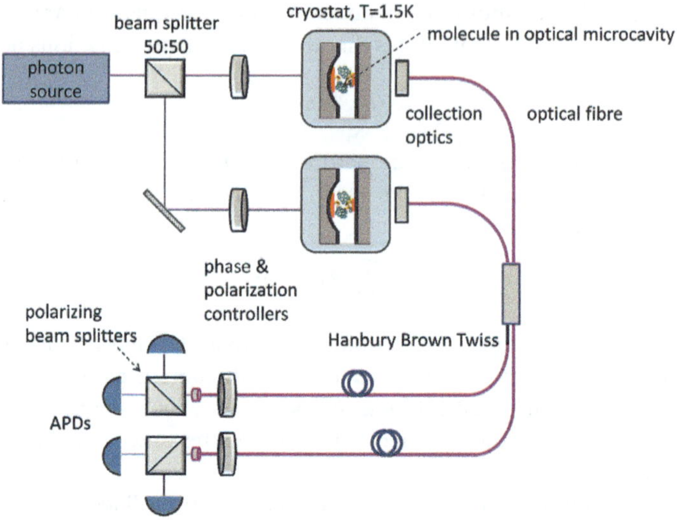

Fig. 1 Generic scheme[10] for entangling two molecules *via* interaction with single-photons. Entanglement has been realised with various systems, including large objects,[11] but not with organic molecules.

The aim of this experiment is primarily to demonstrate that entanglement between complex molecules separate by macroscopic distances (meters) is indeed possible.

The protocol comprises three steps:

The first entails generating a mode-entangled state $|01\rangle + |10\rangle$ consisting of a single photon and a vacuum state by impinging a single-photon on a 50-50 beam splitter.[13]

The second step requires the coupling of the optical mode of the microcavity to the mode-entangled state, or driving field. Treating the optical cavity as an imperfect mirror, we can model the interaction between its optical mode and the field using beam-splitter formalism where we trace over the cavity mode and the external field together, resulting in the cavity field[14]

$$\hat{\rho}_c = Tr_f\left[\hat{B}(\theta)\hat{\rho}_f|0\rangle_c\langle 0|\hat{B}^\dagger(\theta)|\right]$$

where f is the field mode injected into the cavity, $\hat{B}$ and $\hat{B}^\dagger$ are the beam splitter operators and θ is a control parameter of the mirror, *i.e.* transmissivity.

The protocol is completed by a third step whereby the cavity field and molecule are coupled. The rate of coupling is given by the Jaynes–Cummings model[15] where only linear interaction coupling is considered. This is given by the Hamiltonian

$$\hat{H} = \hbar\delta\frac{\hat{\sigma}_z}{2} + \frac{\hbar\Omega}{2}\left(\hat{a}\hat{\sigma}_+ + \hat{a}^\dagger\hat{\sigma}_-\right)$$

where δ is the detuning frequency between the driving field and the two-level molecule, $\hat{a}$ and $\hat{a}^\dagger$ are the cavity field mode ladder operators, and $\hat{\sigma}_\pm$ are the spin raising and lowering operators.

Confirmation of entanglement is possible by measuring the fluorescence fields from the two cavities. Absence of coincidence counts in the two arms of the

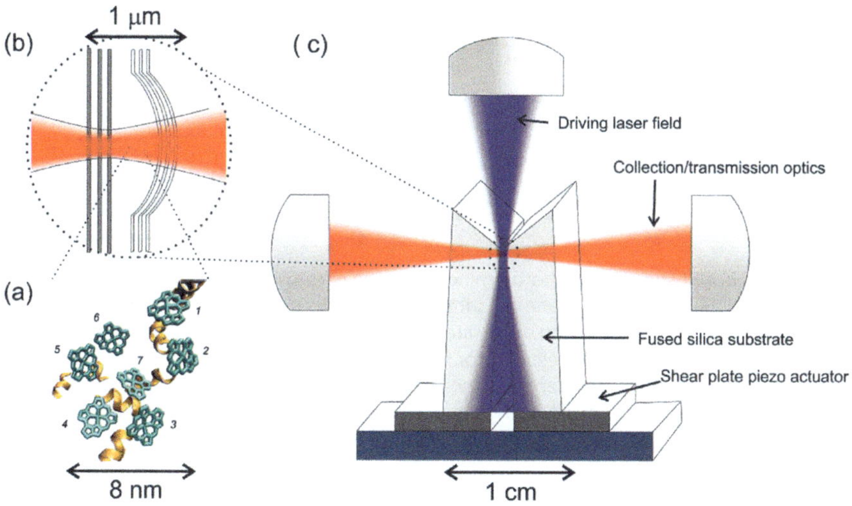

Fig. 2 Schematic of our half-symmetric optical microcavity assembly allowing laser excitation of single molecules either through a mirror into a cavity mode or in the transverse direction. A blow-up of the semi-convex geometry of the cavity is shown (top right). The optical mode is fully tuneable *in situ* with respect to its wavelength and position, as shown in experiments designed by J. M. Smith and R. A. Taylor.[24] The cavities have femtolitre mode volumes and offer high finesses, quality factors >10^4, and coupling efficiencies into detectors of ~30%.[18]

detection apparatus after the decay of the excited state following a characteristic lifetime (approximately 8 ns in the case of DBT in anthracene) finally confirms entanglement between two molecules.

2.3 Single-molecule spectroscopy

How does this translate into the laboratory? Coupling single photons to a molecular transition is hard and places experimental constraints[16] on the photon energy and polarisation, environmental noise and the brightness of the output signal. However, there is cause for optimism thanks to the development of novel tuneable Fabry–Pérot micro-cavities[17] (Fig. 2 below) into which single molecules can be embedded, promising to make the task of interrogating single molecules with single photons easier by increasing the interaction strength and light-collection efficiency.

3 Discussion

Erwin Schrödinger founded quantum mechanics believing that in practise we could never achieve a sufficient degree of experimental precision that would let us experiment on single molecules.[19] It is normal that experiments can only test some aspects of theoretical predictions, while others remain beyond technical reach, for a while. The development and startling success of single-molecule spectroscopy[20–23] in the last 20 years (recognised in last year's Nobel prize for chemistry) today enables us to select the fluorescence of individual molecules from ensembles, realizing experiments that Schrödinger once thought unlikely.

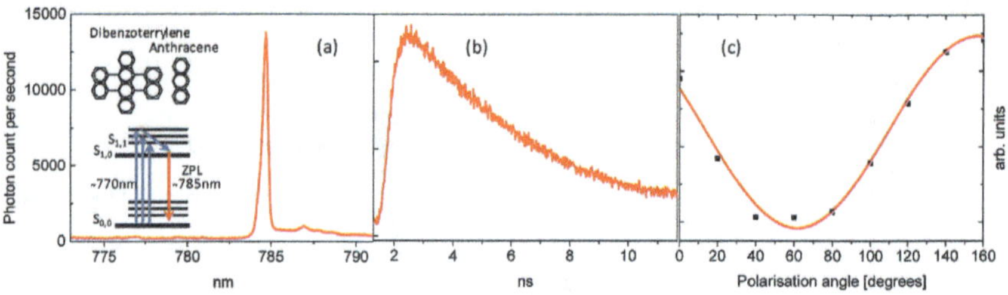

Fig. 3 (a) Bright and sharp photoluminescent peak from organic molecules (dibenzoterrylene – DBT) in an anthracene crystal at 4 K, and (b) its corresponding time-resolved PL trace with characteristic exciton decay lifetime of 5 ns. The molecule is a highly promising source of single IR photons relevant for telecommunications. (c) Polarisation control of photon emission from dibenzoterrylene. Inset in (a) shows the structure of DBT and anthracene comprising benzene rings in 2D corrugated sheets. Samples[26] of DBT were kindly provided by Prof. M. Orrit and his group at Leiden University, Holland.

Single-molecule spectroscopy offers unprecedented access to information on properties masked by the averaging of ensemble populations, and today is the cutting-edge in the study of structure, chemical kinetics, and tracking *in vivo* biological processes in single cells.[24] A series of landmark experiments[22,25] in the last few years demonstrated that single organic molecules compare favourably to quantum dots and nanocrystals as emitters of indistinguishable single photons. Besides this immediate technological benefit, organic molecules also represent one of the frontiers in the study of complex quantum states, and so are ideally married to theoretical objectives in realising entanglement in many-body systems such as two (and possibly more) molecules.

A typical example of an organic molecule that acts as a highly efficient single photon emitter is dibenzoterrylene (DBT). Our preliminary spectroscopic characterisation of DBT in an anthracene crystal is shown in Fig. 3. Optical transitions in most organic dye molecules, including DBT_{AC}, reduces to an effective two-level systems with a dark triplet state with a negligible intersystem crossing yield ($<10^{-7}$). At cryogenic temperatures, the spectral properties of DBT_{AC} in bulk exhibit a bimodal distribution of inhomogeneously broadened spectral peaks centred on 785 nm and 794 nm, corresponding to the two lattice insertion/substitution sites of DBT in anthracene.

The narrow linewidth of the Zero Phonon Line (ZPL) in single DBT molecules (20–40 MHz)[22] and fixed polarisation makes them highly attractive as sources of indistinguishable photons. Below $T \sim 2$ K, phonon modes are suppressed and transitions can be excited resonantly or quasi-resonantly by pumping either the purely electronic state $S_{1,0}$, or the $S_{1,1}$ and higher-order vibrational states. These are non-radiative short-lived states (5–10 picoseconds) that relax into the $S_{1,0}$ level before decaying fluorescently to $S_{0,0}$. This produces an intense and narrow ZPL with wavelength $\sim$785 nm or $\sim$794 nm and a bright source of photons, which makes DBT a near-ideal test-system for experiments on entanglement.

4 Conclusions

There are two motivations driving the search for entanglement. The first is fundamental. We would like to know if complex organic objects can support entanglement and, if so, under what conditions. Conventional wisdom would have it that a system consisting of a large number of subsystems (say 10^{24}, bathing in a hot environment) ought to behave fully classically. Large and hot bodies should not, then, be entangled. Yet mounting experimental evidence suggests otherwise. Studying entanglement in "messy" systems such as organic molecules (even in relatively small ones consisting of tens of atoms) is a step towards probing the quantum-to-classical transition, should any such thing exist at all. The implication is that life itself may rely on non-trivial quantum effects to survive.

The second motivation is technological since entangled molecules could act as efficient encoders of quantum bits, crucial for future applications in quantum information processing. If the power of entanglement is to benefit future quantum computers, we need to be able to handle larger and larger entangled systems[27] in terms of complexity as well as the number of quantum bits.

While atoms, superconductors, quantum dots and nano-diamonds have all been show to exhibit entanglement, a similar milestone remains to be achieved for single organic molecules. The protocol for entangling two molecules presented here, if realised, would showcase the ability of many-body organic systems to compete with their solid state counterparts as viable candidates for applications in quantum information processing.

Acknowledgements

The authors thank Professor Michel Orrit and his group at Leiden University for kindly providing samples of dibenzoterrylene and for helpful discussions. Financial support from the Oxford Martin Programme on Bio-Inspired Quantum Technologies, the Singapore Ministry of Education and National Research Foundation is gratefully acknowledged. V. V. is Fellow of Wolfson College, Oxford, and T. F. is James Martin Fellow of the Oxford Martin School.

Notes and references

1 R. Feynman, R. Leighton and M. Sands, *The Feynman Lectures on Physics*, 1966, vol. 3, p. 1.

2 S. Popescu, Nonlocality beyond quantum mechanics, *Nat. Phys.*, 2014, **10**, 264–270.

3 T. Brixner, *et al.*, Two-dimensional spectroscopy of electronic couplings in photosynthesis, *Nature*, 2005, **434**, 625–628; G. S. Engel, *et al.*, Evidence for wavelike energy transfer through quantum coherence in photosynthetic systems, *Nature*, 2007, **446**, 782–786.

4 T. Farrow and V. Vedral, Scale-estimation of quantum coherent energy transport in multiple-minima systems, *Nature Sci. Rep.*, 2014, **4**, 5520; R. Dorner, J. Goold, L. Heaney, T. Farrow and V. Vedral, Effects of quantum coherence in metalloprotein electron transfer, *Phys. Rev. E: Stat. Phys., Plasmas, Fluids, Relat. Interdiscip. Top.*, 2012, **86**, 031922.

5 V. Vedral, Living in a quantum world, *Sci. Am.*, 2011, **304**, 38–43; V. Vedral, *Nature*, 2008, **453**, 1004–1007; V. Vedral, Quantum physics, Entanglement hits the big time, *Nature*, 2003, **425**, 28; K. C. Lee, *et al.*, *Science*, 2011, **334**, 1253–1256.

6 T. Farrow and V. Vedral, Classification of macroscopic quantum effects, *Opt. Commun.*, 2015, **337**, 22–26.

7 S. Nimmrichter and K. Hornberger, Macroscopicity of Mechanical Quantum Superposition States, *Phys. Rev. Lett.*, 2013, **110**, 160403.

8 M. Wiesniak, V. Vedral and C. Brukner, Heat capacity as an indicator of entanglement, *Phys. Rev. B: Condens. Matter Mater. Phys.*, 2008, **78**, 064108; M. Wiesniak, V. Vedral and C. Brukner, Magnetic Susceptibility as Macroscopic Entanglement Witness, *New J. Phys.*, 2005, **7**, 258.

9 R. Raussendorf and H. Briegel, *Phys. Rev. Lett.*, 2001, **86**, 5188; S. C. Benjamin, *et al.*, *Laser Photonics Rev.*, 2009, **3**, 556.

10 M. Paternostro, W. Son and M. S. Kim, Complete Conditions for Entanglement Transfer, *Phys. Rev. Lett.*, 2004, **92**, 197901.

11 L.-M. Duan, *Nature*, 2001, **414**, 413; C. W. Chou, *et al.*, *Nature*, 2005, **438**, 828; T. Chanelière, *et al.*, *Nature*, 2005, **438**, 833; N. Sangouard, C. Simon, H. de Riedmatten and N. Gisin, *Rev. Mod. Phys.*, 2011, **83**, 33.

12 A. Auffeves, *et al.*, Pure emitter dephasing: A resource for advanced solid-state single-photon sources, *Phys. Rev. A: At., Mol., Opt. Phys.*, 2009, **79**, 053838.

13 M. S. Kim, W. Son, V. Bužek and P. L. Knight, Entanglement by a beam splitter: Nonclassicality as a prerequisite for entanglement, *Phys. Rev. A: At., Mol., Opt. Phys.*, 2002, **65**, 032323.

14 W. Son, M. S. Kim, J. Lee and D. Ahn, Entanglement transfer from continuous variables to qubits, *J. Mod. Opt.*, 2002, **49**(10), 1739–1746.

15 E. T. Jaynes and F. W. Cummings, Comparison of quantum and semiclassical radiation theories with application to the beam maser, *Proc. IEEE*, 1963, **51**(1), 89–109.

16 Y. L. A. Rezus, *et al.*, Single-photon spectroscopy of a single molecule, *Phys. Rev. Lett.*, 2012, **108**, 093601; G. Wrigge, *et al.*, *Nat. Phys.*, 2008, **4**, 60; G. Zumofen, *et al.*, *Phys. Rev. Lett.*, 2008, **101**, 180404; I. Gerhardt, *et al.*, *Phys. Rev. Lett.*, 2007, **98**, 033601.

17 P. R. Dolan, *et al.*, Femtoliter tunable optical cavity arrays, *Opt. Lett.*, 2010, **35**, 3556; Z.-Y. Di, *et al.*, Controlling the emission from semiconductor quantum dots using ultra-small tunable optical microcavities, *New J. Phys.*, 2012, **14**, 103048; D. M. Coles, *et al.*, Strong coupling between chlorosomes of photosynthetic bacteria and a confined optical cavity mode, *Nat. Commun.*, 2014, **5**, 5561.

18 S. C. Benjamin, *et al.*, Prospects for measurement-based quantum computing with solid state spins, *Laser Photonics Rev.*, 2009, **3**, 556.

19 E. Schrödinger, *Br. J. Philos. Sci.*, 1952, **33**, 233.

20 *Single-Molecule Optical Detection, Imaging and Spectroscopy*, ed. T. Basch, W. E. Moerner, M. Orrit and U. P. Wild, VCH, Germany, 1997.

21 W. E. Moerner and M. Orrit, Illuminating Single Molecules in Condensed Matter, *Science*, 1999, **283**, 1670.

22 W. E. Moerner, A Dozen Years of Single-Molecule Spectroscopy in Physics, Chemistry, and Biophysics, *J. Phys. Chem. B*, 2002, **106**(5), 910–927.

23 W. E. Moerner and L. Kador, Optical Detection and Spectroscopy of Single Molecules in a Solid, *Phys. Rev. Lett.*, 1989, **62**(21), 2535; M. Orrit and J. Bernard, Single pentacene molecules detected by fluorescence excitation in a *p*-terphenyl crystal, *Phys. Rev. Lett.*, 1990, **65**, 2716–2719; P. Navarro,

Stable single molecules for quantum optics and all-optical switches, PhDthesis, 2014, Leiden University.

24 S. Weiss, Fluorescence spectroscopy of single biomolecules, *Science*, 1999, **283**(5408), 1676–1683; M. Sotomayor and K. Schulten, Single-Molecule Experiments *in Vitro* and *in Silico*, *Science*, 2007, **316**(5828), 1144–1148.

25 B. Lounis and W. E. Moerner, Single Photons on Demand from a Single Molecule at Room Temperature, *Nature*, 2000, **407**, 491–493; R. Lettow, Realization of two Fourier-limited solid-state single-photon sources, *Opt. Express*, 2007, **15**(24), 15842; V. Ahtee, *et al.*, Molecules as Sources for Indistinguishable Single Photons, *J. Mod. Opt.*, 2009, **56**, 161; C. Toninelli, *et al.*, Near-infrared single-photons from aligned molecules in ultrathin crystalline films at room temperature, *Opt. Express*, 2010, **18**(7), 6577–6582; R. Lettow, *et al.*, Quantum Interference of Tunably Indistinguishable Photons from Remote Organic Molecules, *Phys. Rev. Lett.*, 2010, **104**, 123605.

26 A. A. L. Nicolet, *et al.*, Single Dibenzoterrylene Molecules in an Anthracene Crystal: Spectroscopy and Photophysics, *ChemPhysChem*, 2007, **8**, 1125.

27 R. Raussendorf and H. J. Briegel, A one-way quantum computer, *Phys. Rev. Lett.*, 2001, **86**, 5188–5191.

Single single molecules for quantum optics and ultrafast [illegible], [illegible], [illegible].
[illegible] spectroscopy of single [illegible] molecules, [illegible] and [illegible], Single Molecule [illegible] [illegible].
[illegible] Single [illegible] on [illegible] [illegible]
[illegible]
[illegible] single-photon [illegible]
[illegible]
[illegible]
[illegible] single [illegible] [illegible]
[illegible]
[illegible]
[illegible]
[illegible]
[illegible]
[illegible]

Faraday Discussions

PAPER

Near-field Raman spectroscopy of nanocarbon materials

Zachary J. Lapin,[a] Ryan Beams,[b] Luiz Gustavo Cançado[c] and Lukas Novotny†[a]

Received 1st May 2015, Accepted 26th May 2015
DOI: 10.1039/c5fd00050e

Nanocarbon materials, including sp^2 hybridized two-dimensional graphene and one-dimensional carbon nanotubes, and sp^1 hybridized one-dimensional carbyne, are being considered for the next generation of integrated optoelectronic devices. The strong electron–phonon coupling present in these nanocarbon materials makes Raman spectroscopy an ideal tool to study and characterize the material and device properties. Near-field Raman spectroscopy combines non-destructive chemical, electrical, and structural specificity with nanoscale spatial resolution, making it an ideal tool for studying nanocarbon systems. Here we use near-field Raman spectroscopy to study strain, defects, and doping in different nanocarbon systems.

1. Introduction

Nanocarbon materials, such as carbon nanotubes and graphene, are being considered as a platform for next-generation integrated optoelectronics. Similar to semiconductor device technology, the electronic properties of nanocarbon materials can be engineered by introducing dopants and exploiting interfaces with other materials. For example, in semiconductors, electron concentration and electrical conductivity can be varied over six orders of magnitude by varying the doping densities.[2] Furthermore, optical emission time constants and quantum efficiencies can be varied over four to five orders of magnitude by varying the defect concentrations. Virtually every practical application of semiconductors relies on the effects of dopants, defects, and interfaces. Similarly, it can be expected that defects and dopants will have a strong influence on future nanocarbon devices. For example, graphene transitions from conducting to semiconducting when it is cut into a narrow ribbon, with a bandgap that depends on the ribbon width, the edge chirality, and the nature of the edges, demonstrating

[a]Photonics Laboratory, ETH Zürich, 8093 Zürich, Switzerland
[b]Material Measurement Laboratory, National Institute of Standards and Technology, Gaithersburg, MD, 20899, USA
[c]Departamento de Física, Universidade Federal de Minas Gerais, Belo Horizonte, MG 30123-970, Brazil
† http://www.photonics.ethz.ch.

the strong influence of localized states on the material transport properties. Any nanocarbon-based device engineering relies on a solid understanding of the electronic, structural, and optical properties of the defects or dopants. The ability to zoom in on single defects and measure local physical properties is an important step in realizing these technologies.

A variety of high spatial resolution techniques have been applied for the systematic study of nanocarbon materials.[3] For example, transmission electron microscopy (TEM) has been used to determine nanotube structures[4] and low-temperature scanning tunneling microscopy (STM) has been applied to investigate the electronic density of states.[5] However, none of these techniques provides the ability to simultaneously probe the electronic, chemical, and structural properties of nanocarbons. Raman scattering, on the other hand, uses non-destructive optical excitation to probe characteristic vibrational modes, which are directly linked to unique structural and electronic material properties.[6–10] However, the spatial resolution of Raman scattering is, traditionally, diffraction limited to roughly $\lambda/2$, λ being the wavelength of the exciting radiation, making it difficult to spatially isolate and investigate the effect of defects and dopants on the local material properties.

Here, we use near-field Raman scattering to study local features in different nanocarbon materials, including graphene, carbon nanotubes, and carbyne. Near-field Raman scattering, also referred to as tip-enhanced Raman scattering (TERS), is a spectroscopic technique that makes it possible to perform vibrational analysis with nanoscale spatial resolution, typically in the range of 10–20 nm.[11–13] It uses a laser-irradiated metal tip acting as an optical antenna.[14] The antenna localizes and enhances the incident radiation at its apex, thereby serving as a highly-confined optical excitation source for Raman scattering (see Fig. 1). A near-field hyperspectral image of a sample is acquired by raster-scanning the sample below the tip and recording, pixel-by-pixel, a Raman scattering spectrum. The technique has been used to characterize local strain and structural defects in carbon nanotubes[15,16] and graphene,[17] and to investigate the influence of the local environment on the optoelectronic properties.[18–21] We have extended the technique beyond imaging to measure phonon correlation lengths in graphene samples.[22] Recently, TERS combined with STM has pushed optical spectroscopy to the sub-molecular level, demonstrating atomic-scale resolution.[23]

This article is organized as follows: in Section 2 we introduce the experimental geometry and methodology used in our near-field studies of nanocarbon materials. Sections 3–5 introduce the TERS findings of two-dimensional graphene (Section 3), one-dimensional carbon nanotubes (Section 4), and one-dimensional linear carbon chains (Section 5). Finally, we conclude by discussing the importance of TERS measurements and methodologies in the development of nanocarbon optoelectronic devices in Section 6.

2. Experimental techniques

Our near-field Raman experiments are performed with a home-built microscope,[12,24] schematically shown in Fig. 1. The setup consists of a scanning probe microscope positioned on top of an inverted confocal microscope. First, a continuous-wave laser beam is converted to a radially polarized mode to ensure a

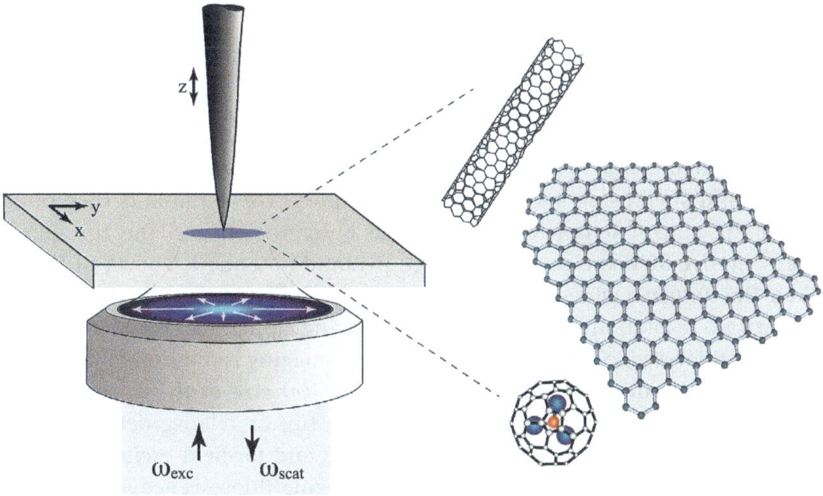

Fig. 1 Schematic diagram of the experimental setup. A radially polarized laser excitation is focused by a high-numerical aperture objective onto a sample from below. An optical antenna, attached to a piezo-electric tuning fork, is positioned into the center of the optical focus and is held in close proximity to the sample surface with shear-force feedback. A scattered optical signal is collected by the same objective, filtered by appropriate interference filters, and detected by either a single photon-counting detector or a spectrograph equipped with a liquid nitrogen cooled CCD. Simultaneous optical and topographic images are formed by raster-scanning the sample between the objective and the tip.

strong longitudinal field component at the laser focus.[25] It is then sent into the inverted microscope where it is reflected by a dichroic beam-splitter and focused from below onto the sample surface by a high-numerical aperture objective. The scanning probe microscope is outfitted with an optical antenna as the probe. The antenna is positioned in the center of the optical focal spot and held in close proximity above the sample surface with a shear-force feedback system[26,27] with a force sensitivity of a few pico-newtons. The laser-irradiated antenna (*e.g.* tip, pyramid, nanoparticle) localizes incident radiation to a subwavelength-sized spot, thereby generating a nanoscale excitation source for local Raman scattering with the sample surface. The locally scattered field is collected by the same objective lens, filtered by a combination of interference filters, and then detected by either a single photon-counting detector or a spectrograph equipped with a liquid nitrogen cooled CCD. The sample is mounted on a $40 \times 40\ \mu m^2$ closed-loop piezo scan-bed. Simultaneous optical and topographic images are formed by raster-scanning the sample in the horizontal plane between the objective and the tip. The control electronics consist of a modified controller for STM. It is used for controlling the scan-bed, tip-sample separation, and image acquisition. The entire setup is placed inside a light-free, acoustically damped enclosure located on a floating optical table.

We have demonstrated the applicability of a range of different optical antenna geometries for near-field microscopy, including a sharp metallic wire,[24,28] a single gold particle,[29,30] cascaded gold particles,[31] and highly reproducible pyramid optical antennas.[32] The pyramid antennas, due to their reproducibly high resolution and large field enhancements, are the geometry we are currently using and will be briefly discussed. The pyramids are prepared by evaporating gold or silver onto a silicon template with inverted pyramidal pits, generally with a base of 20 × 20 μm^2, generating metallic pyramid shells. Following a lift-off of the interstitially deposited metal, the end of a thin tungsten wire is attached to the inside of the pyramid shell using a standard two-part epoxy. The pyramid is then pulled out of the silicon template, exposing the outer region and the tip apex. A representative SEM image of a prepared pyramid antenna is shown in Fig. 2a, where the tungsten wire extends downwards. As shown in Fig. 2b, we determine the spatial resolution of the complete imaging apparatus based on single-molecule fluorescence images. A typical spatial imaging resolution for a pyramid antenna is ~20 nm, and depends on the physical size of the tip apex. Additionally, the fluorescence enhancement of the tip, or relative detected signal intensity between a near-field excited molecule and far-field excited molecule, can be mapped by monitoring the single-molecule fluorescence intensity as a function of the tip-molecule separation. Typical fluorescence enhancement factors are on the order of 50–100-fold, corresponding to a localized electric field enhancement on the order of 10.

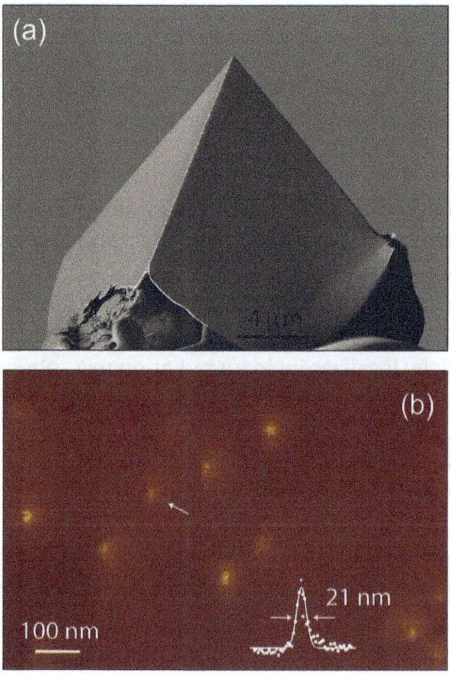

Fig. 2 The pyramid antenna and near-field resolution. (a) An SEM image of a gold pyramid tip. The base of the pyramid is 20 × 20 μm^2 and the tip apex is on the order of 10 nm. (b) A near-field single-molecule fluorescence image used to characterize the local field confinement of the pyramid antenna. The inset shows a cross section of the single-molecule marked with the white arrow. There is an imaging resolution of ~21 nm.

3. Graphene

Raman spectroscopy has been extensively applied to study the material properties of graphene, such as defect density, carrier concentration, strain, and number of layers.[33–39] As such, TERS is an ideal technique for characterizing nanoscale graphene devices. Recently, the high spatial resolution and sensitivity of TERS have been used to measure Raman bands at edges and local strain in graphene.[17,40,41] Additionally, TERS has non-imaging applications where we have utilized the optical antenna as a nanometric, positionable, illumination source to measure the correlation length of optical phonons.[22]

Being an sp^2 carbon system, the Raman spectrum of graphene has three primary features.[1] The most prominent band for single-layer graphene, known as the G′ or 2D band (~2700 cm^{-1}), belongs to the totally symmetric irreducible representation A'_1 and is readily observable in all sp^2 carbon systems.[1] The G′ band is commonly used to determine the number of graphene layers. The second strongest band in pristine graphene is the double degenerate G band (~1600 cm^{-1}) with E_{2g} symmetry in the irreducible representation.[38,42] Both the G and G′ bands are present everywhere on a graphene flake and they are often used to characterize the strain and doping in a graphene flake.[38] The final primary feature is the disorder-induced D band (~1350 cm^{-1}), which has the same symmetry as the G′ band.[38,42] Unlike the G and G′ bands, the D band requires a lattice defect to be Raman active and is therefore highly localized to edges with armchair chirality in pristine graphene flakes.[37,39,40,43,44]

The localized nature of the D band is determined by the phase-breaking length of carriers at optical energies (~1–3 eV), which is important for determining the carrier mobility in optoelectronic devices. The influence of edges is particularly important for the electronic properties of graphene nanoribbons.[45] Understanding and controlling the impact of defects on the electronic properties is crucial for making further progress towards graphene-based optoelectronic devices, such as transparent displays and high-speed photodetectors.

Far-field and near-field Raman intensity images of the G′ band from a graphene flake are shown in Fig. 3a and b, respectively. The color contrast in Fig. 3a is scaled by ×2 relative to Fig. 3b. This data was acquired using a gold pyramid tip illuminated with a 632.8 nm laser. To further characterize the graphene flake, a near-field spectrum was acquired at the edge (square in Fig. 3b) and in the center (circle in Fig. 3b) of the flake, plotted in Fig. 3c in black and red, respectively. The Raman spectra indicate that the flake is single-layer and pristine since the D band is highly localized to the edge.

Fig. 4 shows a hyperspectral line-scan across the edge of another graphene flake. This data was acquired with a 632.8 nm laser and a silver pyramid tip. The sample was scanned below the illuminated tip and a complete Raman spectrum was acquired at each spatial location. The resulting colormap of the near field Raman intensities is presented in Fig. 4a. This illustrates that the G and G′ bands are continuous over the flake, whereas the D band is highly localized at the edge (white arrow). The spectra were fit with single Lorentzians and the resulting amplitudes are plotted in Fig. 4b and c. Fig. 4b shows that the D band is highly localized at the edge. The measured width is due to a convolution of the tip resolution with a D band spatial extent of ~4 nm.[37,40,43,44] The G and G′ bands are

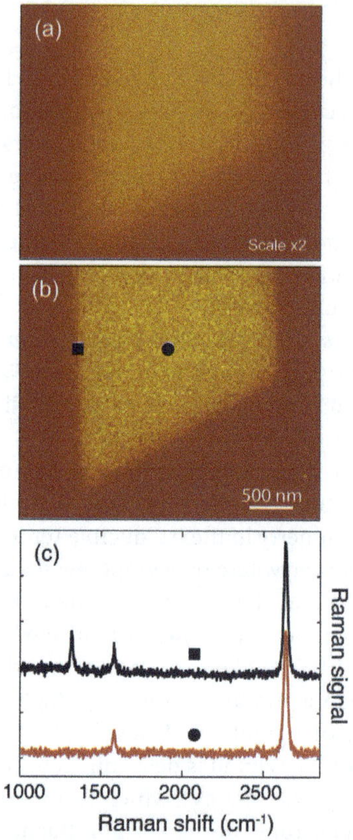

Fig. 3 Near-field Raman imaging of graphene. (a) Far-field and (b) near-field intensity images of the G′ band. The color contrast in (a) is scaled by ×2. (c) Near-field Raman spectra acquired at the edge (black) and center (red) of the flake as indicated by the square and circle in (b), respectively.

shown in Fig. 4c as blue circles and red squares, respectively. The amplitudes of the G and G′ bands are continuous over the flake and sharply decrease as the tip transitions off the flake.

While these measurements were concerned with edges in pristine graphene, the demonstrated resolution and signal enhancement illustrate that TERS is capable of measuring single defects and local strains in a two-dimensional graphene device. Both of these will be necessary as graphene devices transition from academic demonstrations to industrial products.

4. Carbon nanotubes

Carbon nanotubes are formed by rolling graphene into a tubular structure. As such, carbon nanotubes are still sp^2 systems and have Raman active modes that are similar to graphene. Single walled carbon nanotubes (SWNTs) present a combination of two properties that are ideal for near-field Raman experiments: (i) a strong Raman response (strong electron–phonon coupling combined with resonance matching in the optical absorption/emission), and (ii) a truly one

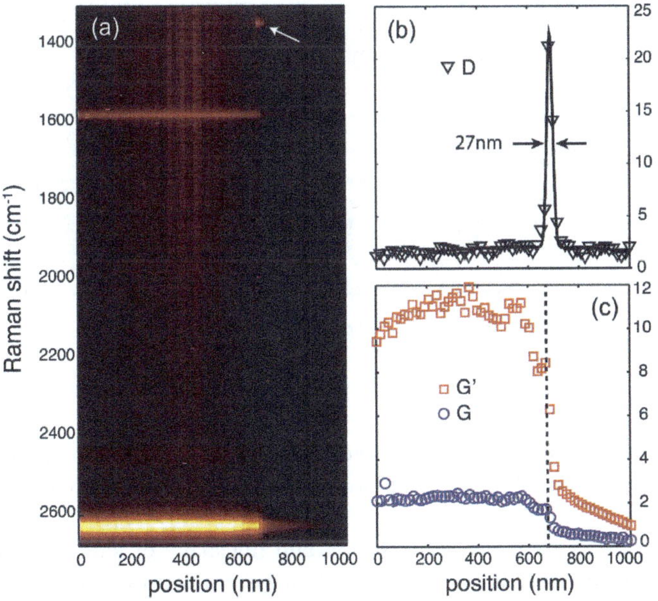

Fig. 4 Hyperspectral line-scan across a graphene flake edge. (a) Colormap of the near-field Raman intensities acquired by scanning the tip from the center of the flake (0 nm) across the edge (~675 nm) and onto the glass substrate. The white arrow indicates the D band localized at the edge. (b and c) Plots of the amplitudes of the D (black triangles), G (blue circles), and G′ (red squares) bands.

dimensional character and nearly one-dimensional physical geometry that can be correlated with the optical response.[1] For this reason, carbon nanotubes are often chosen as prototypes for near-field Raman studies and were naturally involved in the early demonstration of the TERS effect.[12] In this section, we discuss how the innate applicability of the TERS effect on carbon nanotubes advanced the understanding of sp^2 nanocarbon photophysics.

Using near-field Raman scattering, we are able to resolve localized features in the nanotube structure with a resolution of 10–20 nm. We have observed that the intensity of Raman lines along many nanotubes is non-uniform and we tentatively assigned this observation to defects in the different tube structures; however, it was suggested that the observation could be related to interactions of the tube with the supporting substrate. A series of experiments finally confirmed that the substrate has only a minor effect on the Raman lines[21] and that the observed localization of Raman modes was indeed due to nanotube inhomogeneities. The strongest support for these conclusions comes from measurements that were performed on nanotubes grown by different methods but using the same substrates. We found that some tubes are uniform over many tens of microns in length, whereas other tubes show a high density of localized features. Generally, nanotubes grown by the chemical vapor deposition (CVD) method showed the lowest defect density. So-called serpentine nanotubes can be grown by using miscut quartz surfaces.[46] During CVD growth, the nanotubes wind up in the form of serpentines, as shown in Fig. 5. The high quality of the serpentine nanotubes allowed us to quantitatively study the mechanism of near-field Raman

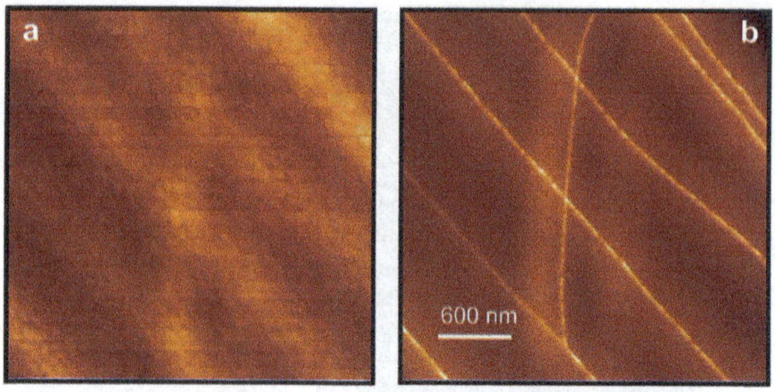

Fig. 5 (a) Confocal Raman image (G band intensity) of CVD grown SWNTs. (b) Corresponding near-field Raman image.

enhancement.[13] Our studies revealed that the near-field Raman intensity is inversely proportional to the 11th power of the separation between the tip and the nanotube.

Analogous to graphene, the position and lineshape of Raman modes reveal much about the structure of the sp^2 carbon system. For example, the electronic properties of SWNTs strongly depend on the nanotube chirality, and Raman spectroscopy is one of the primary techniques used to probe the large variety of possibilities.[1] The tube chirality determines if the nanotube is metallic or semiconducting as well as the optical transition energies. Therefore, the structural characterization (chirality determination) of carbon nanotubes is extremely important for optoelectronic applications of carbon nanotubes. The chirality of the nanotube is determined by the radial breathing mode (RBM) frequency and resonance excitation profile.[1]

The high resolution and sensitivity of TERS allow for the local characterization of SWNT junctions generated by abrupt chirality changes. First, the junction itself is a structural defect, which can be detected by the presence of the disorder-induced D band.[47] Second, changes in the RBM frequency, which strongly depends on the tube diameter and consequently on the tube chirality, can be monitored. Third, the lineshape of the bond-stretching G band is different for metallic and semiconducting nanotubes. Combined, we have used TERS to provide spectroscopic evidence for the existence of local semiconducting–metal SWNT junctions.[19]

Local semiconducting–metal transitions in SWNTs can also be induced by pressure.[48,49] This property was explored in ref. 50, where the authors used TERS to measure the local Raman response of an X-shape crossing of two carbon nanotubes. By comparing the G band shape in the TERS spectra taken at the crossing point and far from it, the authors concluded that the upper tube suffered a semiconducting–metal transition due to a structural deformation (applied pressure) caused by the presence of the tube underneath. The effects of stresses on a nanotube were also explored in ref. 51, where the authors applied controlled forces on an isolated semiconducting nanotube using the near-field probe itself. The authors observed local changes in the G band lineshape and intensity as a function of applied pressure on the nanotube, corresponding to the nanometric deformation of the nanotube itself. In ref. 52, the same group used TERS to measure the effects of strain in SWNTs. The authors used an atomic force microscopy probe to apply lateral forces on isolated SWNTs. Again, TERS was used to extract detailed information regarding the induced mechanical deformations, and therefore the generated strain, by observing shifts in the G band frequency. Besides detecting mechanical distortions, near-field Raman spectroscopy is also very sensitive to local chemical environments, and a recent study has demonstrated chemical variations in SWNTs by TERS with $\sim$2 nm resolution.[53]

Additionally, the effect of defects can be detected by spectral shifts in the Raman bands. Because electrons and phonons are strongly coupled in sp^2 carbon systems, a defect can cause renormalization of electron and phonon energies. For defect-free and undoped SWNTs, the G′ band consists of a single peak centered at $\nu_{G'} = 2676\ cm^{-1}$ (for a 632.8 nm excitation). Upon doping, however, we observe that a new peak appears at a lower/higher frequency for n/p doping. This new peak is slightly shifted from the main G′ peak and the shift is representative of an increase (decrease) in electron velocity near a negatively (positively) charged defect.[16]

5. Carbyne

The newest allotrope to the carbon revolution is one-dimensional carbyne. Carbyne is an infinite linear chain of carbon atoms which, unlike the other discussed carbon allotropes, is characterized by sp^1 hybridized bonds. Although it was proposed originally in the 1960s,[54–57] its realization was subject to much controversy.[58,59] The first indisputable proof of linear carbon chains (LCCs) came from polyynes, or short LCCs that are stabilized by large end-capping groups.[60] Now, it has been demonstrated that carbon nanotubes can be utilized both as nanoreactors for the growth of LCCs and to protect the chain from the ambient environment.[61] Using an optimized fabrication methodology, long linear carbon chains (LLCCs), consisting of thousands of carbon atoms, have been grown inside double walled carbon nanotubes (DWCNTs, with the whole system referred to as LLCCs@DWCNTs),[62] which will be discussed here.

LCCs contained within carbon nanotubes are still a new system with potentially contradicting theoretical and experimental findings. Of primary concern for both Raman studies as well as applications in optoelectronics are Peierls distortions.[63] Theoretical studies have shown that there is a slight energetic preference for an LCC to have a polyynic (alternating single–triple bonds) conformation in free space, meaning that the chain is both semiconducting and Raman active; however, upon insertion into a nanotube, theoretical studies on the bond length alternation (BLA) vary, meaning that it is unclear whether LCCs@CNTs are, in general, metallic or semiconducting.[64–66] Despite these calculations, we find there to be a strong resonance Raman response for LLCCs@DWCNTs,[62] although it is possible that there are additional, undetected, LCCs with zero BLA. This strong Raman response, combined with the well understood nanotube Raman response, makes Raman spectroscopy an ideal tool to study the physical and electrical properties of the combined system.

Theoretical calculations have reproduced the Raman shifts observed in studies of chemically synthesized polyynes of known size, showing that longer chains have a lower-energy Raman shift.[67] Thus far, far-field ensemble measurements have been used to characterize the Raman response of the LLCCs@DWCNTs and have supported this theory; however, ensemble averaging combined with low spatial resolution has made it impossible to directly measure individual chain lengths and corresponding spectra.[62] As shown in Fig. 6, we have used TERS to resolve individual LLCCs that are hosted inside a single DWCNT.

Near-field hyperspectral imaging has revealed a complex local interaction between the LLCCs and the enclosing DWCNT. Fig. 6a shows the superposition of a near-field Raman image (color) of an isolated DWCNT containing multiple LLCCs and its corresponding topographic image (grayscale). We are able to measure the physical length of the individual chains (200–800 nm) and correlate it with the respective Raman shifts (1804–1786 cm^{-1}) across a variety of samples. A near-field hyperspectral image allows us to simultaneously visualize the Raman response of the LLCC and the DWCNT. The spectra shown in Fig. 6b, corresponding to positions just to the left of the diamond, circle, and square in Fig. 6a,

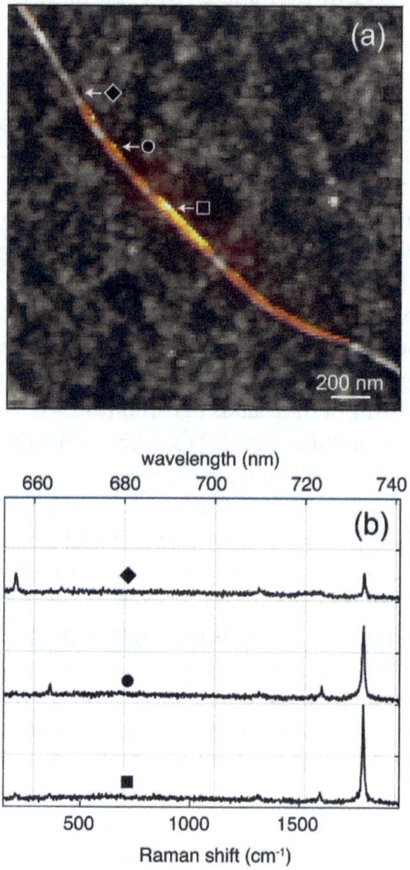

Fig. 6 Near-field Raman imaging of long linear carbon chains (LLCCs) in double walled carbon nanotubes. (a) A superposition of the near-field LLCC mode image (color) acquired with a narrow band-pass filter centered at 735 nm and the corresponding topographic image (grayscale). (b) Near-field spectra of the nanotube just to the left of the positions marked in (a) with the diamond, circle, and square. All spectra are normalized to the LLCC mode at ~1800 cm^{-1} in the bottom spectrum and are offset from each other by 1 for visibility. A complex interplay between the relative intensities of the outer nanotube RBM at ~204 cm^{-1}, the inner nanotube RBM at ~362 cm^{-1}, and the G band at ~1590 cm^{-1}, relative to the LLCC mode, is observed.

are all normalized to the intensity of the LLCC mode at ~1800 cm^{-1} in the bottom spectrum and are offset from each other for visibility. Depending on the sampling location relative to the location of the LLCC, the relative ratios between the outer nanotube RBM at ~204 cm^{-1}, the inner nanotube RBM at ~362 cm^{-1}, and the G band at ~1590 cm^{-1} vary greatly, while the D band component at ~1315 cm^{-1} is rather constant. Studies of LCCs in nanotubes have both predicted[68,69] and measured[70] shifts in the RBM upon filling, while other studies have noticed a dramatic decrease in the RBM of a Mo-filled nanotube.[71] This complex interaction has not been observed in ensemble measurements and highlights the need for the resolution and sensitivity afforded by near-field spectroscopy.

6. Conclusions and outlook

The tunable optoelectric properties of nanocarbon materials make them ideal for the next generation of optoelectronic devices. Combined with other near-field microscopy techniques, TERS allows for a deep understanding of the physical processes taking place. Current optoelectronic studies of carbon nanotubes highlight the synergy between Raman spectroscopy, photoluminescence (PL) microscopy and electroluminescence (EL) microscopy and their importance for real device engineering.

Several works have explored near-field PL as a probe for local changes in carbon nanotubes.[18,20,72–75] Our work on combined Raman and near-field PL measurements of carbon nanotubes revealed the localized nature of PL emission from defects,[18] which may have implications both for nanocarbon-based light-emission and photo-detection. Additionally, TERS has been combined with near-field EL microscopy to study carbon nanotube device junctions.[76] TERS was used to localize a Schottky junction between at least one metallic and one semiconducting nanotube, and confirm that the EL emission originated from the junction. This work further proved the reciprocity theorem, which states that optical antennas are not only able to increase the efficiency of optical absorption, but also optical emission.[77] We have also characterized graphene EL from STM experiments[78] and applied this knowledge to fabricate and characterize an antenna-coupled graphene photodetector.[79]

While the study of sp^2 nanocarbons is still under development, the study of carbyne is only in its infancy. Even still, TERS has already demonstrated the ability to locally sample the complex interaction between overlapping sp^1 and sp^2 orbitals that will be essential to facilitate devices based on this novel material. The chemical sensitivity and non-destructive nature of Raman spectroscopy, combined with the high-spatial resolution and topographic information afforded by near-field microscopy, makes TERS ideal to study nanocarbons.

Acknowledgements

This work was funded by the Swiss National Science Foundation (SNF) under the NCCR-QSIT program and grant CR2212_152944. LGC acknowledges FAPEMIG and CNPq. We thank Ernesto Joselevich for the serpentine nanotube samples, Thomas Pichler, Lei Shi, and Philip Rohringer for the carbyne samples, Sang-Hyun Oh and Tim Johnson for the template stripped gold pyramids, and Ado Jorio for valuable discussions.

References

1 A. Jorio, G. Dresselhaus and M. S. Dresselhaus, *Carbon Nanotubes: Advanced Topics in the Synthesis, Structure, Properties and Applications*, Springer, Berlin, 2008, vol. 111.

2 J. Singh, *Physics of Semiconductors and Their Heterostructures*, McGraw-Hill, 1993.

3 *Carbon Nanotubes*, ed. M. S. Dresselhaus, G. Dresselhaus and P. Avouris, Springer, Berlin, 2001, vol. 80.

4 L.-C. Qin, X. Zhao, K. Hirahara, Y. Miyamoto, Y. Ando and S. Iijima, *Nature*, 2000, **408**, 50.

5 T. W. Odom, J.-L. Huang, P. Kim and C. M. Lieber, *Nature*, 1998, **391**, 62–64.

6 A. Jorio, R. Saito, J. H. Hafner, C. M. Lieber, M. Hunter, T. McClure, G. Dresselhaus and M. S. Dresselhaus, *Phys. Rev. Lett.*, 2001, **86**, 1118.

7 Z. Yu and L. Brus, *J. Phys. Chem. B*, 2001, **105**, 1123–1134.

8 A. Mews, F. Koberling, T. Basché, G. Philipp, G. S. Duesberg, S. Roth and M. Burghard, *Adv. Mater.*, 2000, **12**, 1210.

9 J. Maultzsch, S. Reich, C. Thomsen, S. Webster, R. Czerw, D. L. Carroll, S. Vieira, P. Birkett and C. Rego, *Appl. Phys. Lett.*, 2002, **81**, 2647.

10 G. S. Duesberg, I. Loa, M. Burghard, K. Syassen and S. Roth, *Phys. Rev. Lett.*, 2000, **85**, 5436–5439.

11 R. M. Stöckle, Y. D. Suh, V. Deckert and R. Zenobi, *Chem. Phys. Lett.*, 2000, **318**, 131–136.

12 A. Hartschuh, E. Sanchez, X. Xie and L. Novotny, *Phys. Rev. Lett.*, 2003, **90**, 095503.

13 L. G. Cançado, A. Jorio, A. Ismach, E. Joselevich, A. Hartschuh and L. Novotny, *Phys. Rev. Lett.*, 2009, **103**, 186101.

14 L. Novotny and S. J. Stranick, *Annu. Rev. Phys. Chem.*, 2006, **57**, 303–331.

15 N. Anderson, A. Hartschuh and L. Novotny, *Mater. Today*, 2005, **8**, 50–54.

16 I. O. Maciel, N. Anderson, M. A. Pimenta, A. Hartschuh, H. Qian, M. Terrones, H. Terrones, J. Campos-Delgado, A. M. Rao, L. Novotny and A. Jorio, *Nat. Mater.*, 2008, **7**, 878–883.

17 R. Beams, L. G. Cancado, A. Jorio, A. N. Vamivakas and L. Novotny, *Nanotechnology*, 2015, **26**, 175702.

18 A. Hartschuh, H. Qian, A. J. Meixner, N. Anderson and L. Novotny, *Nano Lett.*, 2005, **5**, 2310.

19 N. Anderson, A. Hartschuh and L. Novotny, *Nano Lett.*, 2007, **7**, 577–582.

20 H. Qian, C. Georgi, N. Anderson, A. A. Green, M. C. Hersam, L. Novotny and A. Hartschuh, *Phys. Status Solidi B*, 2008, **245**, 2243–2246.

21 J. Soares, A. P. Barboza, D. Nakabayashi, P. Araujo, N. B. Neto, N. Shadmi, T. Yarde, A. Ismach, N. Geblinger, E. Joselevich, C. Vilani, L. G. Cancado, L. Novotny, G. Dresselhaus, M. Dresselhaus, B. Neves and M. Mazzoni, *Nano Lett.*, 2010, **10**, 5043–5048.

22 R. Beams, L. G. Cancado, S.-H. Oh, A. Jorio and L. Novotny, *Phys. Rev. Lett.*, 2014, **113**, 186101.

23 R. Zhang, Y. Zhang, Z. Dong, S. Jiang, C. Zhang, L. Chen, L. Zhang, Y. Liao, J. Aizpurua, Y. Luo, *et al.*, *Nature*, 2013, **498**, 82–86.

24 L. G. Cançado, A. Hartschuh and L. Novotny, *J. Raman Spectrosc.*, 2009, **40**, 1420–1426.
25 L. Novotny, M. R. Beversluis, K. S. Youngworth and T. G. Brown, *Phys. Rev. Lett.*, 2001, **86**, 5251.
26 K. Karrai and R. D. Grober, *Appl. Phys. Lett.*, 1995, **66**, 1842–1844.
27 T. R. Albrecht, P. Grütter, D. Horne and D. Rugar, *J. Appl. Phys.*, 1991, **69**, 668.
28 L. Novotny, E. J. Sanchez and X. S. Xie, *Opt. Photonics News*, 1999, **10**, 24.
29 P. Anger, P. Bharadwaj and L. Novotny, *Phys. Rev. Lett.*, 2006, **96**, 113002.
30 P. Bharadwaj, R. Beams and L. Novotny, *Chem. Sci.*, 2011, **2**, 136–140.
31 C. Höppener, Z. Lapin, P. Bharadwaj and L. Novotny, *Phys. Rev. Lett.*, 2012, **109**, 017402.
32 T. W. Johnson, Z. J. Lapin, R. Beams, N. C. Lindquist, S. G. Rodrigo, L. Novotny and S.-H. Oh, *ACS Nano*, 2012, **6**, 9168–9174.
33 L. G. Cançado, M. A. Pimenta, B. R. A. Neves, M. S. S. Dantas and A. Jorio, *Phys. Rev. Lett.*, 2004, **93**, 247401.
34 A. C. Ferrari, J. C. Meyer, V. Scardaci, C. Casiraghi, M. Lazzeri, F. Mauri, S. Piscanec, D. Jiang, K. S. Novoselov, S. Roth and A. K. Geim, *Phys. Rev. Lett.*, 2006, **97**, 187401.
35 A. Das, S. Pisana, B. Chakraborty, S. Piscanec, S. K. Saha, U. V. Waghmare, K. S. Novoselov, H. R. Krishnamurthy, A. K. Geim, A. C. Ferrari and A. K. Sood, *Nat. Nanotechnol.*, 2008, **3**, 210.
36 T. M. G. Mohiuddin, A. Lombardo, R. R. Nair, A. Bonetti, G. Savini, R. Jalil, N. Bonini, D. M. Basko, C. Galiotis, N. Marzari, K. S. Novoselov, A. K. Geim and A. C. Ferrari, *Phys. Rev. B: Condens. Matter Mater. Phys.*, 2009, **79**, 205433.
37 M. M. Lucchese, F. Stavale, E. H. Martins Ferreira, C. Vilahn, M. V. O. Moutinho, R. B. Capaz, C. A. Achete and A. Jorio, *Carbon*, 2010, **48**, 1592–1597.
38 A. C. Ferrari and D. M. Basko, *Nat. Nanotechnol.*, 2013, **8**, 235–246.
39 R. Beams, L. G. Cançado and L. Novotny, *J. Phys.: Condens. Matter*, 2015, **27**, 083002.
40 W. Su and D. Roy, *J. Vac. Sci. Technol., B: Microelectron. Nanometer Struct.*, 2013, **31**, 041808.
41 L. G. Cancado, R. Beams, A. Jorio and L. Novotny, *Phys. Rev. X*, 2014, **4**, 031054.
42 A. Jorio, M. S. Dresselhaus, R. Saito and G. Dresselhaus, *Raman Spectroscopy in Graphene Related System*, Wiley-VCH, Weinheim, Germany, 1st edn, 2011.
43 C. Casiraghi, A. Hartschuh, H. Qian, S. Piscanec, C. Georgi, A. Fasoli, K. S. Novoselov, D. M. Basko and A. C. Ferrari, *Nano Lett.*, 2009, **9**, 1433–1441.
44 R. Beams, L. G. Cancado and L. Novotny, *Nano Lett.*, 2011, **11**, 1177–1181.
45 K. A. Ritter and J. W. Lyding, *Nat. Mater.*, 2009, **8**, 235–242.
46 N. Geblinger, A. Ismach and E. Joselevich, *Nat. Nanotechnol.*, 2008, **3**, 195.
47 C. Georgi and A. Hartschuh, *Appl. Phys. Lett.*, 2010, **97**, 143117.
48 B. Shan, G. W. Lakatos, S. Peng and K. Cho, *Appl. Phys. Lett.*, 2005, **87**, 173109.
49 L. Vitali, M. Burghard, P. Wahl, M. A. Schneider and K. Kern, *Phys. Rev. Lett.*, 2006, **96**, 086804.
50 Y. Okuno, Y. Saito, S. Kawata and P. Verma, *Phys. Rev. Lett.*, 2013, **111**, 216101.
51 T.-a. Yano, P. Verma, Y. Saito, T. Ichimura and S. Kawata, *Nat. Photonics*, 2009, **3**, 473–477.
52 T.-a. Yano, T. Ichimura, S. Kuwahara, F. HDhili, K. Uetsuki, Y. Okuno, P. Verma and S. Kawata, *Nat. Commun.*, 2013, **4**, 2592.

53 C. Chen, N. Hayazawa and S. Kawata, *Nat. Commun.*, 2014, **5**, 2592.
54 V. I. Kasatochkin, A. M. Sladkov, Y. P. Kudryavtsev, N. M. Popov and V. V. Korshak, *Dokl. Akad. Nauk SSSR*, 1967, **177**, 358–360.
55 A. E. Goresy and G. Donnay, *Science*, 1968, **161**, 363–364.
56 A. G. Whittaker and P. L. Kintner, *Science*, 1969, **165**, 589–591.
57 A. Sladkov and Y. Kudryavtsev, *Priroda*, 1969, **5**, 37–44.
58 P. P. Smith and P. R. Buseck, *Science*, 1981, **212**, 322–324.
59 P. P. Smith and P. R. Buseck, *Science*, 1982, **216**, 984–986.
60 W. A. Chalifoux and R. R. Tykwinski, *C. R. Chim.*, 2009, **12**, 341–358.
61 X. Zhao, Y. Ando, Y. Liu, M. Jinno and T. Suzuki, *Phys. Rev. Lett.*, 2003, **90**, 187401.
62 L. Shi, PhD thesis, Universität Wien, 2015.
63 R. E. Peierls, *Quantum theory of solids*, Clarendon, 1955.
64 A. Rusznyak, V. Zolyomi, J. Kurti, S. Yang and M. Kertesz, *Phys. Rev. B: Condens. Matter Mater. Phys.*, 2005, **72**, 155420.
65 X. Fan, L. Liu, J. Lin, Z. Shen and J.-L. Kuo, *ACS Nano*, 2009, **3**, 3788–3794.
66 A. Tapia, L. Aguilera, C. Cab, R. Medina-Esquivel, R. de Coss and G. Canto, *Carbon*, 2010, **48**, 4057–4062.
67 S. Yang, M. Kertesz, V. Zolyomi and J. Kurti, *J. Phys. Chem. A*, 2007, **111**, 2434–2441.
68 Z. L. Hu, X. M. Guo and C. Q. Ru, *Nanotechnology*, 2007, **18**, 485712.
69 Y. Liu, R. O. Jones, X. Zhao and Y. Ando, *Phys. Rev. B: Condens. Matter*, 2003, **68**, 125413.
70 M. Jinno, Y. Ando, S. Bandow, J. Fan, M. Yudasaka and S. Iijima, *Chem. Phys. Lett.*, 2006, **418**, 109–114.
71 H. Muramatsu, T. Hayashi, Y. A. Kim, D. Shimamoto, M. Endo, M. Terrones and M. S. Dresselhaus, *Nano Lett.*, 2008, **8**, 237–240.
72 H. Qian, T. Gokus, N. Anderson, L. Novotny, A. J. Meixner and A. Hartschuh, *Phys. Status Solidi B*, 2006, **243**, 3146–5.
73 H. Qian, C. Georgi, N. Anderson, A. A. Green, M. C. Hersam, L. Novotny and A. Hartschuh, *Nano Lett.*, 2008, **8**, 1363–1367.
74 H. Qian, P. T. Araujo, C. Georgi, T. Gokus, N. Hartmann, A. A. Green, A. Jorio, M. C. Hersam, L. Novotny and A. Hartschuh, *Nano Lett.*, 2008, **8**, 2706–2711.
75 C. Georgi, M. Boehmler, H. Qian, L. Novotny and A. Hartschuh, *Phys. Status Solidi B*, 2009, **246**, 2683–2688.
76 N. Mauser, N. Hartmann, M. S. Hofmann, J. Janik, A. Hošgele and A. Hartschuh, *Nano Lett.*, 2014, **14**, 3773–3778.
77 P. Bharadwaj, B. Deutsch and L. Novotny, *Adv. Opt. Photonics*, 2009, **1**, 438–483.
78 R. Beams, P. Bharadwaj and L. Novotny, *Nanotechnology*, 2014, **25**, 055206.
79 C. Chakraborty, R. Beams, K. M. Goodfellow, G. W. Wicks, L. Novotny and N. A. Vamivakas, *Appl. Phys. Lett.*, 2014, **105**, 241114.

Faraday Discussions

PAPER

Multicolour single molecule emission and excitation spectroscopy reveals extensive spectral shifts

Lukasz Piatkowski,†[a] Esther Gellings†[a] and Niek F. van Hulst*[ab]

Received 10th June 2015, Accepted 16th July 2015
DOI: 10.1039/c5fd00107b

We explore the distribution and shape of single molecule spectra at room temperature, when embedded in a polymer host. Multicolour excitation and emission spectroscopy is implemented to capture the full inhomogeneous distribution. We observe dramatic spectral changes in a distribution of single quaterrylene diimide (QDI) molecules isolated in a PMMA matrix. The molecules are strongly blue shifted with respect to the ensemble absorption maximum and spread over a staggering 200 nm range. Despite these strong shifts, the shape of the emission spectra does not differ much between individual molecules. We demonstrate that a considerable number of molecules may be invisible in single molecule experiments, as they typically rely on only a single excitation wavelength, which predetermines which subensemble is probed in the experiment. Lastly, we make a first step towards single molecule excitation spectroscopy under ambient conditions, which allows us to determine the spectral range at which individual molecules absorb light most efficiently. We show how single molecule emission and excitation spectroscopies can complement each other and a combination of both techniques can help in understanding the origin of underlaying spectral properties of individual molecules.

Introduction

Single molecule (SM) experiments allow us to study a system of interest addressing one molecule at a time. Molecules are not isolated in nature, but are constantly interacting with their surroundings. This, together with an intrinsic molecular diversity, leads to the observation of inhomogeneities and dynamical processes that are usually concealed in the ensemble average.[1,2]

Depending on the system and photophysics of interest, different detection schemes are feasible. With very few exceptions, all SM methods rely on the

[a]ICFO—Institut de Ciences Fotoniques, The Barcelona Institute of Science and Technology, 08860 Castelldefels, Barcelona, Spain. E-mail: Niek.vanHulst@ICFO.eu

[b]ICREA—Institució Catalana de Recerca i Estudis Avançats, 08010 Barcelona, Spain

† These authors contributed equally to this work.

detection of spontaneous emission. In cryogenic studies, the limited interactions between molecules and their local environment are exploited by measuring the typically lifetime-limited zero-phonon absorption lines. Under ambient conditions, the increased interactions with the environment lead to significantly broadened absorption lines as well as a drastic reduction in the photostability and number of photons emitted before photobleaching. As a consequence, at cryogenic temperatures SMs are commonly studied through both their fluorescence excitation and emission spectra,[3–5] while at room temperature SMs are only routinely investigated through their fluorescence emission spectra.[6,7]

Spectral and dynamical variability including spectral shifts and blinking are observed irrespective of the chosen experimental parameters.[8–12] They can be linked to changes in the chromophore itself (intrinsic) or the chromophores' local environment (extrinsic) that are mirrored in changes in the fluorescence emission. Intrinsic variability can for example be induced by intersystem crossing or conformational changes, whereas reasons for extrinsic changes include external fields and changes in the molecules' local environment. Especially in heterogeneous environments like polymers, individual molecules can interact with a manifold of chemically or physically distinct surroundings. Structural differences like the size, shape and the local surface composition of the polymer affect among other things the polarity, acidity and rigidity of the nanoenvironment. Depending on the degree and nature of the interaction, this may be reflected in more or less pronounced deviations from the molecule ensemble average spectral position and shape, quantum yield or lifetime.[13–18] Dynamical changes in the behaviour of the molecules can be photoinduced, for example by a rearrangement of the host matrix around the molecule due to local heating[14] or *cis*/*trans* conformational changes.[19,20] Still, the trapping of molecules in a crystal matrix or a polymer film is generally beneficial. It increases the spectral stability, because it forces the molecules into a certain conformation and hinders the diffusion of oxygen to the molecule, resulting in a lower photobleaching rate.[21–23]

Spectral variability of SMs has been experimentally observed in a number of systems. A spectral distribution of 40 nm has been reported for fluorescent dyes based on the perylene diimide chromophore (structurally similar to QDI).[14,15] In these cases, large spectral shifts have been explained in terms of a twisting of the chromophore core (intrinsic) as well as a varying local polymer environment (extrinsic). Spectral variations of nearly 100 nm have been reported for other systems such as green fluorescent proteins in PVA[19] and carbocyanine dye molecules adsorbed on bare glass.[20]

In this work, we investigated the spectral distribution of individual QDI (quaterrylene diimide)[24] molecules embedded in a polymer host (PMMA) under ambient conditions. QDI molecules belong to the rylene dye family that has been extensively used in SM research over the past decade due to its high quantum efficiency and photostability.[3,25–28] The absorption and emission wavelengths of rylene dyes can be tuned throughout the visible spectrum by adjusting the length of the perylene backbone.[29] This determines the degree of π-electron delocalization and thus the electronic transition strength and energy.[30]

Here, we demonstrate the importance of utilizing a range of excitation wavelengths and detection bandwidths to access and probe the full spectral range of SMs. We show how the choice of excitation and detection wavelengths can put a bias on the measured spectral distribution for both SM emission and excitation spectroscopy.

Results

Environmental effects on the spectral properties of molecules can already be visible in ensemble experiments. That is why it is informative to first compare the ensemble absorption and emission spectra in solution to those in a polymer matrix, and get an indication of the spectral changes we might expect from SM experiments. We dropcast QDI/PMMA/toluene solution with a QDI concentration of $\sim$10 µM onto a cover slip, let the solvent evaporate and subsequently measured the ensemble absorption and emission spectra. We compared the results to the QDI solution spectra measured with the same concentration.

In Fig. 1 we show the absorption and emission spectra of QDI in solution (shaded grey and red, respectively) and in a solidified PMMA film (solid blue and red lines). In both cases the emission spectra were measured using 633 nm excitation light derived from a HeNe laser. The inset shows the chemical structure of QDI.

In the solution absorption spectrum, we observed three bands corresponding to the electronic transition at 750 nm and two vibronic sidebands at 680 nm and 630 nm. The solution emission spectrum is not the exact mirror image of the absorption spectrum as would be expected for chromophores that strictly follow Kasha's rule. The emission spectrum has only two bands at 785 nm and 860 nm and its interband intensity ratio differs from the corresponding interband ratio in the absorption spectrum.

The molecular ratio between the polymer and QDI is roughly 10/1. Given that the polymer molecules are approximately 100 times larger than QDI we could safely assume that at this concentration the QDI molecules mainly interact with the polymer and not with each other. The measured spectra show that the absorption of QDI in a PMMA film significantly differs in shape from the absorption in solution (see Fig. 1). The maximum intensity of the spectrum is shifted towards lower wavelengths by approximately 50 nm and lacks the spectral features of the vibronic progression. Correspondingly, we found that the emission spectrum of the dry film QDI was blue shifted by $\sim$20 nm with respect to the solution emission spectrum. However, its shape was largely unaltered.

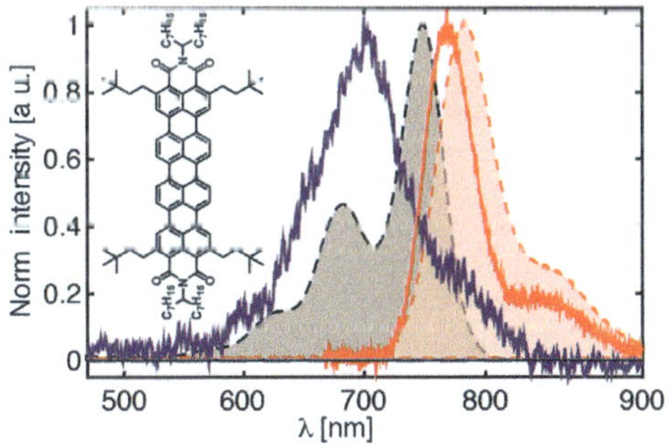

Fig. 1 Absorption and emission spectra of QDI in PMMA/toluene solution (shaded grey and red, respectively). Dry QDI/PMMA film absorption and emission spectra (solid blue and red lines). The chemical structure of QDI is shown alongside the spectra.

It is important to note that the observed changes in the absorption spectra of the QDI film are reversible – when the dry film was dissolved in toluene again, the QDI solution spectrum was recovered.

It is clear that the local nano-environment induced by the proximity of the glass and/or polymer does alter the absorption and emission spectra of the QDI ensemble. The question that arises naturally from these observations is whether comparable changes in spectral shape and position can be found at the SM level. Do individual molecules embedded in PMMA mirror the shape and position of the ensemble dry film absorption and emission spectra, or does an induced spread of molecules with different electronic transition energies and different spectral shapes lead to the observed changes? With these questions in mind we now turn our attention to multicolour SM experiments.

Multicolour excitation confocal spectroscopy

For the SM experiments, samples with a density of about 1 QDI molecule per 2 μm^2 were used to ensure that the molecules are well separated for the confocal detection. As the first step, a confocal image was recorded to locate molecules on the sample. Then the focal position was optimized based on the intensity of the fluorescence signal detected with an APD. We then acquired a series of five confocal images of the same sample area at excitation wavelengths ranging from

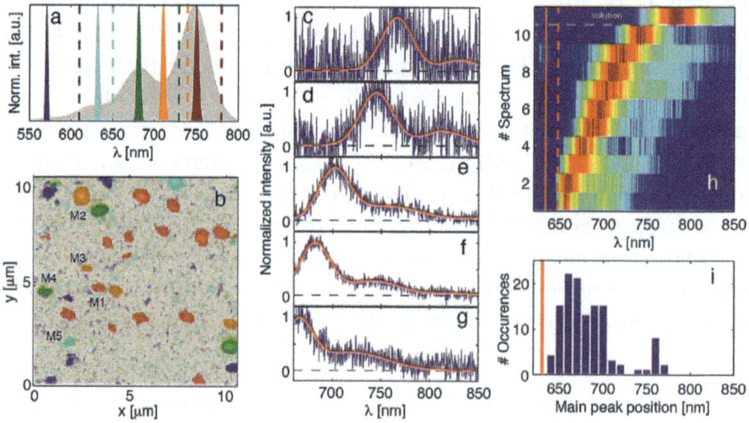

Fig. 2 (a) Concept schematic of the experiment. Coloured profiles and dashed lines represent the excitation wavelengths and the cut-off wavelengths of the corresponding long-pass filters, respectively. The grey shaded area represents the ensemble absorption spectrum of QDI in solution. (b) Zoom in on multicolour excitation confocal image composed of five individual confocal images obtained at the excitation wavelengths indicated in panel a. (c–g) SM emission spectra measured on five distinct molecules (M1 to M5 from top to bottom) numbered on the confocal image in panel b. The solid red lines represent the Gaussian fit to the measured emission spectra. (h) The spread of the emission spectra of individual molecules shown in the form of a 2D map. The top slice shows the emission spectrum of QDI in solution. Solid and dashed red lines indicate excitation (633 nm) and long-pass filter cut-off wavelengths (650 nm), respectively. (i) Histogram of the spectral positions for all (122) measured SM emission spectra. The spectral positions of individual molecules correspond to the position of the maximum intensity in their emission spectrum (maximum of the electronic transition).

750 nm to 570 nm (coloured spectra in Fig. 2a). For each excitation wavelength we used an appropriate long-pass filter with a cut-off wavelength spectrally as close as possible to the excitation wavelength, to maximize the fluorescence detection efficiency. If the molecules have a spread in absorption spectra, a different subset of molecules will get excited at each excitation wavelength so that information on the amount of excitable molecules at each wavelength can be extracted. We started on the red side with a confocal image using 750 nm wavelength excitation light. All molecules whose absorption spectrum overlapped with that wavelength appeared in this image. Subsequently we took a confocal image at 710 nm excitation. As the absorption spectrum of QDI is nearly 200 nm wide, the molecules that were detected at 750 nm excitation will also appear on this image. Additionally, however, we found extra molecules whose absorption spectrum was blue shifted so much that they could be excited at 710 nm, but not yet at 750 nm. We repeated this for subsequent excitation wavelengths towards the blue, to find all the subsets of molecules that start absorbing at each of the excitation wavelengths. We assigned a colour to these sub-ensembles, according to the colour of the excitation wavelength at which they first appear in a confocal image (not really corresponding to the VIS range). The molecules that were visible on a particular image were manually removed from all images acquired at shorter excitation wavelengths. Then, the average background was removed and the data smoothed with a Gaussian filter. Finally, the images were normalized to each other and overlaid. For better visibility of the molecules, the bottom 5% intensity on the images (residual background) has been coloured grey. The resulting multicolour excitation confocal image is shown in Fig. 2b. The variation in size between molecules simply results from the diversity in emission intensity.

We have experimentally checked that there were no molecules significantly redshifted with respect to the ensemble spectrum by measuring a confocal image at 800 nm excitation. Assuming that molecules can be detected when excited at the level of at least 10% of their absorption cross section, they would appear on a confocal image if their maximum absorption peak would reside at most 40 nm below the excitation wavelength (assuming that the spectral shape is largely unaltered). Thus we conclude that there were no molecules shifted by more than 10 nm towards the red.

The continuous appearance of molecules on the images, even as blue shifted as 570 nm, indicates that the sample contains molecules with a large spread of electronic transition energies. To find out whether the observed spread is also associated with changes of the spectral shape and the vibrational band structure, we now turn to SM emission and excitation spectroscopy.

Single molecule emission spectra

We measured emission spectra of individual molecules on the same area of the sample upon excitation at 633 nm. In panels c–g of Fig. 2 we show five exemplary emission spectra measured on the molecules that appeared at different excitation wavelengths and are numbered M1–M5 in Fig. 2b. We found that the emission spectra of molecules emitting on the red side near the QDI solution emission wavelength (Fig. 2c and d) were much weaker and much noisier. This was expected because at 633 nm excitation the absorption cross-section is very low for molecules with maximum absorption around 750 nm (~15% of the maximum).

Still, there was a continuous distribution of molecules absorbing at all intermediates, which became clear when we measured more emission spectra on different areas of the sample. We found that the position of the maxima of the emission spectra ranged from about 760 nm to the cut-off wavelength of our long-pass filter at 647 nm.

In panel h we show a series of 10 SM emission spectra. In the top slice we show the emission spectrum of QDI in solution for comparison. A histogram showing the wavelength of the maximum emission for all measured molecules is shown in panel i. The positions of the maxima were determined using a Gaussian fit to the main peak of the emission spectrum. The distribution clearly peaks towards shorter wavelengths, close to where the excitation wavelength is. This is understandable, as the molecules that have their maximum absorption shifted towards the blue overlap with the excitation laser more strongly and thus appear brighter on the image. In the experiment we are naturally biased towards picking brighter molecules. This observation demonstrates that the choice of excitation and detection bandwidths determines which molecular sub-ensemble is probed most efficiently when the molecules show a large spectral distribution.

We found new molecules appearing when the excitation wavelength is moved towards the blue in the multicolour excitation confocal image (by nearly 200 nm). We also found a distribution of emission spectra with nearly the same shape, while spreading over at least 100 nm. These observations clearly indicate that the individual molecules keep their intrinsic spectral properties but experience an overall electronic shift.

Towards excitation spectra of single molecules

We have learned that the absorption spectrum of QDI molecules shifts together with the emission spectrum. Since the shape of the emission spectra did not appear to depend on the spectral position, we tentatively assumed that the shape of the absorption spectra would not vary significantly between individual molecules either. It is however desirable to find out whether the shape of the absorption spectra resembles that of QDI in solution or whether it is affected by the encapsulation in PMMA.

In the following step we therefore attempted to measure the excitation spectra of individual molecules. Fig. 3f shows the concept of this multicolour experiment. We recorded confocal images at different excitation wavelengths again but kept the long-pass filter in the detection path constant at 776 nm this time. This way the same portion of fluorescence emission was detected at each excitation wavelength. Therefore, for the same excitation power, the detected fluorescence intensity only depended on the molecule's absorption cross-section at the excitation wavelength. Variations in fluorescence intensity between different excitation wavelengths therefore directly yield the excitation spectrum of the molecule in this case. The measured confocal images are shown in Fig. 3a–e. The confocal images measured towards the blue excitation contain fewer and fewer molecules, which results from the fact that the red molecules have a very low absorption cross section at these wavelengths, whereas for the blue molecules the amount of detected fluorescence is very low. Examples of SM excitation spectra are shown in Fig. 3g–j (solid blue lines). The excitation spectrum shown in Fig. 3j corresponds to the molecule marked with a red circle in the confocal images. As a guide to the

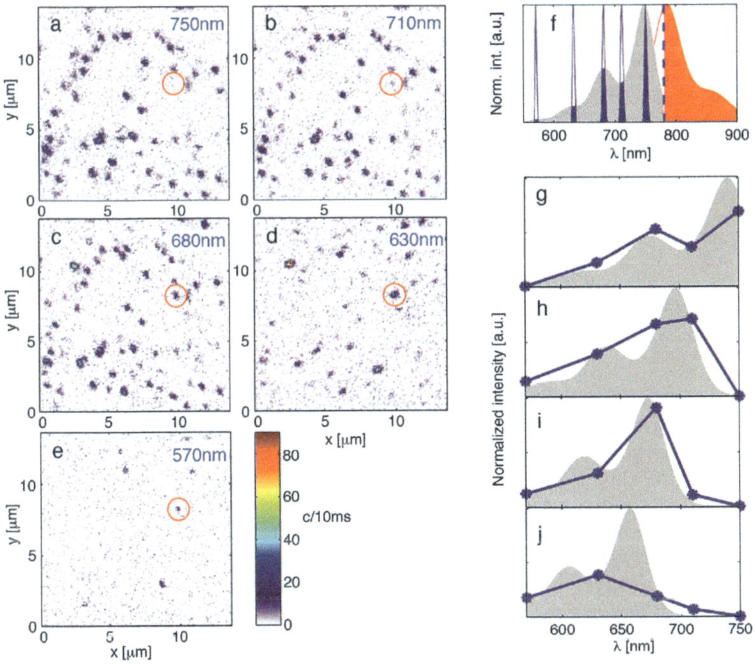

Fig. 3 (a–e) Confocal images measured at different excitation wavelengths, indicated in the right upper corner of each image, with the fluorescence signal detected with the same 776 nm long-pass filter. (f) Concept schematic of the experiment. The blue profiles represent the excitation wavelengths and the blue dotted line the cut-off wavelength of the long-pass filter. The red filled area indicates the detected part of the emission spectrum. (g–j) The blue lines represent the SM excitation spectra reconstructed from the confocal images shown in panels a to e. For comparison we also show the absorption spectrum of QDI in solution (shaded grey) that has been spectrally blue shifted to match the measured SM excitation spectrum. The ensemble absorption spectra were shifted in the frequency domain and then translated back to the wavelength range to account for the difference in photon energy corresponding to different wavelengths.

eye, we added a QDI solution absorption spectrum, which has been spectrally shifted to match the measured excitation spectra (grey, shaded). The five measured experimental points were insufficient to precisely reproduce the spectral shape of the absorption spectrum. However, they were sufficient to determine the spectral position of the excitation spectra with about 20 nm precision and also to verify that the width of the excitation spectra matches that of the solution absorption spectrum. Looking at different molecules we found a large spread of spectral positions of about 100 nm for the excitation spectra.

Quantitative analysis

Upon careful inspection of the SM emission spectra we found that the separation between the electronic transition and the vibronic sideband, as well as their relative intensities, varied between individual molecules. In Fig. 4a we show the correlation between the spectral position of the main peak and the spectral separation between the electronic transition and vibronic sideband. We took into

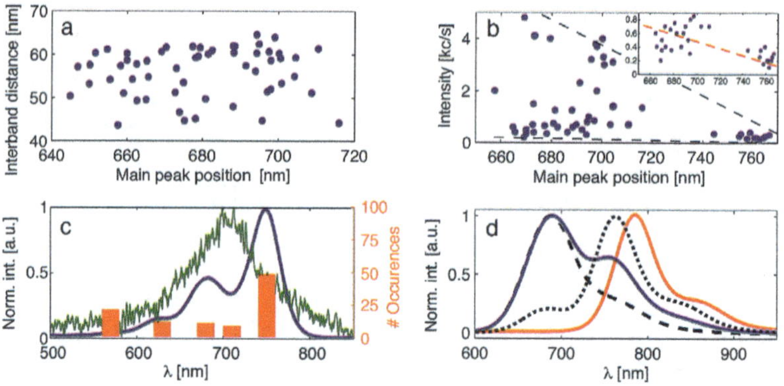

Fig. 4 (a) Correlation between the spectral position of the SM emission spectra and separation between the main electronic transition and the first vibronic band. (b) Correlation between the SM spectral position and intensity of the emission spectrum. The black dashed lines indicate the limits of the plotted distribution. The inset zooms into the low count range of the main panel with the red dashed line acting as guide to the eyes. (c) Comparison between the absorption spectrum of QDI film (green) and the distribution of the single QDI molecule spectral positions (red bars) based on the multicolour excitation confocal image shown in Fig. 2b. (d) Comparison between the QDI solution emission spectrum (red) and the reconstructed SM ensemble emission spectrum for two sub-populations of molecules: bright (dashed line) and dark (dotted line) ones (giving >300 and <300 counts per s, respectively, as shown in the inset of panel b). The solid, blue line represents the SM ensemble emission spectrum, taking into account all the molecules shown in panel b.

account all SM emission spectra for which we could clearly see the two bands and did not find any clear correlation of the interband separation and maximum wavelength at which the molecules emit. The majority of the molecules exhibited the same interband distance of about 50–60 nm even though they absorbed across a wavelength range of nearly 100 nm.

From the analysis of the emission spectra of individual molecules taken from the multicolour excitation confocal image, we learned that picking up darker molecules yields emission spectra at the red side (towards 750 nm), while brighter molecules emit towards the blue side. In Fig. 4b we show the correlation between the maximum intensity of the emission spectrum and the spectral position of the main emission peak, when exciting at 633 nm. Interestingly, the scatter plot has the shape of a wedge, as indicated with the thin, black dashed lines. The intensity spread for red molecules is much smaller than for blue molecules. This can be readily understood in terms of the transition dipole moment orientation and/or the internal quantum efficiency. When exciting at 633 nm we were more sensitive to blue molecules. Even not so efficient molecules and molecules whose transition dipole moment is not aligned well with the polarization of the excitation light would be visible on the image and bright enough to measure their emission spectrum. For the red molecules on the other hand, as they were excited less efficiently in our experiment, we were mainly sensitive to the brightest ones that had high quantum efficiency, and for which the transition dipole moment orientation matched the excitation light polarization, making the distribution much narrower. After looking closely at the low intensity sub-ensemble of molecules (see inset) we indeed found that the spread of intensities is similar for all molecules and we saw a clear linear dependence between the brightness and the spectral position. This dependence illustrates yet again that SM experiments are highly biased by the excitation wavelength, especially when only the bright sub-population is sampled.

Finally, from all our experiments on SMs we have found a large distribution of molecules absorbing and emitting at various wavelengths. We can thus reconstruct ensemble excitation and emission spectra from the SM results and tentatively compare them to the absorption and emission spectra measured on the QDI film. In Fig. 4c we show the absorption spectrum of the dry QDI film (green) together with a histogram (red), which shows occurrences of molecules starting to absorb at the five different excitation wavelengths. The histogram was constructed, taking into account molecules from the multicolour excitation confocal images (see Fig. 2b). The number of molecules that we have measured is insufficient to precisely rebuild the ensemble absorption spectrum of dry film. However, the extent of the distribution is well represented. For comparison, we also show the absorption spectrum of QDI in solution (blue). We note that multicolour excitation confocal microscopy with changing long-pass filters was the only unbiased experiment with respect to excitation/detection wavelength choice that we have performed.

In Fig. 4d we show the reconstructed emission spectrum of the SM ensemble (blue) in comparison to the dry film emission spectrum (red). The SM ensemble spectrum was constructed by adding up normalized solution emission spectra that are shifted according to Fig. 4b. The SM ensemble emission spectrum is clearly blue shifted with respect to the dry film spectrum. This is in line with the fact that in the experiment we are biased towards brighter molecules (633 nm excitation). If we only take into account bright molecules (>300 counts per s), the reconstructed SM ensemble emission spectrum is strongly blue shifted (black dashed line), whereas if we take only the darker molecules (≤300 counts per s) we find that the SM ensemble emission spectrum has nearly the shape of the QDI solution emission (black dotted line).

Discussion

From the multicolour excitation confocal images we have learned that QDI molecules embedded in a PMMA matrix can start absorbing anywhere between 570 and 750 nm. The shifts in the absorption of up to 200 nm away from the solution absorption maximum (750 nm) are quite dramatic, especially when realising that they correspond to changes in the electronic transition energy of about 25%.

Interestingly, we found that the shapes of the emission spectra of molecules absorbing at various wavelengths were nearly identical (Fig. 2c–g). The position of the first vibrational band with respect to the main electronic transition varied between 45 and 65 nm, but there was no clear correlation between the inter-band distance and the wavelength at which the molecules emit. This indicates that the intramolecular coupling between vibrational modes and the electronic transition was largely unaltered. As the vibrational modes are quite sensitive to structural changes of the molecule, we concluded that all molecules must have a very similar conformation.

The spectral distribution in the measured SM excitation spectra shown in Fig. 3 is in full agreement with the spread observed in the measured emission spectra. The excitation spectra comprise only 5 points. While this was not sufficient to precisely determine the shape of the excitation spectra, the overall shape and spectral width still generally agree with the QDI solution absorption spectrum (shaded grey). It would evidently be highly beneficial to be able to measure excitation spectra of individual molecules fast and with high spectral resolution under ambient conditions. Only recently we have developed an experimental approach based on the broadband interferometric excitation that enables fast measurement of excitation spectra of single molecules at ambient conditions with high spectral resolution.[31] Correlations between the excitation and emission spectra will undoubtedly shed new light on many fundamental intra- and intermolecular processes.

Based on our experiments the exact reason for the observed strong spectral variability among the individual molecules is hard to pinpoint. However, we know that the extent of π electron delocalization in the rylene backbone determines the absorption/emission wavelengths and thus one possible explanation is that the conjugation in the backbone is truncated differently for individual molecules. In rylene dyes there is a 100 nm bathochromic shift for each attached naphthalene group. The shorter backbone homologues of QDI, like PDI and TDI, have their absorption maxima in solution at approximately 560 and 660 nm, respectively. Importantly, they all have the same shape of the absorption spectrum. It thus seems plausible that either the change in the conformation and/or interactions with the polymer leads to a different degree of truncation of conjugation and to the observed distribution. It could either be the long hydrocarbon chains folding onto the backbone or the proximity of the polymer interfering with the conjugation. The unchanged emission and excitation spectra (the latter at least in terms of the width) would support this.

To check this hypothesis we repeated the experiments for QDI molecules spin-coated on a glass coverslip without any polymer. Unfortunately, bleaching, strongly reduced observation time and much enhanced blinking dynamics did not allow us to draw any conclusions.

Hofkens *et al.* showed that substitutions in the bay region (middle region of the naphthalene backbone) of PDI based molecules may lead to large twists of the naphthalene backbone. However, twists reaching tens of degrees lead to spectral shifts of the emission spectra of only up to 40 nm.[14] On the one hand, in PDI (shorter homologue of QDI) without any substitutions in the bay area the flat conformation was calculated to be the most favourable one. On the other hand, the twist of the naphthalene backbone is associated with the loss of the fine structure in the vibrational progression. Even though QDI does not have any bay substitutions, we did observe a smoothing of the vibrational progression in the emission spectra of many measured molecules. Perhaps minor intrinsic or polymer induced twisting of the backbone also play a role in our studies.

Our results clearly demonstrate that SM experiments can be highly biased towards sub-ensembles in the case of a large spectral heterogeneity among the molecules. For QDI, the fact that the spread is nearly 200 nm broad combined with the choice of a particular wavelength (excitation or detection) might result in completely failing to capture part of the SM ensemble. It is of fundamental importance to keep this in mind for any SM experiment, because reliance on the ensemble solution spectrum for choosing appropriate excitation and detection wavelengths may lead to highly biased results.

Both the multicolour excitation confocal images and the emission spectroscopy experiments provided information on the relative amount of molecules absorbing and emitting at specific wavelengths. We intuitively built up SM ensemble spectra and compared them to the absorption and emission spectra of QDI in solution, as shown in Fig. 4c and d. The spread of absorption wavelengths was well represented but we did find a discrepancy in the number of molecules absorbing at specific wavelengths. Our statistics were quite limited (~100 molecules) but it seems that there was about a 50/50 ratio of molecules absorbing around 750 nm to molecules absorbing at all other wavelengths, whereas the dry QDI film suggested that the distribution should peak around 700 nm. The origin of this difference is unclear, but we cannot really rule out changes in the absorption cross section for spectrally distinct molecules.

The SM ensemble emission differed significantly from the dry QDI film emission spectrum. This difference is easy to understand because the experiment is highly biased towards molecules absorbing in the blue upon excitation at 633 nm. In fact we could spectrally tune the SM ensemble emission spectrum by simply taking into account molecules of different brightness and consequently of different spectral position (see Fig. 4d). Furthermore, a reasonable number of the molecules whose fluorescence does not significantly stand out against the background might simply be ignored in the analysis. Lastly, changes in the quantum efficiency of the molecule with the spectral position cannot be completely excluded.

Conclusions

In this work we have addressed spectral changes in absorption and emission spectra of SMs embedded in a polymer matrix. We have found that QDI molecules exhibit a nearly 200 nm broad distribution of the position of the absorption and emission spectra, shifted towards lower wavelengths from the maximum absorption of QDI in solution. All molecules show a similar spectral shape but differ significantly in electronic transition energy.

We have demonstrated that any SM experiment may lead to biased results depending on the choice of excitation and detection wavelengths when the investigated molecules exhibit a large spectral variability. In extreme cases, part of the SM population may not be detected at all.

Due to an interplay between numerous intrinsic and extrinsic causes for spectral shifts, the spectral variability among SMs is hard to predict. Multiple wavelength excitation should therefore be utilized to find the full bandwidth of spectral positions of SMs on the sample.

Finally, being able to measure excitation spectra of individual molecules fast and with high spectral resolution would expand our knowledge of fundamental processes driving chemistry at the SM level. In particular, correlations between excitation spectra and other observables such as quantum efficiency, emission spectra, blinking dynamics and conformation might be extremely insightful. Moreover, knowledge of the position of the absorption and emission spectra is of fundamental importance for novel experimental approaches where for instance plasmonic enhancement is used on either absorption or emission of individual molecules.[32,33]

Experimental

Experimental setup

A laser beam was derived from either a white light source (SuperK, NKT) or a HeNe laser. The laser beam was cleaned up using an appropriate band-pass filter and propagated colinearly into an inverted microscope (Observer D1, Zeiss). Inside the microscope, the laser light was reflected from a 50/50 beamsplitter and focused to a diffraction-limited spot onto the sample with a high numerical aperture oil-immersion objective (1.3 NA, 100×, Zeiss Fluar). The fluorescence from the sample was collected through the same objective and then either sent to a spectrometer equipped with an electron-multiplying charge-coupled device (EMCCD, Newton, Andor) camera or focused onto an avalanche photo-diode (APD, Perkin-Elmer) that allows confocal optical imaging of the sample. Long-pass filters were placed in the detection path to block the reflected laser light. The sample was placed onto a piezo-controlled stage that enabled precise positioning and scanning. A shutter was positioned in front of the microscope to prevent sample illumination in between measurements.

The laser power was typically set in the range 1–4 μW for a single beam at the sample position. For an average molecule 2000–3000 counts per s were detected, of which about ~200 counts per s was background count. Background count included the dark counts of the APD (~60 counts per s) and residual stray light above the cut-off wavelength of the long-pass filter.

All the experiments were performed at least twice to ensure reproducibility.

Sample preparation

The experiments were performed on single quaterrylene diimide (QDI) molecules derived from the rylene dye family and obtained from the Müllen group (Max Planck Institute for Polymer Research, Mainz, Germany).

The samples were prepared by dissolving QDI molecules at a ~0.1–1 nM concentration in a poly(methyl methacrylate)–toluene solution (PMMA/toluene mixture ~1% w/v). Approximately 50 μL of the solution was then spin-coated onto a nr 1 microscope cover slip for 60 s at a spinning rate of 2000 rpm. Microscope cover slips were cleaned by immersing them in a piranha solution (1 : 2 ratio hydrogen peroxide to sulfuric acid) for ~30 minutes, then rinsed with deionized water and blow-dried with N_2. We found this procedure yielded no or very little contamination on the cover slips.

Absorption and emission spectra of the QDI ensemble

The absorption spectrum of QDI in solution was measured using a commercial spectrophotometer (NanoDrop 2000, Thermo Scientific). The absorption spectrum of the dry QDI film was measured using the experimental setup described above. A drop of QDI solution was placed onto a cover slip and allowed to dry. White light, derived from the built-in halogen lamp of the microscope, was transmitted through the sample and sent into the spectrometer. A reference white light spectrum was always measured alongside the dried QDI spot.

We used the same setup to measure the emission spectra of both dry QDI and QDI in solution. For the latter we placed a drop of the solution onto the cover slip and measured the emission spectrum directly afterwards. Both emission spectra were measured using 633 nm excitation light derived from a HeNe laser.

SM emission spectra

SM emission spectra were measured exciting at ~633 nm with a HeNe laser. The laser light was cleaned up with a 632/11 band-pass filter and blocked with a 647 nm LP in front of the APD and spectrometer. The filters in front of the spectrometer were chosen as close to the excitation wavelength as possible in order to maximize the detectable spectral range.

After taking the confocal image, the background spectrum was taken with the EMCCD on the sample away from the molecules without changing the focus. Then, individual molecules were placed in the laser focus to record their emission spectra. Typically, 20 background-corrected emission spectra were recorded in a kinetic series with an integration time of 1 s per spectrum. The spectra before photobleaching or spectral jumping were averaged per molecule to improve the signal-to-noise ratio. The gain of the used EMCCD camera was set to 200. The quantum efficiency of our camera reaches its maximum around 700 nm, and its variation in the spectral range of 600–800 nm is less than 8%.

Multicolour excitation confocal images

Multicolour excitation confocal images were acquired in two different schemes. In the first approach the confocal images were measured for a specific set of excitation and detection long-pass filters. In the second approach the long-pass filter was fixed and only the excitation wavelength was tuned. In both schemes, the confocal fluorescence images were recorded with 5 nm broad excitation bands derived from an ultrabroadband laser (SuperK, NKT) equipped with an acousto-optic tunable filter. A narrowband band-pass filter matching the excitation wavelength was placed in the beam to ensure that no other wavelengths were reaching the sample.

In the first experiment, the excitation wavelengths were 570, 630, 680, 710 and 750 nm with respectively a 610 LP, 650 LP, 730 LP, 740 LP and 776 LP filter (schematically indicated in Fig. 2a). In the second experiment we used the same excitation wavelengths while keeping the 776 nm long-pass filter for all acquired images (schematically shown in Fig. 3f). The images have a size of 20 × 20 μm with 200 × 200 pixel resolution and an acquisition time of 10 ms per pixel.

Acknowledgements

L. P. acknowledges financial support from the Marie-Curie International Fellowship COFUND and the ICFOnest program. E. G. acknowledges financial support from the Erasmus+ program. This research was funded by the European Commission (ERC Advanced Grant 247330-Nano Antennas), Fundació CELLEX (Barcelona) and Plan Nacional project FIS2012-35527.

Notes and references

1 W. E. Moerner, *J. Phys. Chem. B*, 2002, **106**, 910–927.
2 B. Kozankiewicz and M. Orrit, *Chem. Soc. Rev.*, 2014, **43**, 1029–1043.

3 R. Korlacki, M. Steiner, H. Qian, A. Hartschuh and A. J. Meixner, *ChemPhysChem*, 2007, **8**, 1049–1055.
4 W. E. Moerner and L. Kador, *Phys. Rev. Lett.*, 1989, **62**, 2535–2538.
5 M. Orrit and J. Bernard, *Phys. Rev. Lett.*, 1990, **65**, 2716–2719.
6 P. Tchenio, A. B. Myers and W. E. Moerner, *Chem. Phys. Lett.*, 1993, **213**, 325–332.
7 P. Tchenio, A. B. Myers and W. E. Moerner, *J. Phys. Chem.*, 1993, **97**, 2491–2493.
8 R. M. Dickson, A. B. Cubitt, R. Y. Tsien and W. E. Moerner, *Nature*, 1997, **388**, 355–358.
9 T. Basche and W. E. Moerner, *Nature*, 1992, **355**, 335–337.
10 A. Zumbusch, L. Fleury, R. Brown, J. Bernard and M. Orrit, *Phys. Rev. Lett.*, 1993, **70**, 3584–3587.
11 W. P. Ambrose and W. E. Moerner, *Nature*, 1991, **349**, 225–227.
12 F. Schleifenbaum, C. Blum, V. Subramaniam and A. J. Meixner, *Mol. Phys.*, 2009, **107**, 1923–1942.
13 H. Wang, A. M. Bardo, M. M. Collinson and D. A. Higgins, *J. Phys. Chem. B*, 1998, **102**, 7231–7237.
14 J. Hofkens, T. Vosch, M. Maus, F. Köhn, M. Cotlet, T. Weil, A. Herrmann, K. Müllen and F. C. D. Schryver, *Chem. Phys. Lett.*, 2001, **333**, 255–263.
15 A. Margineanu, J. Hofkens, M. Cotlet, S. Habuchi, A. Stefan, J. Qu, C. Kohl, K. Müllen, J. Vercammen, Y. Engelborghs, T. Gensch and F. C. de Schryver, *J. Phys. Chem. B*, 2004, **108**, 12242–12251.
16 B. Araoz, D. Täuber, C. v. Borczyskowski and P. F. Aramendia, *J. Phys. Chem. C*, 2012, **116**, 7573–7580.
17 M. A. Bopp, A. J. Meixner, G. Tarrach, I. Zschokke-Gränacher and L. Novotny, *Chem. Phys. Lett.*, 1996, **263**, 721–726.
18 H. Kobayashi, K. Tsuchiya, K. Ogino and M. Vacha, *Phys. Chem. Chem. Phys.*, 2012, **14**, 10114–10118.
19 C. Blum, A. J. Meixner and V. Subramaniam, *Biophys. J.*, 2004, **87**, 4172–4179.
20 K. D. Weston, P. J. Carson, H. Metiu and S. K. Buratto, *J. Chem. Phys.*, 1998, **109**, 7474–7485.
21 Y. J. Lee, D. Y. Kim and P. F. Barbara, *J. Phys. Chem. B*, 2006, **110**, 9739–9742.
22 S. Chong, W. Min and X. S. Xie, *J. Phys. Chem. Lett.*, 2010, **1**, 3316–3322.
23 D. S. English, A. Furube and P. F. Barbara, *Chem. Phys. Lett.*, 2000, **324**, 15–19.
24 H. Quante and K. Müllen, *Angew. Chem., Int. Ed. Engl.*, 1995, **34**, 1323–1325.
25 M. Celebrano, P. Kukura, A. Renn and V. Sandoghdar, *Nat. Photonics*, 2011, **5**, 95–98.
26 R. Hildner, D. Brinks and N. F. van Hulst, *Nat. Phys.*, 2011, **7**, 172–177.
27 J. Hwang, M. M. Fejer and W. E. Moerner, *Phys. Rev. A*, 2006, **73**, 021802.
28 D. Brinks, F. D. Stefani, F. Kulzer, R. Hildner, T. H. Taminiau, Y. Avlasevich, K. Mullen and N. F. van Hulst, *Nature*, 2010, **465**, 905–908.
29 Y. Avlasevich, C. Li and K. Müllen, *J. Mater. Chem.*, 2010, **20**, 3814–3826.
30 F. Würthner, Z. Chen, V. Dehm and V. Stepanenko, *Chem. Commun.*, 2006, 1188–1190.
31 L. Piatkowski, E. Gellings and N. F. van Hulst, Broadband single molecule excitation spectroscopy, *Nat. Commun.*, under consideration
32 E. Wientjes, J. Renger, A. G. Curto, R. Cogdell and N. F. v. Hulst, *Nat. Commun.*, 2014, **5**, 4236.
33 S. Yampolsky, D. A. Fishman, S. Dey, E. Hulkko, M. Banik, E. O. Potma and V. A. Apkarian, *Nat. Photonics*, 2014, **8**, 650–656.

Faraday Discussions

PAPER

Intramolecular photostabilization *via* triplet-state quenching: design principles to make organic fluorophores "self-healing"†

Jasper H. M. van der Velde,[a] Jaakko J. Uusitalo,[b] Lourens-Jan Ugen,[a] Eliza M. Warszawik,[c] Andreas Herrmann,[c] Siewert J. Marrink[b] and Thorben Cordes*[a]

Received 25th June 2015, Accepted 24th July 2015
DOI: 10.1039/c5fd00114e

Covalent linkage of fluorophores and photostabilizers was recently revived as a strategy to make organic fluorophores "self-healing" *via* triplet-state quenching. Although Lüttke and co-workers pioneered this strategy already in the 1980s, the general design principles still remain elusive. In this contribution, we combine experiments and theory to understand what determines the photostabilization efficiency in dye–photostabilizer conjugates. Our results from single-molecule microscopy and molecular dynamics simulations of different Cy5-derivatives suggest that the distance and relative geometry between the fluorophore and photostabilizer are more important than the chemical nature of the photostabilizer, *e.g.* its redox potential, which is known to influence electron-transfer rates. We hypothesize that the efficiency of photostabilization scales directly with the contact rate of the fluorophore and photostabilizer. This study represents an important step in the understanding of the molecular mechanism of intramolecular photostabilization and can pave the way for further development of stable emitters for various applications.

Introduction

Fluorescence has become an essential tool for spectroscopic studies, biomedical sensing, and for imaging of (dynamic) processes and structures across various

[a]Molecular Microscopy Research Group & Single-molecule Biophysics, Zernike Institute for Advanced Materials, University of Groningen, Nijenborgh 4, 9747 AG Groningen, The Netherlands. E-mail: t.m.cordes@rug.nl
[b]Groningen Biomolecular Sciences and Biotechnology Institute, Zernike Institute for Advanced Materials, University of Groningen, Nijenborgh 7, 9747 AG Groningen, The Netherlands
[c]Department of Polymer Chemistry, Zernike Institute for Advanced Materials, University of Groningen, Nijenborgh 4, 9747 AG Groningen, The Netherlands

† Electronic supplementary information (ESI) available: Additional single-molecule and simulation data. See DOI: 10.1039/c5fd00114e

scientific disciplines. The photophysical performance of the employed fluorescent reporters – signal stability, intensity and duration – is still one of the limiting factors for many applications.[1] Recently, intramolecular photostabilization[2–4] was revived[5–9] as a tool to impart organic fluorophores with "self-healing" properties.[7] The covalent linkage of photostabilizers, *e.g.* triplet-state quenchers or redox-active molecules, to organic fluorophores allows suppression of irreversible destruction and transient excursions to dark-states (triplet- or radical-states). While bleaching pathways are not always triplet-based,[10,11] they in fact represent one of the most common reasons for photobleaching.

The underlying mechanism of intramolecular photostabilization *via* triplet-state quenching, along with highly desirable design principles to make any organic fluorophore "self-healing", are still under active debate.[5,7–9,12] Furthermore, all self-healing fluorophores described so far show less efficient photostabilization[5,8,9] compared to the use of the same photostabilizers as a solution-additive[13–17] – an approach that is experimentally much simpler compared to the synthesis of photostabilizer–dye conjugates needed for self-healing.[18,19] It should be emphasized, however, that intramolecular photostabilization overcomes various limitations of solution-based healing approaches since it uses only high local concentrations of the photostabilizer. This reduces the general toxicity, reduces alterations of the biological structure, and makes photostabilization even possible *in vivo* or in situations where diffusion-based collisions between photostabilizer and fluorophore are prohibited. Considering these advantages, it can be suggested that intramolecular photostabilization is a superior method – once as efficient as solution-based approaches. The rational design of photostabilizer–dye conjugates, combinations with solution-based approaches, and mechanistic understanding should hence be primary goals of research in this area.

Collisions between the fluorophore and photostabilizer are essential for triplet-state quenching, independent of the mechanism (inter- or intramolecular). In dye–photostabilizer conjugates, the actual distance dependence seems, however, ambiguous. The performance of triplet-state quenchers, such as cyclooctatetraene (COT), scales directly with proximity,[5,6] however, redox-active compounds, such as nitrophenyl-groups, show a more complicated behaviour.[9,19] Here, cyanine-based photostabilizer–dye conjugates comprising nitrophenyl-groups *via* constrained linkers show a relatively moderate performance with a ~10-fold increase in photostability when using an unnatural amino acid with an aromatic nitro group.[19] A different conjugation between the same photostabilizer and fluorophore, having a larger spacing and increased structural flexibility, shows a significantly higher photostability with a ~100-fold improvement compared to the non-stabilized fluorophore[9] – a value that is close to the results of pristine photostabilizers in solution (~200-fold improvement in the photostability).[9]

To mechanistically understand the differences between triplet-state quenchers and oxidants in intramolecular photostabilization and ultimately to establish design principles for making organic fluorophores "self-healing", a detailed photophysical characterization of cyanine-based photostabilizer–dye conjugates is presented in this paper. Single molecule fluorescence microscopy in combination with molecular dynamics simulations allows us to identify critical parameters that determine the efficiency of intramolecular photostabilization. Therefore, we explored the consequences of structural modifications in oxidizing photostabilizers, *i.e.* different aromatic nitro moieties used for intramolecular photostabilization of cyanine fluorophores, such as Cy5. We further used molecular dynamics simulations to probe the distance and orientation dependence between the fluorophore and photostabilizer. The obtained results suggest that the distance and possible orientations between the fluorophore and photostabilizer are the crucial parameters for efficient intramolecular healing *via* triplet-state quenching rather than the chemical structure of the photostabilizer. We also show that the orientation of the dye and photostabilizer is largely governed by their local chemical environment (in our case DNA), which allows for end-stacking or binding to grooves. This study represents an important step in the understanding of the mechanism and working principles of intramolecular photostabilization and hence paves the way to obtain improved fluorophores for spectroscopic studies and imaging when triplet-based damage is the major route for photobleaching.

Results and discussion

After the pioneering work of Lüttke and co-workers in the 1980s,[2–4] the Blanchard lab started to use both triplet-state quenchers, *e.g.* COT, as well as (anti)oxidants, such as trolox (TX) and nitrobenzoic alcohol (NBA), for intramolecular photostabilization *via* triplet-state quenching.[5,6,12,18] In our studies, we focused on the mechanistic understanding of intramolecular photostabilization *via* (anti) oxidants[7–9] and more recently on its practical implementation. For that purpose, a simple and versatile scaffolding strategy for photostabilizer–dye conjugates based on unnatural amino acids was developed.[19] The resulting conjugates can be readily attached to various biomolecular targets (Fig. 1a, left panel, amino acid scaffolding). More details on the chemical synthesis and applicability of intramolecular photostabilization for various fluorophore classes (rhodamines, carbopyronines and cyanines) *via* amino acid scaffolding are given in ref. 19.

In demanding fluorescence applications, such as single-molecule microscopy, intramolecular photostabilization *via* triplet-state quenching turned out to be less efficient than the use of solution-based stabilizers, *i.e.* the addition of similar compounds to the imaging buffer at high (~mM) concentrations.[7–9] Furthermore, the effect of intramolecular photostabilization for both triplet-state quenchers (COT) as well as (anti)oxidants (TX, NBA) seemed to correlate with their spatial separation. This was shown by using target scaffolding (Fig. 1a, right panel) where the photostabilizer and fluorophore are bound to different complimentary DNA-strands with varying attachment points for the emitter and stabilizer. After hybridization, it was found that a decreasing distance resulted in an increased photostability.[6,12,18] This motivated us to develop more efficient photostabilizers for intramolecular triplet-state quenching to better compete with solution-based approaches.[9] Notably, these experiments showed that a molecule containing an aromatic nitro group can show surprisingly different photostabilization efficiencies for the same fluorophore depending on the attachment geometry: a proximal attachment of the photostabilizer resulted in extremely stable fluorescence (Fig. 1b, bottom, Cy5–NPAA) while direct conjugation of the photostabilizer and fluorophore showed less photostabilization (Fig. 1b, middle, NPA–Cy5); the qualitative increases in the photostability can be seen by a comparison of the trace length of individual fluorophores (Fig. 1b).

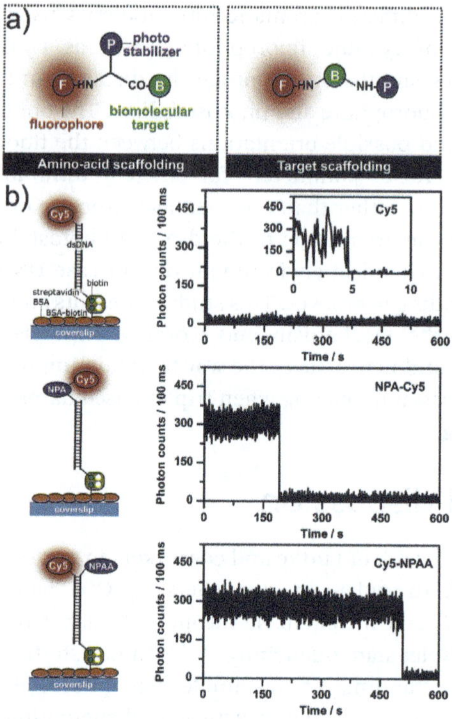

Fig. 1 (a) Chemical strategies to scaffold photostabilizer–dye conjugates using the different residues of (unnatural) amino acids or strategic labelling of the biomolecular target to create permanent proximity. (b) Experimental strategy to immobilize dsDNA on microscope coverslips to study the photophysics of different Cy5 derivatives: parent fluorophore Cy5, photostabilizer and fluorophore linked *via* amino acid scaffolding (NPA–Cy5), and proximal attachment of the photostabilizer and fluorophore (Cy5–NPAA). The figure also shows representative fluorescence time traces of individual molecules from TIRF microscopy of the respective compounds. Additional data for each compound in panel (b) are shown in Fig. S1–S3.†

To understand these effects, the photostability, fluorescence count rate, total number of detected photons and the signal-to-noise ratio (SNR) were determined for different surface-immobilized Cy5-derivatives that were linked to double-stranded DNA (dsDNA, Fig. 1 and 2) using camera-based imaging. Surface-immobilization was done according to published procedures on a BSA-biotin–streptavidin-coated microscope coverslip binding to a 3′-terminal biotin unit of the same dsDNA (Fig. 1b). Experiments were performed in the absence of oxygen and without photostabilizers in solution to minimize the convolution of inter- and intramolecular photostabilization. We performed single-molecule TIRF microscopy with excitation at 640 nm (intensities of ~50–100 W cm^{-2}) with a 100 ms integration time. Typical camera frames of single Cy5 fluorophores are shown in the ESI, *e.g.* Fig. S1–S3.† To quantify the photostability, the number of fluorescent molecules per video frame was determined and the decay in absolute numbers over subsequent image frames was fitted with an exponential decay to obtain the photobleaching lifetime, τ_{bleach} (Fig. 2a). From the videos, we extracted background-corrected single-molecule time traces to determine the fluorescence

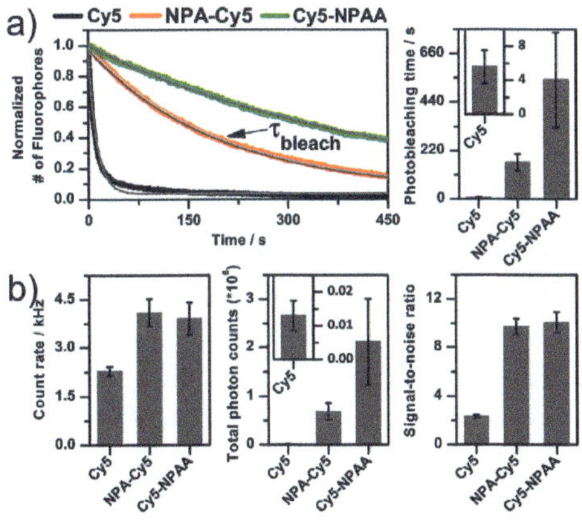

Fig. 2 Quantitative photophysical characterization of photostabilizer–dye conjugates of Cy5 derived from single-molecule TIRF microscopy. Data was recorded in aqueous PBS buffer at pH 7.4 in the absence of oxygen under continuous 640 nm excitation with ~50 W cm^{-2}. (a) Decay curve of the number of fluorophores as a function of time. Fitting an exponential function (grey) to this decay resulted in the photobleaching lifetime. The curves shown were obtained by averaging over >5 TIRF movies. (b) Background-corrected intensity traces were used to derive the total number of photons (N_{total}), the average signal in kHz (brightness) and the signal-to-noise ratio (SNR). For further details of the experimental techniques, data acquisition and analysis see the Experimental section.

count rate in kHz, the signal-to-noise ratio, and the total number of detected photons before photobleaching (Fig. 2b, N_{total} = brightness × τ_{bleach}). The mean and standard deviation of all values was derived from multiple ($n \geq 3$) independent experiments.

The analysis of Cy5 and two of its derivatives, comprising either a photostabilizer covalently attached or in close proximity, is summarized in Fig. 2. It is apparent that Cy5 shows a low photostability (τ_{bleach} of 5.6 ± 1.9 s), count rate (2.3 ± 0.1 kHz), and SNR (2.3 ± 0.1), and hence has little potential for applicability without a photostabilizer (see also traces in Fig. 1b and S1†).

Adding the nitrophenyl-moiety *via* amino acid or target scaffolding improves all values drastically up to a τ_{bleach} value of ~550 s, count rates of ~4 kHz and SNR of ~10, allowing us to collect more than 2 × 10^6 photons from individual molecules (Fig. 2) with little signal fluctuations (see fluorescence time traces in Fig. S2 and S3†). The experiments revealed, however, a substantial difference in photostability, *i.e.* τ_{bleach} and N_{phot} for NPA–Cy5 and Cy5–NPAA (Fig. 1 and 2).

This comparison alone does not, however, allow a satisfying conclusion about the (molecular) reasons for the observed difference. Unfortunately, in NPA–Cy5 and Cy5–NPAA, multiple aspects of the dye–photostabilizer conjugate are altered at the same time: (i) the chemical structure of the aromatic nitro group is different for both compounds, potentially changing its redox-properties[20] *via* the electron-withdrawing or -donating characteristic of the different substituents[21] and hence altering the rates for photo-induced electron transfer;[22] (ii) the fluorophore and the photostabilizer are linked differently, which changes both their spatial separation and biochemical environment. It is known from other studies of cyanine dyes on DNA[23–25] that the local environment has a drastic influence on the photophysical properties, such as the lifetime, spectrum, quantum yield and others, mainly *via* end-stacking or groove-binding with the DNA.[26,27] Our approach in this paper was to de-convolute both effects *via* a combination of experiments on different dye–photostabilizer conjugates and to use molecular dynamics simulations to understand the possible modes of interactions in NPA–Cy5 and Cy5–NPAA.

First, we systematically varied the chemical structure of the aromatic nitro compound while keeping it at a fixed distance to the Cy5-fluorophore. Fig. 3a shows the chemical structures and photophysical properties of three different dye–photostabilizer conjugates that use nitrophenyl-acetic acid (NPAA), nitrophenyl-propanoic acid (NPP) or nitrophenylalanine (NPA) as the photostabilizer. As seen in Fig. 3a, the chemical structure of the conjugates shows variations in the linking element between the nitrophenyl and amide group. In NPAA and in NPP, a methylene group and an ethyl spacer are present, respectively, whereas NPA exhibits an additional amino-group that could have effects on the photophysical properties of fluorophores.[4]

The photophysical properties of all three compounds appeared to be strikingly similar (Fig. 3). Besides variations in the fluorescence count rate from 3–5 kHz (Fig. 3c), all other parameters were found in a very similar range given the experimental errors: the photobleaching lifetimes were in the range of $\sim$550 s, the total detected photon numbers were around $\sim 2 \times 10^6$ and the SNR was $\sim$10. The difference in the count rates indicates that indeed the redox-potential of the compounds might be different when comparing NPAA with NPP/NPA. The lower count-rate of the latter might indicate different degrees of singlet-quenching and hence a reduced quantum yield.[28] Interestingly, the photostabilizing effects are identical for all compounds. Since the photostability is mainly improved *via* depletion of the reactive triplet-state, we propose that the rate of triplet-quenching is governed by interactions of the photostabilizer and fluorophore with the local chemical environment, which dictates the collision rate between the photostabilizer and fluorophore. Our results indicate that the overall performance of intramolecular photostabilization with aromatic nitro groups does not strongly depend on the actual chemical structure but instead on the interactions between the photostabilizer and fluorophore.

To understand the molecular details of these processes, we performed molecular dynamics simulations to (i) map the specific local environment of the fluorophore and photostabilizer, (ii) determine the possible interactions/binding modes of both including (iii) their collision rate. These three factors could account for the differences in the fluorophore brightness, lifetime (i/ii) and photostability (iii).

In Fig. 4, we show the simulation results for proximally attached NPAA (Cy5–NPAA, Fig. 4a), a directly conjugated fluorophore (NPA–Cy5, Fig. 4b), and a derivative of NPA–Cy5 that features a longer PEG-linker of the form $(O{-}CH_2{-}CH_2)_4$ between the fluorophore and photostabilizer (Fig. 4c). The latter should mimic the superior interaction possibilities of Cy5–NPAA but at the same time use the practically more relevant attachment geometry of NPA–Cy5.[19] For each system, the panels from left to right show the type of attachment as a cartoon (left), the most frequently occupied volumes of Cy5 and the photostabilizer around dsDNA, the distances between the different moieties measured during the simulation, and the volume-maps of the photostabilizer around Cy5 (right).

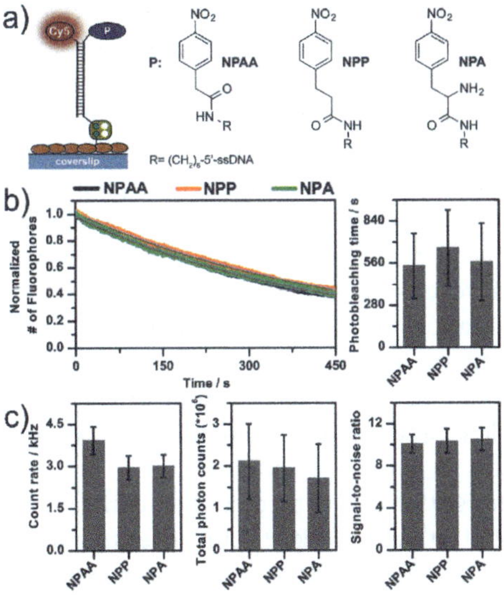

Fig. 3 (a) Scheme of the linking and immobilization strategy of different photostabilizer–dye conjugates of Cy5 including the chemical structure of the aromatic nitro moieties. (b and c) Quantitative photophysical characterization of the photostabilizer–dye conjugates. Data was recorded in aqueous PBS buffer at pH 7.4 in the absence of oxygen under continuous 640 nm excitation with ~50 W cm^{-2}. (b) Decay curve of the number of fluorophores as a function of time. Fitting an exponential function (grey) to this decay resulted in the photobleaching lifetime. The curves shown were obtained by averaging over >5 TIRF movies. (c) Background-corrected intensity traces were used to derive the total number of photons (N_{total}), the average signal in kHz (brightness) and the signal-to-noise ratio (SNR). For further details of the experimental techniques, data acquisition and analysis see the Experimental section.

The second column shows the most frequently occupied volumes of both Cy5 (shown in red) and the photostabilizer (shown in blue). In both cases, the volumes enclose all points relative to the DNA that are occupied in at least 2% of the simulation frames. In the case of Cy5–NPAA, the most likely location for both Cy5 and NPAA is on top of the DNA. The same is true for Cy5 in the absence of the photostabilizer, where pronounced end-stacking can be observed (Fig. S6†). For NPA–Cy5, both NPA and Cy5 are neither in a conformation that is directly on top of DNA nor in a DNA groove. The conformation shown in Fig. 1b is, however, fairly stable. When extending the linker between NPA and Cy5, both Cy5 and NPA are most likely located in a DNA groove. Similar conclusions can be drawn when looking at the distances between Cy5 and the photostabilizer (P–Cy5), as well as the distance of Cy5 from the top of the DNA (stacked DNA–Cy5) and the DNA groove (groove DNA–Cy5). A detailed description of how each distance was measured from the simulations is given in the Experimental section.

In the fourth column, the blue volume map encloses all the points relative to Cy5 that are occupied by the photostabilizer in at least 0.6% of the simulation frames. It is evident from the volume maps that in Cy5–NPAA and NPA–Cy5 the photostabilizer can interact with different parts of Cy5. Whereas in Cy5–NPAA interactions with both the aromatic part and the polymethine chain of Cy5 are observed, in NPA–Cy5 the interactions are limited to a small aromatic region at the linker attachment point of Cy5. Increasing the linker length (NPA–PEG4–Cy5) allows the NPA to interact with a larger surface region of Cy5.

While this data reveals clear differences for the interaction of the photostabilizer and fluorophore as well as their location within the chemical environment, the meaning of these results for photostabilization efficiency remains unclear. Our current model for intramolecular photostabilization *via* photoinduced electron transfer assumes a ping-pong mechanism[7,8] with two subsequent electron transfer steps (Fig. 5). In the case of NPA-based conjugates, fluorophore oxidation results in the formation of two radical species, *i.e.* a fluorophore radical cation and a photostabilizer radical anion. Since both species cannot be separated by diffusive processes, both ultimately recombine in a second redox-reaction. Such a double-electron transfer ("geminate recombination") is also postulated as a mechanism for diffusion-based photostabilization of organic fluorophores *via* thiol-compounds.[29]

Since it is not straightforward to predict whether the exact contact-mode of the fluorophore and photostabilizer (Fig. 4) influences the rate for electron transfer, we determined the "kinetics" of the collisions between Cy5 and the photostabilizer (Fig. 6, collision frequency). Cy5 and the photostabilizer were classified

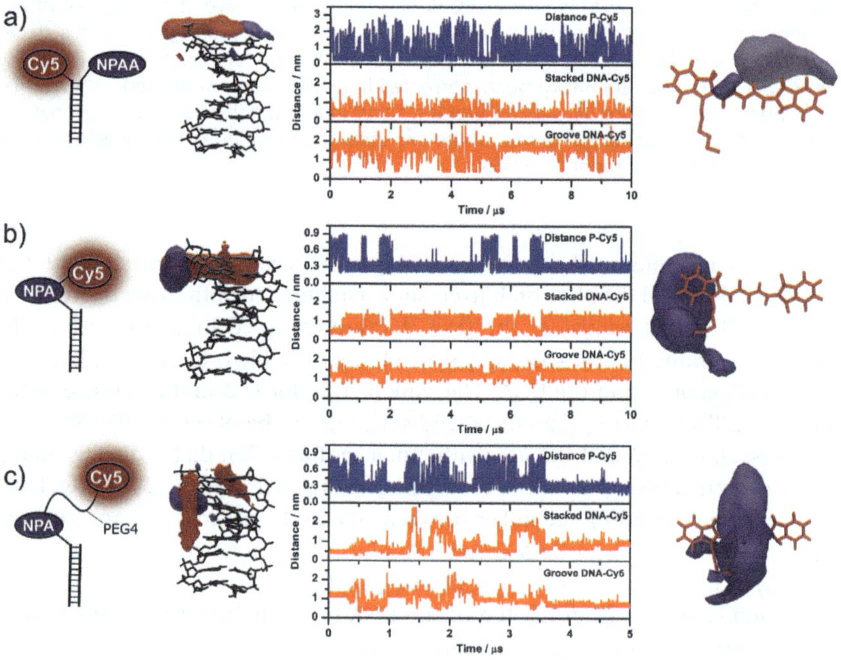

Fig. 4 Relative orientation of Cy5, the photostabilizer and DNA during the simulations of the photostabilizer–dye conjugates: (a) Cy5–NPAA, (b) NPA–Cy5 and (c) NPA–PEG4–Cy5. The first column on the left shows the attachment of the molecules to DNA. The most common locations of Cy5 (shown in red) and the photostabilizer (shown in blue) relative to DNA are shown in the second column. The distances between Cy5 and the photostabilizer as well as Cy5 and the top and groove of DNA are presented in the third column. The fourth column shows the most common location of the photostabilizer relative to the fluorophore Cy5.

as having collided and being in contact when the distance between their centres of mass was smaller than 0.5 nm. Besides the actual collision frequency, and the mean and median times for the duration of being associated (Fig. 6, mean collision time), we also calculated the ratio of time they spent in contact (bound) *versus* unbound during the simulations (Fig. 6, bound/unbound ratio).

For Cy5–NPAA, our data shows a high contact frequency with short association times (Fig. 6 and Table S1†). Although Cy5 and NPAA were frequently in contact, NPAA also explored other regions frequently while Cy5 was mostly located on top of the DNA and was in the groove only in 11% of the simulation frames. The collision frequency is 0.09 ± 0.01 ns^{-1}, the mean and median contact times are 6 ± 1 ns and 2.3 ± 0.2 ns, respectively, and the bound/unbound ratio is 1.4 ± 0.3.

Inverse behaviour was found for NPA–Cy5, which shows a significantly lower collision frequency and a long association time (Fig. 6). Both the photostabilizer and fluorophore remain in contact in 87% of the simulation frames and the duration of the contacts is often long, while the position of Cy5 is somewhere between being on top of the DNA and in the groove of the DNA. In this case, the collision frequency is 0.02 ± 0.01 ns^{-1} and the bound/unbound ratio is 12 ± 11. The mean and median contact times are 40 ± 20 ns and 2 ± 1 ns, respectively. Due to the long contact times, the simulation results shown here are not long enough to accurately determine the exact bound/unbound ratio or the mean or median contact times, although these values are clearly larger than in the case of proximally attached NPAA. This is also reflected in the error estimates.

In the case of NPA–PEG4–Cy5, the increased mobility of NPA with respect to DNA is exhibited in values similar to those for Cy5–NPAA (Fig. 6). Furthermore, the longer linker increases the range of movement of Cy5 and it is frequently located in the grooves of the DNA. The collision frequency between NPA and Cy5 is 0.10 ± 0.04 ns^{-1} and the bound/unbound ratio is 5 ± 4. The mean and median contact times are 11 ± 8 ns and 0.8 ± 0.1 ns, respectively. Again, the simulation time is not long enough to give accurate estimates of the mean and median contact times.

These results indicate that a high collision frequency in combination with short association times of the photostabilizer and fluorophore results in the highest efficiency for photostabilization. This strongly suggests that the binding modes for the different compounds shown in Fig. 4 are not necessarily relevant for triplet-state quenching *via* photo-induced electron transfer since their duration does not scale with photostability. The contact rate, however, seems to follow this trend and only future studies with compounds like NPA–PEG1–Cy5 can show whether our hypothesis holds.

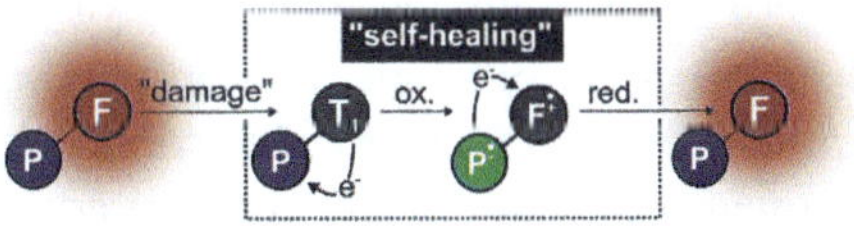

Fig. 5 Proposed model for the depletion of the triplet-state *via* intramolecular photostabilization with aromatic nitro groups: (photo)oxidation of the fluorophore and subsequent recombination of two radicals forms both the fluorophore and photostabilizer in their neutral chemical forms. The redox-cascade is referred to as "self-healing".

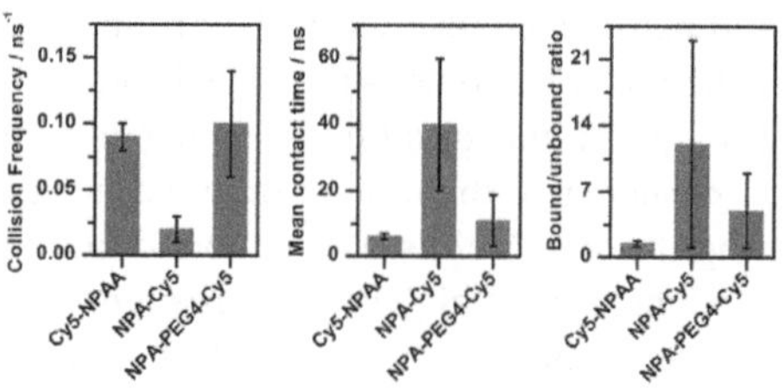

Fig. 6 Interactions of Cy5 and the photostabilizer during the simulations for Cy5–NPAA, NPA–Cy5 and NPA–PEG4–Cy5: collision frequencies (left), the mean contact times for each system (middle) and the bound/unbound ratio (right). All three metrics show that NPA–Cy5 interacts in a different manner compared to the other two derivatives.

Conclusions

This article establishes a mechanistic basis for intramolecular triplet-state quenching and the resulting photostabilization using an oxidizing aromatic nitro group. We investigated the relationship between the chemical structure of the photostabilizer as well as possible interactions with its chemical surrounding and the fluorophore to be protected. Using single-molecule microscopy, we found that the chemical structure and its associated changes in functional properties (redox-potential) play only a minor role in the ability to stabilize organic fluorophores *via* intramolecular triplet-state quenching. Molecular dynamics simulations suggest, however, that interactions of the fluorophore and the photostabilizer with each other and the chemical surroundings govern efficient photostabilization. We suggest that a high collision rate is needed in combination with short interaction times of the photostabilizer and fluorophore. The data indicates that a longer interaction period does not contribute towards an increased photostabilization. Our study represents an important step in understanding the molecular mechanism of intramolecular photostabilization, but also calls for further studies with *e.g.* systematic variation of the linker length and the photostabilizer.

Finally, it has to be stressed that intramolecular photostabilization is less efficient compared to the use of solution additives as photostabilizers. It remains, however, the only possible method for photostabilization when photostabilizers cannot be added to the system (*e.g.* under live-cell conditions), or when the addition of a diffusion-based photostabilizer is ineffective due to missing collisions or not tolerated by the system of interest.

Experimental

Sample preparation and surface immobilization

Immobilization and investigation of single fluorophores was achieved using a dsDNA scaffold comprising two 40-mer oligonucleotides, *i.e.* ssDNA–fluorophore and ssDNA–biotin. Sequences of both oligomers were adapted from ref. 16, 30 and 31. As a non-stabilized control fluorophore, we used ssDNA–fluorophore (Cy5-C6-5′-TAA TAT TCG ATT CCT TAC ACT TAT ATT GCA TAG CTA TAC G-3′, used as received from Eurofins and IBA, Germany). Single immobilized fluorophore molecules were studied in LabTek 8-well chambered cover slides (Nunc/

VWR, The Netherlands) with a volume of 750 μL, as described in ref. 22. After cleaning with 0.1% HF and washing with PBS buffer (one PBS tablet was dissolved in deionized water containing 10 mM phosphate buffer, 2.7 mM potassium chloride and 137 mM sodium chloride at pH 7.4; Sigma-Aldrich, The Netherlands), each chamber was incubated with a mixture of 5 mg/800 mL BSA and 1 mg/800 mL BSA/biotin (Sigma Aldrich, The Netherlands) at 4 °C in PBS buffer overnight. After rinsing with PBS buffer, a 0.2 mg mL^{-1} solution of streptavidin was incubated for 10 minutes and subsequently rinsed with PBS buffer.

The immobilization of dsDNA was achieved *via* a biotin–streptavidin interaction using pre-annealed dsDNA with the aim of observing single emitters for prolonged periods of time and allowing free rotation of the fluorophores. For this, 5–50 μL of a 1 μM solution of ssDNA–fluorophore or ssDNA–NPA–fluorophore was mixed with the complementary ssDNA–biotin at the same concentration (biotin-5′-CGT ATA GCT ATG CAA TAT AAG TGT AAG GAA TCG AAT ATT A-3′, used as received from IBA, Germany). The respective mixtures of the two oligomers were heated to 98 °C for 4 minutes and cooled down to 4 °C with a rate of 1 °C min^{-1} in annealing buffer (500 mM sodium chloride, 20 mM Tris–HCL, and 1 mM EDTA at pH = 8). The treated LabTek cover slides were incubated with a 50–100 pM solution of pre-annealed dsDNA for 1–2 minutes, leading to a typical surface coverage of fluorophore-labelled dsDNA as shown in the ESI, Fig. S1–S5.†

All single-molecule experiments were carried out at room temperature (22 ± 1 °C). Oxygen was removed from the buffer system utilizing an oxygen-scavenging system (PBS, pH = 7.4, containing 10% (wt/vol) glucose and 10% (vol/vol) glycerine, 50 μg mL^{-1} glucose-oxidase, 100–200 μg mL^{-1} catalase, and 0.1 mM tris(2-carboxyethyl)phosphine hydrochloride (TCEP)). As shown before, such low concentrations of the reducer TCEP have no noticeable effect on the photophysics of organic fluorophores[31,32] and hence do not convolute with the effects from intramolecular stabilization. Glucose-oxidase catalase (GOC)[22,33] was used instead of a combination of protocatechuic acid and protocatechuate-3,4-dioxygenase (PCA/PCD[28]) to avoid convolution of inter- and intramolecular photostabilization with PCA.[34]

TIRF-microscopy including data analysis

Wide-field fluorescence microscopy imaging was performed on an inverted microscope (Olympus IX-71 with UPlanSApo 100x, NA 1.49, Olympus, Germany) with an objective type total-internal-reflection fluorescence (TIRF) configuration, as described in ref. 8. Images were collected with a back-illuminated emCCD camera (512 × 512 pixel, C9100-13, Hamamatsu, Japan, filtered by ET700/75, AHF Analysentechnik, Germany). Excitation from a diode laser (Cube, Coherent, Germany, filtered by ZET640/10, Chroma, USA) was at 640 nm with ≈50 W cm^{-2} at the focal plane. To quantitatively characterize the photostability of the different photostabilized Cy5-derivatives, we imaged an area of 42 × 58 μm containing ≥100 molecules. Each area was illuminated for 600 s and recorded with an integration time of 100 ms. Fluorescence time traces of individual molecules were obtained by identifying those pixels that were at least 80–150 counts above the background noise in the standard deviation plot (standard deviation over all pixels over the complete movie) and averaging the intensity within a 3 × 3 pixel area in each recorded frame. Neighbouring peaks within a 5 pixels radius were excluded. The signal-to-noise ratio (SNR) was determined using the obtained fluorescence time traces after background correction by dividing the average fluorescence intensity of the signal before photobleaching by the standard deviation over the signal. The fluorescence intensity of the signal of >900 single emitters per Cy5-derivative was used to obtain quantitative information on the count rate. The number of fluorescent bursts in each frame image was determined using an absolute threshold criterion, *i.e.* the signal had to be higher than 800 counts (arbitrary units in the raw movie). The number of fluorescent emitters per frame was then plotted as a function of time and fitted to a mono-exponential decay: $y(t) = C + Ae^{-c/t}$ (with $c = 1/\tau_{\text{bleach}}$ and τ_{bleach} being the characteristic photobleaching time constant). Photobleaching times and the associated standard deviations were derived from multiple repeats of the experiment on different days with each ≥5 movies. For each Cy5-derivative, the total number of detected photons before photobleaching was calculated by a multiplication of the obtained count rate and photobleaching lifetime.

Functionalization of oligonucleotides

The procedures for the functionalization of oligonucleotides were established and described in detail in ref. 9 and 19. NPA–Cy5 is fully characterized in ref. 19 and Cy5–NPAA in ref. 9.

The synthesis of new ssDNA–photostabilizer conjugates was performed according to the following protocol: single-stranded DNA (ssDNA) was used as received from IBA, Germany (NH_2-C6-5′-TAA TAT TCG ATT CCT CTG GAC GTAA TAT TCG ATT CCT TAC ACT TAT ATT GCA TAG CTA TAC G-3′). The lyophilized ssDNA was resuspended in MilliQ water and adjusted to 80 μM in 0.2 M $NaHCO_3$ buffer (pH 8.35). To this solution, the same volume of a 20 mg mL^{-1} solution of a reactive NHS-ester (see below) dissolved in DMF was added and the mixture was vortexed thoroughly. If necessary, additional DMF was added in 10 μL portions to obtain a clear solution. After reaction at room temperature overnight, the functionalized oligonucleotides were purified on an illustra NAP 5 column with Sephadex G-25 DNA Grade material obtained from GE Healthcare (triethylammonium acetate buffer 50 mM, pH 7.0 was used as the eluent for purification on the illustra NAP 5 columns. MilliQ water was used for ssDNA–NPA–Fmoc to prevent partial deprotection of the Fmoc protecting group). Purification was performed by preparative reversed phase HPLC (rp-HPLC) (*vide supra*) to yield ssDNA–(NPP/NPA–Fmoc). The HPLC time was 18.7 minutes for ssDNA–NPP and 39.8 minutes for NPA–Fmoc.

Fmoc deprotection of ssDNA–NPA–Fmoc was performed as follows. The HPLC-purified and lyophilized oligonucleotide was resuspended in 50 μL of 50 mM triethylammonium acetate buffer (pH 7.0). Deprotection was achieved by the addition of 40 μL of DMF and 10 μL of piperidine. The mixture was vortexed and allowed to react for 2 hours at room temperature. The deprotected oligonucleotide was purified on an illustra NAP 5 column (*vide supra*) and lyophilised to yield ssDNA–NPA–NH_2. The HPLC time was 34.5 minutes for this sample.

rp-HPLC analysis and the preparative purifications (isolations) were performed on a Shimadzu LC-10AD VP machine equipped with a Waters Xterra MS C18 column (3.0 × 150 mm, particle size 3.5 μm) and Waters Xterra MS C18 prep column (7.8 × 150 mm, particle size 10 μm) using a gradient of acetonitrile/

triethylammonium acetate buffer 50 mM, pH 7.0. Gradient: 5/95 for 0 to 10 min, 65/35 at 60 min, 75/25 at 65 min, and 5/95 at 75 min for 15 min.

Synthesis of the reactive photostabilizer-derivatives

(*S*)-2,5-Dioxopyrrolidin-1-yl-2-((((9*H*-fluoren-9-yl)methoxy)carbonyl)amino)-3-(4-nitrophenyl)propanoate ("NPA–NHS"). NPA–Fmoc (173 mg, 0.4 mmol) was added to *N*-hydroxysuccinimide (NHS) (46.0 mg, 0.40 mmol) in 1.5 mL of anhydrous 1,4-dioxane and *N*,*N*′-dicyclohexyl carbodiimide (DCC) (82.8 mg, 0.40 mmol) at 0 °C. After stirring at ambient temperature overnight, the mixture was cooled to 10 °C and the precipitate was filtered off. The filtrate was evaporated *in vacuo*. Residual 1,4-dioxane was removed by subsequent addition and evaporation of anhydrous ethanol. NPA–NHS was used for coupling to ssDNA–NH_2 without further purification.

2,5-Dioxopyrrolidin-1-yl 2-(4-nitrophenyl)acetate ("NPP–NHS"). NPP (72.5 mg, 0.4 mmol) was added to *N*-hydroxysuccinimide (NHS) (48.6 mg, 0.42 mmol) in 1.5 mL of anhydrous 1,4-dioxane and *N*,*N*′-dicyclohexyl carbodiimide (DCC) (87.2 mg, 0.38 mmol) at 0 °C. After stirring at ambient temperature overnight, the mixture was cooled to 10 °C and the precipitate was filtered off. The filtrate was evaporated *in vacuo*. Residual 1,4-dioxane was removed by subsequent addition and evaporation of anhydrous ethanol. NPP–NHS was used for coupling without further purification.

Molecular dynamics simulations

The molecular dynamics simulations were performed using the Gromacs 4.6 simulation package.[35] All simulations used the GROMOS 53A6 force field.[36] The topologies for Cy5 and NPA were created using the Automated Topology Builder[37] then manually curated and attached to the DNA topologies created using the Gromacs tool pdb2gmx. An 8 bp long DNA with a sequence TAATATTC was used. During the simulation, the DNA structure was kept static using position restraints on all its atoms with a force constant of 1000 kJ $(\mathrm{mol\ nm^2})^{-1}$ in each direction. Each system was simulated in a dodecahedron box with around 5400 water molecules, 150 mM NaCl as well as counterions. The SPC water model was used while the ions were from the GROMOS 53A6 force field.[36]

Simulations were run using a 2 fs time step. All bonds were constrained using the LINCS algorithm.[38] A twin-range cut-off of 0.8 nm and 1.4 nm was used. For the electrostatics, a reaction field was used with a 1.4 nm cut-off and an epsilon_rf of 62. Lennard–Jones interactions used a simple cut-off of 1.4 nm with a dispersion correction. The temperature was kept at 298 K using a Berendsen thermostat[39] with $\tau_T = 0.1$ ps. The pressure was kept at a reference pressure of 1 bar with a Berendsen barostat,[39] $\tau_p = 0.5$ ps and compressibility of 4.6×10^{-5} bar.

Each system was first energy minimized for 100 steps using the steepest descent algorithm. They were further equilibrated for 500 ps. The system with Cy5 and NPA attached to opposing sides was then simulated for 1 µs to obtain conformations with Cy5 both on top of DNA as well as in the groove. Two production simulations of 5 µs each were then started, one from each conformation. An identical procedure was followed for the NPA–Cy5 system. The system with a 4 monomer long PEG oligomer added between NPA and Cy5 was simulated first for 100 ns starting from an intermediate conformation and then a 5 µs production run was performed. The system with Cy5 only was simulated for 2 μs starting from each conformation obtained from a 1 μs equilibration simulation.

Before analysis, the simulation trajectories started from different conformations were concatenated and analysed together. The volume maps were produced using the program VMD[40] by root mean square deviation (RMSD) fitting the trajectory on either DNA and Cy5 and then calculating the distribution of Cy5 and NPA or only NPA. When calculating the volume maps around DNA, the volume enclosed points where the molecule is in more than 2% of the simulation frames. For the Cy5 case, this value was 0.6%.

The movements of Cy5 and NPA were studied by measuring the distances between them and DNA. For Cy5 and NPA, the measurements were performed from the center of mass (COM) of the molecule, while the top of the DNA was represented by the COM of the N1 and N3 atoms of the top base pair of DNA. The groove of the DNA was defined by the COM of the N1 and N3 atoms of the fifth base pair from the top.

Acknowledgements

This work was financed by the Zernike Institute for Advanced Materials, the Centre for Synthetic Biology (University of Groningen, start-up grant to T. Cordes). J. H. M. van der Velde acknowledges Ubbo-Emmius funding (University of Groningen). J. J. Uusitalo is supported by a young scientist scholarship from the Emil Aaltonen foundation.

References

1 T. Ha and P. Tinnefeld, *Annu. Rev. Phys. Chem.*, 2012, **63**, 595–617.
2 B. Liphardt, B. Liphardt and W. Lüttke, *Opt. Commun.*, 1981, **38**, 207–210.
3 B. Liphardt and W. Lüttke, *Liebigs Ann. Chem.*, 1981, 1118–1138.
4 B. Liphardt, B. Liphardt and W. Lüttke, *Opt. Commun.*, 1983, **48**, 129–133.
5 R. B. Altman, D. S. Terry, Z. Zhou, Q. Zheng, P. Geggier, R. A. Kolster, Y. Zhao, J. A. Javitch, J. D. Warren and S. C. Blanchard, *Nat. Methods*, 2012, **9**, 68–71.
6 R. B. Altman, Q. Zheng, Z. Zhou, D. S. Terry, J. D. Warren and S. C. Blanchard, *Nat. Methods*, 2012, **9**, 428–429.
7 P. Tinnefeld and T. Cordes, *Nat. Methods*, 2012, **9**, 426–427.
8 J. H. M. van der Velde, E. Ploetz, M. Hiermaier, J. Oelerich, J. W. de Vries, G. Roelfes and T. Cordes, *ChemPhysChem*, 2013, **14**, 4084–4093.
9 J. H. M. van der Velde, J. Oelerich, J. Huang, J. H. Smit, M. Hiermaier, E. Ploetz, A. Herrmann, G. Roelfes and T. Cordes, *J. Phys. Chem. Lett.*, 2014, **5**, 3792–3798.
10 R. Gronheid, J. Hofkens, F. Köhn, T. Weil, E. Reuther, K. Müllen and F. C. De Schryver, *J. Am. Chem. Soc.*, 2002, **124**, 2418–2419.
11 P. Tinnefeld, J. Hofkens, D.-P. Herten, S. Masuo, T. Vosch, M. Cotlet, S. Habuchi, K. Müllen, F. C. De Schryver and M. Sauer, *ChemPhysChem*, 2004, **5**, 1786–1790.
12 Q. Zheng, S. Jockusch, Z. Zhou, R. B. Altman, J. D. Warren, N. J. Turro and S. C. Blanchard, *J. Phys. Chem. Lett.*, 2012, **3**, 2200–2203.
13 I. Rasnik, S. A. McKinney and T. Ha, *Nat. Methods*, 2006, **3**, 891–893.
14 C. Eggeling, J. Widengren, L. Brand, J. Schaffer, S. Felekyan and C. A. Seidel, *J. Phys. Chem. A*, 2006, **110**, 2979–2995.

15 J. Widengren, A. Chmyrov, C. Eggeling, P.-A. Löfdahl and C. A. M. Seidel, *J. Phys. Chem. A*, 2007, **111**, 429–440.

16 T. Cordes, J. Vogelsang and P. Tinnefeld, *J. Am. Chem. Soc.*, 2009, **131**, 5018–5019.

17 L. A. Campos, J. Liu, X. Wang, R. Ramanathan, D. S. English and V. Muñoz, *Nat. Methods*, 2011, **8**, 143–146.

18 Q. Zheng, M. F. Juette, S. Jockusch, M. R. Wasserman, Z. Zhou, R. B. Altman and S. C. Blanchard, *Chem. Soc. Rev.*, 2014, **43**, 1044–1056.

19 J. H. M. van der Velde, J. Oelerich, J. Huang, A. Aminian, J. H. Smit, S. Galiani, K. Kolmakov, G. Gouridis, C. Eggeling, A. Herrmann, G. Roelfes and T. Cordes, 2015, paper in revision.

20 W. H. Smith and A. J. Bard, *J. Am. Chem. Soc.*, 1975, **97**, 5203–5210.

21 C. A. Briehn, M.-S. Schiedel, E. M. Bonsen, W. Schuhmann and P. Bäuerle, *Angew. Chem., Int. Ed.*, 2001, **40**, 4680–4683.

22 J. Vogelsang, R. Kasper, C. Steinhauer, B. Person, M. Heilemann, M. Sauer and P. Tinnefeld, *Angew. Chem., Int. Ed.*, 2008, **47**, 5465–5469.

23 M. E. Sanborn, B. K. Connolly, K. Gurunathan and M. Levitus, *J. Phys. Chem. B*, 2007, **111**, 11064–11074.

24 B. J. Harvey, C. Perez and M. Levitus, *Photochem. Photobiol. Sci.*, 2009, **8**, 1105–1110.

25 J. Spiriti, J. K. Binder, M. Levitus and A. van der Vaart, *Biophys. J.*, 2011, **100**, 1049–1057.

26 M. Levitus and S. Ranjit, *Q. Rev. Biophys.*, 2011, **44**, 123–151.

27 E. M. S. Stennett, M. A. Ciuba and M. Levitus, *Chem. Soc. Rev.*, 2014, **43**, 1057–1075.

28 J. Hu and C. Zhang, *Anal. Chem.*, 2013, **85**, 2000–2004.

29 P. Holzmeister, A. Gietl and P. Tinnefeld, *Angew. Chem., Int. Ed.*, 2014, **126**, 5792–5796.

30 J. Vogelsang, T. Cordes, C. Forthmann, C. Steinhauer and P. Tinnefeld, *Proc. Natl. Acad. Sci. U. S. A.*, 2009, **106**, 8107–8112.

31 J. Vogelsang, T. Cordes and P. Tinnefeld, *Photochem. Photobiol. Sci.*, 2009, **8**, 486–496.

32 J. C. Vaughan, G. T. Dempsey, E. Sun and X. Zhuang, *J. Am. Chem. Soc.*, 2013, **135**, 1197–1200.

33 C. E. Aitken, R. A. Marshall and J. D. Puglisi, *Biophys. J.*, 2008, **94**, 1826–1835.

34 A. Le Gall, D. Dulin, G. Clavier, R. Méallet-Renault, P. Bouyer, K. Perronet and N. Westbrook, *ChemPhysChem*, 2011, **12**, 1657–1660.

35 S. Pronk, S. Páll, R. Schulz, P. Larsson, P. Bjelkmar, R. Apostolov, M. R. Shirts, J. C. Smith, P. M. Kasson, D. van der Spoel, B. Hess and E. Lindahl, *Bioinformatics*, 2013, **29**, 845–854.

36 C. Oostenbrink, A. Villa, A. E. Mark and W. F. Van Gunsteren, *J. Comput. Chem.*, 2004, **25**, 1656–1676.

37 K. B. Koziara, M. Stroet, A. K. Malde and A. E. Mark, *J. Comput.-Aided Mol. Des.*, 2014, **28**, 221–233.

38 B. Hess, H. Bekker, H. J. C. Berendsen and J. G. E. M. Fraaije, *J. Comput. Chem.*, 1997, **18**, 1463–1472.

39 H. J. C. Berendsen, J. P. M. Postma, W. F. van Gunsteren, A. DiNola and J. R. Haak, *J. Chem. Phys.*, 1984, **81**, 3684–3690.

40 W. Humphrey, A. Dalke and K. Schulten, *J. Mol. Graphics*, 1996, **14**, 33–38.

Faraday Discussions

PAPER

Fluorescence microscopy and spectroscopy of subsurface layer dynamics of polymers with nanometer resolution in the axial direction†

Yuri G. Vainer,‡*[ab] Yaroslav I. Sobolev,‡[a] Andrei V. Naumov,[abc] Igor S. Osad'ko[a] and Lothar Kador[d]

Received 3rd May 2015, Accepted 14th July 2015

DOI: 10.1039/c5fd00055f

We studied the dynamics in ultrathin subsurface layers of an amorphous polymer by the spectra of single fluorescent molecules embedded into the layer by vapor deposition and subsequent controlled diffusion to the desired depth in ≈0.5 nm steps. The spectral trails of single molecules were recorded at 4.5 K as a function of diffusion depth. In depths shallower than 20 nm, the spectral dynamics deviate from those deep in the bulk. Less than 5 nm deep, the linewidths increase rapidly, whereas the number of detected molecules decreases. No zero-phonon lines were observed closer than 0.5 nm to the polymer surface. Possible physical reasons of the observed phenomena are discussed.

1 Introduction

For most macroscopic objects, the surface is composed of a tiny fraction of the total count of the object's atoms or molecules. Still, the properties of the surface play an important role in many physical effects. As we turn to ever smaller – nanoscale – objects, the surface-to-volume ratio increases by many orders of magnitude and the contribution of the surface grows accordingly. In spite of extensive research, our knowledge of the microscopic nature of surface processes is limited and many fundamental questions are still open. One of the difficulties

[a]*Institute for Spectroscopy, Russian Academy of Sciences, Fizicheskaya str. 5, Troitsk, 142190 Moscow, Russia. E-mail: vainer@isan.troitsk.ru*

[b]*Moscow Institute of Physics and Technology, Institutskiy lane 9, 141700 Dolgoprudny, Russia*

[c]*Moscow Pedagogical State University, M. Pirogovskaya str. 29, 119992 Moscow, Russia*

[d]*University of Bayreuth, Institute of Physics and BIMF, 95440 Bayreuth, Germany*

† Electronic supplementary information (ESI) available: More details on the experimental procedure and accuracy of determining the depth of the performed diffusion; possible experimental artifacts that may be the real source of the observed effect; dependence of the number of detected molecules on the distance from the surface. See DOI: 10.1039/c5fd00055f

‡ These authors contributed equally to this work.

is that most experimental methods have access to the surface only, while surface properties are determined by what happens below the surface as well. Furthermore, it is often unclear how deep the subsurface layer is affected by the surface and what parameters define this depth.

In the present work, we developed a method that yields information about local dynamics in transparent disordered solids on the nanolevel. It is based on introducing fluorescent molecules into a subsurface layer at a desired depth. We studied single-molecule spectra slightly below the surface of thicker polymer films and we report the results. One of the purposes was to determine experimentally the thickness of a subsurface layer where local parameters differ from those in the bulk volume.

At cryogenic temperatures, individual fluorescent chromophore molecules in solids often feature strong and narrow zero-phonon lines (ZPLs) with widths on the order of 10^{-5} to 10^{-6} nm due to weak coupling with the phonon bath. Hence, they can serve as sensitive probes for small spectral shifts caused by *e.g.* the coupling to flipping tunneling systems in disordered materials. The investigation of these lines and their variation with time has been the subject of single-molecule spectroscopy for many years. In almost all cases, however, only molecules in the bulk of solids were recorded. Fleury *et al.*[1] investigated single terrylene molecules in crystalline *n*-hexadecane and found that the linewidth strongly increased close to the interface between the organic crystal and the silica substrate. On, or close to, the free surface of a polymer, ZPLs of single molecules have never been detected to our knowledge.

Unexpected fast and rich spectral dynamics of single chromophore lines at low temperatures were previously found in low-molecular-weight glass formers, such as toluene,[2] cumene, and propylene carbonate, and in oligomers of amorphous polyisobutylene (PIB).[3] The spectral diffusion was sometimes so fast that it was impossible to record single-molecule lines; only short fluorescence flashes were detected instead. The most probable explanation is that in these materials a lack of polymer entanglements leads to a large number of tunneling systems with low barriers, which are still active at temperatures of a few K.

Recently, our group reported single-molecule studies in films of PIB with thicknesses between 5–300 nm.[4] These films were weakly and homogeneously doped with tetra-*tert*-butylterrylene (TBT). The detectable ZPLs and their dynamics were investigated as a function of film thickness at $T = 4.5$ K. The basic finding was that in films thicker than about 100 nm, the majority of the spectra showed the usual behavior, which is known from bulk PIB and is consistent with standard tunneling two-level systems (TLSs) model, *i.e.* well-defined and reversible jumps between a small number (2^n; n integer) of spectral positions. In thinner films, an increasing fraction of chromophores feature non-standard spectral trails with richer dynamics. The largest number of them was found in the thinnest films (5 nm). Moreover, the fraction of non-standard lines increased with laser intensity, and also this influence was strongest in the thinnest films.[4] In ultrathin films, both the substrate and the free surface of the polymer are close to the chromophores, and it is not clear which of the interfaces is mainly responsible for the altered spectral dynamics. This ambiguity is also resolved in the present work, as we now address only the free surface interface.

We again chose the fluorescent dye TBT and deposited it on top of the surface of PIB films by evaporation at low temperature. Then, the chromophores were allowed to diffuse into well-defined depths up to about 100 nm by heating to suitable temperatures. All spectroscopic measurements were performed at $T = 4.5$ K.

Our study adds new experimental data characterizing the low-temperature dynamics in the first few nanometers below a polymer surface. This can be viewed as a subdomain of molecular spectroscopy that lies between the well known spectroscopy of impurity molecules fixed inside a solid and the spectroscopy of nearly free molecules in gas or vacuum.

Low-temperature studies of surface dynamics are scarce; in the temperature range 250–400 K, the dynamics of polymer surfaces have been studied much more (for a review see ref. 5 and references therein; for single-molecule studies see *e.g.* ref. 6). It is established that, in most cases, free surface layers are more mobile than the bulk, *i.e.* more "liquid-like", with faster relaxation processes.[7,8] But even in this temperature region, experimental results are often controversial: for example, it is still unclear whether the glass transition temperature in ultra-thin polystyrene films is higher or lower than in the bulk material.[9] Therefore, there is a demand for new experimental data of polymer surfaces. The low temperatures used in our approach reduce the complexity of the dynamics to a large degree.

2 Experimental

PIB of high molecular weight (M_w = 4 700 000 g mol^{-1}) was purchased from Sigma Aldrich and was used without further purification. The material was dissolved in toluene and thin films were prepared by spin-coating on microscope glass slides. The thickness of the polymer films was adjusted *via* the concentration of the solution and the rotation rate, and it was measured with a profilometer (Veeco DEKTAK 150) with a vertical resolution of about 3 nm.

The slide with the undoped polymer film was inserted in an optical He-4 cryostat (Cryovac) and cooled down to temperatures between 30 and 50 K, at which the translation diffusion of molecules can be neglected. A thin layer of TBT was placed on the surface of a ceramic cylinder carrying a coil of resistance wire for electrical heating. The ceramic cylinder with the dye was also placed in the cryostat, first at some distance from the polymer film, and the sample chamber of the cryostat was evacuated to about 10^{-5} mbar. The cylinder was heated to a temperature slightly below the sublimation temperature of TBT to remove impurities. Subsequently, it was placed about 5 mm from the polymer film and heated to 100–200 °C for some seconds. The exact temperature of the cylinder could not be measured; the heating power and sublimation time were adjusted such that the desired amount of dye molecules were deposited on the polymer surface (below one molecule per 100 nm^2 surface area within a spectral interval of 1 cm^{-1} inside the inhomogeneous absorption spectrum).

After evaporation at low temperature (30–50 K), the dye molecules were immobile on the polymer surface. Thus, heating to elevated temperatures was required to induce their diffusive penetration into the material. At room temperature, the diffusion coefficient of large organic dye molecules in PIB is about 1000 nm^2 s^{-1}, and it decreases at lower temperatures. By keeping the temperature of the sample low enough so that the dye molecules travelled less than 1 nanometer per minute of conditioning, we were able to dye the ultrathin

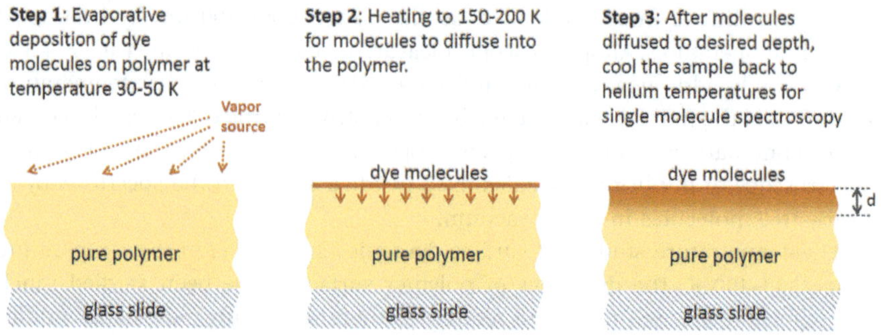

Fig. 1 Procedure for preparing samples of the polymer with fluorescent dye molecules in the ultrathin subsurface layers at a desired depth. Thicknesses of the polymer and glass slide are not to scale. The temperature was tracked and recorded every second during the entire procedure (see ESI Fig. 1†) in order to integrate diffusion and calculate the total distance *d* travelled by the dye molecules, thus obtaining the effective thickness of the dyed subsurface layer.

surface layer only (see Fig. 1 and ESI Fig. 1†). To quantitatively estimate the diffusion depths from the well-known equation

$$d = 2[\int D[T(\tau)]\mathrm{d}\tau]^{1/2}, \tag{1}$$

we used the temperature dependence of the translational diffusion coefficient $D(T)$ of rubrene in PIB[10] and inserted the measured time-dependent temperature excursion $T(\tau)$ during our heating cycles. The actual diffusion depths may be 1–2 nm larger, since the diffusion coefficient in the first 1–2 nm below a polymer surface is known to be higher than in the bulk, at least at the glass transition temperature (see *e.g.* ref. 7).

The ZPLs of single fluorescing TBT molecules were recorded with a low-temperature wide-field microscope, as described in detail previously.[11] The excitation light source was a tunable single-mode dye laser COHERENT CR 599-21, operating with rhodamine 6G in the spectral range 565–625 nm. Its linewidth was roughly 2–3 MHz and the continuous tuning range was 1 cm^{-1} (0.03 nm). The laser frequency was repeatedly scanned over the same spectral interval of 1 cm^{-1} width, and the single fluorescing chromophores in the illuminated area of the sample of about 100 μm diameter were imaged onto the chip of a sensitive Peltier-cooled CCD camera (either PCO Sensicam or PCO Luca-S) with a microscope objective (Microthek, Hamburg, Germany, NA = 0.85) inside the cryostat. The duration of each laser scan was 75 s, and the time between two scans was about 1 s. During the scans, Doppler-broadened absorption lines of an iodine cell were recorded to control the tuning range of the laser. The signals of individual molecules were extracted from the database of the CCD camera images and plotted in two-dimensional representation as a function of excitation frequency and time, with the fluorescence intensity encoded on a gray scale – so called "spectral trails".

3 Results

The first interesting result was that immediately after evaporating the dye molecules onto the surface of the cold PIB films, no fluorescence images of individual

TBT molecules were visible in any of the experiments. Instead, the films showed only spatially featureless fluorescence emission from the entire area illuminated by the laser light. Varying the temperature of the PIB substrate between about 10 and 30 K had no influence on this behavior. Spectral analysis with a polychromator revealed that the fluorescence emission was close to the typical spectrum of TBT. Hence, we can conclude that the homogeneous emission of the samples must be mainly due to TBT and not to an unknown fluorescing impurity.

The situation changed completely after a brief temperature cycle to an excursion temperature slightly above the glass point of PIB (180 K). After cooling the sample back down to 4.5 K, fluorescence images of single TBT molecules appeared immediately and the background fluorescence level decreased significantly. From the temperature-dependent diffusion coefficient, we can estimate that the average diffusion depth reached in this temperature cycle is no larger than about 0.12 nm. Also, the fluorescence emission spectrum changed distinctly during the cycle: after the diffusion process it developed a band around 620 nm, which is also present in the room temperature spectrum of TBT in liquid toluene solution. Photobleaching by prolonged laser irradiation did not cause any additional spectral changes but only a decrease in amplitude (see Fig. 2).

The observed change in the emission spectrum can be explained in the following manner. After molecules collide with the cold surface, their spatial structure is far from equilibrium; heating the sample above the glass transition allows them to relax to equilibrium.

The spectral dynamics of single TBT molecules close to the surface of PIB films are significantly different from those in the bulk and this does, for many molecules, not agree with the standard tunneling model. Specifically, spectral jumps are more frequent and do not occur reversibly between a small number (2^n, n integer) of well-defined positions. More importantly, we see continuous spectral drifts and abrupt changes in linewidth and in the rate of spectral jumps. Fig. 3

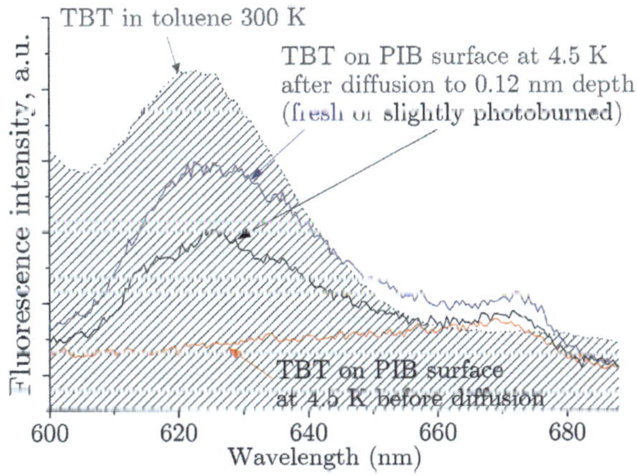

Fig. 2 Ensemble fluorescence emission spectra of TBT evaporated onto the surface of a cold PIB film before (red) and after (blue) a heating cycle causing diffusion to a depth of 0.12 nm. For comparison we show the ensemble fluorescence emission spectra of TBT in toluene solution at room temperature (hatched). The black line shows the spectrum after some degree of photobleaching by the laser. The excitation wavelength for the low-temperature spectra is 575 nm. All spectra are ensemble spectra.

shows some examples. Continuous spectral drifts, as in panels (a) and (d), are possible, although uncommon, in the standard tunneling mode.[12] They require the weak coupling of the fluorescing molecule to numerous TLSs. Interactions between TLSs [panel (b)] and temporal variations of the TLS or coupling parameters [panel (c)] do not agree with the tunneling model. Spectral trails similar to those presented in Fig. 3 are also typical for low-molecular-weight glass formers[3] and ultrathin PIB films.[4]

The fraction of single chromophore molecules featuring non-standard spectral dynamics decreases strongly with increasing diffusion depth. Fig. 4 compares this variation (filled circles) with the relative number of non-standard spectral trails recorded in ultrathin PIB films (open circles).[4] The behavior in both kinds of samples is similar. The offset in the depth dependence is most likely due to the proximity of the second interface (*i.e.* a polymer–substrate one) in the ultrathin films.

Also, the spectral width of the single-molecule lines varies with diffusion depth. The increase in linewidth begins at a depth of about 10 nm and proceeds rapidly towards the surface, as Fig. 5 illustrates. Some of the spectral trails of single molecules at those depths are presented in Fig. 7. At a depth of 0.6 nm, the average linewidth is about 20 times broader than that in the bulk. For diffusion depths smaller than 0.6 nm, no ZPLs could be observed. The number of detected ZPLs increases with depth and reaches a constant value at roughly 7 nm.

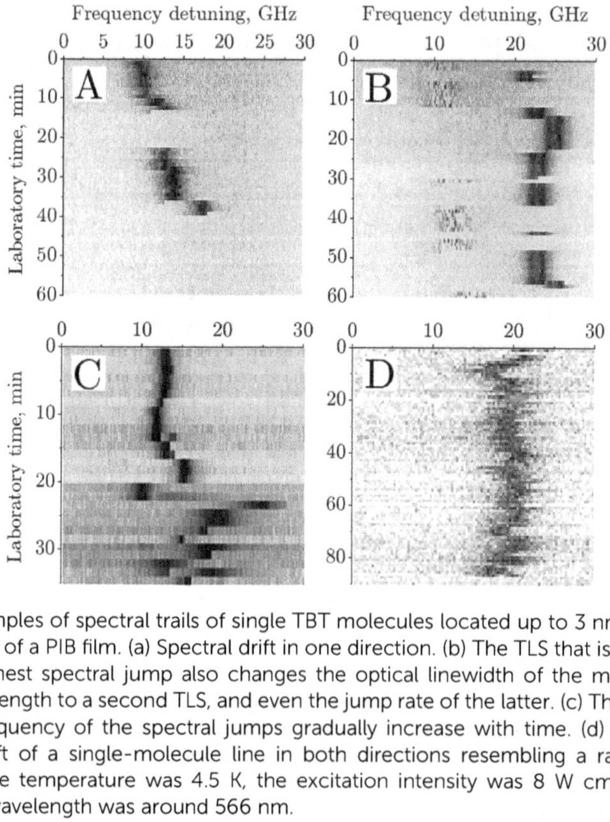

Fig. 3 Examples of spectral trails of single TBT molecules located up to 3 nm below the free surface of a PIB film. (a) Spectral drift in one direction. (b) The TLS that is responsible for the farthest spectral jump also changes the optical linewidth of the molecule, the coupling strength to a second TLS, and even the jump rate of the latter. (c) The amplitude and the frequency of the spectral jumps gradually increase with time. (d) Continuous spectral drift of a single-molecule line in both directions resembling a random-walk process. The temperature was 4.5 K, the excitation intensity was 8 W cm^{-2}, and the excitation wavelength was around 566 nm.

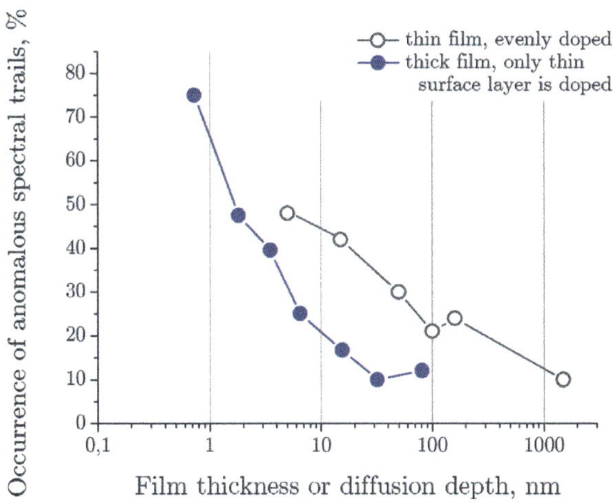

Fig. 4 Relative number of single TBT molecules in PIB exhibiting anomalous spectral dynamics as a function of film thickness or diffusion depth. The figure compares data measured in ultrathin films (ref. 4; empty circles) and in thicker films of which only the surface layer has been doped by diffusion (this work; filled circles). Spectral trails were considered non-standard if they featured any of the following effects: continuous spectral drifts in one direction; abrupt changes in the linewidth or jump rate; spectral jumps that were irreversible during the measuring time. Temperature = 4.5 K; excitation intensity = 4 W cm^{-2} (ultrathin films) or 8 W cm^{-2} (surface doping).

4 Discussion

The fast and complex spectral dynamics of single-molecule lines close to the surface indicate the presence of a high density of tunneling systems, in particular ones with low barriers. Since the TLSs have a certain spatial extension, the high density is expected to give rise to a correspondingly large number of interactions between them.[13] Their jump rates may be additionally enhanced by the interaction with extra (*i.e.* non-Debye) low-frequency surface vibrational modes. We can assume that the penetration depth of the surface modes is roughly given by their wavelength. If we set the average excitation energy of the modes equal to the thermal energy at 4.5 K and insert the sound velocity of PIB (1.5 km s^{-1}), we obtain a wavelength of about 16 nm, which is in good agreement with the depth in which the number of anomalous line shapes begins to increase (Fig. 4).

The TLS in PIB causes a significant shift in the electronic transition frequency of a nearby fluorescent molecule when this TLS is closer than about 10 nm to the molecule in question. Then, if we suppose that the spatial density of TLSs is constant everywhere in the sample, one can expect that when a fluorescent molecule is less than 10 nm below the surface, it is influenced by fewer TLSs. Furthermore, when a molecule is at the surface, the entire half-space is missing and this fact should halve the number of TLSs influencing the molecule. This is not what happens in reality. If anything, the spatial density of TLSs near a surface is larger than in the rest of the sample. To illustrate this, we performed computer simulations of single-molecule spectra by implementing the approach of Geva and Skinner.[14] Fig. 6 shows how the spatial density of TLSs influences the simulated spectral trails of a single fluorescent molecule. Halving the number of

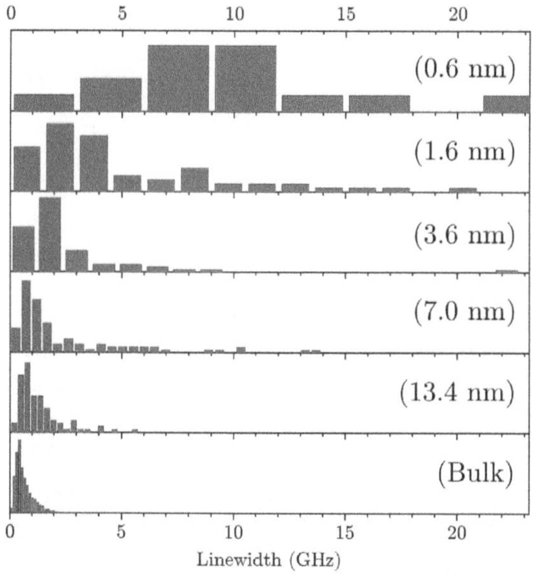

Fig. 5 Distribution of the spectral linewidths of single TBT molecules in PIB with different diffusion depths between 0.6 nm and 13.4 nm, and a comparison with the bulk data. Temperature = 4.5 K; excitation intensity = 8 W cm^{-2}; excitation wavelength between 565 and 568 nm.

TLSs interacting with a molecule means going from trails in the second row to trails in the first row. This is not what is observed experimentally. Comparing these simulations with the experimentally recorded spectral trails in Fig. 7, it seems that going from 3.5 nm to 0.7 nm below the surface modifies the spectral trails as if the density of the TLSs has increased by about an order of magnitude.

Another possible cause of the strongly increasing linewidths of molecules located 1–2 nm below the surface (Fig. 5) may, in part, be due to direct dephasing by the surface quasi-localized modes. In bulk PIB, quasi-localized low-frequency modes affect the linewidths of single chromophore molecules only at temperatures above 5–7 K,[16,17] since they have frequencies between about 300 and 600 GHz (15–30 K).[18] Close to the surface, their energies may be distinctly lower. If so, the absence of detectable single-molecule lines directly on the surface is probably caused either by their extremely broad linewidths or by strong linear electron–phonon coupling, so the lines consist only of phonon sidebands.

The absence of detectable single-molecule ZPLs directly on the surface is unclear. This may happen because the spectral position of a ZPL constantly changes over a wide range of wavelengths. These changes may have several possible sources: coupling to low-energy degrees of freedom (*e.g.* surface vibrational modes), continuous transitions between the larger number of accessible conformation states an impurity molecule obtains on a surface, and continuous translational motions of an impurity molecule along the surface that can be induced by excitation laser light. A high enough magnitude and frequency of such changes in wavelength can make the detection of ZPLs impossible at a given temporal resolution and signal-to-noise ratio. In experiments, there is usual a

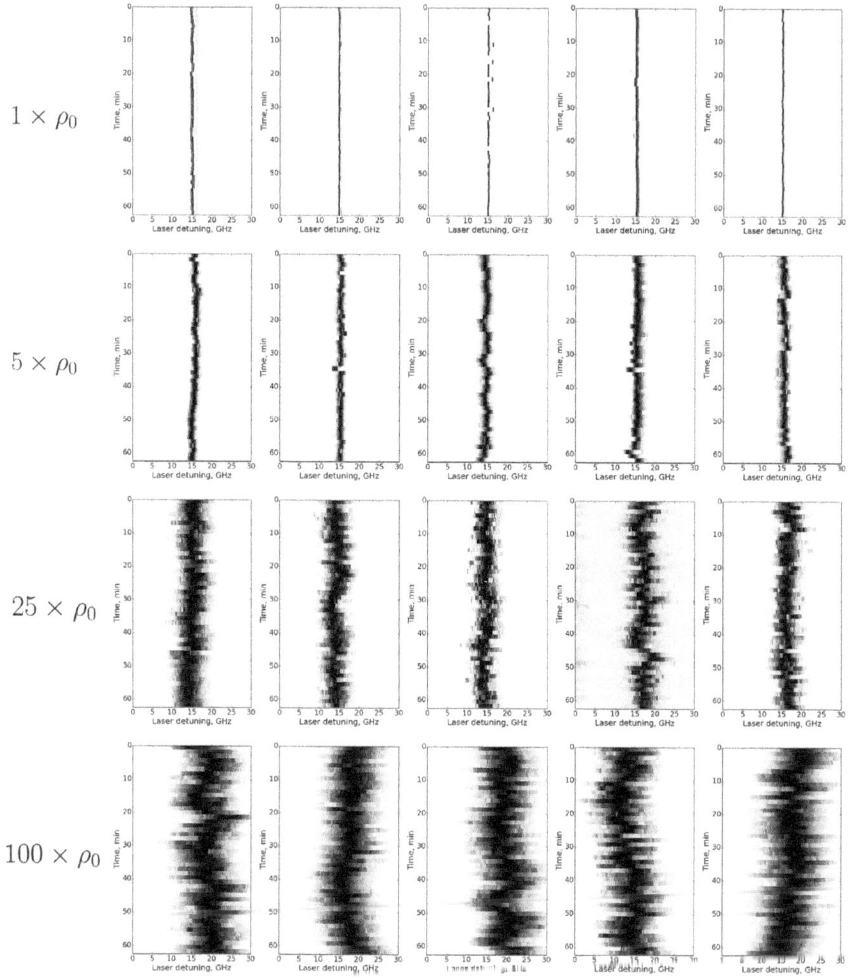

Fig. 6 Simulated spectral trails of single TBT molecules in PIB for different spatial densities of the TLSs, as indicated for each row of images. Here, ρ_0 means the density of the TLS used by Geva and Skinner, who have shown that it gives good agreement with photon echo experiments and distributions of single molecule spectral linewidths.[14] For each density, we sampled 5 instantiations. For all trails, the distributions of the TLS parameters were kept the same (but the parameters were generated anew for each trail) and these distributions are valid for the bulk PIB sample, as shown in ref. 15. Interactions between the TLSs, however, were not taken into account in the simulations. For all trails, the spectral range scanned by the excitation laser was 30 GHz, and the time span of the experiment (vertical axis) was 62.5 minutes.

trade-off between the signal-to-noise ratio and temporal resolution – both are defined by the exposure time at a given laser excitation power density, and increasing the excitation power density may itself induce additional spectral dynamics. Since, in cases where we can still observe ZPLs, their spectral dynamics intensify as the impurity gets closer to the surface, the dynamical explanation of the disappearance of ZPLs for impurity molecules on the surface seems more likely than ascribing this disappearance to quenching of the fluorescence itself.

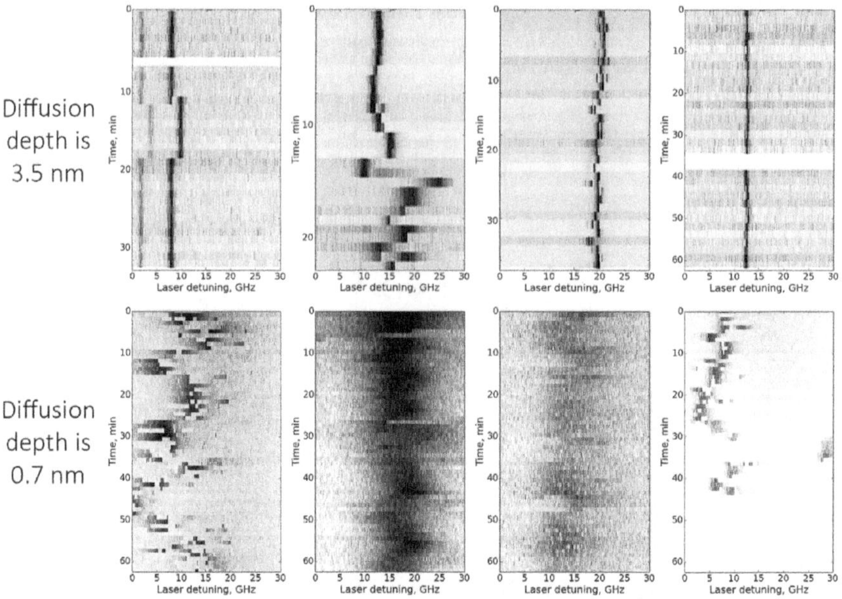

Fig. 7 Experimentally recorded spectral trails of single TBT molecules at diffusion depths of 3.5 nm (top row) and 0.7 nm (bottom row) below the free surface of a PIB film. For all trails, the spectral range scanned by the excitation laser was 30 GHz.

Spectral broadening alone does not seem to be sufficient for explaining the total disappearance of detectable single molecule ZPLs on the polymer surface. Let's turn to a more detailed analysis of an impurity molecule interacting with light. Upon electronic excitation of the molecule, the vibrational part of the full Hamiltonian changes and the inharmonic approximation Hamiltonians of the ground and excited states are:

$$H^{g} = T(R) + R\frac{U}{2}R, \quad H^{e} = T(R) + (R+a)\frac{U+W}{2}(R+a). \tag{2}$$

Here, $T(R)$ is the molecule kinetic energy operator, R is the multidimensional vector of the molecule's deviation from the ground state equilibrium position, and U and $U + W$ are force matrices for the molecule's ground and electronically excited state, respectively. The electronic excitation causes changes to the force matrix, as well as inducing a shift a of the equilibrium position. The electron–phonon coupling operator is the difference between the vibrational Hamiltonians:

$$H^{e} - H^{g} = a\frac{U}{2}a + aUR + R\frac{W}{2}R. \tag{3}$$

The second and third terms describe the linear and quadratic coupling, respectively. Fig. 5 shows that the ZPL width of a single molecule is larger for molecules that are closer to the surface. Since the ZPL width is defined solely by the magnitude of the electron–phonon coupling W that is quadratic with respect to the creation and annihilation operators (quadratic coupling), an increase in the

ZPL-widths means that the quadratic coupling gets stronger as the distance to the surface decreases. The same conclusion was drawn by Fleury *et al.*[1]

Chapter 4 of the book in ref. 19 features a detailed calculation of the shape of an entire optical band under the influence of interaction (3). According to these calculations, the half-width at half-maximum is given by:

$$\gamma_{\rm ph} = \int_0^\infty \frac{d\omega}{2\pi} \ln\left[1 + W^2 \Gamma^{\rm e}(\omega)\Gamma^{\rm g}(\omega)/\sinh^2(\beta\omega/2)\right] \tag{4}$$

where $\Gamma^{\rm e}(\omega)$ is the density of the phonon states in the ground (g) and excited (e) electronic states of the impurity molecule:

$$\Gamma^{\rm e}(\omega) = \frac{\Gamma^{\rm g}(\omega)}{(1 - W\Delta^{\rm g}(\omega))^2 + (W\Gamma^{\rm g}(\omega))^2} \tag{5}$$

where

$$\Delta^{\rm g}(\omega) = \int_0^\infty \frac{d\nu}{\pi} \Gamma^{\rm g}(\nu) \frac{2\nu}{\omega^2 - \nu^2}. \tag{6}$$

According to (4), when W becomes very large, the half-width of the ZPL stops growing with W (one example where this 'saturation' of the ZPL width manifests itself is the optical line shape of paramagnetic crystals upon increasing the temperature).[20]

However, when a molecule gets spatially close to the sample surface, there can be an increase in coupling $V = aUR$ as well. This coupling is not quadratic, but linear in phonon operators and thus does not affect the width of the ZPL. Nonetheless, linear coupling V is capable of broadening the overall optical band much more strongly than quadratic coupling. This broadening effect of strong linear coupling is worked out in detail in ref. 19 and the result is the optical shape

$$I^{\rm g,e}(\Delta) = {\rm e}^{-\phi(0,T)} \frac{1/T_2\pi}{(\omega - \omega_0)^2 + (1/T_2)^2} + \frac{1}{\sqrt{2\pi B}} {\rm e}^{-\frac{(\omega-\omega_0\pm A)^2}{2B}} \tag{7}$$

where the function

$$\phi(t, T) = \sum_q \frac{{a_q}^2}{2} \left[n_q \exp(i\nu_q t) + (n_q + 1)\exp(-i\nu_q t)\right] \tag{8}$$

depends on the phonon frequencies ν_q and shifts a_q in the equilibrium position under electronic excitation, and describes the influence of the linear electron–phonon coupling on the optical line shape. Here, n_q is

$$n_q = \frac{1}{\exp(\hbar\nu_q/kT) - 1}. \tag{9}$$

At large shifts a_q, the Lorentz shape of the ZPL disappears entirely as the Debye–Waller factor ${\rm e}^{-\phi(0,T)}$ gets infinitely small, while the Stokes shift

$$2A = \sum_{q=1}^{N} {a_q}^2 \nu_q \tag{10}$$

of the phonon side-band (in absorption and fluorescence) and the half-width $\delta\omega_{1/2} = \sqrt{B2 \ln 2}$ of the Gauss-shaped phonon side-band, where

$$B = \sum_{q=1}^{N} \frac{{a_q}^2}{2} \nu_q \coth(\beta\nu_q/2), \tag{11}$$

can be orders of magnitude greater than the broadening induced by quadratic coupling. For example, in spectral hole burning experiments it can be at least tens of thousands of GHz. In experimental schemes normally used to detect molecules with narrow ZPLs, the fluorescence of a single molecule with such a wide optical line is practically impossible to detect on background light. We can only speculate what real microscopic motions possibly underlie this increased linear coupling, but from Fig. 5 and from seeing no molecules directly on the surface, we conclude that while this linear coupling may increase gradually starting at a distance of several nanometers from the surface, it sharply increases when the surface is reached.

5 Conclusion

To conclude, our single-molecule studies of a model disordered solid (amorphous polymer) have shown that the dynamical properties of ultrathin subsurface layers differ from the properties of the polymer bulk. We found that this deviation becomes noticeable at distances less than about 20 nm from the surface. The spectral dynamics of single chromophores become faster and present irreversible features. At depths less than 5 nm, along with a drastic increase in the complexity of the spectral dynamics, we also observed an additional strong increase in the individual spectral linewidths. Finally, less than 0.6 nm below the surface, no zero-phonon lines were detected at all.

Acknowledgements

The authors would like to thank A. Krekhov and A. Vagov for fruitful discussions and the members of the chair EP IV in Bayreuth for invaluable support. This work was funded by the Deutsche Forshungsgemeinshaft (DFG) and the Russian Foundation for Basic Research (14-02-00834).

References

1 L. Fleury, A. Gruber, A. Dräbenstedt, J. Wrachtrup and C. von Borczyskowski, *J. Phys. Chem. B*, 1997, **101**, 7933–7938.
2 A. V. Naumov, Y. G. Vainer and L. Kador, *Phys. Rev. Lett.*, 2007, **98**, 145501–145504.
3 I. Y. Eremchev, Y. G. Vainer, A. V. Naumov and L. Kador, *Phys. Chem. Chem. Phys.*, 2011, **13**, 1843–1848.
4 Y. I. Sobolev, Y. G. Vainer, A. V. Naumov and L. Kador, *J. Chem. Phys.*, 2014, **140**, 204907–204910.
5 G. McKenna, *Eur. Phys. J.: Spec. Top.*, 2010, **189**, 285–302.
6 N. Tomczak, R. A. L. Vallée, E. M. H. P. van Dijk, L. Kuipers, N. F. van Hulst and G. J. Vancso, *J. Am. Chem. Soc.*, 2004, **126**, 4748–4749.

7 C. J. Ellison and J. M. Torkelson, *Nat. Mater.*, 2003, **2**, 695–700.
8 J. L. Keddie, R. A. L. Jones and R. A. R. Cory, *Europhys. Lett.*, 1994, **27**, 59–64.
9 M. Tress, M. Erber, E. U. Mapesa, H. Huth, J. Müller, A. Serghei, C. Schick, K.-J. Eichhorn, B. Voit and F. Kremer, *Macromolecules*, 2010, **43**, 9937–9944.
10 D. Bainbridge and M. D. Ediger, *Rheol. Acta*, 1997, **36**, 209–216.
11 Y. G. Vainer, A. V. Naumov, M. Bauer and L. Kador, *J. Lumin.*, 2007, **127**, 213–217.
12 V. Lubchenko and R. J. Silbey, *J. Chem. Phys.*, 2007, **126**, 064701.
13 A. Burin and Y. Kagan, *Phys. Lett. A*, 1996, **215**, 191–196.
14 E. Geva and J. L. Skinner, *J. Chem. Phys.*, 1998, **109**, 4920.
15 A. V. Naumov, Y. G. Vainer, M. Bauer and L. Kador, *J. Chem. Phys.*, 2002, **116**, 8132–8138.
16 A. V. Naumov, Y. G. Vainer, M. Bauer and L. Kador, *J. Chem. Phys.*, 2003, **119**, 6296–6301.
17 Y. G. Vainer, A. V. Naumov, M. Bauer and L. Kador, *J. Lumin.*, 2004, **107**, 287–297.
18 Y. G. Vainer, A. V. Naumov, M. Bauer and L. Kador, *J. Chem. Phys.*, 2005, **122**, 244705.
19 I. S. Osadḱo, *Selective Spectroscopy of Single Molecules*, Springer, Berlin, 2003.
20 S. Gladenkova and I. Osad'ko, *Chem. Phys. Lett.*, 1991, **187**, 628–632.

Faraday Discussions

PAPER

Design and synthesis of aromatic molecules for probing electric fields at the nanoscale

Sanli Faez,†*[a] Nico R. Verhart,[a] Marios Markoulides,[b] Francesco Buda,[c] André Gourdon[b] and Michel Orrit[a]

Received 7th May 2015, Accepted 2nd June 2015
DOI: 10.1039/c5fd00065c

We propose using halogenated organic dyes as nanoprobes for electric fields and show their greatly enhanced Stark coefficients using density functional theory (DFT) calculations. We analyse halogenated variants of three molecules that have been of interest for cryogenic single molecule spectroscopy: perylene, terrylene, and dibenzoterrylene, with the zero-phonon optical transitions at blue, red, and near-infrared. Out of all the combinations of halides and binding sites that are calculated, we have found that fluorination of the optimum binding site induces a dipole difference between the ground and excited states larger than 0.5 D for all three molecules with the highest value of 0.69 D for fluoroperylene. We also report on the synthesis of 3-fluoroterrylene and the bulk spectroscopy of this compound in liquid and solid organic environments.

Introduction

Single organic molecules in solid state host media have found a vast variety of applications in physics, biophysics, and chemistry.[1] Under cryogenic conditions, some organic molecules acquire a very strong zero-phonon line (ZPL) on their electronic transitions at optical frequencies (about 300 THz) with a linewidth of merely 30 MHz, which is limited by spontaneous emission. The exact value of the transition energy is strongly influenced by the nano-environment. Given the high quality factor of this transition ($\sim 10^7$), each molecule can be seen as a highly sensitive probe of its local environment.[2,3]

In 2001, one of us has suggested using organic molecules as nanoprobes to measure charge distributions in organic semiconductors.[4] In follow-up

[a]Huygens-Kamerlingh Onnes Laboratorium, Leiden University, PO Box 9504, 2300 RA Leiden, The Netherlands
[b]CEMES-CNRS, 29 rue J. Marvig, P.O. Box 94347, F-31055 Toulouse Cedex, France
[c]Leiden Institute of Chemistry, Leiden University, PO Box 9502, 2300 RA Leiden, The Netherlands
† Current address: Department of Physics and Debye Institute for Nanomaterials Science, Utrecht University, Princetonplein 5, 3584 CC Utrecht, Netherlands. E-mail: S.Faez@uu.nl

experiments, we demonstrated the optical detection of trapped charges in organic crystals.[5] Other theoretical modeling has shown the possibility of detecting charge displacements of merely 10 pm by triangulation methods.[6] More recently, we have shown that only 10 probe molecules are enough to yield an accuracy better than 3 nm for tracking up to 6 electrons simultaneously.[7] Puller, Lounis, and Pistolesi have investigated the possibility of displacement detection of nanomechanical oscillators based on electrostatic interactions[8] and have shown that regimes of strong back-action between the two-level system of a molecule and a clamped carbon nanotube can be realized.

These proposals, experimental advances in embedding these organic molecule in nanostructures,[9] and recent progress in circuit-QED have motivated Das, Faez and Sørensen to propose and theoretically investigate a hybrid superconducting qubit (SCQ)-molecule-photon quantum system that enables direct state transfer from a stationary SCQ to a flying photon qubit.[10] State transfer is the key enabling step towards building large quantum networks, connecting individual electron-based nodes with light channels, while preserving quantum properties.[11] Their proposal extends the use of "molecular nanoprobes" from the effectively-classical detection of single electrons towards the quantum inspection of the electron's wavefunction, which is a crucial enabling technology for hybrid quantum information processing.

Direct probing of electrons with light in solid-state devices is challenging because the electronic state can be easily disrupted, considering the large energy gap between a single visible photon and the conduction electrons (few eV for photons *versus* few meV for electrons). Elastic scattering from organic molecules that are influenced by the electron field can act as the interfacing mechanism between electrons and photons with minimal energy dissipation: a molecule's internal states are influenced by the electron's electric field at distances of hundreds of nm. As the molecular states can be read out optically, the molecule can act as a transducer between optical and electronic energy scales.

The most important requirement for realizing these proposals is to obtain molecules with necessary Stark shifts, *i.e.*, the spectral shift of the ZPL in response to an external electric field, without losing the crucial property of exhibiting a narrow zero-phonon electronic transition. Previous experiments have shown that this is indeed possible. For example, the dipole moment difference between electronic ground and excited states of some terrylene molecules in a *para*-terphenyl crystal reaches 1 Debye,[12] corresponding to a linear shift of more than 3 MHz (kV^{-1} m^{-1}).[13] This high sensitivity to electric fields is induced by deformations of the otherwise centrosymmetric molecular orbitals by the shell of adjacent molecules in the crystal.[14] The electrostatic field of a single electron is roughly 150 kV m^{-1} at a distance of 100 nm, which is enough to shift the ZPL frequency of those terrylene molecules by three times their 42 MHz linewidth.[15]

Other molecular guest–host systems with equivalently high Stark factor and simultaneously narrow ZPL are terrylene in a polyethylene matrix[12] and dibenzoterrylene (DBT) in an anthracene crystal.[16] However, in both of those systems, the Stark coefficients of the molecules were randomly set by their local environments. To pursue more quantitative measurement and to realize intriguing proposals such as mapping charge transport in nano-electronic devices, it is necessary to obtain several molecules with consistently large Stark coefficients.

In this article, we pursue one strategy for designing new organic dye molecules for electric field sensing based on the Stark shift of ZPL. To this end, we have considered targeted halogenation of otherwise centrosymmetric conjugated molecules and have calculated their enhanced Stark coefficients using density functional theory (DFT) modelling. We have analyzed halogenated variants of three molecules that have been investigated before by cryogenic single molecule spectroscopy: perylene, terrylene, and dibenzoterrylene, with their ZPL optical transitions at blue, red, and near infrared frequencies. Out of all the combinations of halides and binding sites that are calculated, we have found that fluorination of the optimum binding site induces a change of dipole moment larger than 0.5 D for all three molecules with the highest value of 0.69 D for fluoroperylene. We have synthesized targeted 3-fluoroterrylene molecules and characterized the solvent-dependent Stokes shift by bulk and fluorescence spectroscopy. Preliminary cryogenic measurements indicate that a Shpol'skii matrix such as tetradecane is not a suitable host for obtaining narrow and bright ZPL from this molecule.

Experimental

The electric field response (Stark shift) of the zero-phonon optical transition is due to the difference between the ground state and excited state dipole moments. Most of the previously investigated dyes, however, are centrosymmetric and hence present no intrinsic dipole moment. By embedding these molecules in a solid matrix, their symmetry is broken and they can obtain a permanent dipole moment. This dipole moment is strongly influenced by the slight deformation of each molecule and hence is highly inhomogeneous. The simplest way of directing the inhomogeneity toward larger values is to synthesize molecules with intrinsically broken symmetries. Synthesis of all of these molecules, however, is a cumbersome and expensive procedure. Therefore, we first used DFT calculations to choose the most promising candidates. It is also important not to fully disturb the orbital structure of the molecules and to keep some spatial overlap between the ground and excited electronic states, else the ZPL transition will be too weak or completely absent.

Previously, methylation of terrylene has been tested experimentally and the possibility of obtaining a narrow ZPL has been demonstrated by hole-burning experiments.[17] The ZPLs of the molecules in a hexadecane matrix, however, were not completely stable and were prone to spectral jumps presumably because of the rotational freedom of the methyl group, which could tunnel between two eigen-states. Based on that experience, we chose to replace one hydrogen atom with a single electronegative halogen atom for this project.

The structures of perylene, terrylene, and DBT are shown in Scheme 1. For terrylene, we have considered replacement of a H-atom in the corner (position 3) with each of the four halogen atoms and found out that fluorine substitution provides the highest dipole change. Therefore, for perylene and DBT we have only considered halogenation with fluorine. For all three molecules the effect of modifying the position of Fluorine substitution has been investigated for all nonequivalent sites of each. The DFT results for absorption and emission energies and the direct transition are summarized in Tables 1 and 2. The computed absorption energy for pure terrylene is in relatively good agreement with our experimental data (see Fig. 1) and results reported in the literature.[13] The absorption energy for perylene is also in a very good agreement with the

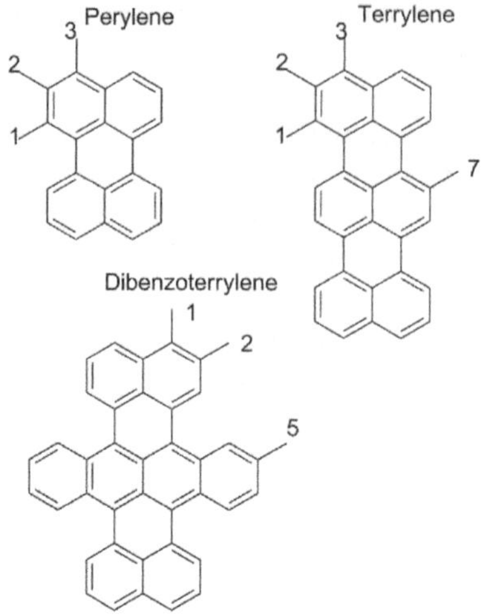

Scheme 1 Model systems considered in this study. The numbers refer to the positions where we have considered halogen substitution.

experimental value of ~434 nm. Note that solvent-specific interactions have not been considered in our DFT calculations. This showed that singlet excitation in terrylene has a strong HOMO–LUMO character. As a representative example we show the HOMO and LUMO for the 3-fluoroterrylene in Scheme 2. According to the calculation, the highest dipole moment difference between the ground and the excited states for fluorine-substituted perylene, terrylene, and DBT are for positions 3, 7, and 2, respectively. The calculations also indicate that the main contribution to the dipole difference between the two states is the rotation of the dipole moment orientation rather than a change in its absolute value.

Table 1 DFT/TDDFT results for dipole moments of terrylene with different halogen substitutions in position 3 (see Scheme 1). The first column is the absorption energy calculated at the optimized geometry in the ground state (GS) S0 (R0); the second column is the first excitation energy computed at the optimized geometry in the first excited state (ES) S1 (R1). E-ZPL is the energy difference between the two minima in the GS and ES. The difference in dipole moment is calculated as the modulus of the difference vector between the dipole moment in the GS at R0 and the dipole moment in the ES at R1

Compound	Absorption (nm) [eV]	Emission (nm) [eV]	E-ZPL (eV)	GS dipole (Debye)	ES dipole (Debye)	Dipole change (Debye)
Terrylene	566.0 [2.191]	628.1 [1.974]	2.078	0	0	0
Terrylene + F p1	559.6 [2.216]	619.6 [2.001]	2.105	1.10	1.07	0.20
Terrylene + F p2	566.8 [2.187]	628.6 [1.972]	2.076	1.97	2.14	0.22
Terrylene + F p3	568.1 [2.182]	630.2 [1.967]	2.071	1.38	1.36	0.36
Terrylene + F p7	559.6 [2.215]	619.8 [2.000]	2.104	1.37	1.35	0.66
Terrylene + Cl p2	567.6 [2.185]	629.7 [1.969]	2.073	2.47	2.70	0.23
Terrylene + Cl p3	573.1 [2.163]	635.6 [1.951]	2.053	2.2	2.21	0.08
Terrylene + Br p3	574.6 [2.158]	637.2 [1.946]	2.048	2.23	2.21	0.05
Terrylene + I p3	586.5 [2.114]	653.0 [1.899]	2.002	2.69	2.64	0.05

Table 2 Same as Table 1 for perylene and DBT

Compound	Absorption (nm) [eV]	Emission (nm) [eV]	E-ZPL (eV)	GS dipole (Debye)	ES dipole (Debye)	Dipole change (Debye)
Perylene	436.3 [2.84]	491.2 [2.52]	2.68	0	0	0
Perylene + F p1	429.5 [2.89]	481.9 [2.57]	2.73	1.17	1.13	0.40
Perylene + F p2	437.4 [2.83]	492.4 [2.52]	2.67	1.72	1.94	0.37
Perylene + F p3	439.0 [2.82]	494.9 [2.51]	2.66	1.29	1.28	0.69
DBT	836.4 [1.48]	975.4 [1.27]	1.376	0.18	0.32	0.24
DBT + F p1	837.9 [1.48]	975.9 [1.27]	1.374	1.25	1.20	0.13
DBT + F p2	839.1 [1.48]	978.7 [1.27]	1.372	1.94	2.34	0.51
DBT + F p5	837.4 [1.48]	976.6 [1.27]	1.374	1.58	1.52	0.08

Based on the results of the DFT calculations, the relative yield of chemical synthesis, and considering the accessible range of our laser sources for high-resolution spectroscopy, 3-fluoroterrylene (3FT) was chosen as the most proper candidate for testing our design strategy and to test the agreement between calculations and experimental outcome. We therefore, synthesized this substance and performed absorption and fluorescence emission spectroscopy on it. Bulk emission and absorption spectra of terrylene and 3FT in various liquid solvents, chosen to yield a large span of refractive indices, are depicted in Fig. 1a and b. The peak absorption wavelength for terrylene dissolved in toluene is at 560 nm which is in relatively good agreement with the DFT result. The corresponding value for 3FT is 540 nm, exhibiting a significant blue-shift of 20 nm, as opposed to the DFT calculations that indicated a light red-shift. This discrepancy might be due to the solvent effect missing in the DFT result. The absorption spectrum of 3FT, showing a main peak followed by two characteristic vibronic components, although much broader, very much resembles the spectrum of terrylene. The emission spectrum of 3FT, on the other hand, does not resemble the mirror image of the absorption spectrum. Most of the emission seems to be in one of the vibronic components suggesting that the electron cloud in the excited state is heavily perturbed by the presence of the fluorine atom. Upon cooling the sample down to the liquid nitrogen temperatures, no sharp zero phonon line appears in the spectrum (see Fig. 1c). The position of the emission peak is very close to the value predicted by TDDFT/B3LYP.

Optical spectroscopy

Absorption and fluorescence emission spectra were recorded for 3-fluoroterrylene and terrylene in a variety of solvents at ambient conditions. For the absorption measurements the sample was illuminated with white light from a xenon lamp (Mikropack, HPX-2000, High power xenon light source). Transmitted light was collected and sent to an imaging spectrometer (Horiba Jobin Yvon iHR320, f = 320 mm, 150 grooves per mm grating, LN2 cooled CCD (Symphony II detection system)). To record the fluorescence emission spectra the sample was excited with 532 nm laser light (Coherent, Verdi V-5, diode pumped solid-state laser). The fluorescence emission was collected in reflection mode and sent to the spectrometer. The laser light was blocked by two 532 nm notch filters.

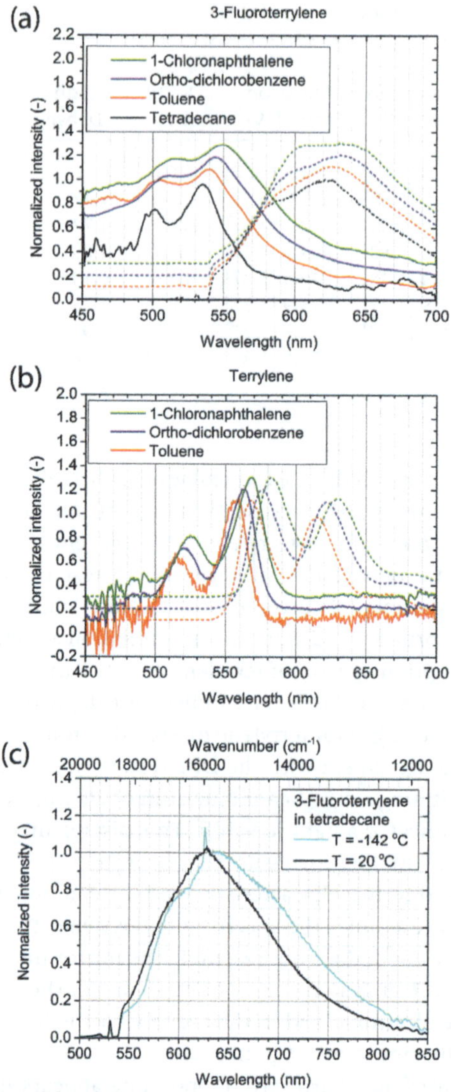

Fig. 1 The normalized absorption (solid curves) and emission (dashed curves) spectra of 3-fluoroterrylene (a) and terrylene (b) in several solvents. The spectra for tetradecane has zero offset; the other spectra have been cumulatively offset by 0.1 in the vertical direction for clarity. (c) Emission spectrum of 3-fluoroterrylene in tetradecane at 131 K and 293 K. The sharp peak at around 625.8 nm is Raman scattering of C–H stretching. There is no clear signature indicating to the presence of a narrow zero phonon line.

Computational details

All DFT and time-dependent DFT (TDDFT) calculations have been performed with the Gaussian 09 program.[18] We used the B3LYP exchange-correlation functional. The basis set was cc-pVDZ for all calculations except for the case with I (iodine) since this basis set is not available for this element. Instead, for I we used the DGDZVP basis set, which should be of similar accuracy. TDDFT/B3LYP has been used to also optimize the geometry in the first excited state.

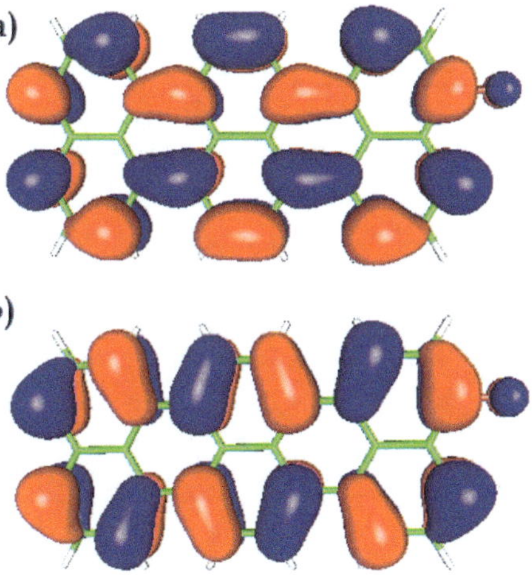

Scheme 2 (a) LUMO and (b) HOMO for 3-fluoroterrylene.

Synthesis

The synthesis of the targeted 3-fluoroterrylene **6** is shown in Scheme 3.[19] The synthetic route started with the preparation of 2-(4-fluoronaphthalen-1-yl)-4,4,5,5-tetramethyl-1,3,2-dioxaborolane **2**, which was obtained in 58% yield by Miyaura borylation of 1-bromo-4-fluoronaphthalene **1**, using a modified procedure to that reported by Miyaura *et al.*[20] 3-Bromoperylene **4** was obtained in 87% yield by NBS bromination of perylene **3**, following the procedure reported by Mitchell *et al.*[21,22] in combination with our previously reported method.[23] Subjecting 2-(4-fluoronaphthalen-1-yl)-4,4,5,5-tetramethyl-1,3,2-dioxaborolane **2** with 3-bromoperylene **4** to modified Suzuki cross-coupling conditions[23,24] afforded the key intermediate 3-(4-fluoronaphthalen-1-yl)perylene **5** in 75% yield. The structure of **5** was confirmed by ^{1}H, ^{13}C and ^{19}F NMR spectroscopy, and by UV and mass spectrometry.[1] Having obtained the key intermediate 3-(4-fluoronaphthalen-1-yl) perylene **5**, attention was then turned to selective cyclodehydrogenation reaction conditions. A representative cyclodehydrogenation procedure of **5** involved the use of eight equivalents of aluminium chloride in chlorobenzene and the mixture was stirred at 80 °C for 4 h. Following a standard workup the precipitate was then washed thoroughly with dichloromethane and extracted in a Soxhlet apparatus with toluene. The orange solution was decanted and the solid product was collected from the flask walls, rinsed with petroleum ether and dried in vacuum to give the targeted product 3-fluoroterrylene **6** as a purple solid in 31% yield. The structure of **6** was confirmed by ^{19}F NMR spectroscopy, UV vis and mass spectrometry.[19]

Discussion

Over the past 25 years, experimentalists have discovered and studied a variety of organic fluorescent molecules with excellent quantum properties.[25] Embedding these molecules in carefully chosen host crystals confers a high degree of coherence to their optical ZPL electronic transition already at moderately low

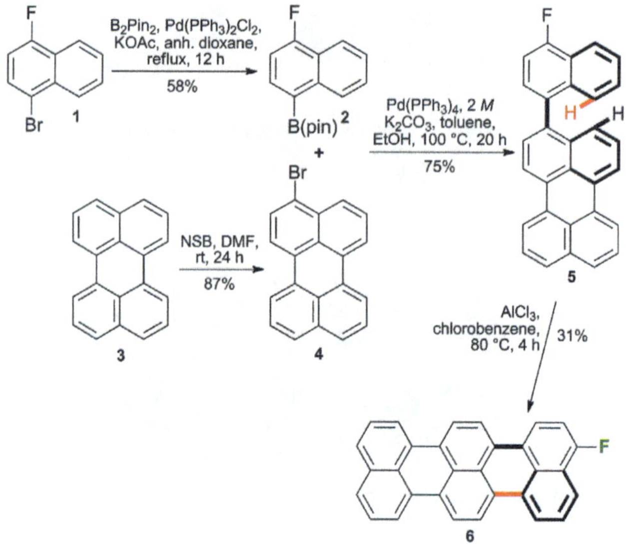

Scheme 3 Synthesis of 3-fluoroterrylene.

temperatures (around 2 K). Other emitters have also been suggested for optical single electron sensing, most notably the nitrogen vacancy center in diamond.[26] All these emitters, however, must be embedded in extremely pure inorganic crystals which puts stringent conditions on their fabrication and makes them harder to combine with other types of electronic devices. Besides their narrow coherent transition and large Stark shift, organic molecules have three great advantages over other solid-state emitters: they are

(a) Small: less than 1 nm. They can be placed very close to the investigated device and can be used in concentrations as high as 1000 per cubic micrometer. Current optical super-resolution techniques with bright and stable emitters routinely provide position accuracies close to 1 nm.

(b) Flexible: their structure can be tailored according to the specific needs of the experiment.

(c) Reproducible: each molecule of a given substance is completely identical to any other molecule. Using molecules, we avoid structural inhomogeneity, which is a big hindrance in nanofabricated structures such as nanocrystals or self-assembled quantum dots.

In order to realize some intriguing proposals such as electric charge mapping or hybrid quantum interfaces between electrons and photons, it is necessary to invent new molecules with high electric field sensitivity. Given the long and expensive procedure of synthesizing new organic substances, we have first performed DFT modeling on a class of substances and have found single fluorination a plausible strategy.

Based on our calculations, the magnitude of the dipole moment stays almost constant for the ground and excited states of the lowest energy excitation in all three systems. By looking at the components of the dipole moment, we see that in all three molecules, the large change in dipole moment between ground state and excited state is due to the reorientation of the dipole. The orientation of the dipole moment is shown in Scheme 4 for one of these molecules. An interesting case is DBT with a F substitution in position 2 with a change in dipole moment of about 0.5 Debye. Also for perylene with F substitution in position 3 there is a large change of about 0.7 D, which again is mainly due to a rotation of the dipole moment upon excitation.

To this end, we have synthesized only one of the simulated compounds, 3-fluoroterrylene, which was seen as the most promising candidate. Since we could not find a strong ZPL for this molecule at cryogenic temperatures, we used the method of solvent-induced shifts for measuring the Stark factor.

Previously, it has been shown that by measuring the solvent induced shift of the absorption and emission peak, it is in some cases possible to estimate the dipole moments by plotting the Stokes shift $\Delta\nu$ *versus* the difference between the static and optical Clausius–Mossotti factors $\Delta f(\epsilon, n^2)$, where ϵ is the static permittivity and n is the refractive index at optical frequencies. The change of the Stokes shift with solvent consists of two parts: a universal part proportional to the squared difference in dipole moments of the ground and excited states $\Delta\mu^2 = (\mu_e - \mu_g)^2$ given by:

$$\Delta\nu_{\text{universal}} = \frac{2\Delta\mu^2}{hc\varepsilon_0 a^3}\Delta f\left(\epsilon, n^2\right) = \frac{2\left(\mu_e - \mu_g\right)^2}{hc\varepsilon_0 a^3}\left[\frac{\epsilon - 1}{2\epsilon + 1} - \frac{n^2 - 1}{2n^2 + 1}\right],$$

plus a solvent specific term.[27–29] Unfortunately, although the dipole moments for fluoroterrylene are large for an aromatic molecule (see Table 1), the difference $\Delta\mu$ is still very small compared to those of the dyes that were successfully investigated before with this method, and which all present strong charge rearrangements in the excited state. Therefore, it appears that the solvent-induced shifts are dominated by the specific interactions with the solvent, and therefore that the systematic trend with solvent dielectric properties is not large enough to appear in

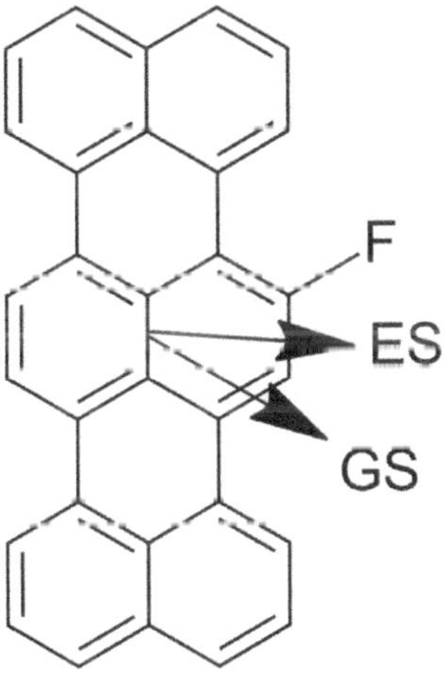

Scheme 4 The electric dipole moments of 7-fluoroterrylene in the ground (GS) and excited (ES) states indicating that the change in the moments between the two states is mainly due to the re-orientation and not because of the change in the dipole moment magnitude.

Table 3 Solvent shifts for 1-fluoroterrylene and terrylene. Reported are, from left to right, the solvent, the static dielectric constant, the refractive index of the solvent, the difference between the static and optical Clausius–Mossotti factor, the peak absorption and emission wavelengths of the fluorophore in the specific solvent and the difference (Stokes shift) and mean of the absorption and emission wavenumbers calculated from those values. The reported errors are the statistical errors obtained from the fit. The errors for the peak wavelengths of terrylene were <0.1. Note that the statistical errors are smaller than the spectral resolution of ~0.4 nm of the spectrometer and the pixel resolution of ~0.5 nm of the spectrometer CCD

Solvent	ε	n	Δf	$\lambda_{abs.}$ (nm)	$\lambda_{em.}$ (nm)	$\Delta\tilde{\nu}$ (cm^{-1})	$\tilde{\nu}$ (cm^{-1})
3-Fluoroterrylene							
1-Chloronaphthalene	5.00[b]	1.633[a]	0.100	548.7 ± 0.1	634.0 ± 0.2	2451 ± 7	16 998 ± 4
Ortho-dichlorobenzene	9.93[a]	1.551[a]	0.186	544.3 ± 0.1	631.8 ± 0.1	2544 ± 4	17 100 ± 2
Toluene	2.379[a]	1.496[a]	0.013	539.5 ± 0.1	626.4 ± 0.1	2570 ± 3	17 250 ± 2
Tetradecane	2.039[c]	1.429[a]	~0	535.2 ± 0.1	621.5 ± 0.2	2595 ± 6	17 386 ± 3
Terrylene							
1-Chloronaphthalene	5.00[b]	1.633[a]	0.100	567.9 ± 0.1	582.6 ± 0.1	443 ± 2	17 387 ± 1
Ortho-dichlorobenzene	9.93[a]	1.551[a]	0.186	562.6 ± 0.1	576.1 ± 0.1	418 ± 1	17 567 ± 1
Toluene	2.379[a]	1.496[a]	0.013	557.1 ± 0.1	568.3 ± 0.1	354 ± 1	17 773 ± 1

[a] ed. Robert C. Weast, *CRC Handbook of chemistry and physics*, 49th edn, Chemical Rubber Publishing Company, Cleveland, Ohio, 1968. [b] Arthur A. Maryott and Edgar R. Smith, *Table of dielectric constants of pure liquids*, National Bureau of standards circular 514, 1951. [c] Extrapolated from data for other *n*-alkanes (pentane, hexane, heptane, octane, *etc.* up to dodecane) as reported in footnote b.

the plot of the shift *versus* the solvent polarizability. Hence we were unable to determine the dipole moment change by this method (Table 3).

Before trying the other halogenated substances that we have modeled, it is possibly necessary to perform new calculations on, for example, other substituents like nitrile groups that might induce less distortion on the electronic orbitals of the dye molecule. More detailed molecular dynamics calculations that also include positioning of the molecule in the host crystal might be necessary to predict the best molecular design for single molecule electric field sensing. An alternative experimental route can be to use other chromophores such as a single chain of poly(2-methoxy-5-(2-ethyl-hexyloxy)-1,4-phenylene-vinylene) (MEH-PPV) that exhibit electric field response[30] and a sizable ZPL[31] or to use semiconductor nanocrystals[32] or vacancy centers in diamond[26], which have comparable properties. Much work, however, will be necessary to obtain a natural linewidth-limited spectrum in these alternative systems, which is required for using them as quantum transducers. Given the vast variety of applications for such quantum nanoprobes and the numerous advantages of organic molecules over other types of emitters, this investigation is certainly worth the effort.

Conclusions

In summary, we have investigated one possibility for designing new dye molecules specifically as sensitive electric field probes. To our knowledge, no previous report in single molecule spectroscopy has been devoted to this topic and all the previous reports on sensing charges and electric fields with single molecules have used the matrix-induced dipole moment difference for this purpose. We have calculated the Stark factor of the ZPL for variants of three widely used molecules under cryogenic conditions in which a single hydrogen atom has been substituted by a halogen atom. Our results indicate that fluorine could be the best substituent and we have synthesized 3-fluoroterrylene for our experiment. While room temperature spectroscopy on solutions of this substance shows a large spectral shift relative to terrylene in both absorption and emission spectra, we could not find an indication of a strong ZPL for 3FT in tetradecane. While choosing other host matrices might still be helpful in obtaining a narrow ZPL, we suspect that other design strategies might be necessary to find new candidate molecules with both strong Stark factor and strong ZPL emission.

Acknowledgements

Nico Verhart acknowledges support from NanoNextNL (project 06D.12). Marios Markoulides acknowledges funding from the FP7 project PAMS (contract: 610441). The use of supercomputer facilities was sponsored by NWO Physical Sciences, with financial support from the Netherlands Organization for Scientific Research (NWO).

Notes and references

1 M. Orrit, T. Ha and V. Sandoghdar, *Chem. Soc. Rev.*, 2014, **43**, 973–976.
2 F. Kulzer, T. Xia and M. Orrit, *Angew. Chem., Int. Ed.*, 2010, **49**, 854–866.
3 Y. Tian, P. Navarro and M. Orrit, *Phys. Rev. Lett.*, 2014, **113**, 135505.
4 J.-M. Caruge and M. Orrit, *Phys. Rev. B: Condens. Matter Mater. Phys.*, 2001, **64**, 205202.
5 A. A. L. Nicolet, M. A. Kol'chenko, C. Hofmann, B. Kozankiewicz and M. Orrit, *Phys. Chem. Chem. Phys.*, 2013, **15**, 4415–4421.
6 T. Plakhotnik, *ChemPhysChem*, 2006, **7**, 1699–1704.
7 S. Faez, S. J. van der Molen and M. Orrit, *Phys. Rev. B: Condens. Matter Mater. Phys.*, 2014, **90**, 205405.
8 V. Puller, B. Lounis and F. Pistolesi, *Phys. Rev. Lett.*, 2013, **110**, 125501.
9 S. Faez, P. Türschmann, H. R. Haakh, S. Götzinger and V. Sandoghdar, *Phys. Rev. Lett.*, 2014, **113**, 213601.
10 S. Das, S. Faez and A. S. Sørensen, *Proc. SPIE*, 2014, 9136–9189.
11 H. J. Kimble, *Nature*, 2008, **453**, 1023–1030.
12 M. Orrit, J. Bernard, A. Zumbusch and R. I. Personov, *Chem. Phys. Lett.*, 1992, **196**, 595; M. Orrit, J. Bernard, A. Zumbusch and R. I. Personov, *Chem. Phys. Lett.*, 1992, **199**, 408.
13 F. Kulzer, R. Matzke, C. Bräuchle and T. Basché, *J. Phys. Chem. A*, 1999, **103**, 2408–2411.
14 P. Bordat, M. Orrit, R. Brown and A. Würger, *Chem. Phys.*, 2000, **258**, 63–72.
15 S. Kummer, T. Basché and C. Bräuchle, *Chem. Phys. Lett.*, 1994, **229**, 309–316.
16 C. Hofmann, A. Nicolet, M. A. Kol'chenko and M. Orrit, *Chem. Phys.*, 2005, **318**, 1–6.
17 A. Sigl, C. Scharnagl, J. Friedrich, A. Gourdon and M. Orrit, *J. Chem. Phys.*, 2008, **128**, 044508.

18 M. J. Frisch, G. W. Trucks, H. B. Schlegel, G. E. Scuseria, M. A. Robb, J. R. Cheeseman, G. Scalmani, V. Barone, B. Mennucci, G. A. Petersson, H. Nakatsuji, M. Caricato, X. Li, H. P. Hratchian, A. F. Izmaylov, J. Bloino, G. Zheng, J. L. Sonnenberg, M. Hada, M. Ehara, K. Toyota, R. Fukuda, J. Hasegawa, M. Ishida, T. Nakajima, Y. Honda, O. Kitao, H. Nakai, T. Vreven, J. A. Montgomery Jr, J. E. Peralta, F. Ogliaro, M. Bearpark, J. J. Heyd, E. Brothers, K. N. Kudin, V. N. Staroverov, R. Kobayashi, J. Normand, K. Raghavachari, A. Rendell, J. C. Burant, S. S. Iyengar, J. Tomasi, M. Cossi, N. Rega, J. M. Millam, M. Klene, J. E. Knox, J. B. Cross, V. Bakken, C. Adamo, J. Jaramillo, R. Gomperts, R. E. Stratmann, O. Yazyev, A. J. Austin, R. Cammi, C. Pomelli, J. W. Ochterski, R. L. Martin, K. Morokuma, V. G. Zakrzewski, G. A. Voth, P. Salvador, J. J. Dannenberg, S. Dapprich, A. D. Daniels, Ö. Farkas, J. B. Foresman, J. V. Ortiz, J. Cioslowski and D. J. Fox, *Gaussian 09, Revision D.01*, Gaussian, Inc., Wallingford CT, 2009.
19 M. S. Markoulides, C. Venturini, D. Neumeyer and A. Gourdon, *New J. Chem.*, 2015, **39**, 6498–6503.
20 T. Ishiyama, M. Murata and N. Miyaura, *J. Org. Chem.*, 1995, **60**, 7508–7510.
21 R. H. Mitchell, Y.-H. Lai and R. V. Williams, *J. Org. Chem.*, 1979, **44**, 4733–4735.
22 H. Maeda, Y. Nanai, K. Mizuno, J. Chiba, S. Takeshima and M. Inouye, *J. Org. Chem.*, 2007, **72**, 8990–8993.
23 S. Nagarajan, C. Barthes, N. K. Girdhar, T. T. Dang and A. Gourdon, *Tetrahedron*, 2012, **68**, 9371–9375.
24 For a review, see: A. Suzuki, *J. Organomet. Chem.*, 1999, **576**, 147–168.
25 P. Tamarat, A. Maali, B. Lounis and M. Orrit, *J. Phys. Chem. A*, 2000, **104**, 1–16.
26 F. Dolde, H. Fedder, M. W. Doherty, T. Nöbauer, F. Rempp, G. Balasubramanian, T. Wolf, F. Reinhard, L. C. L. Hollenberg, F. Jelezko, *et al.*, *Nat. Phys.*, 2011, **7**, 459–463.
27 *Photophysics of aromatic molecules*, ed. J. B. Birks, John Wiley and Sons Ltd., London, 1970, section 4.12, pp. 109–119.
28 E. G. McRae, *J. Phys. Chem.*, 1957, **61**, 562–572.
29 K. K. Rohatgi-Mukherjee, *Fundamentals of Photochemistry*, New Age International Publishers, New Delhi, 1978, section 4.6 and 4.7, pp. 101–106.
30 F. Schindler, J. M. Lupton, J. Müller, J. Feldmann and U. Scherf, *Nat. Mater.*, 2006, **5**, 141–146.
31 F. A. Feist, G. Tommaseo and T. Basché, *Phys. Rev. Lett.*, 2007, **98**, 208301.
32 S. A. Empedocles and M. G. Bawendi, *Science*, 1997, **278**, 2114–2117.

Faraday Discussions

PAPER

Single-molecule spectromicroscopy: a route towards sub-wavelength refractometry

T. A. Anikushina,[ab] M. G. Gladush,[a] A. A. Gorshelev[a] and A. V. Naumov*[ab]

Received 20th May 2015, Accepted 10th July 2015
DOI: 10.1039/c5fd00086f

We suggest a novel approach for spatially resolved probing of local fluctuations of the refractive index n in solids by means of single-molecule (SM) spectroscopy. It is based on the dependence $T_1(n)$ of the effective radiative lifetime T_1 of dye centres in solids on n due to the local-field effects. Detection of SM zero-phonon lines at low temperatures gives the values of the SM natural spectral linewidth (which is inversely proportional to T_1) and makes it possible to reveal the distribution of the local n values in solids. Here we demonstrate this possibility on the example of amorphous polyethylene and polycrystalline naphthalene doped with terrylene. In particular, we show that the obtained distributions of lifetime limited spectral linewidths of terrylene molecules embedded into these matrices are due to the spatial fluctuations of the refractive index local values.

1. Introduction

One of the most important characteristics of a material that determines many of its macroscopic properties is the refractive index, n. It is included as a parameter in classical equations (*e.g.* Fresnel's and Maxwell's), is linked to the dielectric permittivity and magnetic permeability, and is known to depend on temperature, pressure and wavelength.[1] In recent years, special attention has been given to the refractive index in relation to metamaterials, which are assumed to have a negative n value.[2,3]

Since Ernst Abbe invented the refractometer in 1874 (ref. 4), the most precise techniques for measuring n have been based on the principle of the total internal reflection (TIR) on the interface of two media. There are also some other practical techniques for direct and indirect measurements of n (goniometry, ellipsometry, interferometry, frustrated TIR, see *e.g.* ref. 1 and 5–8), however, most of them

[a]*Molecular Spectroscopy Dept., Institute for Spectroscopy of the Russian Academy of Sciences, Fizicheskaya Str 5, Troitsk, Moscow, 142190 Russia; Web: http://www.single-molecule.ru. E-mail: a_v_naumov@mail.ru; Fax: +7 495 8510886; Tel: +7 495 8510236*

[b]*Chair of the Theoretical Physics, Institute of Physics, Technology and Information Systems, Moscow State Pedagogical University, Malaya Pirigivskaya Str.29, Moscow, 119991, Russia*

require averaging over a large macroscopic volume of the sample. As a result, the question of local fluctuations of n in real solids remains unanswered. Moreover, in view of the growing interest in nanotechnology and nanophotonics, it is important to know whether there is a relationship between n and the micro- and nanoscopic structure of solids. In order to find the answer, some special instruments for probing the local n values on the sub-wavelength (nanometer) level must be developed.

Topical papers describe a great number of very promising "nanoinstruments" for probing various local parameters of a sample based on single quantum light emitters embedded into the sample's material. For example, single-molecule (SM), single quantum dot (QD) and single nanocrystal/nanoparticle spectroscopy and imaging provide an outstanding basis for various kinds of nanoinstrumentation, like nanoscale thermometry,[9] nanodetectors of acoustic strain,[10] single charge detectors,[11] instruments for probing low-energy excitations in disordered solids,[12–17] *etc.* Besides, we see a possibility to use single-molecule spectroscopy (SMS) for probing the local-field effects in dye-doped solids.

It is well known that the local-field effects cause the dependence of the excited state lifetime of a light emitter on the refractive index of the hosting medium $T_1(n)$.[18–25] Thus, the study of the photo-physical properties of a single dye center is the way to probe the effective local n values. For example, in ref. 26 such an idea was realised by analysing the fluorescence decay of QDs.

In this work we propose a method for probing the local (nm-scale) fluctuations of n in solids by the analysis of zero-phonon spectral lines (ZPL) of single impurity dye-molecules at ultra-low temperatures (Fig. 1).

2. Theoretical

The possibility to measure the index of refraction on the sub-wavelength scale is extremely important because of the explosive growth of interest in nanophotonics and related fields. The refractive index n is taken into account in all model calculations and theoretical analyses (*e.g.* the development of single-photon sources, hybrid nanostructures, nanodevices, nanoscale metamaterials), but scientists usually consider just the average n value for the specific material. At the same time, the local n values in a macroscopically homogeneous material can be distributed over a very broad range for the different microscopic volumes. This fact should be taken into account in any nanophotonic application.

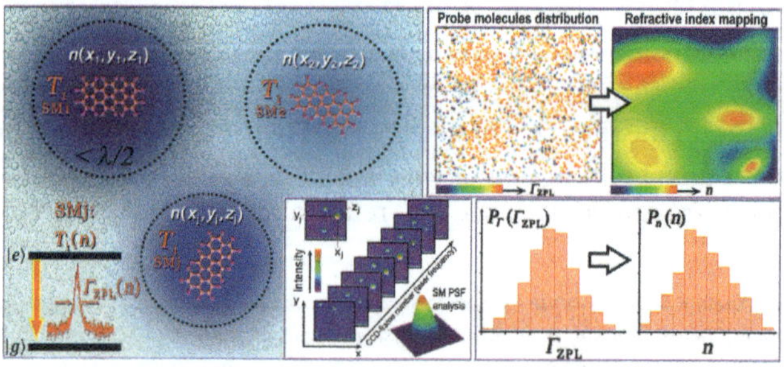

Fig. 1 An illustrative sketch of the developed approach for probing local values of the refractive index n in solids.

A ZPL, which corresponds to a purely electronic transition in an impurity molecule,[27] is a unique source of information about dye–matrix interactions.[12,13] ZPL parameters (frequency, intensity, width, *etc.*) are very sensitive to the local environment of the corresponding chromophore SM. This fact makes SMs good candidates for spectral probes to obtain data on the structure and the internal dynamics of solids.

As we know from numerous studies, the temperature T dependence of the homogeneous spectral width of ZPL is determined by three main contributions,[12–16] *i.e.*

$$\Gamma_{ZPL}(T) = \Gamma_0 + \Delta\Gamma_{e\text{-tunn}}(T,t_m) + \Delta\Gamma_{e\text{-phon}}(T), \tag{1}$$

where the natural (lifetime limited) linewidth Γ_0 is

$$\Gamma_0 = \frac{1}{2\pi T_1}; \tag{2}$$

the ZPL broadening due to the interaction of electronic transitions in impurity molecules with tunneling excitations in a matrix is

$$\Delta\Gamma_{e\text{-tunn}}(T,t_m) \sim T^\alpha \ln(t_m); \tag{3}$$

and the ZPL broadening due to the quadratic electron–phonon interaction, which, in the simplest case of an interaction with a single quasilocalized low-frequency vibrational mode (LFM), is expressed as

$$\Delta\Gamma_{e\text{-phon}}(T) = w\frac{\exp(-\Delta E/kT)}{[1-\exp(-\Delta E/kT)]^2}. \tag{4}$$

In these equations, t_m is the total time of measurement, $1 \le \alpha < 2$ and its exact value depends on the dye–matrix system, w is the constant of the quadratic electron–phonon coupling, ΔE is the energy of LFM and k is Boltzmann constant.

The ZPL width is also a function of the laser excitation intensity P_{LAS}:

$$\Gamma_{ZPL}(P_{LAS}) = \Gamma_{ZPL}(0)\sqrt{1+\frac{P_{LAS}}{P_S}}, \tag{5}$$

where P_S is the saturation intensity and $\Gamma_{ZPL}(0)$ is the unsaturated ZPL spectral width. Note the saturation behavior depends on the relative orientation[12] of SM transition dipole moment and laser excitation polarization. However this dependence leads only to "recalibration" of curve, and does not change the value of unsaturated linewidth $\Gamma(0)$.

From eqn (1)–(5), it follows that if $P_{LAS} \ll P_S$ and the temperature is in the range of ultra-low values, then the additional broadenings $\Delta\Gamma_{e\text{-tunn}}$ and $\Delta\Gamma_{e\text{-phon}}$ can be neglected and for each SM it is possible to measure the lifetime limited ZPL spectral width related to T_1. Numerous experiments have shown that these conditions are achieved at $P_{LAS} < 0.1$–10.0 W cm^{-2} and at $T < 1$–5 K (depending on the type of matrix, see *e.g.* ref. 28 and 29). Thus, SMs can be used as very sensitive probes of T_1 fluctuations due to the local-field effects.

As a primary step, light emitters can be viewed as if they are (quasi)homogeneously distributed over some macroscopic space, while the space that separates their localities is an n-valued "continuous medium". We assume that this n is the averaged value and there are variations in the effective index in the regions that contain a single emitter (due to local matrix inhomogeneities). In reality, a specific host material may require a more accurate theory because of numerous types of complexity. The local environments around an emitter can have their own n value contrasting with the bulk host matrix, or be a discrete or structured region where the continuous medium approximation is not applicable. If we leave these cases for a later discussion in this section (*e.g.* recent work in ref. 30–32), the correction factors for T_1 can be summarized using five models.[20–24] They all consider the situation when a guest molecule emits light into the continuous medium of the host matrix, that is weakly absorbing, isotropic and characterized by n. Most generally, the effective T_1 is written as:[19,20]

$$T_1(n) = \frac{\tau_0}{nf(n)}, \tag{6}$$

where τ_0 is the "vacuum" value of the excited state lifetime; n is the refractive index (if precisely at the wavelength of the emitted light); function $f(n)$ reflects the local-field contributions and includes either a squared or straight n-dependent coefficient $l(n)$ being the proportionality between the local-field acting on the emitter E_L and the average Maxwellian field E_M, *i.e.* $E_L = l(n)E_M$.

To date, the choice of $l(n)$ has been among two concepts of the local-field. One is the Lorentz local-field, which is classically calculated under the assumption that the field due to polarized molecules inside a small sphere centered at the site of a light emitter may be neglected. The other is a simplified variant of Onsager's approach,[33] where there is a small empty cavity around the emitter. These cases are often distinguished by the concept of interstitial and substitutional guest molecules.[25] The most popular models for the lifetime correction are the virtual-cavity model[21,22] based on the Lorentz local-field:

$$f(n) = \left(\frac{n^2+2}{3}\right)^2, \tag{7}$$

and the real- or empty-cavity model[23] based on the Onsager model for the local-field:

$$f(n) = \left(\frac{3n^2}{2n^2+1}\right)^2. \tag{8}$$

Another approach is called the fully microscopic model,[24] in which

$$f(n) = \frac{1}{n}\left(\frac{n^2+2}{3}\right). \tag{9}$$

Similar to eqn (7), this model uses the Lorentz field but here the $l(n)$ function is not squared. At the same time, it misses the n factor in the final form of eqn (6). The remaining models imply $f(n) = l(n)$,[20] which brings back the missing n and provides agreement with the fully microscopic model (eqn (9)) in terms of the fact that the local-field coefficients $l(n)$ are not squared. Thus, for the likely-to-be-interstitial emitters:

$$f(n)=\frac{n^2+2}{3},\tag{10}$$

while for the likely-to-be-substitutional emitters:

$$f(n)=\frac{3n^2}{2n^2+1}.\tag{11}$$

There are the models that describe specific materials and systems and, consequently, include a greater number of parameters, we note that in ref. 30, 32 and 34 they also agree with writing the concept (6)–(8) as a starting point. Now some sophistications follow. In ref. 32, they study the decay dynamics of ions in nanocrystals in various solvents with different n values. A more appropriate model for their case was shown to be:

$$f(n)=\left(\frac{3n^2}{2n^2+n_{\mathrm{NC}}}\right)^2[1+f(n,n_{\mathrm{NC}},\alpha)]\tag{12}$$

where, if compared to eqn (8), the unit in the denominator (being the index of the empty cavity) is substituted with the refractive index of the nanocrystal n_{NC}, the term in the square brackets describes the second-order correction including the radius,[35] and n is the index of the solvent.

In the studies of polymers using the SM lifetime fluctuations,[34] they used:

$$f(n)=\left|\frac{\mu_{\mathrm{tot}}}{\mu}\right|^2\left(\frac{n^2+2}{3}\right)^2\tag{13}$$

as the final working formula. Here, the difference compared to eqn (7) is the factor on the right hand side. This factor follows on from the polymer structure theory. It is the ratio between the total dipole μ_{tot} in the cavity and the dipole μ associated with the charge distribution of the probe molecule, and is calculated numerically. It must be noted that the simulated lifetime distributions for this case are strongly dependent on this ratio in terms of broadening and asymmetry. This behavior of distributions was shown to be in good agreement with the one experimentally observed in ref. 34 and 36, where SM lifetime fluctuations were used to reveal the segmental dynamics in polymers at room and higher temperatures.

The calculations of the spontaneous decay of an excited atom in a dense atomic cloud were performed numerically for the model when the continuous medium approximation does not apply.[31] It was shown that direct calculations give different results from the n-value concept for both the similar and foreign atom cases.

Speaking generally, the models in eqn (7) and (8) are based on the field quantization procedures in dielectrics and other macroscopic concepts for treating the problem of spontaneous emission. The models in eqn (9)–(11) appeared from the derivation of the Maxwell–Bloch equations describing the light–matter interaction using the Heisenberg operator formalism for (9) and the Bogoliubov–Born–Green–Kirkwood–Yvon equations for the reduced density matrices and correlation operators of material particles and modes of the quantized radiation field for eqn (10) and (11). In both approaches, the field operators were initially written for material vacuum, providing interactions between the guest and host particles.

Nevertheless, each of these models (7)–(11) has found "verification" during the analyses of the experimental data.[19,20] In all cases, the so-called vacuum excitation lifetime τ_0 was subjected to variation during the fitting procedure. Depending on the model used for data fitting, for each emitting system, the preferable $T_1(n)$ dependence was obtained. It must be noted, however, that the verifications of the models were based on the data from the excited state lifetime measurements for dopants in crystals and glasses as well as quantum dots in various solutions. This approach has never been used for impurity systems, such as organic molecules in solid matrices with different degrees of disorder and, specifically, has never been tried on SMs. Here we make the first attempt to test this approach using several sets of unique data obtained by different research groups.

3. Experimental results and methods

In this work we have chosen terrylene (Tr)[37] as the object of analysis. It has been widely studied in different matrices by laser selective spectroscopy methods, and is one of the most used fluorophores in SMS. For this compound, the T_1 values were measured for a set of solid matrices:[38] polyethylene (PE), polystyrene (PS), polyvinyl-butyral (PVB), polymethyl-methacrylate (PMMA), and solid *n*-hexadecane (Hex). In the context of the present work, we also had to find the refractive indices for these matrices. The *n* values for PE, PS, and PVB were found in ref. 39, whereas the PMMA and HEX refractive indexes were measured by our team using the laboratory Abbe refractometer (see the table in Fig. 2).

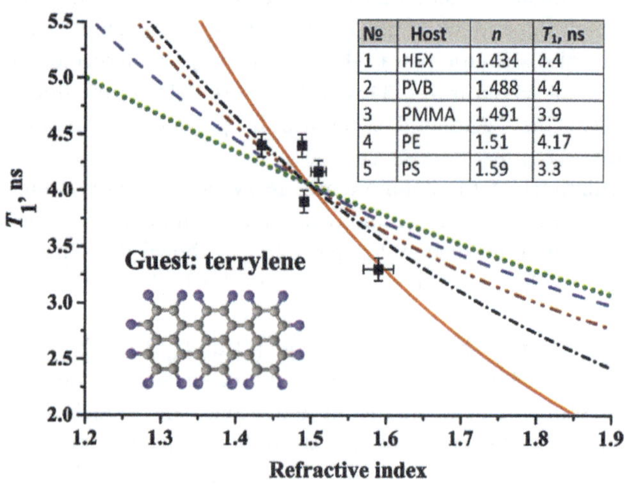

№	Host	n	T_1, ns
1	HEX	1.434	4.4
2	PVB	1.488	4.4
3	PMMA	1.491	3.9
4	PE	1.51	4.17
5	PS	1.59	3.3

Fig. 2 The dependence of the excited state lifetime $T_1(n)$ of terrylene on the refractive index of its host matrix (squares). The curves show the best fits of the $T_1(n)$ dependence using different models: the virtual-cavity model in Eqn. (7) (red solid); the empty-cavity model in Eqn. (8) (brown dash-dot-dot); the fully microscopic models in Eqn. (9) (green dot); microscopic model in Eqn. (10) (black dash-dot) and in Eqn. (11) (blue dashed).

A. Refractive index measurements

To obtain the refractive indices of the different polymer matrices as well as of liquid solutions, the laboratory Abbe refractometer URL-1 was used. The measurements were performed at a temperature of 20 ± 0.1 °C.

(a) In the case of the liquids, a drop of the solution under study was placed between the faces of the illuminating and the measuring prisms.

(b) In the case of the solids, the optically uniform sample should have two mutually perpendicular polished planes, one of which must be equal in size to the entrance face of the measuring prism. The sample under study was placed on the entrance face of the measuring prism. To achieve a good optical contact, an immersion oil was used. The second polished plane of the sample was facing towards the refractometer illuminator.

(c) During the measurements, the critical angle of the total internal reflection on the sample/prism interface was obtained and the sample refractive index n value was found.

B. Excited state lifetime measurements in the bulk dye-doped sample

The values of the terrylene fluorescence lifetimes in different matrices were obtained in ref. 38 by measuring the fluorescence decays excited by picosecond pulses of a Nd:YAG laser and were recorded using time-correlated single photon counting followed by specific data analysis.

C. Lifetime-limited single-molecule ZPL width measurements

The spectra of single terrylene molecules in a naphthalene crystal and in polyethylene were obtained in unique measurements performed in ETH Zurich in 1999–2000.[40,41] There they used optical ^{3}He/^{4}He cryostat (Oxford instruments). It had allowed SM ZPL detection at 30 mK using the fluorescence excitation technique. Each registered SM ZPL was then fit with the Lorentzian function, and thus the values of the SM linewidths were obtained. The power broadening of the SM spectra was taken into account and the appropriate corrections of the SM linewidth values were performed (see details in ref. 37 and 38).

D. Materials and samples under study

Terrylene (Tr) is a fluorescent molecule that belongs to the family of rylene dyes, which are extremely persistent against photo-bleaching and are characterized by weaker blinking (compared to other types of fluorophores) over a broad range of temperatures. Tr is characterized by a strong absorption, high fluorescence quantum yield, negligible population of triplet states, weak electron–phonon coupling at low temperatures, and high photo-stability.

Samples for excited state lifetime measurements were prepared by adding a 0.1 to 1 μM solution of terrylene dissolved in dichloromethane into about a 0.1% (w/w) solution of the polymers or semi-crystalline forming solids. The hosts utilized were polyvinyl-butyral, polymethyl-methacrylate, polystyrene, and hexadecane. Approximately 12 μL of the combination was transferred *via* a micropipette onto quartz cover glasses, where the drop size averaged about 5 mm in diameter. All samples were baked for evaporation of the solvent and for structural stability. The standard sample thickness was about 1 to 5 mm (see details in ref.

38). The preparation of the samples for lifetime-limited ZPL width measurements was as follows:

(a) Mixed crystals of terrylene in naphthalene were prepared by co-sublimation under a nitrogen atmosphere, which produced thin crystals of excellent optical quality. A single crystal was used as a sample, which had a thickness of 10 μm and a diameter of 1 mm. The concentration of terrylene molecules was 10^{-8} M (see details in ref. 40).

(b) The sample was prepared by means of diffusion, by heating a small amount of terrylene together with some polyethylene at 180 °C under vacuum for several hours. A small amount of the hot mixture was pressed between two polished stainless steel flats to produce a 2 μm thick film. To avoid surface broadening, the doped film was then melted between two undoped PE films, so that no SM was within 10 μm from the sample surface (see details in ref. 41).

E. Analysis of experimental data

In Fig. 2 we have plotted the T_1 values for the Tr molecules in all of the above mentioned matrices against the values of n in these systems. The $T_1(n)$ experimental dependencies were fitted using different models (eqn (7)–(11)) of the local-field correction factors, including the virtual-cavity model,[21,22] the real- or empty-cavity model,[23] the fully microscopic model,[24] and the two models from a quantum-kinetic approach for the derivation of the effective Maxwell–Bloch equations.[20]

4. Results and discussion

The best fit was proved to correspond to the virtual-cavity model Eq. 7 (red curve in Fig. 2), with the value of $\tau_0 = 12.1$ ns (i.e., $\Gamma_0 = 13.1$ MHz).

Given the value of τ_0 and the best dependence $T_1(n)$ for the Tr molecule, we can calculate the local value of n at the position of a SM from its lifetime-limited ZPL width measured in the matrix involved (Fig. 1). To do this, we have taken the unique experimental data on the SM ZPL width distributions (inset in Fig. 3), reported in refs. 40, 41. The authors had performed unprecedented and complicated measurements at milliKelvin temperatures on the SMS setup equipped with the $^3He/^4He$ dilution cryostat. Under these conditions, the broadening contributions $\Delta\Gamma_{e\text{-tunn}}$ and $\Delta\Gamma_{e\text{-phon}}$ were obviously negligible. The measurements were performed at the laser excitation intensities well below the saturation intensity. Thus, in accordance with eqn (5), the obtained data were the distributions of lifetime-limited SM ZPL widths Γ_0 directly related to the SM effective excited state lifetimes T_1.

These distributions of $P_{\Gamma_0}(\Gamma_0)$ can be converted into the distributions of $P_{T_1}(T_1)$ using eqn (2). Afterwards, distributions of $P_{T_1}(T_1)$ can be transformed into the distributions of the refraction indices $P_n(n)$ by numerical solving of the corresponding equations ((7)–(11) with core eqn (6)). For example, Fig. 3 demonstrates the results of such conversions with the virtual-cavity model (eqn (7), $\tau_0 = 12.1$ ns), the empty-cavity model (eqn (8)) and the microscopic model (eqn (10)).

Considering these distributions of n obtained for the polycrystalline and the polymer media, one should note the following: (a) there are significant fluctuations of the refractive index local values in real materials; (b) the value of n

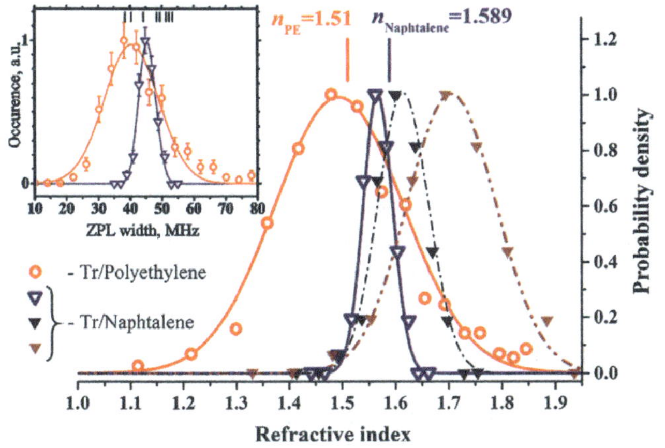

Fig. 3 The distribution of the local values of the refractive index in amorphous polyethylene (red circles) and polycrystalline naphthalene (blue triangles). The data are normalized to the maximum value. The sticks mark the average (for bulk media) values of the refractive index of polyethylene (red)[39] and naphthalene (blue).[43] The black and brown triangles describe the distributions in naphthalene using the satisfactory fit (eqn (10)) and the fit with eqn (8). The inset shows the spectral linewidth distribution for terrylene SMs in polyethylene at 30 mK with a Gaussian fit of the data.[41] The same is shown for Tr in naphthalene crystals.[40] Vertical sticks on the inset show values of unsaturated (lifetime-limited) spectral widths for Tr molecules in ortho-dichlorobenzene at T = 1.5 K, ref. 42.

measured in a bulk sample using classical methods corresponds to the peak of $P_n(n)$ obtained in the sample. Fig. 3 shows that this is valid for polyethylene and naphthalene. This implies the applicability of the proposed approach for probing the local fluctuations of the refractive index; (c) the local fluctuations of n in the amorphous polymer are substantially greater than those in the molecular polycrystal.

One can note that the experimental data points of the initial dependence $T_1(n)$ lie in the region of the n-values (1.4–1.6), which is narrower than the found region of the n-value distributions $P_n(n)$. This means, in principle, that for a broader range of n, another model may be more appropriate (or even a new model should be developed). In order to answer this question, the experimental data for $T_1(n)$ for a broader range of n should be obtained.

If one looks again at Fig. 2, another satisfactory fit is seen (the black dash-dot curve), which corresponds to the microscopic model (eqn (10)). Note that it includes the Lorentz local-field factor as the best fit with eqn (7). The other three models result in much worse approximations. We have recalculated the distribution of the lifetime-limited Γ_0 in naphthalene into the distributions of refraction indices $P_n(n)$ using all of the other models. Fig. 3 shows that the distributions using the fits for eqn (8) and (10) are subjected to significant broadening and, what is most important, their peaks are notably shifted from the value of n in the bulk naphthalene. The other two $P_n(n)$ distributions not shown are neither good. We treat this as further evidence of the correct choice of model and the validity of the developed approach. Note that naphthalene was absent among the systems (Fig. 2) in which T_1 was directly measured.

It is necessary to note that for some materials the proposed method can be realized at liquid ^{4}He temperatures (1.5 K and higher). For example, in ref. 42 it was shown that the measurement of unsaturated ZPL widths Γ_0 for Tr molecules in *ortho*-dichlorobenzene at $T = 1.5$ K gives the distribution of lifetime-limited spectral width values (see vertical sticks on the inset of Fig. 3). A similar result was obtained in ref. 29 for *para*-dichlorobenzene doped with Tr. It is important for the prospective applications of the developed method that quite simple ^{4}He based cryostats can be used for nanoscale mapping of local n values instead of the complicated ^{3}He cryogenic technique.

5. Conclusions

To conclude, in our study we propose a unique approach for probing the local refractive index fluctuations in solids. The method is based on the detection of SM ZPLs at conditions allowing lifetime-limited spectral line widths. The potential of this approach is demonstrated. In particular, simultaneous reconstruction of SM spatial coordinates with nanometre accuracy by super-resolution fluorescence microscopy[44] opens the way to perform sub-wavelength refractometry. In the most advanced version, probing the local refractive index values in solids can be performed in all 3 dimensions using any 3D luminescence microscopy technique (*e.g.* double helix point spread function[44,45]). It was found that there are significant fluctuations of the local n values in amorphous polymer and molecular polycrystalline media. These fluctuations are substantially greater in more disordered medium. The peak of the distribution $P_n(n)$ corresponds to the value of n, averaged over the bulk sample, which is usually obtained by traditional methods.

Acknowledgements

The authors acknowledge support from the Russian Foundation for Basic Researches (project # 14-29-07270 – development of the methods for probing of local fields; project # 13-02-01303 – analysis of life-time limited SM ZPL linewidths), and from Russian Science Foundation (project # 14-12-01415 "Statistical methods for single nanoobjects luminescence study: fundamental aspects and applications of spectromicroscopy of single molecules and quantum dots in material science and nanotechnology").

Notes and references

1 B. V. Ioffe, *Russ. Chem. Rev.*, 1960, **29**, 53–66.
2 S. Foteinopoulou, E. N. Economou and C. M. Soukoulis, *Phys. Rev. Lett.*, 2003, **90**, 107402.
3 J. Kastel, M. Fleischhauer and G. Juzeliunas, *Phys. Rev. A*, 2007, **76**, 062509.
4 E. Abbe, *Neue Apparate zur Bestimmung des Brechungs – und Zerstreuungsvermögens fester und fluüssiger Körper*, Mauke's Verlag, Jena, 1874.
5 J. E. Chamberlain, H. A. Gebbie and J. E. Gibbs, *Nature*, 1963, **198**, 874–875.
6 K. G. Muller, S. Sainov, S. MittlerNeher and W. Knoll, *Appl. Opt.*, 1996, **35**, 708–710.
7 A. Y. Nassif, *Appl. Opt.*, 1997, **36**, 779–785.

8 A. V. Zvyagin, K. Silva, S. A. Alexandrov, T. R. Hillman, J. J. Armstrong, T. Tsuzuki and D. D. Sampson, *Opt. Express*, 2003, **11**, 3503–3517.
9 G. Kucsko, P. C. Maurer, N. Y. Yao, M. Kubo, H. J. Noh, P. K. Lo, H. Park and M. D. Lukin, *Nature*, 2013, **500**, 54–U71.
10 Y. X. Tian, P. Navarro and M. Orrit, *Phys. Rev. Lett.*, 2014, **113**, 135505.
11 S. Faez, S. J. van der Molen and M. Orrit, *Phys. Rev. B: Condens. Matter Mater. Phys.*, 2014, **90**, 205405.
12 T. Basché, W. E. Moerner, M. Orrit and U. P. Wild, *Single-molecule optical detection, imaging and spectroscopy*, VCH, Weinheim, Cambridge, 1997.
13 A. V. Naumov, *Phys.-Usp.*, 2013, **56**, 605–622.
14 I. Osad'ko, *Selective Spectroscopy of Single Molecules*, Springer, 2010.
15 E. Geva and J. L. Skinner, *J. Phys. Chem. B*, 1997, **101**, 8920–8932.
16 Y. G. Vainer, A. V. Naumov, M. Bauer and L. Kador, *Phys. Rev. Lett.*, 2006, **97**, 185501.
17 P. Tamarat, A. Maali, B. Lounis and M. Orrit, *J. Phys. Chem. A*, 2000, **104**, 1–16.
18 R. S. Meltzer, S. P. Feofilov, B. Tissue and H. B. Yuan, *Phys. Rev. B: Condens. Matter Mater. Phys.*, 1999, **60**, 14012–14015.
19 K. Dolgaleva and R. W. Boyd, *Adv. Opt. Photonics*, 2012, **4**, 1–77.
20 D. V. Kuznetsov, V. K. Roerich and M. G. Gladush, *J. Exp. Theor. Phys.*, 2011, **113**, 647–658.
21 S. M. Barnett, B. Huttner and R. Loudon, *Phys. Rev. Lett.*, 1992, **68**, 3698–3701.
22 M. Fleischhauer, *Phys. Rev. A*, 1999, **60**, 2534–2539.
23 R. J. Glauber and M. Lewenstein, *Phys. Rev. A*, 1991, **43**, 467–491.
24 M. E. Crenshaw, *Phys. Rev. A*, 2008, **78**, 053827.
25 P. de Vries and A. Lagendijk, *Phys. Rev. Lett.*, 1998, **81**, 1381–1384.
26 A. Pillonnet, P. Fleury, A. I. Chizhik, A. M. Chizhik, D. Amans, G. Ledoux, F. Kulzer, A. J. Meixner and C. Dujardin, *Opt. Express*, 2012, **20**, 3200–3208.
27 K. K. Rebane, *J. Lumin.*, 2002, **100**, 219–232.
28 T. Plakhotnik, W. E. Moerner, V. Palm and U. P. Wild, *Opt. Commun.*, 1995, **114**, 83–88.
29 P. Navarro, Y. X. Tian, M. van Stee and M. Orrit, *ChemPhysChem*, 2014, **15**, 3032–3039.
30 R. A. L. Vallee, M. van der Auweraer, F. C. de Schryver, D. Beljonne and M. Orrit, *ChemPhysChem*, 2005, **6**, 81–91.
31 A. S. Kuraptsev and I. M. Sokolov, *Phys. Rev. A*, 2014, **90**, 012511.
32 T. Senden, F. T. Rabouw and A. Meijerink, *ACS Nano*, 2015, **9**, 1801–1808.
33 L. D. Landau, E. M. Lifésic and L. P. Pitaevskij, *Electrodynamics of continuous media*, Pergamon Press, Oxford *etc.*, 2nd edn, 1984.
34 R. A. L. Vallee, N. Tomczak, G. J. Vancso, L. Kuipers and N. F. van Hulst, *J. Chem. Phys.*, 2005, **122**, 114704.
35 H. T. Dung, L. Knoll and D. G. Welsch, *Phys. Rev. A*, 2000, **62**, 053804.
36 R. A. L. Vallee, N. Tomczak, L. Kuipers, G. J. Vancso and N. F. van Hulst, *Phys. Rev. Lett.*, 2003, **91**, 038301.
37 A. Bohnen, K. H. Koch, W. Luttke and K. Mullen, *Angew. Chem., Int. Ed. Engl.*, 1990, **29**, 525–527.
38 G. S. Harms, T. Irngartinger, D. Reiss, A. Renn and U. P. Wild, *Chem. Phys. Lett.*, 1999, **313**, 533–538.
39 *US Pat.*, 4310584, *Multilayer light-reflection film*, S. A. Cooper, R. Shetty, J. Pinsky, 1982.

40 E. A. Donley, V. Burzomato, U. P. Wild and T. Plakhotnik, *J. Lumin.*, 1999, **83–84**, 255–259.
41 E. A. Donley, S. Bonsma, V. Palm, V. Burzomato, U. P. Wild and T. Plakhotnik, *J. Lumin.*, 2000, **87–89**, 109–114.
42 A. A. Gorshelev, A. V. Naumov, I. Y. Eremchev, Y. G. Vainer, L. Kador and J. Koehler, *ChemPhysChem*, 2010, **11**, 182–187.
43 D. R. Lide, *CRC Handbook of Chemistry and Physics*, CRC Press, Boca Raton, Florida, 90th edn, 2009.
44 A. Naumov, I. Y. Eremchev and A. A. Gorshelev, *Eur. Phys. J. D*, 2014, **68**, 348.
45 S. R. P. Pavani, M. A. Thompson, J. S. Biteen, S. J. Lord, N. Liu, R. J. Twieg, R. Piestun and W. E. Moerner, *Proc. Natl. Acad. Sci. U. S. A.*, 2009, **106**, 2995–2999.

ROYAL SOCIETY OF CHEMISTRY

DISCUSSIONS

Quantum optics, molecular spectroscopy and low-temperature spectroscopy: general discussion

Michel Orrit, Geraint Evans, Thorben Cordes, Irena Kratochvilova, William Moerner, Lisa-Maria Needham, Sergey Sekatskii, Yuri Vainer, Sanli Faez, Vlatko Vedral, Himangshu Prabal Goswami, Alex Clark, Alfred J. Meixner, Lukasz Piatkowski, Victoria Birkedal, Vahid Sandoghdar, Gary M. Skinner, Wolfgang Langbein, Jiangfeng Du, Felix Koberling, Jens Michaelis, Fazhan Shi, Robert Taylor, Arindam Chowdhury, Brahim Lounis, Niek van Hulst, Patrick El-Khoury, Lukas Novotny, Jörg Wrachtrup, Tristan Farrow, Andrei Naumov, Maxim Gladush and Ronald Hanson

DOI: 10.1039/c5fd90088c

Himangshu Prabal Goswami opened the discussion of the paper by Jörg Wrachtrup: When you describe the Hamiltonian, the coupling between the ancillary and sensor spin is g and the coupling between the sensor and sample spin is J_i. Are these frequency dependent? If not, what is the quantitative range of g and J_i under which they can be treated as system parameters? This would give a better explanation to understand the two regimes $g \gg J_i$ and $g \ll J_i$.

Jörg Wrachtrup responded: The couplings between the ancillary and sensor spin g and between the sensor and sample spin J are DC coupling strengths as both systems are treated as permanent static dipole moments. Typical coupling strengths for g are on the order of 200 kHz and for J are around 10 kHz.

Robert Taylor asked: How do you engineer the NV centre so close to the surface and how do you determine the depth at which it is placed? What kind of yield do you get in your samples?

Jörg Wrachtrup replied: In most cases we implant nitrogen with low energy so that it is positioned a few nm below the surface. Then we use the implanted defect as a sensor for the NMR signal of water on the surface to determine the distance.[1] This is the most reliable method so far. The depths we use in the experiments are between 10 nm and 2 nm. The yield to get a NV per implanted N is unfortunately very low for such a shallow NV, *i.e.* around 1–10%.[2]

1 T. Staudacher *et al.*, *Science*, 2013, **339**, 561–563.
2 F. F. de Oliveira *et al.*, *Appl. Phys. Lett.*, 2015, **107**, 073107.

Vahid Sandoghdar said: Could you please comment on the stability of the microwave transitions? One would think that at some level there must be some spectral diffusion or instability somewhat similar to what one knows from the optical domain. If there is such an effect, could you say how large this can be? What about the temperature stability requirement?

Jörg Wrachtrup answered: Indeed, the electron spin resonance transitions undergo spectral diffusion. This spectral diffusion arises because the electron spin is coupled to ^{13}C nuclear spins, which change the spin projection causing a randomly varying magnetic field at the location of the electron spin. This would be similar to the spectral diffusion of low temperature optical lines coupled to two-level systems. In our case, the magnitude of the spectral diffusion depends on the ^{13}C concentration and ranges from kHz to some 100 kHz in the case of a large ^{13}C concentration.[1] Indeed, temperature changes also have an impact on the electron spin resonance transitions. Shifts are around 70 kHz K^{-1}.[2] However, they can be eliminated by using proper spin transitions.

1 N. Mizuochi *et al.*, *Phys. Rev. B*, 2009, **80**, 041201.
2 P. Neumann *et al.*, *Nano Lett.*, 2013, **13**, 2738.

Jens Michaelis remarked: You discussed a lot about sensitivity at low temperatures. Could you please elaborate on how the signal is compromised when going to room temperature.

Jörg Wrachtrup replied: Indeed, we are aiming at room temperature NMR experiments on single proteins. A challenge is proper immobilization as the detection volume is on the order of the size of the protein. We tested a couple of approaches. In a recent study,[1] we found, for example, that one can slow down water molecules by three orders of magnitude close to the surface. In this case they stay for a sufficiently long time in the observation volume that their NMR linewidth is narrow enough to get high resolution spectra. In any case, even an ideally immobilized molecule will undergo dynamics, which will impact the structure we see. As a result, we will detect an average structure similar to other methods. In addition, time resolved NMR and line shape analysis will allow us to conclude on the dynamics.

1 T. Staudacher *et al.*, *Nat. Commun.*, 2015, **6**, 8527.

Vlatko Vedral commented: What would you say is quantum in your quantum sensor? This could also be useful for the algorithm known as DQC1. Here a single pure system, like your electron spin, interacts with many mixed spins (like *e.g.* nuclear ones). Could something like this be implemented in your experiment?

Jörg Wrachtrup answered: In the specific scheme we discuss in the paper, the sensing electron spin gets entangled with the nuclear spin. By this we achieve enhanced sensitivity and this is the genuine quantum aspect of our work. Indeed, one can consider our system as being an ideal DQC1 model system. Implementation is straight forward.

Vahid Sandoghdar said: How would you define a "quantum" effect? Would you really need quantum mechanics to describe T_2 and defacing?

Jörg Wrachtrup responded: Indeed, T_2 in itself can be mapped onto classical analogs as it appears as a parameter in *e.g.* the Bloch equations. There is no doubt though, that the electron spin we use is a quantum object as it, for example, has been experimentally demonstrated that it violates temporal Bell's inequality.[1] Nevertheless, in a lot of applications those quantum properties are not used. In our paper, however, the sensing electron spin gets entangled with the memory spin, which is key to achieving the sensitivity enhancement we propose.

1 G. Waldherr *et al.*, *Phys. Rev. Lett.*, 2011, **107**, 090401.

Jiangfeng Du commented: What is the limitation of the sensing range by ancilla nuclear spin? 100 nm for the electron spin and 10 nm for the nuclear spin? During laser pumping, how much is T_2 of the ancilla nuclear spin? Is it the limitation for NV sensing?

Jörg Wrachtrup replied: The method we propose in the paper is similar to earlier work where a quantum memory was used. The sensing range scales with the T_1 of the memory spin used. That means that it is extended by a factor of 10. Indeed, it is the T_2 (or T_1 in the case of dynamical decoupling) of the memory limiting the sensing range.

Vahid Sandoghdar asked: Would there be any advantage to using concepts like adiabatic passage?

Jörg Wrachtrup answered: Indeed, adiabatic passage is another way of achieving a similar function of the memory spin. Under certain conditions, when the memory spin is even weaker coupled to the electron spin, then we assume in the paper that adiabatic passage might be more effective.

Michel Orrit said: The prospect of single-molecule NMR spectroscopy is really fascinating. But what is your vision of the new science that this would open? Would you search mainly for structural information (protein conformations, associations, reorganisation upon substrate or cofactor binding), or do you think that dynamic information would also be accessible? And, if yes, in which range of times?

Jörg Wrachtrup responded: Certainly the prospects for structural resolution are exciting as, for example, rare protein species, or deviation in structures can be explored. Also, a whole world of molecules can not be investigated by conventional techniques as either they are too large for solution NMR spectroscopy or they can not be crystallized. Dynamics can be derived from line shapes, making use of the full potential of NMR spectroscopy, like 2D methods. We recently published a paper giving the first experimental evidence.[1] I personally think that combining NMR spectroscopy with high resolution optical microscopy is most promising as, for example, interacting proteins can be visualized in unprecedented detail.

1 T. Staudacher *et al.*, *Nat. Commun.*, 2015, **6**, 8527.

Felix Koberling asked: What can we learn about the protein from the spin measurements (compared, for example, to single-molecule FRET measurements, which tell us something about the distances in the protein and therefore its conformation and 3D structure)? Are you preparing a single spin in the protein by a specific labeling? Can we decorate the protein, which spins at specific positions?

Jörg Wrachtrup answered: Single-protein NMR and electron spin resonance complement other single-molecule methods like FRET. NMR spectroscopy, however, is label-free and ideally yields structure data with atomic resolution rather than distances between labeled parts. We hope to get far enough to do these kinds of experiments in, as close as possible, natural environments. Electron spin resonance (EPR) requires spin labeling similar to dyes for FRET. Also, there is a whole catalog of compounds available. The advantage of EPR over NMR is that it allows for detection at greater stand-off distances between the sensor and molecule. Ideally, both methods should be combined with high resolution optical microscopy to yield high resolution structural data inside an optical microscope.

Fazhan Shi opened the discussion of the paper by Ronald Hanson† by communicating: What is the most promising way towards a scalable system based on single spins? A hybrid with superconductivity or photons or others? Please comment on both of the hybrid systems, benefits and disadvantages.

Ronald Hanson communicated in reply: My perspective is that spins can be coupled both *via* direct interactions (magnetic or electric) and *via* photons. The advantage of photons is that one can cover large distances (this appears the only feasible path towards long-range networks) and that one can create modular structures, where only small local cells need to be fully coherently controlled instead of one large chip with millions of spins.

Fazhan Shi communicated: In five years, how many qubits can be entanglement together with NVs *via* photon medium?

Ronald Hanson communicated in reply: This will depend very strongly on the progress on cavities and other technologies. I am aiming for at least 4 cells, each containing something between 5 and 10 spins.

Sanli Faez opened the discussion of the paper by Tristan Farrow: Measurements of indistinguishable photons from two separate organic molecules have been done previously (without using cavities).[1] To what extent does the use of cavities help when measuring new effects that could not have been done with the previous settings? Please make a quantitative estimate.

1 R. Lettow, Y. L. A. Rezus, A. Renn, G. Zumofen, E. Ikonen, S. Götzinger, and V. Sandoghdar, *Phys. Rev. Lett.*, 2010, **104**, 123605.

Tristan Farrow answered: Quantum information processing technologies would immensely benefit from the availability of reliable, easily controllable, individually

† The paper by Ronald Hanson was not presented at the meeting. Subsequently, any questions regarding this paper were submitted after the meeting.

addressable, coherent systems. Organic molecules represent one completely unexplored candidate. The above-mentioned paper[1] shows that it is possible to excite one molecule by a single photon emitted by another molecule. This is an important first step. However, what would be valuable would be to entangle the two molecules as this would create resource for quantum information processing. The above experiment does not do this since once the first molecule has emitted the photon, it actually goes into the ground state, rather than being in a superposition. To do that, we proposed (DOI: 10.1039/C5FD00101C) to inject a single photon (already in a mode entangled state) into two trapped molecules in optical micro-cavities. The cavities are instrumental because it is critical that our signal remains well above the classical noise. Our preliminary tests on these cavities show strong coupling in bulk on complex proteins, and are characterized by quality factors ranging from 10^4 to 10^6. Witnessing entanglement is conceptually simple as it follows the logic of standard Mach–Zehnder interferometry (as described in any standard quantum optics text). Beyond our entanglement protocol to create |eg> + |ge>, we believe also that this ability to address individual molecules would allow us to prepare and verify all four Bell states of the two organic molecules using phase shifts and pi-pulses. In essence, this would be the first elementary unit of an organic quantum computer.

1 Y. L. A. Rezus, S. G. Walt, R. Lettow, A. Renn, G. Zumofen, S. Götzinger and V. Sandoghdar, *Phys. Rev. Lett.*, 2012, **108**, 093601.

Vahid Sandoghdar remarked: How do you plan to read out the entanglement that results from when you do a two-photon interference experiment? The entanglement only lasts within the lifetime of the excited state, which is a few nanoseconds.

Tristan Farrow responded: This protocol relies on single photon mode entanglement rather than on two photon interference. However, it could be adapted to the two photon interference variant (for three level systems where path information could be erased), in which case the read out would be generated on a beam-splitter and entanglement would be confirmed by post-selection in the same manner as in our protocol.

Alex Clark addressed Tristan Farrow and Vlatko Vedral: I am interested to know what kind of entangled states can be created in this type of system. There is an issue in that after the decay has occurred and you have measured the photon statistics, the system is not entangled anymore. It would be good to have a measurement that can confirm the entanglement without destroying it. Is that possible? The other issue is the fact that the entanglement is so short lived, so, as above, it is hard to check whether you had the entanglement you wanted to allow some measurements, manipulations, or feed forwards. Is that an issue for scalability if you wanted to make larger entangled states?

Vlatko Vedral replied: Yes, ideally you would like to execute some gates in order to manipulate the initial entangled state |eg> + |ge>. One thing you can do is apply a phase gate to one molecule thereby obtaining |eg> − |ge>. You could also do a NOT gate, resulting in |gg> + |ee>. So you can prepare all Bell states. All these operations would be orders of magnitude faster than ns. However, doing a verification of entanglement without destroying it is harder. I could see how to do it by utilizing say light inside cavities (roughly an analogue of what Haroche does with atoms in cavities),[1] but I can not think of anything much simpler.

1 S. Haroche *et al.*, *Rev. Mod. Phys.*, 2013, **85**, 1083.

Tristan Farrow responded: We believe that the 5 ns lifetime is ample time to perform other logical operations on the molecules while they are in an entangled state. For example, we can implement phase shifts and pi-pulses on the individual molecules as qubits. Hence, it is irrelevant that the entanglement dissipates after we have exploited this resource. If one wants to perform a non-demolition verification of entanglement, one would have to implement a scheme analogous to that used in, for example, the Haroche type cavity experiments.[1] This however presents experimental challenges for molecules as it would require probing a single molecule with a single photon in the strong coupling regime.

1 S. Haroche *et al.*, *Rev. Mod. Phys.*, 2013, **85**, 1083.

Vlatko Vedral addressed Alex Clark and Vahid Sandoghdar: There are two ways in which you can entangle two systems, each of which is a qubit (two level system). 1. Excite them both, then observe a single photon emission but in a way that does not reveal which system it came from. 2. Inject a single photon superposed over two spatial modes, which leads to |eg> + |ge>. One way to witness that entanglement is created is to look at a single photon emission, which involves another beam-splitter. A mixed state of |eg> and |ge> would give us a random output at the two ports, while an entangled state would always lead to the same output. This is a crude way as it does not discriminate all separable states. To do so, you would have to measure, for instance, two complementary observables, x and z, on each of the systems. This would be challenging. We can also think about doing the full state tomography, but this would involve measuring y on them in addition. How can we do that?

Alex Clark answered: These methods to entangle matter systems have the problem that you are just measuring if your system *was* entangled, rather than if it *is* entangled. As these are singlet transitions we can not think to manipulate the spin of the electron, which could perhaps give us access to tomographic measurements, as has been shown with NV centres.[1] It is intriguing to think about whether tomographic measurements would be possible with access to another state of the molecule. Also, it would be interesting to investigate if dibenzoterrylene is the best molecule of choice for these experiments, or if another system would be preferable.

1 H. Bernien *et al.*, *Nature*, 2013, **497**, 86–90.

Vahid Sandoghdar responded: The dye molecules that are usually used for low-temperature single-molecule spectroscopy have a strong zero phonon line between the electronic singlet ground and singlet excited states. Other available transitions are either to the triplet state or to the vibrational levels of the ground state. The former have been used in ODMR measurements and there might be schemes to exploit them for entanglement purposes. The latter ones are very broad because of a fast non-radiative decay on the order of picoseconds to the ground state.

Himangshu Prabal Goswami addressed Vlatko Vedral and Tristan Farrow: How do you rule out the possibility of sudden death of entanglement?

Vlatko Vedral replied: If the singlet fraction is higher than a half, then we know that the state must be entangled, *i.e.* it can not be separable. This is a very simple witness of entanglement that would work in the system we presented (DOI: 10.1039/C5FD00101C).

Tristan Farrow responded: This is not relevant as our protocol intends to discriminate between entangled and separable states. It is immaterial exactly how entanglement vanishes.

Vlatko Vedral commented: A 5 ns window in which to complete the entangling and witnessing protocol is more than sufficient for the molecular entanglement.

Himangshu Prabal Goswami responded: How did you quantify the time window?

Vlatko Vedral answered: With regard to the relevant time window, this is determined by the relevant decoherence time of the transition we intend to use for storing a qubit.

Vahid Sandoghdar said: The paper would benefit from more elaborate discussions on how you plan to use the information about the emitted photon statistics to learn about the transition from classical to quantum behavior in molecules.

Tristan Farrow answered: We agree as the paper focuses on a single family of quantum states in the hierarchy of quantum states, which includes discordant, entangled and just classically correlated. The protocol presented would only be able to identify entangled states, however, we are intending to extend this to a methodology to further discriminate between all the subsets. This is uncharted territory, which offers an exciting avenue to pursue in the context of organic molecules. While DBTs are too simple and well understood, they offer useful test systems from which we can learn how to then probe more complex structures (many body systems interacting with heat baths). These molecules could be used as probes of local phonon modes, which push quantum systems over the classical boundary.

Vahid Sandoghdar asked: How do you relate the process of light emission to the overall quantum state of a large object/molecule? For example, if you consider a diamond nanocrystal, would information on the emission from a color center say something about the whole nanoparticle?

Tristan Farrow replied: For example, in the case of some polymers, the emission of a single photon that undergoes double slit interference can tell us about the delocalisation of the exciton that was created in that polymer under the action of a weak optical pump. Specifically, it tells us that the exciton has spatial coherence over "macroscopic" distances over 10 microns (the length of the polymer). In turn, this tells us that the polymer behaves as a quantum wire whose delocalized pi-bond acts as a uniform one dimensional confining potential.

Wolfgang Langbein said: Using single organic molecules in a matrix for long term stable devices is an issue due to the more accessible surroundings compared to epitaxial quantum dots (*e.g.* a GaAs/InGaAs system), and more photobleaching. How do you plan to overcome this stability issue?

Tristan Farrow answered: Yes, although this depends on how we define "long term". The devices incorporating single organic molecules would certainly be sufficiently stable on the time scale of the relevant measurements and could potentially be re-used if handled with care (low laser powers, low pressure conditions, *etc.*). It is true that they are unlikely to ever be as stable as a solid state system, such as quantum dots. But their great advantage over such systems is that they are almost perfectly reproducible, much brighter, and cheaper in comparison.

Vahid Sandoghdar commented: I do not agree with the statement that molecules are not good quantum systems or are somehow "dirty". Interestingly, these days, many physicists immediately associate dye molecules with biology and systems that are not reproducible. This is somehow a misconception. The molecular systems that are used for low-temperature spectroscopy are embedded in an organic crystal and are extremely well behaved. Semiconductor physicists seem to often think that structures like quantum dots are well engineered and molecules are somehow sloppy. In fact, in molecules, every atom sits at a well-defined position, whereas we still do not have any technique that can make a quantum dot where every atom counts. Moreover, every molecule is the same, while there is certainly a huge inhomogeneity in the quality of quantum dots.

Wolfgang Langbein replied: I agree that molecules themselves are atomistically well defined, while the distribution of *e.g.* In atoms in InGaAs/GaAs quantum dots is not. On the other hand, InGaAs QDs are sharing the crystal lattice with the host, so that there is no interface as such. The organic molecules used today are not replacing a lattice site of the embedding crystalline structure in the same way but they have a large number of possible embedding sites. Furthermore, the exciton in these crystals has a smaller extension, which increases the potential for photochemistry. Presently, semiconductor quantum dots are less affected by photobleaching and allow us to make reliable single photon emitters, which function over many years, with $>10^{12}$ photons emitted. I guess that combining the advantages of the two systems would be the best way forward, *i.e.* designing fluorescent molecules in a way that they are embedded in a host lattice in a well-defined lattice site, and extending the exciton wavefunction to achieve stronger radiative emission and less photochemistry.

Himangshu Prabal Goswami stated: Photon emitting sources have different statistical regimes depending on the second order intensity correlation function *viz.* bunched, anti-bunched and totally random. Which of these statistics gives rise to the maximum entanglement and why? Do any of these statistics never give entanglement?

Tristan Farrow responded: Entanglement, in this context, relies on using an anti-bunched source, specifically indistinguishable single photons. Using a single photon source to generate an entanglement resource is more efficient as this can be swapped with the molecules rather than relying on the simultaneous emission of two indistinguishable single photons interfered on a beam-splitter. This adds another layer of complexity and relies on post-selecting a small subset of relevant events.

Sergey Sekatskii queried: In relation to the possible existence of quantum effects in complex biomolecules at room temperature, it seems rather improbable that electronic degrees of freedom might be involved, the decoherence time should be too short. The situation can be different if magnetic (spin) degrees of freedom are at work here. Is there some experimental or at least theoretical evidence in real biological molecules for the transfer of quantum properties from electronic to magnetic degrees of freedom?

Tristan Farrow replied: Yes, it is harder for electronic degrees of freedom to be coherent in a complex bio molecule at room temperature. In addition to the magnetic degrees of freedom, vibronic degrees of freedom provide an alternative handle on coherence. There is some speculation that the recent experiments on photosynthetic FMO complexes benefit from the vibronic contributions. It seems plausible, at least theoretically, that these contributions couple to the magnetic degrees of freedom. In fact, it would be surprising if this were not the case, however, it is a moot point if this contributes or detracts from coherence. Experiments aiming to ascertain the relevant regimes and parameters would be a valuable contribution to this field.

Wolfgang Langbein remarked: In the discussion about quantum weirdness and entanglement, how large the involved objects are is considered important. This is somehow related to the perception that a larger extension means a more classical behaviour. However, this is not the case in general. The decoherence mechanisms which are coupling to the quantum states that are probed are important. In a NV defect in a diamond, the electron carrying the quantum state of interest is extended over only about a unit cell of the crystal. Recent reports on the entanglement of the LO vibrational excitations in two diamonds have to consider that these vibrational excitations are also localized to a few unit cells in the crystal due to their short lifetime of a few picoseconds and the slow propagation speed of these modes.

Tristan Farrow answered: We second this viewpoint concerning the fuzziness of the quantum to classical boundary. "Macroscopicity" in quantum physics is a hotly contested concept currently lacking a clear definition – is the right figure of merit mass, or spatial extent of the quantum state, or the number of superposed qubits, or indeed the scaling of the phase estimation with the number of qubits (in which case the GHZ state with lots of qubits is the most macroscopic). For a fuller discussion see ref. 1. Entanglement in the vibronic degrees of freedom in diamonds is said to extend over macroscopic distances, however, this still involves a single mode in each diamond (and therefore would not be macroscopic if we were to apply any of the above definitions). Furthermore, their lifetime is picoseconds, which does not permit measurements to create or witness the state.

1 T. Farrow and V. Vedral, *Opt. Commun.*, 2015, **337**, 22–26.

Vlatko Vedral noted: How many atoms can you claim participate in the phononic (vibrational) entanglement in the diamond experiment? Certainly only a small fraction contribute to a single phononic mode in each diamond. In addition, these phonons live for only a picosecond or so. However, it takes light a nanosecond to arrive at the detector in order to even create that entanglement (let alone witness it afterwards). How do we demonstrate that the whole thing is entangled? I do not think that there is any evidence of this in such and related experiments.

Sanli Faez opened the discussion of the paper by Lukas Novotny: Is the effect of measurement back-action (because of light illumination) negligible in all cases? Especially for the measurements on carbyne, how can one make sure that the carbon atoms are not pushed aside during the measurement because of the illumination?

Lukas Novotny answered: DFT calculations show that the interaction between carbyne and the surrounding nanotube is 0.5 eV per two carbon atoms. This interaction stabilizes the carbyne chain inside the nanotube. There are, indeed, configurations (larger tubes) for which the carbyne has a negligible interaction, but these configurations prove to be unstable, that is, carbyne can not be grown. The interaction and stabilization with the outer nanotube is a key ingredient for growing carbynes. From a practical consideration, we do not observe any shifting of the carbyne, that is, it can be imaged many times, yielding the same images without noticeable changes.

Wolfgang Langbein commented: You stated that the high electric fields at the tip during the measurements would not give any observable effect – can you summarize which effects you see when increasing the excitation intensity, and how these intensities compare with the ones used in your experiments?

Lukas Novotny responded: In the case of CW excitation and increased excitation, intensity leads to heating. This is mostly reflected in melting of the tip and a deterioration of the resolution.

Brahim Lounis asked: You have reported that long chains of carbyne have been grown in DWCNTs. Is this possible only in this CNT system? Do you have any explanation for this? I would expect that it is only the properties of the first NT in contact with the carbyne chain that matter.

Lukas Novotny replied: The synthesis of carbyne in nanotubes is done by collaborators so I cannot answer this question accurately. However, I suspect that if single-wall nanotubes were used then outside potentials (*e.g.* from patch charges on surfaces) would affect the stability of the carbyne inside the nanotube. Thus, in double-wall nanotubes the outer tube shields the inner tube and ensures the stability of the carbyne chain.

Michel Orrit remarked: The shielding of the linear carbon (carbyne) chain by the nanotube opens the possibility to have excitons or polaritons with very large delocalization lengths. This system reminds me of 1D conjugated polymers in 3D crystal matrixes of the monomer, investigated by Schott and coworkers.[1,2] It may be interesting to compare with the results from that group.

1 F. Dubin *et al.*, *Nat. Phys.*, 2006, **2**, 32.
2 L. Legrand *et al.*, *Phys. Stat. Sol. B*, 2008, **245**, 2702.

Lukas Novotny replied: Thanks. I have downloaded these papers and will follow up on them.

Himangshu Prabal Goswami asked: Graphene has special points called the Dirac points around which the energy dispersion is linear. If you have a single molecule around these points, how will its optical response change? Also, how does the phonon correlation function behave around the Dirac points?

Lukas Novotny answered: We typically have doped graphene, that is, the Fermi level is not at the Dirac point. The doping arises from the surface (contact doping). Also, we did not do any Raman scattering measurements on single molecules deposited on graphene, so I can not answer the question.

Patrick El-Khoury commented: In RRS, where the incident photon is resonant with an excited electronic state of the material system, one often needs to account for the various properties of that (upper) state to rationalize the observables. For RRS, Bloembergen's 48 terms[1–3] are reduced to 3, which retain three resonant factors in their respective denominators. One of the RRS diagrams is reminiscent of its familiar non-resonant Raman analogue – with the exception that resonant polarization is achieved (yet no excited state population is achieved). The other two diagrams involve the population of excited vibronic states – and real evolution therein – prior to scattering. This is the molecular realm that we are all familiar with. In this work, the excited state is excitonic. With the above in mind, the question asked is: in what way (if any) does resonant polarization affect the observables (frequency shifts, relative intensities of the observable vibrational states, vibrational dephasing times)?

1 N. Bloembergen, *Encounters in Nonlinear Optics: Selected Papers of Nicolaas Bloembergen, with Commentary*, World Scientific Publishing Co. Pte. Ltd., New Jersey, 1996.
2 A. C. Albrecht, On the Theory of Raman Intensities, *J. Chem. Phys.*, 1961, **34**, 1476–1484.
3 S. Mukamel, *The Principles of Nonlinear Optical Spectroscopy*, Oxford University Press, New York, 1995.

Wolfgang Langbein responded: This is a rather broad question – I will try to give an approximate answer. Just to clarify, you refer to Resonant Raman scattering, not Resonant Rayleigh scattering. Your question is concerning the influence of high excitation fields on the lineshape of RRS. These effects are beyond the terms considered in the discussion of RRS mentioned above as I recall, which are in all orders for the vibrations, but only first order in the driving field. Any changes in the lineshape by the excitation needs to be of a higher order in the driving field – at which point the description gets even more complex. Concerning the role of the resonant polarization, this is the easiest of the processes as it only involves interactions with the molecular vibrations. Once a real evolution in the excited state is important, this evolution involves the interaction with the environment, effectively multiple scattering with vibrations of the environment (bath), which tends to thermalize the excited state distribution.

Felix Koberling remarked: The bright spot on the carbon nanotube, this is not an additional signal but it is a defect that is enhancing the existing signal. Have I got this right? Do we want to have a lot of defects or do we want to have more or less no defects? What is the atomic configuration that is causing this defect and that enhances the signal?

Lukas Novotny replied: We want to have controlled defects. Defects are localized states, which can lead to local doping, band renormalization, and chemical reactions. A defect relaxes the light-matter selection rules and allows "forbidden" transitions to become optically active.

Lisa-Maria Needham opened the discussion of the paper by Lukasz Piatkowski: Do you believe that these effects could be observed with smaller chromophores in a similar environment? These observations could have significant implications for discriminating between the diverse structural and environmental features of biological systems.

Lukasz Piatkowski responded: We have not done the experiments with different, smaller chromophores. The measured spectral heterogeneity is always a combination of the specific chromophore and the specific matrix/environment. It is therefore hard to predict what the effect of just changing the size of the chromophore would be. It is certainly true that both the excitation spectrum and emission spectrum of a chromophore could be used to detect locally, structurally different features in complex biological systems. Yet, this would probably require prior understanding of how a specific macroscopic structure affects the spectral properties of the chromophoric probe.

Thorben Cordes asked: Could you also explain the large spectral shifts by changes in the local environment of the fluorophores due to the use of polymer embedding? Do you expect the same (large) shifts for molecules that can rotate freely in (aqueous) surroundings and, let's say, biophysically relevant environments (DNA, proteins, *etc.*)?

Lukasz Piatkowski replied: Yes, the local environment definitely plays a role in the observed spectral heterogeneity of the studied fluorophore. The polymer matrix is a heterogeneous film with various pores and pockets, which may have different local charges and may trap and force the molecules into various, distinct conformations. In fact, there might be more than just one reason for the observed spectral shifts. Fluorophores in aqueous media generally exhibit the same spectral properties, as any local dynamics are faster than the observation time and on average they all experience the same local environment. However, in the case of biophysically relevant conditions, the local environment (due to the presence of large molecules like DNA, fats or proteins) can also induce spectral heterogeneity. If the fluorophore of interest is exposed to a more hydrophobic or hydrophilic environment or if it interacts with the local bio-environment in a different way then spectral heterogeneity is expected. Whether it is feasible to detect such heterogeneity depends on the timescale on which the environment changes. If spectral diffusion is much faster than the measurement time, one would observe average spectral properties for all the chromophores.

Lisa-Maria Needham said: It may be interesting to try a system in which the chromophore could undergo π–π stacking interactions with the surrounding matrix. This could potentially be used as a model environment for dyes binding to amyloid proteins though this mode.

Lukasz Piatkowski answered: This is a very interesting idea. If, somehow, the specificity of the binding of the chromophore to amyloid proteins could be achieved and providing that π–π stacking for a specific chromophore leads to a certain spectral change, then the idea seems feasible, indeed.

Sanli Faez addressed Lukasz Piatkowski and Niek van Hulst: When the excitation spectra are taken at both high spectral and spatial resolution, is it possible to identify the spatial correlation between the magnitude of the spectral shift and relate it to the local charge distribution inside the polymer? How many scans can be made from a single molecule (on average) before it bleaches? Is there enough sensitivity to measure the influence of applying an external electric field on the spectrum (to measure the Stark shift)?

Lukasz Piatkowski responded: This is an interesting suggestion. In our experiment we spin-coat a solution of the QDI molecules and the polymer. The polymer makes a heterogeneous film and we do not have any knowledge either on the local structure or on the charge distribution. The QDI molecules used are quite robust against bleaching. Typically, 5–10 fluorescence interferograms could be measured (total time of 5–10 min). At this stage, the sensitivity of determining the position of an excitation spectrum is better than 10 nm. This, of course, can be further improved by measuring the excitation spectra at a higher spectral resolution (now 4 nm) and for a longer time.

Niek van Hulst answered: Typically, an induced 10 nm spectral shift would be observable on a single molecule, with a simultaneous localization accuracy of 20 nm. Detection of the Stark shift due to an external field would be an interesting exercise indeed. Yet we have shown spectral shifts over 100 nm due to polymer heterogeneity, much larger than any Stark effect, making such a suggested experiment challenging.

Alfred J. Meixner remarked: Your system has many features that can undergo conformational changes. I do not think that it is too unexpected for you to see shifts as large as you have.

Lukasz Piatkowski replied: We fully agree that given the structure of QDI, its size – long and flexible hydrocarbon chains, it is not that surprising that they exhibit a large spectral heterogeneity when encapsulated in a polymer matrix. What is, in our opinion, somewhat unexpected is the large extent of the observed shifts (nearly 200 nm) and their continuous character. This suggests a continuous truncation of the electron delocalization in the perylene backbone, rather than, for instance, a bimodal distribution originating from two distinct conformers.

Vahid Sandoghdar queried: What is the fundamental advantage of recording the signal in Fourier space, in this case time domain, compared to direct detection of absorption, where you use a camera and grating to record all wavelengths at the same time? What kind of artefacts would one have to worry about in this method? I would like to comment that my question, raised at the meeting, came from the fact that I had not realized that they detect fluorescence and not the excitation light. So, this part of the question is somewhat ill posed.

Lukasz Piatkowski responded: In addition to the work published (DOI: 10.1039/C5FD00107B), we have presented the results of an additional experiment in which we have used a broadband interferometric approach to measure the excitation of an individual molecule at room temperature. In this experiment, a broadband interfering pulse pair excites a molecule and its fluorescence interferogram is detected. Upon Fourier transformation and normalization to the laser spectrum, the excitation spectrum of the molecule is obtained.

The main difference between our experiment and the one to which the questioner referred to[1] is that we detected the *fluorescence*, while Celebrano *et al.* detected the *absorption*. Yet, in their experiment no spectral information was obtained as only a single wavelength CW laser was used. We use a broadband interferometric excitation so the acquired fluorescence interferogram contains spectral information. Obviously, fluorescence detection is a background-free approach, as opposed to absorption detection, and offers a much higher signal to noise ratio.

From our experience, there are two experimental conditions that one has to pay particular attention to. Firstly, as our approach relies on broadband (>120 nm) laser pulses, one has to make sure that the chromatic aberration of the used microscope objective is minimal in the wavelength range of the excitation pulse so that the molecule of interest is excited with the whole bandwidth. Secondly, in order to extract the excitation spectrum of the molecule, one has to Fourier transform the acquired fluorescence interferogram into the frequency domain and then normalize to the excitation laser spectrum. Care needs to be taken so that the used laser spectrum reflects the true laser spectrum that the molecule experiences.

1 Celebrano *et al.*, Single-molecule imaging by optical absorption, *Nat. Photonics*, 2011, **5**, 95–98.

Thorben Cordes asked: The long hydrophobic carbon chains – can they alter the local environment of the fluorophore to account for the observed spectral shifts? Can you perform the same experiments with a fluorophore without those groups? The chemists should be able to make such a compound easily.

Lukasz Piatkowski replied: This is a valid remark. The long and floppy hydrocarbon chains will definitely play a role here. It is hard to imagine that for all the molecules they would take up the same conformation. They may, for instance, fold onto the perylene backbone of the QDI molecule and alter the degree of electron delocalization along the backbone. It would be very interesting to perform the same experiment on a molecule that does not have long hydrocarbon chains. We will look into such a possibility.

Alfred J. Meixner commented: You have two nitrogen atoms which are sp2 hybridized, but could also be sp3. This would give a blue shift, shortening the pi system. If the branches rotate, you lose the vibronic structure. Can you comment on this?

Lukasz Piatkowski responded: The flexible hydrocarbon chains might, indeed, be responsible for the shortening of the pi delocalization in the perylene backbone. They are long enough so that if folded onto the backbone, they may alter the pi system. Alternatively, by twisting and bending, the hydrocarbon chains may bend the nitrogen atoms and the backbone end terminals out of plane, which also could lead to truncation of the pi system. It would thus be very instructive to perform our experiments on a similar chromophore that does not have the hydrocarbon chains.

Sergey Sekatskii opened the discussion of the paper by Thorben Cordes: What is the maximum attainable number of absorption–emission cycles that the molecule can survive after the chemical modification you propose is done? And what is the increase in this number in comparison with the bare molecule case, for different types of dye molecules?

Thorben Cordes replied: We found that the total number of detected photons (which is equivalent to the cycle number but reduced by the detection efficiency of the microscope) increased from 15 000 in bare cyanines (Cy5) up to 2 million in some conjugates. Similar results were found in recent unpublished work for rhodamines, carbopyronines and other dyes.

Niek van Hulst said: You mention how spectacular lengthening of the bleach times of Cy5 can be achieved by attachment of a photostabilizer. The approach seems very promising for nanoscale imaging. Yet, how specific is the method for certain types of dyes, or would it be functional more generally?

Thorben Cordes answered: We think that the approach is generally applicable provided that the major damage routes are associated with the triplet-state. As shown earlier by Luettke and coworkers,[1] other dye classes also benefit from covalent linkage of the photostabilizer. As mentioned before, we have unpublished results that show the positive effects of different covalently-linked photostabilizers on rhodamines, carbopyronines and cyanines.

1 B. Liphardt and W. Luettke, *Opt. Commun.*, 1981, **38**, 207–210.

Lisa-Maria Needham commented: Could the use of these photostabilizers have any adverse effects on a living biological system, for example if one were to use these for live cell single-molecule tracking experiments?

Thorben Cordes replied: I am convinced that the small photostabilizer unit (which is typically smaller than the dye) has no larger effects on the fluorophore toxicity (with some exceptions). Of course, you need to test whether the new photostabilizer–dye conjugate has negative effects on your system of interest but the use of our conjugates should be tolerated more easily than the addition of millimolar concentrations of photostabilizer, as in other experiments.

William Moerner remarked: Congratulations on achieving a considerable increase in the number of photons detected from a single-molecule of Cy5 in the absence of oxygen. Many of us look forward to the use of these derivatives in cells.

Thorben Cordes answered: This is a good remark. Oxygen still represents a major problem since its involvement automatically provides two additional damage routes (besides the intrinsic triplet-based bleaching of the fluorophore): photooxidation and generation of reactive-oxygen species *via* reactions with the fluorophore. This behaviour makes it, so far, complicated to achieve efficient intramolecular photostabilization in the presence of oxygen – often we observe no effect of covalently linked photostabilizers when oxygen is present. We are, however, in the process of solving this problem.

Felix Koberling asked: The self-healing mechanism is working well if the fluorophore has a significant triplet fraction, which can be reduced. If I have a labelled cell, can I measure the triplet fraction of my fluorophores in this specific environment? To get an idea whether it would make sense to apply the self healing mechanism. Do you have a measurement for the triplet fraction in mind? I want to know, in advance, whether it is worth spending time to check the newly developed conjugates.

Thorben Cordes responded: Most organic fluorophores have a significant yield of triplet formation and hence suffer from its population. But it is correct that some fluorophores, *e.g.* oxazines, with a low triplet yield do not show effects as reported in this contribution. These results are yet unpublished but in conjugates of such fluorophores with NPA, we do not find beneficial effects upon conjugation, but rather strong quenching of the fluorescence signal itself. The triplet formation yield could therefore be used as a general guideline to predict if it is worthwhile to make photostabilizer–dye conjugates. We have, however, not considered how to do this in the environment of a cell.

Michel Orrit commented: In your discussion of the mechanism of bleaching reduction, you assume that the triplet quencher "collides" with the dye during the triplet excited state lifetime. How critical is this collision mechanism? Wouldn't it be more efficient to directly link the quencher to the dye so that the triplet would be quenched even more quickly? Or would you then open up the possibility of harmful singlet quenching? In the latter case, I imagine it would be important to optimize the collision rate, so that collisions are too infrequent to quench the singlet, but frequent enough to quench the longer-lived triplet.

Thorben Cordes answered: We and others[1] consider collisions essential for efficient quenching of the triplet state. It is correct that the proximity increases the efficiency of photostabilization for direct triplet-state quenchers (*e.g.* cylcooctatetraene, COT) – so here, close conjugation is optimal. But for stabilizers using photo-induced electron transfer, such as the aromatic nitro group, we find a more complicated behaviour (see Fig. 1 and 2 in DOI: 10.1039/C5FD00114E). I fully agree with your latter comment that the local concentration of the photostabilizer (varied by linker length) has to be balanced to allow efficient depopulation of the triplet but not the singlet state.

1 R. B. Altman, D. S. Terry, Z. Zhou, Q. S. Zheng, P. Geggier, R. A. Kolster, Y. F. Zhao, J. A. Javitch, J. D. Warren and S. C. Blanchard, *Nat. Methods*, 2012, **9**, 68–71.

Gary M. Skinner remarked: How do these new kinds of fluorophores perform in super resolution applications? Do they still exhibit the spontaneous chemical driven switching such as that used in dSTORM?

Thorben Cordes responded: As shown by van der Velde *et al.*,[1] self-healing Cy5 fluorophores still exhibit blinking in the presence of the switching agent MEA. We did not test super resolution imaging (STORM) here since the off/on-ratio was not optimized, but I am convinced that photoswitching *via* MEA and increases in the photostability by intramolecular healing can be combined. We also recently tested the performance of rhodamine-based conjugates in STED-based super resolution imaging and saw significant increases in the number of images that can be obtained when using our photostabilization approach.

1 van der Velde *et al.*, *ChemPhysChem*, 2013, **14**, 4084–4093.

Geraint Evans asked: Does the conjugated triplet state quencher itself eventually become damaged?

Thorben Cordes replied: It seems that "photostabilizer-bleaching" is rare. Most frequently, the fluorophore signal vanishes before any other effects are seen that could be interpreted as destruction of the photostabilizer. In some cases, however, we observe a stable fluorophore emission for a few seconds (photo stabilizer active) with a sudden transition into an emission signal that is characteristic for the fluorophore in the absence of the stabilizer. Often, this transition is preceded by a dim-state that we interpret as a chemical change in the stabilizer.

Victoria Birkedal communicated: How would the dye (with the triplet quenching molecule attached) perform if it was not attached to the DNA molecule or attached to a different place on the DNA molecule? Do you expect that it can perform as well as when the triplet quenching molecules are in the solution?

Thorben Cordes communicated in reply: If the molecules are not attached to DNA then we still expect a significant increase in the photostability. I am convinced, however, that the (bio)chemical environment plays a huge role in the observed quantitative photostabilization effect – especially considering the data in this paper (DOI: 10.1039/C5FD00114E). Regarding the comparison of inter- and intramolecular photostabilization, as we observed earlier,[1] a covalently attached photostabilizer can be nearly as good as its equivalent in solution.

1 van der Velde *et al.*, *J. Phys. Chem. Lett.*, 2014, **5**, 3792–3798.

Sergey Sekatskii opened the discussion of the paper by Yuri Vainer: In your paper you mention that the different characteristics of the surface and volume phonons can influence the optical spectra of molecules in matrices. To what depth does it make sense to speak about surface phonons in your system, and also in other systems?

Yuri Vainer answered: We can assume that the low-frequency surface vibrational modes penetrate to a subsurface layer at a depth comparable to their wavelength. Taking into account that the sound velocity value in polyisobutylene is about 1.5×10^3 m s^{-1} and evaluating the mean excitation energy of the surface modes as equal to the average value of the thermal energy, we obtain a depth of about 16 nm. This is close to the depth of the subsurface layer in polyisobutylene, as evaluated by us from experimental data.

William Moerner asked: You noted that you do not observe single molecules extremely close to the surface of your polymer. It is interesting that single molecules can be observed on a surface at room temperature, although they are often unstable, being relatively unprotected compared to molecules surrounded by host material. Would you say that in your low temperature, high resolution experiments, the reason you do not see molecules is likely due to extremely large and rapid spectral shifts that may be occurring?

Yuri Vainer responded: Thank you for the comment. I think that you are right. The exact reason for the absence of detectable zero phonon lines of single molecules directly on the surface is unclear. Taking into account that the observed spectral dynamics intensify strongly as the chromophore molecule gets closer to the surface, we believe that the dynamical explanation for the disappearance of ZPLs for single molecules on the surface seems more likely than ascribing this disappearance to quenching of the fluorescence itself. This may happen because the spectral position of the ZPL constantly changes over a wide range of wavelengths. These changes may have several possible sources: coupling to low-energy surface vibrational modes, continuous transitions between the larger number of accessible conformation states a single molecule obtains on a surface, and continuous translational motions of a single molecule along the surface that can be induced by excitation laser radiation. It is interesting to note that, in spite of that, the ZPLs of single molecules sputtered on the surface were not detectable and the spatially featureless broadband fluorescence emission from the entire area illuminated by the excitation laser was observed. However, the shape of this spectrum differed from that known from the literature.

The situation changed completely after briefly heating the sample to a temperature slightly above the glass transition point for PIB (~180 K). After cooling the sample back down to 4.5 K, the fluorescence emission spectrum changed distinctly. It developed a band around 620 nm, which is also present in the room temperature spectrum of tetra-*tert*-butylterrylene solution in liquid toluene. The observed change can be explained in the following manner. After molecules collide with the cold surface, their spatial structure is far from equilibrium; heating the sample allows them to relax to equilibrium.

Michel Orrit remarked: Would it be possible to contact the polymer's surface to some solid substrate after the diffusion of the dye? This way, you could test whether the rearrangements at the surface are due to freer motion of the polymer molecules at the interface with vacuum. These motions would be prevented by the solid substrate.

Yuri Vainer replied: Thank you for the interesting and good idea. In principle it is possible, but I do not think so by using an additional substrate. The polymer surface is not plain and contains specks of dust, which will prevent good contact between the polymer surface and a substrate. So, for this purpose, one could spray some additional organic molecules on the polymer surface immediately after sputtering TBT molecules. It should be a substance close to the polymer properties.

An experimental investigation of a surface is a very difficult task. One of the problems is that atoms and molecules from the surroundings are quickly absorbed on the surface (for example, water molecules). This makes it necessary to support a very high vacuum. The main purpose of our experiments was to study the spectral dynamics of chromophore molecules in subsurface layers of a polymer, but not on the surface. We only report the interesting effects observed in our experiments and discuss the possible reasons.

Sanli Faez said: When using crystalline host matrices, such as anthracene, guest molecules are often repelled from the crystal during the annealing process. This is, in some sense, opposite to the behavior observed in DOI: 10.1039/C5FD00055F for polymer host matrices. How should one perform these debt-dependent broadening measurements for such crystalline matrices?

Yuri Vainer answered: Thank you for the good question. You are right, in the case of crystalline matrices, impurity molecules are repelled by a host. This makes our procedure inapplicable for studying subsurface layers of perfect crystals. In our SMS experiments with polycrystalline matrices prepared by spin-coating, we observed that probe molecules were collected in the region of the boundaries between the crystalline grains. However, some parts of the observed single molecules were placed inside the grains, which may be because of local defects or due to not having enough time to move outside the grains.

Arindam Chowdhury commented: Some single molecules in polymer matrices display continuous spectral diffusion while some show rather discrete spectral jumps over time. What is the reason for such contrasting behaviors? Is it possible to perform polarization resolved spectral measurements to understand the possible mechanisms of such processes?

Yuri Vainer responded: Thank you for the good question. One of the reasons for different single molecule spectral behavior is the temporal resolution of our measurements. The very fast spectral jumps appear as line-broadening, whereas slow jumps appear as frequency jumps. The explanation of continuous spectral drifts is more unclear. It can be caused by continuous structural relaxations in the nearby surroundings of a given single molecule (for example due to interactions between TLS).

You are right. Polarization resolved spectral measurements yield more information about the dynamics of probe molecules. Furthermore, we tried to do such measurements, but, due to poor optical quality of the cryostat windows, our attempts were unsuccessful.

Vahid Sandoghdar asked: Could you please give a brief overview of similar work in different systems?

Yuri Vainer replied: We only know of one paper of a similar type.[1]

1 L. Fleury, A. Gruber, A. Dräbenstedt, J. Wrachtrup and C. von Borczyskowski, *J. Phys. Chem. B*, 1997, **101**, 7933–7938.

William Moerner said: You noted that the spectral diffusion or wandering you observe does not follow standard models of glass dynamics. Is this surprising? Each individual molecule may only feel a specific subset of two-level systems (TLSs) that are nearby, rather than the entire distribution of TLSs that may be present throughout the polymer. The situation is not unlike the case of pentacene molecules in *p*-terphenyl, where a specific set of TLSs in a planar arrangement was needed to understand the spectral diffusion.[1,2]

1 P. D. Reilly and J. L. Skinner, *Phys. Rev. Lett.*, 1993, **71**, 4257.
2 P. D. Reilly and J. D. Skinner, *J. Chem. Phys.*, 1995, **102**, 1540.

Yuri Vainer answered: Thank you for the comment. I agree that the deviation of the observed spectral dynamics near the surface from the predictions of the standard model of glasses is not surprising. It was only surprising by the fact that ZPLs of all single molecules on the surface were not detectable. So, we will omit the word "first" in the sentence: "The first surprising result was....". In our experiments we chose, as a system for study, a polyisobutylene with a very high molecular weight (4 200 000 g mol^{-1}), doped with tetra-*tert*-butylterrylene molecules, because (as we know from our numerous measurements) this system demonstrates, in the case of the bulk, excellent agreement with the predictions from the standard tunneling model of low-temperature glasses (much better than in the case of polyisobutylene with a lower molecular weight, which is usually used in most experiments). The results of our model calculations (discussed in DOI: 10.1039/C5FD00055F) show that it is impossible to explain the observed data by simple modifications of the assumptions of the standard model of glasses (increase in the TLS density, decrease of the potential barriers, modifications of the TLS parameters distributions, *etc.*).

Alfred J. Meixner commented: With respect to Professor Moerner's question, I would like to comment that a molecule at the surface of the organic glass substrate is not embedded in a confined volume. The molecule has another environment than in the bulk glass; this is what you observe. You are investigating the surface.

Yuri Vainer responded: Thank you for the comment. You are right. A molecule at the surface of the polymer film can be considered as partially "free". It is especially true for the case of tetra-*tert*-butylterrylene molecules sputtered on the surface of a polyisobutylene film. As a result, the excitation laser radiation can cause diffusion motion (jumps) of TBT molecules along the surface, even at cryogenic temperatures. During this motion, molecules can change their conformation states and the electronic transition frequencies of the molecules will also change.

Himangshu Prabal Goswami opened a discussion of the paper by Sanli Faez: Since a high degree of coherence exists in these molecules, I expect them to exhibit electromagnetically induced transparency (EIT) in the presence of a field. If it really does exhibit EIT, will the molecules' utility as sensors be limited to a specific regime of the electromagnetic spectrum?

Sanli Faez replied: The method suggested here is only for measuring quasi static-electric fields, up to a few GHz in bandwidth. The optical excitation frequency of each molecule is slightly different from that of neighboring molecules because of its local environment, and we probe a single two or three level system. The conventional formulation of EIT, which describes quantum interference in a dense ensemble of atoms, is not relevant. In cases of achieving a high extinction from a single molecule by means of nano-antennas (*e.g.* the experiment of Hwang *et al.*),[1] one can switch the extinction by ground-state depletion with an optical pump field, but that is a totally different configuration than what is described in our paper (DOI: 10.1039/C5FD00065C).

1 J. Hwang *et al.*, *Nature*, 2009, **460**, 76–80

Michel Orrit commented: Indeed, these molecules at a low temperature can present lifetime-limited linewidths, meaning that they can present maximal coherence properties in a range of times limited to a few nanoseconds by spontaneous emission. Electromagnetically induced transparency for a single molecule corresponds to saturation of the absorption, and even to amplification or stimulated emission by the molecule if the population can be inverted.

Irena Kratochvilova asked: Describe in more detail the general properties of the molecules for electric field sensing. Explain why these properties are important for such a specific case.

Sanli Faez replied: For an organic molecule to be a good candidate for electric field sensing based on the method we have proposed, its 00-zero phonon transition must be narrower than the eventual line shift due to the Stark shift caused by the hopping electrons. Ideally, the 00-ZPL should be lifetime limited, but also spectrally strong. For the calculations, this means that there must be significant overlap between the ground state and excited state molecular orbitals in the absence of vibronic excitation so that the matrix element for this transition is large and hence the 00-ZPL is recognizable in the excitation spectrum.

Vahid Sandoghdar remarked: Could you please say whether your DFT calculations can predict whether the lines are narrow or not? How good is your Stark effect sensitivity with the existing symmetric molecules?

Sanli Faez answered: In our basic DFT calculations, we do not consider the molecules of the host medium and therefore it will not be able to predict the exact position of a transition line or its broadening. Using DFT calculations, one can check the strength of the 00-ZPL transition with those of the vibronic bands for an isolated molecule to make sure that 00 is at least stronger than 01, 02, *etc.*, but that still does not guarantee that coupling to matrix phonons will be weak enough so that 00-ZPL can be experimentally isolated.

The experimentally reported Stark values for a fraction of terrylene in *para*-terphenyl (0.3 Debye) and some DBT in polyethylene (1 Debye) are more than enough for the proposed experiments, but one should note that only a fraction (~10%) of the molecules in those systems show the large field sensitivity that we need, so it will require some trial and error to achieve a working system.

Wolfgang Langbein asked: It might not be a coincidence that the field sensing dyes are broad. What other mechanism is expected for broadening of the lines? Can it be due to fast field fluctuations that you are sensing?

Sanli Faez replied: True. High sensitivity to an external electric field usually entails a broad excitation line because of the fluctuations in the environment. However, observation of molecules with lifetime limited narrow lines and a very large Stark factor is not unprecedented and has been reported for two systems of terrylene in *para*-terphenyl and dibenzoterrylene in polymer host matrices. However, only a fraction of all the molecules in those systems show such a desired behavior. Therefore, we suggest that it is possible to engineer new molecules that are predictably and consistently sensitive to an external field, yet preserve their narrow excitation line.

Brahim Lounis said: When a heavy atom is substituted into an aromatic molecule, the intersystem crossing rate increases and the fluorescence yield decreases because of an enhancement in the spin–orbit coupling. Can you comment on this?

Sanli Faez answered: True. We have not measured the fluorescence yield or the intersystem crossing of fluoroterrylene dye so I can not make a quantitative statement about this issue.

Lukasz Piatkowski asked: You have discussed the possibility of using suitably engineered aromatic molecules as very sensitive probes for electric fields. The high sensitivity comes from a spectral shift of the narrow zero phonon line at cryogenic temperatures. How do you envision using such a system as an actual probe as it requires specific (cryogenic) conditions to ensure high sensitivity?

Sanli Faez replied: Our long term goal is to use this molecule in hybrid quantum systems to probe the coherence of electron qubits. Such measurements have to be done at cryogenic temperatures even lower than the liquid helium temperature necessary for obtaining narrow lines for these organic molecules. To perform field sensing at elevated temperatures, one can still utilize the concept of triangulation but one needs molecules with a much larger response to an electric field (such as push–pull systems) so that the line shift is larger than the (much broader) excitation spectrum of these molecules.

Michel Orrit said: I think that neither the virtual cavity (or Clausius–Mossotti) model nor the empty cavity models apply to your case. The Clausius–Mossotti model assumes that the molecule placed at the center of the cavity has the same polarizability as the matrix molecules. However, dye molecules being conjugated are generally more polarizable than matrix molecules. The empty cavity model completely ignores the molecular polarizability and is, therefore, even worse. The molecular polarizability should enter the expression of the local field strength, as was demonstrated by Böttcher[1] and as was done in the paper you cite[2] for fluctuations of fluorescence lifetimes in polymers. The molecular polarizability should be an adjustable parameter in your theory, that you could vary to fit your measurements in different matrixes.

1 C. J. F. Böttcher, *Theory of electric polarization*, Elsevier, Amsterdam, 1973, vol. 1.
2 R. A. L. Vallee *et al.*, *ChemPhysChem*, 2005, **6**, 81–91.

Maxim Gladush responded: Thank you very much for the comment and the suggestions towards further research and development of the theory. The necessity to adjust to different types of host matrices is on the agenda. The limitations and drawbacks of the local-field corrections to the radiative lifetime, which implement the concepts of either Lorentz (Clausius–Mossotti) or the so-called empty cavity local fields, are known and must be dealt with. The enhancement (or correction) factors that contain the local-field contribution follow from different theories. Some theories suggest local-field corrections for foreign guest particles with a transition resonant to the incident light while the host absorption lines are way too far from the corresponding wavelength. The Clausius–Mossotti-like expressions may appear within the correction factors for T_1 as a result of the need to exclude the host particles from the "position" of the guest emitter while performing the integration over the material volume. Alternatively, the Glauber–Lewenstein procedure (empty cavity) may be performed to do the same thing in integration but offers the guest a greater empty space around. This fact by no means hides the need for trickier models but it certainly argues the following: "the Clausius–Mossotti model assumes that the molecule placed at the center of the cavity has the same polarizability as the matrix molecules". In other words, this statement is not necessarily true when connected to the T_1 corrections. Nevertheless, a more rigorous model is to be incorporated into the method.

Brahim Lounis asked: The lifetime of terrylene in *p*-terphenyl (index of refraction 1.65) at room temperature is 3.8 ns.[1] One should add this data point to Fig. 2 in the manuscript (DOI: 10.1039/C5FD00086F) and comment.

1 B. Lounis and W. E. Moerner, *Nature*, 2000, **407**, 491–493.

Maxim Gladush answered: In the article cited in the question, the authors measured the radiative lifetime of a single terrylene molecule in *p*-terphenyl at room temperature, and report a value of 3.8 ns. From the article it is not clear for how many different single molecules such measurements were performed and what the dispersion of the obtained lifetime values was. Having found the lifetime for a single terrylene molecule in *p*-terphenyl to be 3.8 ns, the authors said that it is "consistent with the lifetime of terrylene in *p*-terphenyl as deduced from the homogeneous width of approximately 40 MHz at low temperature" (citing the article of S. Kummer *et al.*[1]).

However, in the latter article, the authors reported the significant dispersion of lifetime limited linewidth values – from 37 MHz up to 53 MHz for terrylene single molecules in *p*-terphenyl. Next they write the following: "The fluorescence emission lifetime for terrylene in a poly(ethylene) matrix was determined to be 3.8 ns.[2] Making the reasonable assumption that the fluorescence lifetime does not differ between a poly(ethylene) or a *p*-terthenyl matrix, we expect a lifetime limited linewidth around 42 MHz".

So, answering the question, I would like to say: 1) there is a dispersion of the excitation lifetime values for terrylene in *p*-terphenyl – from about 3 ns to about 4.3 ns. We show this dispersion by adding the corresponding bar to the calibration curve in Fig. 1 in this Discussion, corresponding to Fig. 2 of our article (DOI: 10.1039/C5FD00086F). 2) Besides, from the article by S. Kummer *et al.*,[1] we may take the terrylene lifetime in polyethylene (3.8 ns), which is more favorable for the virtual cavity model (see the empty square data point in Fig. 1 in this Discussion). To conclude, I would like to note that the calibration curve we show in Fig. 2 (DOI: 10.1039/C5FD00086F) is based on the data for ensemble averaged lifetimes

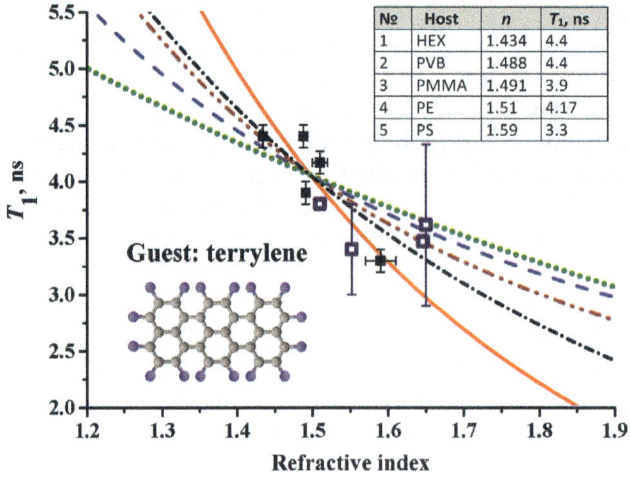

Fig. 1 The dependence of the excited state lifetime *T*1(*n*) of terrylene on the refractive index of its host matrix (squares). The curves show the best fits of the *T*1(*n*) dependence using different models: the virtual-cavity model (red solid); the empty-cavity model (brown dash-dot-dot); the fully microscopic models (green dot); microscopic models (black dash-dot and blue dashed).

obtained from the literature. All data points were reported as measured for bulk samples. We use it to find connections with the lifetime distributions observed in independent experiments with single molecules. Besides, this curve is not intended to defend a particular model for the local-field correction. As soon as we collect more points, some new model for a class of materials may arise. Nevertheless, even the data you suggest fits the general trend seen in Fig. 1 in this discussion or Fig. 2 (see DOI: 10.1039/C5FD00086F). Even if one decides that the extra data point in Fig. 1 is due to be around the middle of the bar on the right end, then the black dash-dot curve may be considered as a preferable calibration variant. But, this is exactly the case shown in Fig. 3 (see DOI: 10.1039/C5FD00086F) by the black dash-dot curve distribution which gives a good matching with the bulk refractive index of naphthalene. This is as well supported by the recently obtained data point for terrylene in 2,3-dichloronaphthalene as reported by M. Białkowska, *et al.*[3] We have also added the point for terrylene in *ortho*-dichlorobenzene evaluated from the unsaturated spectral linewidths shown in the inset of Fig. 3 (see DOI: 10.1039/C5FD00086F). It is seen in Fig. 1 as the point with a tall error bar on the left side. One may note that it covers both curves we have used.

1 S. Kummer, Th. Basché and C. Bräuchle, Terrylene in *p*-terphenyl: a novel single crystalline system for single molecule spectroscopy at low temperatures, *Chem. Phys. Lett.*, 1994, **229**, 309–316.
2 T. Plakhotnik, W. E. Moerner, T. Irngartinger and Urs P. Wild, Single Molecule Spectroscopy in Shpol'skii Matrices, *Chimia*, 1994, **48**, 31.
3 M. Białkowska, W. Chaładaj, A. Makarewicz and B. Kozankiewicz, *ACTA PHYSICA POLONICA SERIES A*, 2015, **128**, RK.128.3.1-1–RK.128.3.1-4.

Yuri Vainer opened a discussion of the paper by Maxim Gladush: I have several questions which are more related to your paper, but not to the introduction, because of some differences between them. In your paper (DOI: 10.1039/C5FD00086F), you mention a route towards sub-wavelength refractometry. Can you explain what you mean when you say "sub-wavelength refractometry"? A refractive index is known as a macroscopic parameter averaged over some volume much larger than the single molecule size and is used when light propagation effects are considered. For example, if we consider materials composed from two components: transparent dielectric media characterized by refractive index n1 and very small balls made from a transparent dielectric material with index n2, the light will "sense" this material as a material with *n* averaged over the layer with thickness lambda divided by four. So, in my opinion, you should at least explain what you mean when you say about sub-wavelength refractometry and what sizes of the local "*n*" heterogeneity do you propose to measure with probe single molecules. In the paper I do not see any reference about coordinates or distances at all. The quantum dot in a drop is not a good example, which is directly related to the considered case.

Maxim Gladush replied: I would like to express my firm conviction that the content of the article (DOI: 10.1039/C5FD00086F) is fully consistent with the presentation. The presentation slides showed the complete track of all the formulas, figures, hypotheses, approximations, and results strictly and carefully, exactly as they appear in the text of the manuscript. Neither in the paper's title nor in the text have we tried to introduce any new concepts or connotations for the terminology.

A refractive index is certainly a macroscopic parameter. It follows from averaging over a volume much larger than the size of a single molecule. It is used when light propagation effects are considered. If measured as a sole parameter of a bulk sample under study, it is written in the wave equation for light which propagates through this sample. However, if one goes down from the wave's domain to the realms of photons it often becomes crucial to take account for the polarizable neighborhood of a molecule which gives birth to these photons. This neighborhood is certainly larger than the size of the molecule and can be attributed as an effective or regional value of refraction following from the indigenous dielectric susceptibility. The connection between the bulk n and the effective zonal n_{eff}, may be under question. However, it is n_{eff} that works in the formulas which express, for instance, the rate of spontaneous emission of a single molecule (quantum emitter) or the rate of FRET in a single donor–acceptor pair. Besides, in the issue of light–matter interaction, we note that in many cases of modern nanophysics, one ought to think very carefully what refractive index should be used in the working formulae.

Andrei Naumov communicated in reply: We do not introduce any new concepts. Probing with a single molecule gives information about the "effective" refractive index value in the region of the sample where the probe is. Spatial coordinates of a single molecule can be reconstructed with nanometer accuracy (a fact fully recognized among members of this discussion community). It is this circumstance that is reflected in the title, that is, sub-wavelength reconstruction of the probe position. The question about the size of the region in the material around the probe, generating that very effect of the local field, which leads to a T_1 change – that is a problem which can be solved with the use of the approach proposed in our paper.

Yuri Vainer remarked: You have considered only one possible contribution to the observed distribution of T_1 values and calculated, on this basis, parameters of the refractive index distribution for systems under study. For these purposes you used very simple models of the local field, which are valid only for atoms or small molecules and in the case of locally isotropic, non-absorbing, continuous dielectric media. However, in the paper (DOI: 10.1039/C5FD00086F), you deal with polymers and polycrystalline solids with very small grains, which can not be considered as locally isotropic and continuous media. The large and plain in the free state TBT molecules incorporated in these media also can not be considered as simple dotted molecules. So, the above mentioned approximations are not valid in the considered systems. In particular, the plain TBT molecules frozen in a cold polymer matrix could be in different conformation states and as a result they may have different orientations and different values of the transition dipole moments. The chromophore molecules can be concentrated on the boundaries between grains in polycrystalline media, which can be anisotropic. The quantum yield values for the TBT molecules also could be different. This means that the T_1 values for these molecules are expected be different. However, the above mentioned important effects were not taken into consideration. As a result, you do not know what the relative value of the local field effect is in the observed T_1 distributions. So, I think that the calculated refractive index distributions in your paper are not proven.

The main result you obtained (the refractive index distribution for polyethylene) looks physically not reasonable. Indeed, if the refractive index values for polyethylene are really so different, as is indicated in your paper (from 1.1 to 1.8), the film of PE should be not optically transparent and the light beam reflected from such a film should be strongly blurred in the incident plane. However, it is not so. In spite of some scattering, the PE film looks transparent and the reflected beam does not look blurred.

Maxim Gladush responded: The key point of this article was to show the agreement between the series of experiments to measure the ensemble-averaged excitation lifetime and two trusted single molecule experiments to reveal the lifetime-limited T_1 distributions in terms of choosing the proper (although case-limited) function $T_1(n)$. The local-field correction models (we choose from), namely the functions $T_1(n)$, were listed and cited carefully according to the up-to-date physics. Despite the "simplicity", the local-field correction model used in calculations has already demonstrated interesting and not obvious matchings. At the same time, I feel it must be noted and emphasized that the necessity for more sophisticated maths has been clearly and literally stated in the text of the paper.

Andrei Naumov communicated in reply: In our article (DOI: 10.1039/C5FD00086F), we clearly showed that at least in two materials the maxima of the refractive index distribution coincides with the n value obtained by classical refractometry for a large volume of the material. Getting ahead, we note that the same coincidence was found for two other materials. This, at least, does not contradict the possibility of probing the local n values by single molecules, but actually indicates that we are on the right path. As for the dispersion (spread of values) of n, the obtained qualitative result is also logical, that is, in a more disordered medium we have a larger spread of values. If, however, it is up for discussion of what other factors may lead to a change of T_1 time, we again note that the proposed approach for the first time opens up the possibility of investigating such local effects.

In this paper we did not consider the spatial distribution of the areas (regions) with different refractive indices, as well as the sizes of the respective areas of the same (almost the same) n localization. We (or anyone) also can not say anything about the characteristics of the light scattering in those films in which the experiments were performed – it was not the subject of the study done by E. A. Donley, S. Bonsma, V. Palm, V. Burzomato, U. P. Wild and T. Plakhotnik,[1,2] and in their papers there is no information about it. In this regard, it would be inappropriate to discuss the question about scattering in the context of our paper. When the relevant experimental results with a spatial resolution appear, it will be possible to discuss the scattering as well.

1 E. A. Donley, V. Burzomato, U. P. Wild and T. Plakhotnik, *J. Lumines.*, 1999, **83–84**, 255–259.
2 E. A. Donley, S. Bonsma, V. Palm, V. Burzomato, U. P. Wild and T. Plakhotnik, *J. Lumines.*, 2000, **87–89**, 109–114.

Yuri Vainer asked: Your paper is written as an experimental paper and includes both theoretical and experimental parts. But really you have measured only two parameters: the refractive indexes for PMMA and for n-hexadecane. All other parameters you used are from the literature. It is not clear why and for what reason you measured the above two values, they both are well known from handbooks and the literature. May be your measurements were more accurate? But no, you measured with the same accuracy. May be you have measured the same material that was used in experiments? But also no. So, it looks like you included the measurements simply to make it look like your paper is experimental, whereas it is really pure theoretical.

Maxim Gladush replied: I am surprised at this interpretation of our work. This article has been positioned as neither experimental nor purely theoretical. This is an in-depth analysis that has bonded a number of recognized theories and trusted experiments (some of them are *sui generis*). All the corresponding references, citations and descriptions are given in the text. The two trivial extra measurements were carried out to clarify the discrepancies found in the literature.

Andrei Naumov communicated in reply: The paper describes all the details of the analysis performed. In particular, as it is important for understanding the essence of the proposed approach, Section 3 (DOI: 10.1039/C5FD00086F) describes, in detail, all peculiarities of the unique experiments carried out at ETH, Zurich. Unfortunately, reproducing such experiments at the present time is not possible, not only in our laboratory, but, to our knowledge, anywhere in the world. All references to the relevant literature sources are given correctly. Additional measurements of the refractive index for the two systems were carried out because in different sources one finds different n values. Concerning your impression, we are sorry, however it is not related to the scientific content of our paper.

Vahid Sandoghdar commented: It seems to me that one first has to decide on a suitable definition for the index of refraction, which is usually a macroscopic quantity. It would be great if one could do an ideal Gedanken experiment, where one starts with an atom in vacuum, measures its transition wavelength and emission lifetime, and then starts to fill its surroundings with host molecules/atoms and sees how these measurements change. One might think that the near-field coupling of the probe atom/molecule with its immediate surroundings would dominate over long-range effects caused by an index of refraction.

Maxim Gladush answered: In our research, we stay with the common understanding that the index of refraction is a macroscopic quantity. There is some small space around a single quantum emitter of light that may be treated as polarizable continuous medium, *i.e.* not a discrete ensemble of particles. The minimum size of this space can be different but it is, certainly, sufficiently large to contain the number of particles suitable for the continuous medium approximation to work. For us it is not the time and not the case yet to decide on a specifying definition for the index of refraction. We are on the way to obtaining the data on the spatial distribution of T_1 values. As soon as they have been processed, we might go into this issue.

Concerning the great idea of an ideal Gedanken experiment, we see this challenge as extremely important, but still unconquered by thinkers. However, the mathematical modelling of how a particle with resonant polarizability interacts with one, two, *etc.* particles with non-resonant polarizabilities is to be put into practice. We look at it as a way to add some clarity to the question of whether there is a competition of the dominance between the nearest-range, the *meso*-range and the far-range effects. Meanwhile, in our article, we show that for at least two materials the maxima of the "refractive index distributions" match the n values obtained by classical refractometry. We have recently obtained the data for two extra materials, which give similar matchings. These facts support the existence of a connection between radiative lifetimes of single molecules and the refractive indices of the hosting regions.

Wolfgang Langbein addressed Yuri Vainer and Maxim Gladush: Molecules in their nanoscopic environment are sensitive to a variety of different external parameters. For a sensor for a single parameter, we need a set of molecules with different sensitivites to these parameters, so that we can deduce the variations of the parameter of interest unaffected by others. We need a strategy to implement such a method in order to use molecules as reliable sensors.

Yuri Vainer responded: Thank you for the comment. I think that you are right. In order to study the properties of subsurface layers more correctly, it is necessary to measure more probe molecules and to develop a corresponding strategy of data analysis. We consider our results as the first step in this direction.

Maxim Gladush replied: Thank you very much for the comment. We have been thinking about implementing a similar idea of multiple sensors. The concept for the matter is being worked on.

Faraday Discussions

PAPER

Optical micro-spectroscopy of single metallic nanoparticles: quantitative extinction and transient resonant four-wave mixing†

Lukas Payne,[a] George Zoriniants,[a] Francesco Masia,[b] Kenton P. Arkill,[c] Paul Verkade,[c] Darren Rowles,[d] Wolfgang Langbein[b] and Paola Borri*[a]

Received 14th May 2015, Accepted 29th May 2015
DOI: 10.1039/c5fd00079c

We report a wide-field imaging method to rapidly and quantitatively measure the optical extinction cross-section σ_{ext} (also polarisation resolved) of a large number of individual gold nanoparticles, for statistically-relevant single particle analysis. We demonstrate a sensitivity of 5 nm^2 in σ_{ext}, enabling detection of single 5 nm gold nanoparticles with total acquisition times in the 1 min range. Moreover, we have developed an analytical model of the polarisation resolved σ_{ext}, which enabled us to extract geometrical particle aspect ratios from the measured σ_{ext}. Using this method, we have characterized a large number of nominally-spherical gold nanoparticles in the 10–100 nm size range. Furthermore, the method provided measurements of in-house fabricated nanoparticle conjugates, allowing distinction of individual dimers from single particles and larger aggregates. The same particle conjugates were investigated correlatively by phase-resolved transient resonant four-wave mixing micro-spectroscopy. A direct comparison of the phase-resolved response between single gold nanoparticles and dimers highlighted the promise of the four-wave mixing technique for sensing applications with dimers as plasmon rulers.

1 Introduction

In the last decade, increasing attention has been devoted to methods for the optical detection of single nano-objects, which do not rely on the object's fluorescence emission, but on its absorption and scattering properties (for a recent review see ref. 1). These characteristics are particularly relevant for weakly

[a]Cardiff University School of Biosciences, Museum Avenue, Cardiff CF10 3AX, United Kingdom. E-mail: borrip@cf.ac.uk
[b]Cardiff University School of Physics and Astronomy, The Parade, Cardiff CF24 3AA, United Kingdom
[c]University of Bristol, Wolfson Bioimaging Facility, University Walk, Bristol BS8 1TD, United Kingdom
[d]BBI Solutions OEM Ltd, Golden Gate, Ty-Glas Avenue, Llanishen, Cardiff CF14 5DX, United Kingdom

† Electronic supplementary information (ESI) available. See DOI: 10.1039/c5fd00079c

luminescent systems, such as metallic nanoparticles (NPs), whose morphology-dependent localized surface plasmon resonances (LSPR) produce large particle polarisabilities and, in turn, large absorption and scattering cross-sections at the LSPR frequency. Notably, these resonances can be exploited to image metallic NPs with high sensitivity, and to probe nanoscale regions in the NP vicinity *via* the local field enhancement effect. Possible applications range from photonics devices,[2] catalysis[3] and photovoltaics[4] to biomedical imaging,[5,6] sensing[7–9] and medical therapy.[10] For many of these applications, it is essential to have a quantitative knowledge of the linear and nonlinear NP optical properties at the single particle level, beyond traditional ensemble-averaged measurements.

A number of approaches have been developed recently to measure the optical extinction cross-section σ_{ext} (defined as $\sigma_{ext} = P_{ext}/I_i$ with I_i the incident intensity and P_{ext} the total power lost, due to both absorption and scattering) of single NPs, and to separate this into the absorption and the scattering cross-section, *via* $\sigma_{ext} = \sigma_{abs} + \sigma_{sca}$. The scattering cross-section σ_{sca} is most commonly measured using dark-field microscopy, which offers the advantage of a wide-field illumination with commercial easy-to-use optical set-ups and detector cameras and, in turn, the rapid observation of a large number of individual NPs simultaneously for statistical analysis. However, the quantification of σ_{sca} in absolute units in this technique is usually not reported, since it requires the detailed knowledge of the excitation and collection geometries, or a calibration reference. Moreover, for particles much smaller than the wavelength of light (dipole approximation), σ_{sca} decreases with the square of the particle volume, limiting the practical applicability of dark-field microscopy to NPs of moderately large sizes (>20 nm diameter for gold and silver nanospheres[1]).

A sensitive quantitative method to measure σ_{ext} uses a spatial modulation micro-spectroscopy (SMS) technique.[11] The method was shown to achieve detection of single gold NPs as small as 5 nm in diameter, corresponding to a sensitivity in terms of σ_{ext} down to about 5 nm^2. However, for quantitative determination of σ_{ext} this method needs a precise knowledge of the modulation amplitude and beam profile at the sample. Moreover it is a modulation-based laser-scanning technique, therefore less amenable to the rapid characterization of a large number of NPs compared to wide-field techniques. Additionally, it requires rather costly and specialized equipment. Similar considerations hold for photothermal imaging (PI),[12] another laser-scanning modulation-based technique. Moreover, in PI quantifying the absorption cross-section amplitude in absolute units is not trivial, since absorption is only indirectly detected in photothermal signals through the temperature change in the surrounding medium following NP heating.

We have recently shown a simple wide-field microscopy method with which we measured σ_{ext} on a large number (>100) of single gold NPs rapidly, quantitatively in absolute units, and without the need for calibration standards or complex modulation techniques.[13] In this first demonstration, we showed measurements on gold NPs of 40 nm and 100 nm diameter for proof-of-principle, and a sensitivity in σ_{ext} of around 500 nm^2, limited by the acquisition speed of the consumer color camera used for detection. We also showed that we could characterize the NP asphericity by using linearly polarized illumination, however the measured polarisation dependence was not analyzed quantitatively to extract the geometrical aspect ratio of the NP.

Beyond linear optical properties, we also developed a nonlinear optical technique, which detects single gold NPs *via* their transient degenerate four-wave mixing (FWM) resonant at the LSPR.[6,14] Owing to an interferometric and time-resolved detection, this technique is specific to gold NPs and background-free (contrary to the above mentioned methods), even in highly scattering and fluorescing environments, such as cells and tissue. Notably, rather than providing the absolute value of σ_{ext}, the technique is sensitive to the change in the NP extinction induced by a short pump pulse at the LSPR. In turn, owing to the phase-sensitivity of the interferometric detection, we showed that we could measure the pump-induced ultrafast change in the particle polarisability as a complex quantity, *i.e.* separating its real and imaginary parts.[14] The phase of this complex quantity was shown to be sensitive to the relative shift between the LSPR and the optical wavelength in the FWM experiment. Hence, beyond fundamental interest, this background-free, phase-resolved, intrinsic ratiometric readout, at the single-particle level, could be exploited for LSPR-based sensing applications, for example using NP dimers as plasmon rulers.[9]

In this work, we report a major improvement on our previous wide-field extinction microscopy method, which now enables us to reach a sensitivity in σ_{ext} of about 5 nm^2, hence allowing detection of single 5 nm diameter gold NPs. The technique remains rapid, wide-field, and easy to implement on any conventional microscope. We also present a method to extract the nanoparticle geometrical aspect ratio from the measured polarisation-resolved σ_{ext} using analytical formulas for elliptical particles. With this wide-field extinction technique, we have characterized a statistically significant number of gold NP conjugates fabricated in house, and used the measurements to distinguish single dimers from isolated particles and larger aggregates. We then performed FWM measurements on the same NP conjugates, allowing direct correlation with the optical extinction method. The results highlighted the promise of the FWM technique for sensing applications.

2 Wide-field extinction microscopy

2.1 Isolated quasi-spherical nanoparticles

The experimental set-up consists of an inverted microscope (Nikon Ti-U) equipped with a white-light illumination (halogen lamp 100 W with Nikon neutral color balance filter or light balancing LB200 optical blue Hoya filter), an oil condenser of 1.4 numerical aperture (NA) with a removable home built dark field illumination of 1.1–1.4 NA, a choice between a 40× 0.95 NA dry objective and a 100× 1.45 NA oil objective, 1× or 1.5× intermediate (tube lens) magnification, a consumer Canon EOS 40D color camera attached to the left port of the microscope and the PCO Edge 5.5, a low noise, high frame-rate, grayscale (b/w), water-cooled, scientific-CMOS camera (sCMOS), attached to the eye-piece side port. With the Canon camera, images were taken in Canon 14-bit RAW format with 10.1 megapixel resolution. The RAW images were converted using the DCRAW plugin in ImageJ, providing 16 bit RGB images with a linear response to intensity and no color balancing. The color camera enables a coarse spectroscopic detection, separating the three wavelength ranges of red (R) 570–650 nm, green (G) 480–580 nm, and blue (B) 420–510 nm. With the sCMOS camera, spectroscopic detection was achieved using bandpass filters (Semrock) in the excitation beam path, in the wavelength ranges of 470 ± 11 nm for blue, 530 ± 20 nm for green and 607 ± 18 nm for red.

The investigated samples were nominally spherical gold NPs of 10 nm, 30 nm, 40 nm, 60 nm, 80 nm and 100 nm diameter (BBI Solutions) drop cast onto a glass coverslip, covered in silicone oil (refractive index $n = 1.518$) and sealed with a glass slide. Before use, glass slides and coverslips were cleaned thoroughly to avoid debris. Cleaning was performed first with acetone and high-quality clean-room wipes, followed by a chemical etch called Caro's etch, or more commonly, Piranha etch.

Extinction microscopy was performed as described in ref. 13. Briefly, the condenser NA was adjusted to match that of the objective. Two bright-field images were taken; one with the NPs in focus, and another with the NPs either out-of-focus (moving the objective by approximately $d = 15$ μm axially away from the sample) or laterally shifted by approximately 1 μm. Background images were taken for blocked illumination. To achieve the lowest shot noise with the Canon camera, the lowest camera sensitivity was used (100 ISO), for which the full-well capacity of the pixels of about $N_{fw} = 4 \times 10^4$ electrons occurs at 70% of the digitizer range (3.4 electrons per count). An exposure time in the order of 10 ms was chosen to reach N_{fw}. Averaging over $N_a = 36$ acquisitions was performed for each image set. With the sCMOS, we had $N_{fw} = 30\,000$ electrons (0.54 electrons per count, 16 bit dynamic range), an exposure time of 1 ms and $N_a = 25\,600$ at 400 frames per s, resulting in a total acquisition time of 64 s for an image set.

We call the background-subtracted, transmitted intensity of the bright-field image with NPs in focus I_f, and the defocused, or shifted, intensity I_0. In the defocused image, a NP distributes its effect over a radius of about $r_d = \mathrm{NA}d$, making I_0 similar to the intensity I_f in the absence of the NP. In the shifted image, we directly measure the intensity in the absence of the NP, which has been sideways shifted from its location. The extinction cross-section of a NP located within the area A_i in the image can then be expressed as $\sigma_{ext} = \int_{A_i} \Delta \mathrm{d}A$, with the relative extinction $\Delta = (I_0 - I_f)/I_0$. To account for the slight mismatch between I_0 and I_f without a NP due to, for example, a drift of the illumination intensity, the residual influence of the NP out-of-focus, and/or a non-perfect optical quality of the glass surface, we determine a local background extinction $\Delta_b = {A_b}^{-1}\int_{A_b} \Delta \mathrm{d}A$ in the area A_b between the radius r_i and $2r_i$, yielding the background-corrected $\sigma_{ext} = \int_{A_i} (\Delta - \Delta_b)\mathrm{d}A$. As discussed in ref. 13, a good choice of r_i, for which saturation of σ_{ext} is reached, is given by $\sim 3\lambda/(2\ \mathrm{NA})$, approximately at the second Airy ring of the objective point-spread function.

In order to determine σ_{ext} of many particles from an extinction image, we developed an image analysis programme written in ImageJ macro language, as described in ref. 13 (updated version available). Isolated spherical NPs are distinguished using their extinction color. NPs with σ_{ext} largest in the color channel corresponding to the expected plasmon resonance are retained, *i.e.* green (red) for 10–40 nm (100 nm) spherical gold NPs having a LSPR dominated by the absorption (scattering) part of the extinction at around 540 nm (~590 nm) in a surrounding medium of 1.5 refractive index. NPs which likely correspond to aggregates, debris, or largely non-spherical NPs are excluded in this way. NPs of 60 nm and 80 nm nominal diameter have a size range distribution which allows for isolated particles to have a σ_{ext} largest in either the green or the red channel of the Canon camera. This was taken into account in the image analysis automated routine. The statistical results over 54 individual gold NPs of nominally 10 nm diameter are summarised in Fig. 1, measured using the 100× 1.45 NA oil objective, the sCMOS camera, and by laterally shifting the sample for reference. An example of an extinction image is shown in Fig. 1a, demonstrating the high contrast in visualizing single 10 nm NPs (which appear as two spots, one bright and one dark, due to the lateral shifting). We adopted a "time-symmetric referencing" procedure, namely I_0 was measured before and after I_f and its average value was used for reference. This reduces the effect of slow drifts. The distribution of σ_{ext} in the G channel corresponding to the plasmon resonance has a mean of $\bar{\sigma}_{ext} = 69$ nm^2, which is consistent with experimental and theoretical values found in the literature.[15,16] The measurement noise in the experiment, $\hat{\sigma}_{noise}$, is obtained from the standard deviation of the zero-mean distribution of σ_{ext} in image regions free of NPs. As shown in Fig. 1, we find $\hat{\sigma}_{noise} = 5.9$ nm^2 for

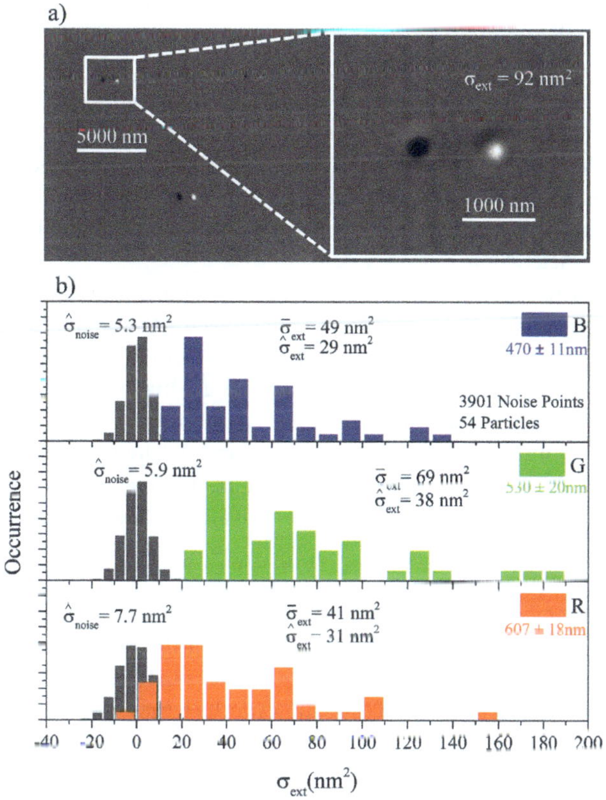

Fig. 1 (a) Wide-field extinction image of single gold NPs of nominal 10 nm diameter in the G channel, using a 100× 1.45 NA oil objective and a sCMOS camera (see text). The right panel is a zoom over the region indicated by the white frame. Grey scale min = −0.005, max = 0.00055. (b) Distributions of σ_{ext} for 54 gold NPs of nominal 10 nm diameter in the R, G, and B color channels as indicated. The grey histograms show the noise distribution obtained by measuring σ_{ext} in 3901 randomly selected regions without NPs.

the green channel, and down to $\hat{\sigma}_{\text{noise}} = 5.3\ \text{nm}^2$ in the blue channel. Notably, this number is two orders of magnitude lower than what was shown in our previous work using the consumer Canon camera. The main reason for such reduction of $\hat{\sigma}_{\text{noise}}$ is the large number of averages ($N_a = 25\,600$) possible with the high-speed sCMOS camera, while retaining a practically feasible total acquisition time over which setup drifts are acceptable. The experimental $\hat{\sigma}_{\text{noise}}$ is, however, still larger than the theoretical shot-noise limit of $3\ \text{nm}^2$, as estimated using the expression in ref. 13. This is likely due to a non-perfect referencing in the measurement of the extinction image Δ, namely I_0 not being identical to the intensity I_f in the absence of the NP, even with local background correction. To that end, we find that defocusing the objective gives $\hat{\sigma}_{\text{noise}} = 25\ \text{nm}^2$ compared to $\hat{\sigma}_{\text{noise}} = 5.9\ \text{nm}^2$ when laterally shifting the sample (under otherwise identical conditions), most likely because a lateral shift does not alter the beam path geometry and thus enables a better referencing. We also find $\hat{\sigma}_{\text{noise}} = 7\ \text{nm}^2$ when laterally shifting but without time symmetric referencing, indicating that sample/objective/camera drifts play a role. Furthermore, at this sensitivity, the spatial and optical quality of the glass surface provides a spatial extinction structure which contributes to $\hat{\sigma}_{\text{noise}}$.

In the histogram in Fig. 1, the standard deviation of σ_{ext} is much larger than $\hat{\sigma}_{\text{noise}}$ and is dominated by the NP size/shape/environment distribution. The NP-related standard deviation, $\hat{\sigma}_{\text{ext}}$, is determined using $\hat{\sigma}_{\text{ext}}{}^2 = \hat{\sigma}_{\text{total}}{}^2 - \hat{\sigma}_{\text{noise}}{}^2$, where $\hat{\sigma}_{\text{total}}$ is the total standard deviation of σ_{ext} as measured. Using $\delta R/R = \hat{\sigma}_{\text{ext}}/(3\bar{\sigma}_{\text{ext}})$, we then inferred a size distribution $\delta R/R = 0.18$ (R is the NP radius). This value is larger than the manufacturer specified distribution $\delta R/R = 0.10$, as determined by electron microscopy. This suggests, consistent with the literature, that additional factors influence the optical $\hat{\sigma}_{\text{ext}}$, beyond size fluctuations, such as the NP asphericity,[17] fluctuations in the local dielectric environment, and the electron-surface scattering damping parameter.[16]

An overview of $\bar{\sigma}_{\text{ext}}$ and $\hat{\sigma}_{\text{ext}}$ in the green color channel for NPs in the 10–60 nm diameter range, and both green and red channels for 60 nm, 80 nm and 100 nm sized NPs, is shown in Fig. 2. Measurements for NPs in the 30–100 nm size range were performed using the 40× 0.95 NA objective, 1.5 magnifier and the consumer Canon camera, as described in ref. 13. For each nominal diameter, a statistical analysis on >50 individual gold NPs was performed. The bars indicate the standard deviation $\hat{\sigma}_{\text{ext}}$. We also measured σ_{sca} using dark-field, as described in ref. 13, and in turn deduced σ_{abs} from σ_{ext}, as shown in Fig. 2. Note that for these data the calibration constant η (such that $\sigma_{\text{sca}} = \eta \int_{A_i} I_{\text{df}} \frac{dA}{I_0}$ with I_{df} scattered intensity in dark-field) was measured by using nanodiamonds,[18] which are scattering-only particles, where $\sigma_{\text{sca}} = \sigma_{\text{ext}}$. A comparison between the experimental values and calculations from Mie theory[15] of σ_{sca} and σ_{abs} at the LSPR wavelength is shown in Fig. 2, indicating good agreement (note how for the largest particles the scattering cross-section in the R channel dominates). Deviations between experiment and theory could be due to the spectral averaging of the color channel bandwidth, especially when using the consumer Canon camera. Furthermore, as mentioned in ref. 13, the finite angular range of the objective implies an underestimate of the extinction. In Fig. 2 we also compare $\hat{\sigma}_{\text{ext}}$ with the manufacturer specified size distribution (black error bars) using $\delta R/R = \hat{\sigma}_{\text{ext}}/(\bar{\sigma}_{\text{ext}}\gamma)$ with $\gamma = 3$ for small particles, where the extinction is dominated by absorption. For particle sizes

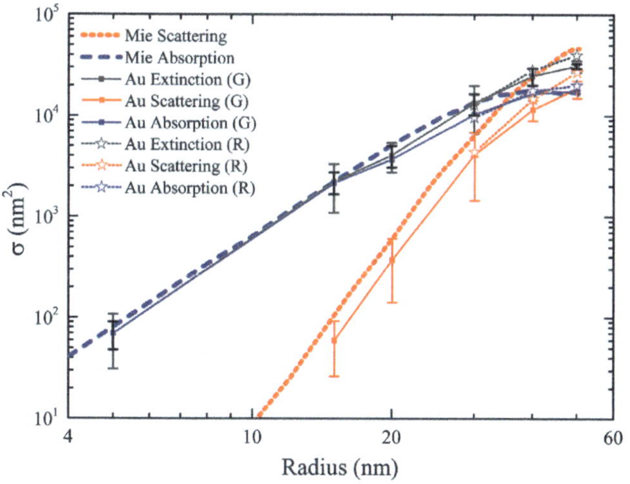

Fig. 2 Extinction, scattering and absorption cross section measured in the G channel on a series of nominally spherical gold NPs with a nominal diameter from 10 nm to 100 nm (solid squares). For 60 nm, 80 nm and 100 nm diameters, the values measured in the R channel are also given (empty stars). The thick blue dashed (red dotted) line is the calculated absorption (scattering) cross section from Mie theory at the LSPR. Symbols (connected by thin lines as a guide to the eye) give the experimental cross-sections as an average over >50 individually measured NPs in the wide-field image, and bars give the standard deviation due to NP size/shape/environment distribution. Black bars are the manufacturer specified size distributions (see text). Note that 10 nm gold NPs cannot be measured in dark-field due to their small σ_{sca}.

larger than 40 nm diameter, γ, such that $\sigma_{ext} \propto R^{\gamma}$, was extracted from σ_{ext} calculated from Mie theory using the sum of σ_{sca} and σ_{abs} at the corresponding LSPR wavelength. We find that $\hat{\sigma}_{ext}$ is generally larger than the manufacturer specifications, again suggesting that additional factors influence the optical $\hat{\sigma}_{ext}$, beyond size fluctuations.

2.2 Nanoparticle asphericity

In our previous work, we showed that by measuring the dependence of σ_{ext} on the in-plane linear polarisation angle θ of the exciting light we could extract information on the NP asphericity.[13] In particular, we fitted the experimental σ_{ext} with the expression $\sigma_{ext}(\theta) = \sigma_0(1 + \alpha_P \cos(2(\theta - \theta_0)))$, where σ_0 is the polarisation-averaged σ_{ext}, $\alpha_P \geq 0$ is the amplitude of the polarisation dependence, and $0 \geq \theta_0 \geq \pi$ is an angular offset, indicating the orientation (in-plane) of the NP asymmetry. We found that, within the measurement noise, aspherical NPs could be identified as those having α_P larger than about 0.1 (in the red channel). However, the relationship between α_P and the particle aspect-ratio was not derived, hence this parameter remained a phenomenological value rather than a useful structural quantity. Here we have developed a theoretical model for aspherical particles, with which we have simulated $\sigma_{ext}(\theta)$ and, in turn, calculated α_P as a function of the particle aspect ratio.

We describe a non-spherical particle as a metallic ellipsoid with three orthogonal semi-axes a, b and c. In the particle reference system, the polarisability tensor $\hat{\alpha}$ is diagonal, and its eigenvalues (in the limit of particles with a radius much smaller than the light wavelength) are given by[19] $\alpha_i = 4\pi abc\frac{\varepsilon - \varepsilon_m}{3\varepsilon_m + 3L_i(\varepsilon - \varepsilon_m)}$. ε is the dielectric constant of gold, ε_m is the dielectric constant of the medium surrounding the NP, and L_i with $i = a, b, c$ are dimensionless quantities defined by the particle geometry (see ESI†). We then expressed σ_{ext} as a function of $\hat{\alpha}$, taking into account that for an arbitrary particle orientation in the laboratory system, the polarisability tensor is $\hat{\alpha} = \hat{A}^{-1}\hat{\alpha}'\hat{A}$, where $\hat{A}$ is the rotation matrix transforming the particle reference system into the laboratory system, and α' is the polarisability tensor in the laboratory system. The absorption, scattering, and extinction cross-sections are calculated following ref. 19 (for more details see ESI†). If the polarisation of light is parallel to one of the symmetry axes of the particle, the extinction cross-section is given by:

$$\sigma_{ext,m} = k\Im(\alpha_m) + \frac{k^4|\alpha_m|^2}{6\pi}, \quad (1)$$

where $m = 1, 2, 3$ and $k = 2\pi n/\lambda$ with λ the wavelength in vacuum, n is the refractive index of the surrounding medium, and $\Im$ denotes the imaginary part. If the particle is oriented in-plane at an angle γ with respect to the laboratory x-axis, the extinction cross-section for an incident polarized light in-plane at an angle θ with respect to the laboratory x-axis can be accordingly calculated using the rotation matrix $\hat{A}$, and is found to be (for details see ESI†):

$$\sigma_{ext}(\theta) = \sigma_{ext,1}\cos^2(\gamma - \theta) + \sigma_{ext,2}\sin^2(\gamma - \theta) \quad (2)$$

This represents a sinusoidal dependence of the measured extinction cross-section *versus* θ, which can be rewritten as:

$$\sigma_{ext}(\theta) = \bar{\sigma}(1 + \alpha_P\cos[2(\theta - \gamma)]) \quad (3)$$

with

$$\bar{\sigma} = \frac{\sigma_{ext,1} + \sigma_{ext,2}}{2}, \quad \alpha_P = \frac{\sigma_{ext,1} - \sigma_{ext,2}}{\sigma_{ext,1} + \sigma_{ext,2}} \quad (4)$$

With these formulas, for a non-spherical particle with in-plane asymmetry $a > b$ (note that in the experiment we are not sensitive to the asymmetry in the z-axis), we calculated α_P and $\bar{\sigma}$ as a function of the particle ellipticity parameter $1 - b/a$ and the particle volume $(4/3)\pi abc$ (where we used $c = \sqrt{ab}$). We find that in the limit of small ellipticity ($\alpha_P < 0.5$), α_P is proportional to the ellipticity parameter and is almost independent of the particle volume (see ESI†). Thus, the ellipticity can be simply determined as $1 - b/a = K\alpha_P$ with the parameter K being a function of the wavelength range used in the experiment. For the G and R channels of the Canon 40D camera used in ref. 13, we find $K_G = 0.568$ and $K_R = 0.339$. Therefore a measured $\alpha_P = 0.1$ in the R channel corresponds to an ellipticity $1 - b/a$ of 3.39%.

2.3 Nanoparticle dimers

We investigated the applicability of this wide-field imaging method to the characterization of gold NP dimers fabricated in-house. Nominally spherical 40 nm diameter gold NPs coated with a nominal 6 nm thick silica shell were fabricated by BBI Solutions. We covalently bound them to bare 40 nm gold NPs in solution using (3-mercaptopropyl)trimethoxysilane (3-MT). Briefly, this was achieved by mixing equal volumes (250 μl each) of bare and silica-coated NPs stock solutions (each at a concentration of approximately 9×10^9 NP per ml) with 0.985 ml of water (Millipore) and 15 μl of 3-MT. The dimerisation reaction was monitored in real-time through the occurrence of a shifted LSPR peak using absorbance spectroscopy and the reaction was stopped (typically after 2 min) using *N*-ethylmaleimide. Examples of the type of dimers obtained in this way are shown in the transmission electron microscopy (TEM) images in Fig. 3. For TEM imaging, 5 μl of the solution containing gold NP dimers was drop cast onto a TEM grid (3 mm diameter, 200 mesh amorphous holey carbon-coated copper grid (EM Resolutions)). The grid was then imaged using a 200 keV LaB_6 filament transmission electron microscope (Tecnai T20, FEI) with a 4096 × 4096 pixel 16 bit bottom-mounted camera (BM Eagle, FEI). For optical measurements, solutions with gold NP dimers were covalently bound onto a glass coverslip functionalised with 3-MT, covered in silicone oil, and sealed with a glass slide. The solution contained a mixture of isolated NPs, dimers and larger aggregates (see overview TEM image – Fig. 3a). We therefore investigated the applicability of our wide-field microscopy method to characterize such a heterogeneous ensemble, and specifically, to distinguish dimers from single particles and larger aggregates.

A cloud plot representing a combination of the extinction cross-section σ_{ext} measured for these NPs on glass in the G and R channels (using the 40× 0.95 NA objective, 1.5 magnifier and the Canon camera) alongside the asymmetry parameter α_P (for σ_{sca}) in the corresponding channels is shown in Fig. 4. We chose

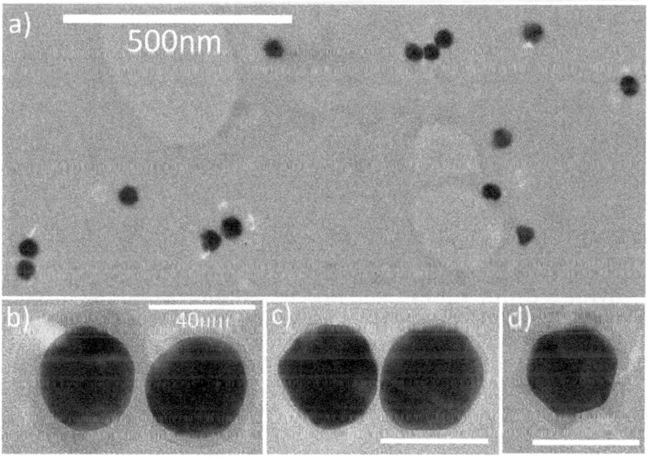

Fig. 3 Transmission electron microscopy (TEM) images of gold NPs drop-cast onto a TEM grid, from a solution containing nominal 40 nm NPs coated with a nominal 6 nm thick silica shell, covalently bound to bare 40 nm NPs. (a) Overview showing single particles, dimers and trimers. (b) Example of a dimer between a silica-coated NP (left) and a bare NP (right). (c) Example of a dimer between two bare NPs. (d) Example of a single 40 nm NP coated with a 6 nm silica shell. The thin coating visible on the nominally bare NPs is likely due to the (3-mercaptopropyl)trimethoxysilane forming a polymerized structure.

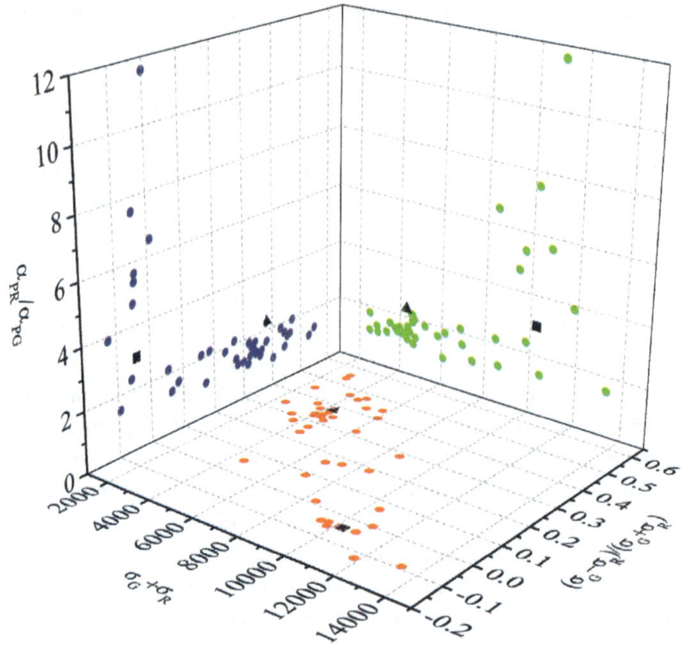

Fig. 4 Optical properties measured on individual NPs covalently bound onto glass, from a solution containing nominal 40 nm NPs coated with a 6 nm thick silica shell and covalently bound to bare 40 nm NPs (same as in Fig. 3). Measurements are shown as a trivariate distribution using the cross-section $\sigma_R + \sigma_G$, the contrast $(\sigma_G - \sigma_R)/(\sigma_G + \sigma_R)$, and the asymmetry parameter ratio α_{PR}/α_{PG} (see text). Black symbols indicate the two particles shown in Fig. 5.

to represent α_P extracted from σ_{sca} due to its lower uncertainty compared to α_P from σ_{ext} when measuring with the Canon color camera.[13] The coordinates were chosen as (i) the extinction cross-section represented by the sum of R and G channels $\sigma_R + \sigma_G$, (ii) the G to R cross-section contrast $(\sigma_G - \sigma_R)/(\sigma_G + \sigma_R)$, and (iii) the asymmetry ratio α_{PR}/α_{PG}. The cross-section indicates whether particles are single or aggregates, since this value will scale roughly proportional to the number of particles in the focal volume. The differential quantity $(\sigma_G - \sigma_R)/(\sigma_G + \sigma_R)$ gives information on the spectral position of the LSPR, where we expect isolated spherical particles to have $\sigma_G > \sigma_R$, while dimers and aggregates will have $\sigma_R > \sigma_G$ due to the red shift of the LSPR upon plasmonic coupling in a dimer with a gap much smaller than the NP diameter.[20] The ratio α_{PR}/α_{PG} is used to emphasize the asymmetry due to dimer formation, where we expect $\alpha_{PR}/\alpha_{PG} \gtrsim 1$ for a single slightly aspherical particle and $\alpha_{PR}/\alpha_{PG} \gg 1$ for dimers with small gaps. We can see a systematic trend in Fig. 4 with isolated particles being the group where $(\sigma_G - \sigma_R)/(\sigma_G + \sigma_R) > 0$, $\alpha_{PR}/\alpha_{PG} < 2$ and $\sigma_R + \sigma_G < 6000$ nm^2 ($\bar{\sigma}_{ext} = 4000$ nm^2 in the G channel and $\bar{\sigma}_{ext} = 2000$ nm^2 in the R channel for an isolated 40 nm NP), while strongly coupled dimers are being identified as the group where $(\sigma_G - \sigma_R)/(\sigma_G + \sigma_R) < 0$, $\alpha_{PR}/\alpha_{PG} > 2$ and $12\,000 > \sigma_R + \sigma_G > 6000$ nm^2. For larger aggregates $\sigma_R + \sigma_G > 12\,000$ nm^2, $(\sigma_G - \sigma_R)/(\sigma_G + \sigma_R) < 0$ but α_{PR}/α_{PG} varies, depending on the anisotropy of the assembly. In fact, we see in the cloud plot a large aggregate with $\alpha_{PR}/\alpha_{PG} < 2$ and $\sigma_R + \sigma_G \sim 14\,000$ nm^2, consistent with a closely-packed trimer.

To confirm these attributions, for the two particles highlighted by black symbols in the cloud plot in Fig. 4, we have measured the spectrally-resolved and polarisation-resolved extinction. We inserted a linear polariser in the white-light illumination beam path before the condenser, and coupled the transmitted light collected by the microscope objective (100× 1.45 NA oil objective, 1× intermediate magnification) into an imaging spectrometer (Horiba Jobin-Yvon iHR 550) equipped with a 100 lines per mm grating (input slit opening of 60 μm corresponding to ~1 nm spectral resolution) and a CCD camera (Andor Newton DU-971N). Referencing was performed similarly to the extinction microscopy discussed in the previous section. Briefly, a stripe-like region of the sample containing a NP in the center was imaged onto the input slit of the spectrometer and, in turn, onto the CCD camera. Within this stripe, image regions above and below the NP were used as a reference under identical spectral illumination conditions. A measurement without the NP (by lateral shifting) was also acquired to correct for irregularities in the camera pixels. Fig. 5a shows the extinction spectra for the particle highlighted as a black triangle in Fig. 4. The spectra exhibit a single LSPR at around 540 nm, almost independent of the polarisation angle, supporting the attribution that this particle is an isolate quasi-spherical NP. Fig. 5b shows the extinction spectra for the particle highlighted as a black square in Fig. 4, which exhibit a strong polarisation dependence and two distinct LSPRs (one at 535 nm and the second one at 575 nm) for two orthogonal polarisations, supporting the attribution that this particle is a dimer. Numerical simulations, using the software COMSOL Multiphysics, of the extinction spectra of a gold dimer surrounded by a homogeneous medium of refractive index 1.5, calculated for the two in-plane polarisations along and perpendicular to the interparticle axis, are shown for comparison (dotted lines in Fig. 5). The best agreement between simulations and the experiment is found assuming that the two NPs

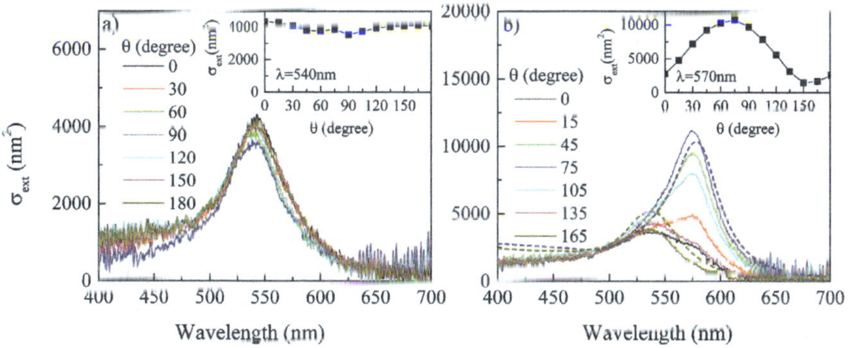

Fig. 5 Extinction spectra on the particles indicated as black symbols in Fig. 4 for various in-plane linear polarisation angles θ of the white-light illumination as indicated. The insets show the extinction *versus* polarizer angle at the indicated wavelength. (a) Spectra on the particle indicated by the black triangle in Fig. 4. (b) Spectra on the particle indicated by the black square in Fig. 4. Dotted lines are numerical simulations of the extinction spectra of a dimer consisting of a spherical NP of 32 nm diameter with a gap distance of 5.5 nm from a second particle of 33 nm diameter, calculated using COMSOL Multiphysics for the two in-plane polarisations along and perpendicular to the interparticle axis. Measured spectra were acquired using an exposure time of 0.1 s, and 200 frames average.

forming the dimer have slightly different diameters of 32 nm and 33 nm (heterodimer model) and an interparticle gap of 5.5 nm. Larger diameters result in larger peak cross-sections, while different gaps do not reproduce the red-shifted LSPR wavelength at 575 nm. Using NPs with different diameters enabled us to optimize the amplitude ratio between the longitudinal and transverse plasmon peaks. The deviation between simulations and experiment might be due to the imperfect sphericity of the particles, not included in the calculations. Furthermore, simulations assume an incident plane wave, and do not account for the experimental angular excitation and detection range.

3 Resonant four-wave mixing microscopy

Transient resonant four-wave mixing micro-spectroscopy was performed with a heterodyne interferometric detection very similar to the one discussed in our previous works,[6,14] the main difference being that we used an epi-detection geometry instead of a transmission geometry. Briefly, optical pulses of ~100 fs duration at 80 MHz repetition rate and 550 nm center wavelength were provided by the signal output of an optical parametric oscillator (OPO). The pulse train is split into three beams (pump, probe and reference), having all the same center wavelength (degenerate FWM scheme). The pump beam excites the sample with an intensity which is temporally modulated by an acousto-optic modulator (AOM) driven with a square wave amplitude modulation of about 1 MHz frequency. The change in the sample optical properties induced by this excitation is probed by the probe pulse at an adjustable delay time τ after the pump pulse. Pump and probe pulses of fields E_1 and E_2, respectively, are recombined into the same spatial mode and focused onto the sample by a microscope objective (for the data shown a 40× 0.95 NA) mounted into the same inverted microscope stand used for the extinction microscopy discussed in the previous section. The sample position is moved relative to the focal volume of the objective by an *xyz* piezoelectric stage. A FWM signal (proportional to $E_1E_1^*E_2$) is collected in reflection (epi-direction) by the same objective, and recombined in a beam splitter with the reference pulse of adjustable delay. The resulting interference is detected by balanced photodiodes (for common-mode noise rejection) and a dual-channel high-frequency lock-in amplifier. We used a heterodyne scheme where the probe optical frequency is upshifted by a radio-frequency amount using an AOM and its interference with the unshifted reference is detected at the appropriate beat frequency, as in the scheme published in ref. 14. This enables us to discriminate the FWM field from the pump and probe pulses and to detect the amplitude and phase of the field. Full control of the pump, probe and reference polarisation was achieved using waveplates in each beam. Note that this interferometric time-resolved FWM detection enables background-free imaging of single gold NPs even in highly scattering and fluorescing environments.[6]

In Fig. 6a we first recall an example of FWM dynamics *versus* τ measured on a single NP of nominal 40 nm diameter drop cast onto glass (from ref. 14) alongside the theoretical model discussed in detail in ref. 14. A key point is the ability of the technique to distinguish the real and imaginary part of the pump-induced change in the particle complex polarisability, indicated as $\Delta\hat{\varepsilon}_R$ and $\Delta\hat{\varepsilon}_I$, respectively (where we defined an effective dielectric constant $\hat{\varepsilon}$ proportional to the NP polarisability $\alpha = 4\pi R^3\hat{\varepsilon}$ for spherical particles, and denoted the pump-induced

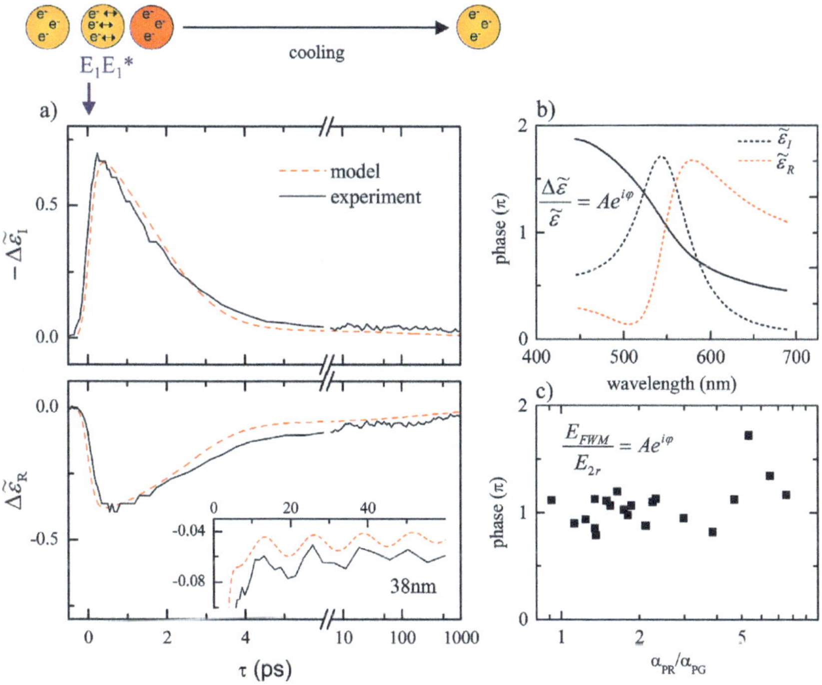

Fig. 6 a) Transient changes of the real ($\Delta\hat{\varepsilon}_R$) and imaginary ($\Delta\hat{\varepsilon}_I$) part of the dielectric function of a single 38 nm diameter gold NP resonantly excited and probed at 550 nm. The pump (probe) fluence is 0.65 J m^{-2} (0.05 J m^{-2}). The acquisition time per point is 200 ms. The dashed lines are the corresponding calculations. The top sketch illustrates the LSPR excitation by the pump (E_1) and the subsequent heating and cooling dynamics monitored by the probe pulse. The inset shows coherent phonon oscillations from which the 38 nm diameter is deduced (reproduced from ref. 14). (b) Spectrally resolved real and imaginary part of the dielectric constant $\hat{\varepsilon}$ (dotted lines) simulated for the 38 nm NP in (a) prior to the arrival of the pump, together with the phase φ (solid line) of the complex quantity $\Delta\hat{\varepsilon}/\hat{\varepsilon}$ with the pump-induced change $\Delta\hat{\varepsilon}$ calculated at $\tau = 0.5$ ps. (c) Measured phase of the ratio between the FWM field and the reflected probe at $\tau = 0.5$ ps on a group of NPs from the cloud plot in Fig. 4 as a function of the asymmetry parameter α_{PR}/α_{PG}.

change as $\Delta\hat{\varepsilon}$). Both $\Delta\hat{\varepsilon}_R$ and $\Delta\hat{\varepsilon}_I$ exhibit dynamics that are quantitatively explained by the model. These are due to the pump-induced increase in the electron temperature (which, in turn, broadens and shifts the LSPR on a ~100 fs time scale), followed by a recovery to equilibrium *via* the thermalization of the hot electron gas with the cooler lattice by electron–phonon coupling (~1 ps time scale), and the thermalization of the heated NP with the cooler surroundings (~100 ps time scale). We also observe coherent phonon oscillations in $\Delta\hat{\varepsilon}_R$ (see inset in Fig. 6a). These are due to the modulation of the plasma frequency by the breathing vibrational mode of the NP with an oscillation period proportional to the particle radius[21–23] from which we infer a diameter of 38 nm for the specific NP shown.

Besides these dynamics, we showed in ref. 14 that the phase of $\Delta\hat{\varepsilon}$ is sensitive to the spectral position of the LSPR relative to the probe optical frequency in the experiment. The phase is a ratiometric quantity independent of the signal strength, and therefore could be used for robust LSPR-based sensing applications. Owing to the red-shift of the LSPR occurring in a NP dimer (increasing with decreasing interparticle gap distance[20]), we would thus expect that this phase measured at 550 nm wavelength in our FWM experiment would gradually change, going from a single quasi-spherical NP to a plasmonically coupled dimer (in a gap-dependent manner). To address this expectation, we measured such phase on the group of NPs identified as single quasi-spherical or dimers in Fig. 4.

Since the experiment is performed in epi-geometry, we are detecting the complex quantity $\Delta\hat{\varepsilon}/\hat{\varepsilon}$ *via* the ratio between the FWM field $E_{FWM} \propto \Delta\hat{\varepsilon}E_2$ and the reflected probe field $E_{2r} \propto \hat{\varepsilon}E_2$ (as opposed to the transmitted probe field E_2 in ref. 14). The phase φ of $\Delta\hat{\varepsilon}/\hat{\varepsilon}$ calculated from $\Delta\hat{\varepsilon}$ at $\tau = 0.5$ ps for a spherical 40 nm NP (see Fig. 5 in ref. 14) is plotted in Fig. 6b together with the real and imaginary parts of $\hat{\varepsilon}$. From this calculation, we expect to observe $\varphi \sim \pi$ for a NP at resonance and $\varphi > \pi$ when probing at shorter wavelengths compared to the LSPR. As shown in Fig. 5, the LSPR of the coupled (longitudinal) plasmon mode in a dimer is significantly red shifted compared to the wavelength at 550 nm used in the FWM experiment. Hence, based on these qualitative considerations, we would expect to observe $\varphi > \pi$ in a plasmonically coupled dimer compared to $\varphi \sim \pi$ in a single spherical NP. Fig. 6c shows the measured phase of the ratio E_{FWM}/E_{2r} in a group of NPs from the plot in Fig. 4, as a function of the asymmetry ratio α_{PR}/α_{PG}. Measurements were performed using a circularly polarized pump and probe and a co-circularly polarized reference, rather than linear polarisation as in ref. 14, to enable the excitation of dimers without prior knowledge of their orientation. We observe that NPs with a large asymmetry parameter ratio $\alpha_{PR}/\alpha_{PG} > 2$, indicating dimers according to Fig. 4, tend to exhibit $\varphi > \pi$ compared to single quasi-spherical NPs with $\alpha_{PR}/\alpha_{PG} < 2$ and $\varphi \sim \pi$, in agreement with our qualitative expectation. A more in-depth understanding of the measured phase both experimentally and theoretically, including a careful calibration against instrument-related phase offsets in the set-up,[14] a confirmation that the measured particles are dimers *via* correlative TEM, and a more quantitative description of the expected dimer response *versus* interparticle distance and polarisation configuration is in progress and will be published in a future manuscript.

4 Conclusions

In conclusion, we have developed a rapid and sensitive method to quantitatively measure the optical extinction cross-section of individual nanoparticles using wide-field imaging. The method enables simultaneous acquisition of hundreds of nanoparticles for statistical analysis with a sensitivity limit down to 5 nm^2, corresponding to a single 5 nm gold NP, using acquisition times in the 1 min range. An experimental demonstration is shown on single nominally-spherical gold nanoparticles with diameters ranging from 10 nm to 100 nm. Importantly, the method can be implemented cost-effectively on any conventional wide-field microscope and is applicable to any nanoparticle. Furthermore, we have investigated the effect of the particle asphericity on the optical extinction cross-section measured with in-plane linearly polarised light as a function of the polarisation direction, and derived, from theory, a simple expression linking these measurements to the particle aspect ratio. We also fabricated gold NP dimers and showed that the optical extinction cross-section analysis can be used to distinguish plasmonically-coupled dimers from single quasi-spherical NPs and larger aggregates. Finally, we performed transient resonant FWM heterodyne interferometry on the same group of NPs characterized by the optical extinction cross-section analysis. We specifically investigated the phase of the ratio between the FWM field and reflected probe field, which is expected to be sensitive to the relative shift between the nanoparticle LSPR and the optical probe wavelength. A comparison between single NPs and dimers identified from the optical extinction analysis supported this expectation and highlighted the potential of these phase-resolved measurements for LSPR-based sensing applications.

Data access

Data access Information about the data created during this research, including how to access it, is available from Cardiff University data archive at http://dx.doi.org/10.17035/d.2015.100095.

Acknowledgements

This work was funded by the EPSRC UK Research Council under the grant agreement EP/L001470/1. P.B. acknowledges the EPSRC for her Leadership fellowship award EP/I005072/1 and the Royal Society for her Wolfson Research Merit Award. W.L. acknowledges support by a Leverhulme Royal Society Research Fellowship (Grant no. LT20085).

References

1 A. Crut, P. Maioli, N. D. Fatti and F. Vallée, *Chem. Soc. Rev.*, 2014, **43**, 3921–3956.
2 W. Barnes, A. Deraux and T. Ebbesen, *Nature*, 2003, **424**, 824–830.
3 K. Awazu, M. Fujimaki, C. Rockstuhl, J. Tominaga, H. Murakami, Y. Ohki, N. Yoshida and T. Watanabe, *J. Am. Chem. Soc.*, 2008, **130**, 1676–1680.
4 H. Atwater and A. Polman, *Nat. Mater.*, 2010, **9**, 205–213.
5 L. Cognet, S. Berciaud, D. Lasne and B. Lounis, *Anal. Chem.*, 2008, **80**, 2288–2294.
6 F. Masia, W. Langbein, P. Watson and P. Borri, *Opt. Lett.*, 2009, **34**, 1816–1818.
7 K. Kneipp, H. Kneipp, I. Itzkan, R. Dasari and M. Feld, *J. Phys.: Condens. Matter*, 2002, **14**, R597–R624.
8 A. McFarland and R. Van Duyne, *Nano Lett.*, 2003, **3**, 1057–1062.
9 C. Sönnichsen, B. M. Reinhard, J. Liphardt and A. P. Alivisatos, *Nat. Biotechnol.*, 2005, **23**, 741–745.
10 C. Ayala-Orozco, C. Urban, M. W. Knight, A. S. Urban, O. Neumann, S. W. Bishnoi, S. Mukherjee, A. M. Goodman, H. Charron, T. Mitchell, M. Shea, R. Roy, S. Nanda, R. Schiff, N. J. Halas and A. Joshi, *ACS Nano*, 2014, **8**, 6372–6381.
11 A. Arbouet, D. Christofilos, N. D. Fatti, F. Vallée, J. R. Huntzinger, L. Arnaud, P. Billaud and M. Broyer, *Phys. Rev. Lett.*, 2004, **93**, 127401.
12 D. Boyer, P. Tamarat, A. Maali, B. Lounis and M. Orrit, *Science*, 2002, **297**, 1160–1163.
13 L. M. Payne, W. Langbein and P. Borri, *Appl. Phys. Lett.*, 2013, **102**, 131107.

14 F. Masia, W. Langbein and P. Borri, *Phys. Rev. B: Condens. Matter Mater. Phys.*, 2012, **85**, 235403.

15 M. van Dijk, A. Tchebotareva, M. Orrit, M. Lippitz, S. Berciaud, D. Lasne, L. Cognet and B. Lounis, *Phys. Chem. Chem. Phys.*, 2006, **8**, 3486–3495.

16 O. L. Muskens, P. Billaud, M. Broyer, N. D. Fatti and F. Vallée, *Phys. Rev. B: Condens. Matter Mater. Phys.*, 2008, **78**, 205410.

17 A. Tcherniak, J. Ha, S. Dominguez-Medina, L. Slaughter and S. Link, *Nano Lett.*, 2010, **10**, 1398–1404.

18 I. Pope, L. Payne, G. Zoriniants, E. Thomas, O. Williams, P. Watson, W. Langbein and P. Borri, *Nat. Nanotechnol.*, 2014, **9**, 940–946.

19 C. F. Bohren and D. R. Huffman, *Absorption and Scattering of Light by Small Particles*, Wiley-VCH, 1998.

20 P. K. Jain, W. Huang and M. A. El-Sayed, *Nano Lett.*, 2007, **7**, 2080–2088.

21 C. Voisin, N. D. Fatti, D. Christofilos and F. Vallee, *J. Phys. Chem. B*, 2001, **105**, 2264–2280.

22 M. A. van Dijk, M. Lippitz and M. Orrit, *Phys. Rev. Lett.*, 2005, **95**, 267406.

23 G. V. Hartland, *Chem. Rev.*, 2011, **111**, 3858–3887.

Faraday Discussions

ROYAL SOCIETY OF CHEMISTRY

PAPER

Coupling single quantum dots to plasmonic nanocones: optical properties

Alfred J. Meixner,*[a] Regina Jäger,*[a] Sebastian Jäger,[a] Annika Bräuer,[b] Kerstin Scherzinger,[a] Julia Fulmes,[b] Sven zur Oven Krockhaus,[a] Dominik A. Gollmer,[b] Dieter P. Kern[b] and Monika Fleischer[b]

Received 12th May 2015, Accepted 15th July 2015

DOI: 10.1039/c5fd00074b

Coupling a single quantum emitter, such as a fluorescent molecule or a quantum dot (QD), to a plasmonic nanostructure is an important issue in nano-optics and nano-spectroscopy, relevant for a wide range of applications, including tip-enhanced near-field optical microscopy, plasmon enhanced molecular sensing and spectroscopy, and nanophotonic amplifiers or nanolasers, to mention only a few. While the field enhancement of a sharp nanoantenna increasing the excitation rate of a very closely positioned single molecule or QD has been well investigated, the detailed physical mechanisms involved in the emission of a photon from such a system are, by far, less investigated. In one of our ongoing research projects, we try to address these issues by constructing and spectroscopically analysing geometrically simple hybrid heterostructures consisting of sharp gold cones with single quantum dots attached to the very tip apex. An important goal of this work is to tune the longitudinal plasmon resonance by adjusting the cones' geometry to the emission maximum of the core–shell CdSe/ZnS QDs at nominally 650 nm. Luminescence spectra of the bare cones, pure QDs and hybrid systems were distinguished successfully. In the next steps we will further investigate, experimentally and theoretically, the optical properties of the coupled systems in more detail, such as the fluorescence spectra, blinking statistics, and the current results on the fluorescence lifetimes, and compare them with uncoupled QDs to obtain a clearer picture of the radiative and non-radiative processes.

I. Introduction

Various nanoantennas have been designed to study their influence on the emission properties of single emitters, such as scanning near-field tips,[1–3] single gold nanoparticles,[4,5] discs,[6] arrays like optical Yagi Uda configurations,[7] and

[a]Center for Light-Matter Interaction, Sensors & Analytics (LISA[+]), Institute of Physical and Theoretical Chemistry, University of Tübingen, Auf der Morgenstelle 18, 72076 Tübingen, Germany. E-mail: alfred.meixner@uni-tuebingen.de; regina.jaeger@uni-tuebingen.de

[b]Center for Light-Matter Interaction, Sensors & Analytics (LISA[+]), Institute for Applied Physics, University of Tübingen, Auf der Morgenstelle 10, 72076 Tübingen, Germany

bowtie antennas.[8] The positioning of single emitters with high accuracy is a key issue for photonic crystal cavities,[9] waveguides for single-photon generation,[10] quantum dot lasers[11] or nonlinear optical devices.[12] The interaction of quantum emitters in the high near-field region of such metal structures results in energy transfer between the two objects. It is well known that the luminescence and Raman signals of molecules can be enhanced on metal nanostructures, but on the other hand, the radiation of a quantum emitter can also be quenched.[4,5] Tuned to the right wavelength, a plasmonic nanostructure acts as an antenna, leading to an enhancement of the incident exciting radiation, and it can also amplify the reemission from the quantum system.[13,14] In this context, the directionality of coupled systems has also been investigated for more complex nanoantennas.[7]

Nanoscale placement of *e.g.* quantum dots (QDs) has been demonstrated in the vicinity of metamaterial surfaces[15] or metallic nanostructures.[16] The placement of emitters within the high near-field regions of plasmonic structures requires manipulation with a lateral accuracy of < 10 nm *e.g.* by optical tweezers[17] or a scanning probe microscope.[18–20] Recently, we developed a straight-forward method, which fulfills the coupling conditions, while all hybrid structures are manufactured at the same time.[21]

The unique optical properties of QDs, such as broad absorption and narrow emission bands, and bright and stable luminescence, make them promising for various applications.[22] QDs are very interesting for applications in nanophotonics and quantum optics, in which they serve as single photon sources.[23] They also play an important role as tags and markers in biological sensing.[24] Therefore, placing QDs in the hot spot of a gold nanocone tip to investigate the coupling mechanisms between the antenna and the emitter may allow one to understand the basic effects concerning the interaction between a single quantum system and a plasmonic antenna, and moreover, to use gold nanocones as sensor arrays, *e.g.* for biochemical analysis.[25]

A. Luminescence and plasmons of gold nanostructures

Luminescence of gold was first reported by Mooradian in 1969, although with a low quantum efficiency of about 10^{-10}.[26] According to the band structure, the luminescence can be explained by electron–hole-pair recombination involving the 5d and 6sp bands of gold.[27] For rough films and gold nanoparticles, the efficiency increases to 10^{-4}.[28–30] This behavior has been correlated with the particle size, giving evidence for a relationship with the coherent electron oscillations in metals, the plasmonic modes.[31] Luminescence can originate directly from the radiative decay of a plasmon oscillation[32] or an electron–hole-pair can recombine and release its energy *via* the plasmon oscillation.[33]

Surface-plasmon polaritons are electromagnetic waves in the visible or near infrared region, involving both a coherent oscillation of the conducting electrons, leading to an oscillating charge separation, and a respective electromagnetic wave at the dielectric interface. Both must fulfill the phase-matching condition dictated by the dielectric functions of the respective materials.[34,35] When the particles are smaller than the wavelength, surface-plasmon polaritons cannot propagate any longer and the phase matching condition does not apply. Then, the conducting electron oscillations are defined by the size and shape of the structure, and the particle can act as a resonator forming a standing wave, creating localized surface plasmons or particle plasmons, as they involve the whole nanoparticle.

In the case of a gold cone, two types of plasmonic modes can be distinguished: (1) in the transversal direction at the cone base, and (2) in the longitudinal direction, concentrating the electrical field at the tip due to the lightning rod effect as in a SNOM (scanning near-field optical microscopy) tip.[36] Two different conditions are illustrated in Fig. 1. Depending on the polarization of the exciting electrical field, either mode can be induced. Note that only one possibility is sketched in Fig. 1 for the transversal plasmonic modes at the cone base.

B. Coupling effects between single emitters and metallic structures

Placing a single molecule or QD in the vicinity of a metallic structure can alter the emission behavior of the emitter. The influence on the radiative and non-radiative decay rates of a QD by a nanoantenna observed *via* fluorescence intermittency has been a topic of intense study[1,2,37] and several mechanisms are discussed in the literature.[38,39] For describing the interplay between the enhanced excitation due to coupling to the enhanced density of radiative local optical states (LDOS) and quenching due to coupling to the local density of evanescent states, we follow the formalism outlined in ref. 40. We describe the QD's absorption and emission by a four-level system, as shown in Fig. 2.

The excitation rate between the electronic ground-state and excited-state, X_0' and X_1', respectively, is according to Fermi's golden rule proportional to the square of the electric field $\vec{E}(\vec{r})$ amplitude at the position $\vec{r}$ of the QD:

$$k_{\text{exc}}(\vec{r}, \omega_0) = \frac{2\pi}{\hbar}\left|\vec{\mu}\vec{E}(\vec{r})\right|^2 \rho(\vec{r}, \omega_0),$$

with the local density of modes $\rho(\vec{r}, \omega_0)$ of the incident photon with energy $\omega_0\hbar$ matching the gap between X_0' and X_1', and the transition dipole moment $\vec{\mu}$ connecting the electronic ground state and the excited state. Similarly, the

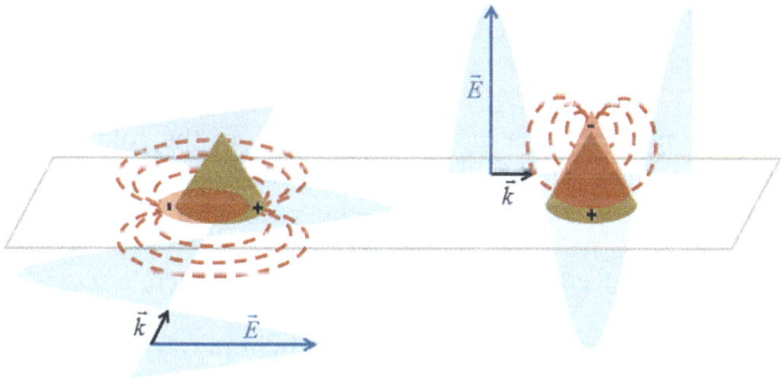

Fig. 1 Scheme of the gold nanocones excited by the electric field component of an incident electromagnetic wave with different propagation directions. In the left cone, a transversal plasmon mode is excited in the base, while a longitudinal plasmon mode in the right cone is induced, leading to a concentration of the field at the tip of the cone. The directional re-radiation is depicted in a dipole-like fashion.

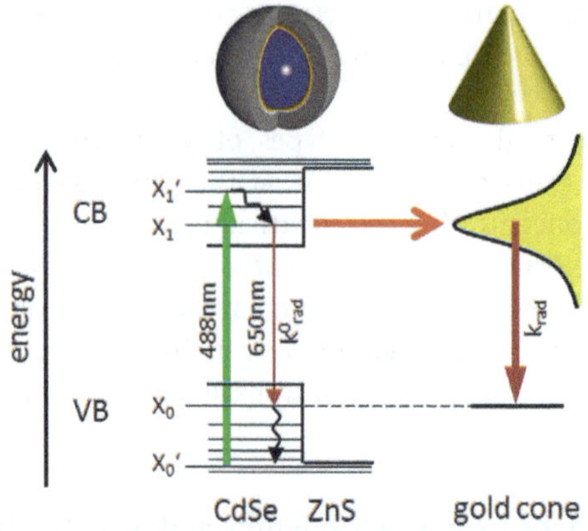

Fig. 2 Schematic energy level illustration of a CdSe/ZnS-QD coupled to a gold nanocone. First, an electron is excited from the ground state X_0' of the valence band (VB) with 488 nm radiation to a short lived level X_1' in the conduction band (CB), from where it relaxes thermally to the state X_1. A QD in free space can either relax radiatively to the ground state X_0 by emitting a photon of about 650 nm or by dissipating its energy to the crystal. If the QD is in close proximity to the tip, its optical near-field couples efficiently to the local optical mode density of a gold cone, which can transfer its energy either to a radiative plasmon mode, which can emit a photon with the rate constant k^0_{rad}, or by coupling to evanescent modes, dissipating the energy *via* ohmic losses in the metal. The non-radiative relaxation paths are not shown in this scheme. The illustration above shows the shell system of the QD, consisting of the CdSe core (blue), the ZnS shell (yellow) and the organic surfactant shell (gray), and a gold cone.

frequency dependent emission rate of a QD between its electronically excited state X_1 and its electronic ground state X_0 is:

$$k_{rad}(\vec{r}, \omega_{em}) = \frac{2\pi}{\hbar}\left|W_{fi}(\vec{r}, \omega_{em})\right|^2 \rho(\vec{r}, \omega_{em}),$$

where $W_{fi}(\vec{r}, \omega_{em})$ is the transition matrix element, *i.e.* the magnitude of the QD's emission dipole moment. Assuming that the non-radiative relaxation processes from X_1' to X_1 and X_0 to X_0' are fast, and X_1' and X_0 are only occupied for infinitely short times, the relationship $X_1 + X_0' = 1$ holds. Hence, the dynamics of the radiative excited state in free space can be described as:

$$\dot{X} = (1 - X_1)\cdot k^0_{exc} - X_1\left(k^0_{nr} + k^0_{rad}\right),$$

where k^0_{nr} is the non-radiative rate in the absence of the gold cone. Thus, the enhanced and localized electric field close to the cone apex can increase the excitation rate of the QD linked to the cone's tip with respect to a flat glass substrate by the enhancement factor $F_{exc} = \frac{\rho(\vec{r}, \omega_0)}{\rho^0(\omega_0)}$, where $\rho^0(\omega_0)$ is the density of the optical modes populated by the incident light in free space and $\rho(\vec{r}, \omega_0)$ is the density of the radiative modes at the tip of the cone. Similarly, the emission of a QD attached to the tip of a cone can be enhanced by the factor $F_{rad} = \frac{\rho(\vec{r}, \omega_{em})}{\rho^0(\omega_{em})}$, where $\rho(\vec{r}, \omega_{em})$ is the radiative mode density in close proximity to the cone and

$\hbar\omega_{em}$ is the energy of a photon emitted from the QD. In the near-field of the cone, evanescent parts $\rho_{ev}(\omega_{em})$ of the mode also exist, which decay within a fraction of the wavelength. These modes are also populated by photons emitted from the QD, leading to energy dissipation in the near-field, and may induce additional radiative losses[41] known as quenching.[38] Hence, similar to the enhancement factor, one can also define the quenching factor $F_q = \frac{\rho_{ev}(\vec{r}, \omega_{em})}{\rho^0(\omega_{em})}$.

With the modified rate constants and stationary conditions, the total photoluminescence (PL) emission rate of the hybrid system can therefore be expressed as:

$$\Gamma_{rad} = X_1 \cdot F_{rad} \cdot k_{rad}^0 = \frac{F_{rad} \cdot k_{rad}^0 \cdot F_{exc} \cdot k_{exc}^0}{k_{exc}^0 \cdot F_{exc} + X_1 \cdot k_{rad}^0 \cdot \left(F_{exc} + F_q + \Phi_0{}^{-1} - 1\right)}$$

where $\Phi_0 = \frac{k_{rad}^0}{k_{rad}^0 + k_{nr}^0}$ is the free space quantum yield.

II. Optical characterization of pure gold nanocones

In this work we investigate arrays of gold nanocones, which have been fabricated on glass substrates with a thin indium tin oxide (ITO) layer by a procedure similar to those in ref. 42 or ref. 43 and 44. A detailed description of our fabrication process is given in ref. 21.

For studying the emission enhancement of a QD placed on a cone tip, an important issue is to tailor the cone's geometry such that the longitudinal plasmon resonance overlaps with the QD emission which is, according to the manufacturer, around 650 nm.[45,46] For this purpose we have fabricated two kinds of cones: smaller ones with heights of about 111 ± 10 nm and base diameters of 162 ± 10 nm (due to minor variations in the fabrication process), and larger ones with heights of 142 ± 10 nm and base diameters of 164 ± 10 nm. Spectroscopic measurements of single cones were performed with a home-built inverted confocal microscope equipped with a feed-back controlled sample-scanning stage (Physik Instrumente, Karlsruhe, Germany) and an NA 1.46 immersion objective lens (Zeiss, Jena, Germany) for optical excitation and signal collection. Hence, after recording an overview image, we could reproducibly address individual cones or QDs for various spectroscopic measurements. For continuous-wave excitation, either an Ar^+Kr^+ laser at $\lambda = 488$ nm or a HeNe laser at $\lambda = 632.8$ nm was used, focusing the laser spot with a power of 20 μW onto the sample. For recording raster-images, the position dependent PL intensity signal generated in the focus was recorded by a single-photon detector (APD) (Perkin Elmer, SPCM-AQR-14), while the laser excitation is blocked by a long pass filter. Spectra were recorded with an Acton 300 pro spectrometer equipped with a nitrogen cooled SPEC-10 CCD-detector (Princeton Instruments, Trenton, USA).

We prefer azimuthally or radially polarized excitation beams, known also as cylindrical vector beams (CVBs)[47,48] over *e.g.* linearly polarized Gaussian beams, since they have cylindrical symmetry and give distinct PL excitation patterns when a single molecule or a sub-wavelength nanoparticle is raster scanned through the focal volume, which allow us to visualize directly the orientation of the transition dipole moment or the induced dipole moment. As shown in Fig. 3a (top), an

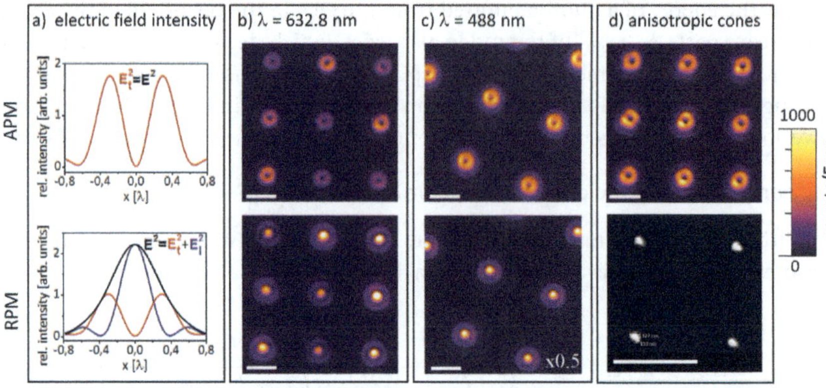

Fig. 3 Confocal PL images obtained by raster-scanning gold nanocones (111 nm height, 162 nm base diameter) through the focus of an APM (top) or RPM (bottom). (a) shows the field intensity distribution in the focal volume of an APM (upper panel) with an exclusively transversal electric field component, E_t. In the lower panel, the focal field of an RPM is shown with a longitudinal field component in the center, E_l, surrounded by a radially polarized transversal field, E_t. (b) and (c) show typical PL patterns acquired with two different wavelengths, 632.8 nm and 488 nm, respectively. As the nanocones are smaller than the focus, the sizes of their patterns reflect the electric field distribution in the respective focus. (d) illustrates inhomogeneous PL patterns obtained by an APM for nanocones with a non-circular base, as can be seen in the SEM image below. All white scale bars represent 1 μm.

azimuthally polarized laser mode (APM) has a doughnut-shaped field distribution in the focal volume with the electric field component being present exclusively in the focal plane.

Raster-scanning a cone with a circular base through such an azimuthal field distribution results in a ring-shaped PL pattern, as shown in Fig. 3b and c (top). A radially polarized laser mode (RPM) forms a focal field with a dominant longitudinal electric field component in the center surrounded by a radially polarized ring. Hence, raster-scanning a sub-wavelength cone through the focal volume results in a bright PL spot in the center surrounded by a less intense ring, as shown in Fig. 3b and c (bottom).

When the nanostructure has a non-circular base, the ring patterns are inhomogeneous, as presented in Fig. 3d. Here, gold nanocones scanned through a focused APM are shown, and they exhibit deviations from the usual uniform patterns. The scanning electron microscope (SEM) image below was acquired by viewing this particular sample from the top to see the shapes of the bases of the structures, which are clearly distorted and look triangular. After acquisition of the SEM images however, the samples are generally contaminated with carbon, showing an intense surface-enhanced carbon Raman spectrum,[49,50] thus obscuring the PL spectra of the single QDs fixed on the cone tips when the hybrid structure is complete. Therefore, the possibility to visualize deviations from the cylindrical shape of the gold nanocones by CVBs without contaminating them is a clear advantage over SEM imaging. The optical characterization *via* CVBs is absolutely non-destructive as long as low laser powers are used. A comparison of the PL intensities of the patterns obtained with a similar excitation power at 632.8 nm and 488 nm shows that the latter results in a higher maximum intensity. For

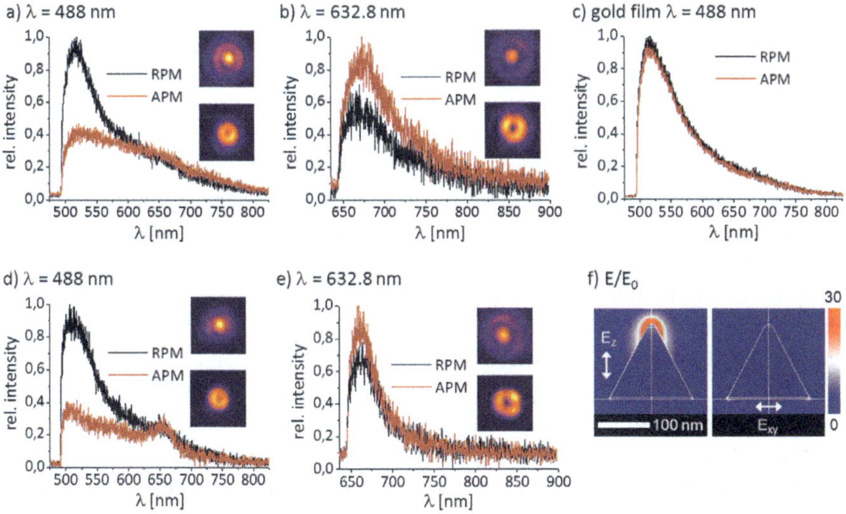

Fig. 4 (a) and (b) PL spectra of a smaller cone (115 nm height, 160 nm base diameter) excited with RPM and APM with 488 nm or 632.8 nm wavelength radiation, respectively, normalized to the respective intensity maxima. The insets show the corresponding PL excitation patterns. (c) PL spectra recorded with RPM and APM of a flat gold film. (d) and (e) PL spectra of a taller cone (142 nm height, 160 nm base diameter). (f) Numerical simulations of the longitudinal and transversal plasmon modes performed with COMSOL Multiphysics, showing the location of a bright intensity maximum at the tip for longitudinal excitation and a significantly weaker intensity at the base for transversal excitation. The cone geometry corresponds to the cone presented in (d) and (e) excited by 632.8 nm radiation.

the cones presented in Fig. 3b, the bright spot in the center excited by the RPM shows twice the intensity compared to the ring from exciting the base of the cone by the APM. These intensity differences between APM and RPM can be explained by examining the spectra in Fig. 4.

Depending on the aspect ratios and tip radii of the gold nanocones, the intensity ratio between the PL excited by the focused APM and RPM varies. The measurements presented in Fig. 4a and b were performed with the same gold nanocone (111 nm height) excited with the same acquisition time (10 s) and a similar laser power (20 μW), but with different wavelengths and modes. The insets show the corresponding PL excitation patterns. For 488 nm excitation with the RPM, the PL spectrum of this nanocone shows a dominant band at about 520 nm, in agreement with the spectrum of a flat gold film, and a decreasing tail extending from 600 nm to the near infrared region with a barely visible shoulder at 650 nm. Exciting with the APM at 488 nm, the band at 520 nm drops to 40% intensity while the red wavelength tail displays a similar intensity for both polarization modes.

Scanning the same gold nanocone through the focus of an APM and RPM at 632.8 nm (*cf.* Fig. 4b), we notice for both cases a band at about 660 nm. The spectra of a continuous gold film resulting from excitation by 488 nm radiation reveal no significant difference for excitation by APM or RPM (*cf.* Fig. 4c). The dominant peak at 520 nm can be assigned to the electron–hole-pair recombination of gold, resulting in a PL emission as described in Section I (A). As for metallic nanostructures where the shape alters the plasmon resonance in contrast to bulk samples, the spectra observed for the gold nanocones differ. The highest peak at 520 nm can be observed by excitation with the RPM, but may shift for different cone heights and aspect ratios, as studied in ref. 36. Upon illuminating the gold nanocone with 488 nm radiation, the strong longitudinal electric field in the focus of an RPM creates an enhanced local field at the cone tip, hence most electron–hole pairs are created right there. The lifetime of electron–hole pairs in bulk gold is only on the order of 10 fs,[32] therefore the quantum efficiency for radiative decay is extremely low. However, it is well known that the PL quantum efficiency excited with 632.8 nm radiation on rough Au surfaces and sharp tips is orders of magnitude higher due to the restricted geometry and the longer plasmon lifetime and hence, radiative recombination prevails at the very tip.[51,52] In Fig. 4d and e, respective PL spectra are shown for a taller cone (142 nm height, sharp tip). While the spectra have a similar qualitative shape as in Fig. 4a and b, a clear band appears at 660 nm. This band also appears for 632.8 nm excitation and is definitively missing in the PL spectra of the flat gold film. Therefore, this band may be attributed to the PL decay *via* the longitudinal plasmon mode of the cone.[53,54]

III. Hybrid structures

Core–shell CdSe/ZnS QDs with an emission maximum at nominally 650 nm (purchased from PlasmaChem) were bound to the tips of the cones. The QDs are capped with trioctylphosphine oxide (TOPO) as a stabilizing agent and hexadecylamine (HDA) ligands to link the QDs to the 3-mercaptopropionic acid layer on the cone surface. Binding by thiol chemistry immobilizes the QDs at the cone tips and provides, together with the QD ligands, a spacer of up to 3 nm between the QDs and gold surface. The chemical spacer in combination with the 0.7 nm thick ZnS shell is supposed to provide a sufficient distance to suppress strong PL quenching.[4] A major challenge is to prevent QDs from binding to the sides of the cone. The selective positioning of QDs at the tips requires temporary embedding of the cones in a resist, which is subsequently removed again. A thin residual resist layer may still be observed on the sample surface. Placing single QDs on the very tips of the cones is by far not trivial and is described in detail in ref. 21.

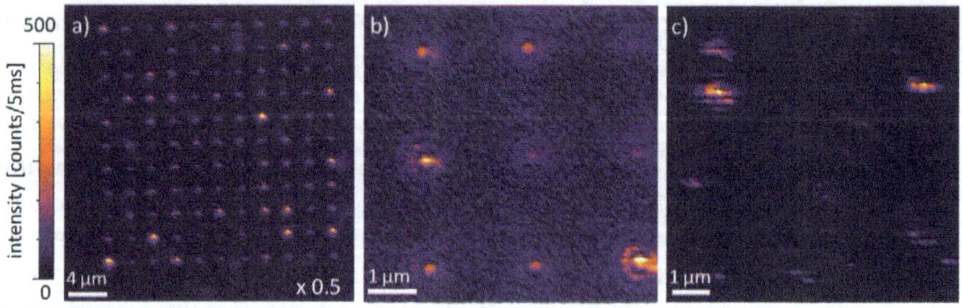

Fig. 5 Confocal PL excitation images recorded by raster scanning the sample through the focus of an RPM at 488 nm under identical experimental conditions. (a) and (b) show typical patterns of an array of gold cones with QDs at their tips, while (c) exhibits pure QDs on glass, some are bright and show intensity fluctuations typically observed for single quantum systems.

A. Imaging of hybrid structures

The hybrid structures are arranged in rectangular fields and have been scanned directly after the preparation through the diffraction limited focus of an RPM at 488 nm. According to the optical investigations of all arrays, the number of QDs successfully linked to the gold nanocones can be estimated to be about 70%. The arrays of the hybrid structure exhibit different PL intensities, as presented in Fig. 5a. We can clearly distinguish dim spots and bright spots, the latter ones originating from cones with QDs on the tips.

A close-up scan (Fig. 5b) shows seven cones with a weak PL intensity and two structures with bright signals. The image (6 × 6 μm, 100 × 100 pixels) has been recorded with an excitation laser power of 1 μW and an acquisition time of 5 ms per pixel. Usually, for these experimental conditions, the pure gold luminescence from the nanocones is too weak to be detected, but due to the luminescence of the remaining thin layer of the resist, it is also possible to observe the cones without linked QDs. Two structures in Fig. 5b are significantly brighter and show fluorescence intermittency, giving evidence for the presence of single or a few QDs attached to the tips of the cones. Single QDs on a glass substrate (Fig. 5c) typically show significant blinking and an isotropic excitation behavior,[55] while most of them can barely be observed. Thus, for the majority of QDs, only blurred patterns are detected. In the presented image (Fig. 5c), only three excitation patterns can clearly be identified, showing bright and dark lines typical for line by line scanning of single QDs through a laser focus. In comparison, the bright PL patterns of the hybrid structures exhibit less fluorescence intermittency than single QDs on a pure glass substrate (Fig. 5c). This

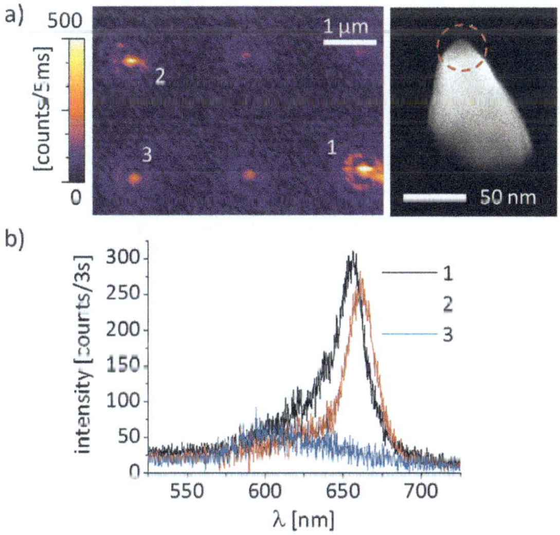

Fig. 6 (a) Confocal image of the hybrid structure (left) excited in the focus of a 488 nm RPM. The SEM image (right) shows a typical hybrid structure. Due to the weak contrast, the QD appearing as a small bump at the tip is marked by a dashed circle. The corresponding spectra of individual cones labeled 1, 2 and 3 are presented in (b), where the difference between cones with CdSe/ZnS QDs (spectra 1 and 2) and the cone without a QD (spectrum 3) can clearly be recognized.

difference indicates a change in the radiative and non-radiative decay rates of the QDs linked to a gold nanocone.

B. Spectroscopic characterization

As typical examples, the background corrected spectra from two bright cones displayed in Fig. 6a (labeled 1 and 2) are shown in Fig. 6b. The patterns in Fig. 6a exhibit clear fluorescence intermittency, while a weaker PL pattern (labeled 3) shows no blinking behavior. Cones 1 and 2 feature sharp Lorentzian PL peaks at the wavelength of the QD emission (~660 nm), indicating that they originate from the luminescence of single or at most a few QDs. Cone 3 features a broad peak at shorter wavelengths with a spectral maximum at about 610 nm, which also appears for cones where no QDs have been fixed and is typical for a thin remaining layer of the resist on the cone. Note that this feature also appears in the spectra of cone 1 and cone 2 as dim short wavelength wings.

A similar very weak PL spectrum of the remaining resist can also be seen on the substrate in between the cones. Indeed, one challenging issue in preparing gold cones with single QDs is to carefully and completely remove traces of the remaining resist since its luminescence may obscure the QD emission. However, such traces are easy to recognize since their spectra are broader and blue shifted with respect to those of QDs and, most importantly, they do not blink. The SEM image of a typical hybrid structure is presented in Fig. 6a (right). To see the QDs linked to the tips of the gold nanocones in the SEM images, the sample has been tilted. Due to the weak material contrast in SEM images, it is difficult to quantify

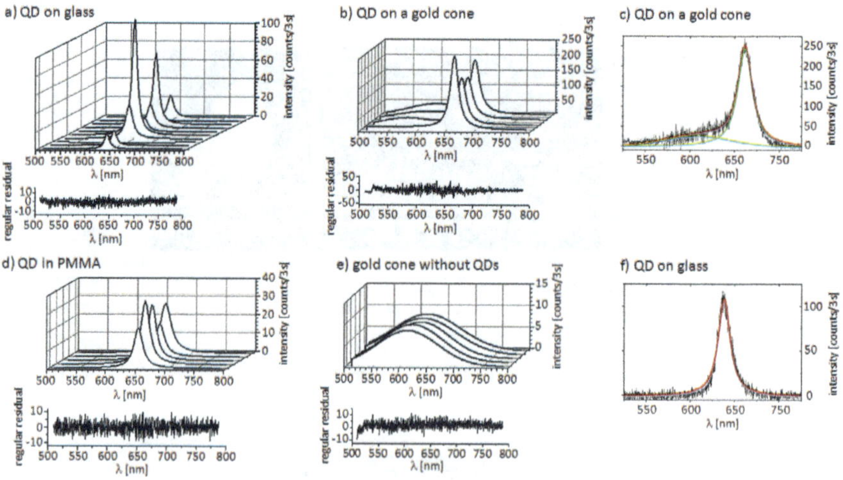

Fig. 7 Series of spectra of (a) one single QD on a glass cover slide, (b) a single QD linked to the tip of a gold nanocone, (d) a single QD embedded in a PMMA matrix and (e) a gold cone without linked QDs where the PL originates from some residual resist. All spectra were recorded with the same experimental conditions (excitation power of 1 μW at 488 nm, 3 seconds integration time) and fitted to Lorentzian line profiles. For each spectral series, the residuals are shown for one spectrum and they demonstrate the high quality of the fits. A typical fit of a Lorentzian to the spectrum of a single QD on a gold cone is shown in (c) with a Gaussian accounting for the background of the resist and for the spectrum of a single QD on glass, as shown in (f).

the number of QDs on the cones directly, and the images exhibit only a bump with weak contrast, indicating a QD. One way to estimate the number of QDs coupled to a nanocone is by comparing the spectra of different hybrid structures, such as 1 and 2 in Fig. 6. A typical fit of a cone with most likely only one QD attached is shown in Fig. 7c, which is the spectrum for hybrid 2 in Fig. 6b. It can be very well described by a single Lorentzian with a full-width-at-half-maximum (FWHM) of 19.8 nm, which is in good agreement with the average value we find for single QDs on glass (*cf.* Fig. 7f). The spectrum for hybrid structure 1 in Fig. 6b shows significant broadening and can only be fitted nicely with two slightly separated, superimposed Lorentzians (not shown), indicating that it originates from more than one QD. For both cases 1 and 2, an additional Gaussian centered at ~610 nm accounts for the short wavelength tail of the spectra of the cones without QDs, as indicated by spectrum 3 in Fig. 6b.

In Fig. 7a, a typical spectral sequence of one single QD deposited on a glass cover slide is shown, together with a spectral sequence of one single QD attached to the tip of a gold cone in Fig. 7b. For comparison, a spectral sequence of one single QD embedded in a thin PMMA film is shown in Fig. 7d. As a reference, a sequence from a gold cone without QDs showing only the background of the resist is displayed in Fig. 7e. All spectra were obtained by exciting with the focus of an RPM at 488 nm and an excitation power of 1 μW with an integration time of 3 seconds. Under these weak excitation conditions, the luminescence of bare gold cones is below the detection limit (see also Fig. 4). For the QD on the gold cone, we observe the highest average spectral intensity of 175 ± 47 counts and an average FWHM of the spectra of 18.84 ± 0.96 nm. The spectra are centered at 650 ± 10 nm, *i.e.* in resonance with the gold cones. For the QD deposited on the glass surface, the spectra also display Lorentzian line profiles. However, we observe large fluctuations in the intensity maxima from 10 counts to almost 120 counts, with an average of 38 ± 38 counts within one spectral series of the same QD. Here the spectral intensity maximum lies at 638 ± 10 nm and the FWHM is 13.91 ± 2.47 nm. We have also recorded spectra of QDs embedded in a PMMA matrix to protect them from air, as shown in Fig. 7d, and we observe an average intensity of 27 ± 7 counts and a FWHM of 18.81 ± 2.11 nm. Here we see less dramatic intensity fluctuations than for the example on the glass slide without a protective layer. To confirm this different spectral behavior, we have measured and analyzed the spectra for 51 QDs attached to the tips of gold cones that indicate single QD behavior and 78 QDs deposited on glass cover slides.

The statistical analysis is shown in Fig. 8. On glass, the PL-maximum of the QDs is symmetrically distributed around 639 ± 13 nm and the spectra have a FWHM of 19 ± 6 nm. For the QDs attached to the gold cones, the PL is red-shifted by about 10 nm, distributed around 648 ± 11 nm, and the FWHM is slightly larger, *i.e.* 21 ± 7 nm. Furthermore, the average spectrally integrated intensity of the QDs on the gold cones is 32 ± 15 kcounts/3 s and for the bare QDs on glass cover slides we find 7 ± 6 kcounts/3 s. Comparing the two distributions in Fig. 8b, we see that the QDs on gold cones can be up to an order of magnitude brighter than the QDs on glass. This lets us conclude that the gold cones enhance the PL of the QDs, which agrees well with a similar PL enhancement found in the ensemble of CdSe/ZnS-QDs on rough gold films.[56,57]

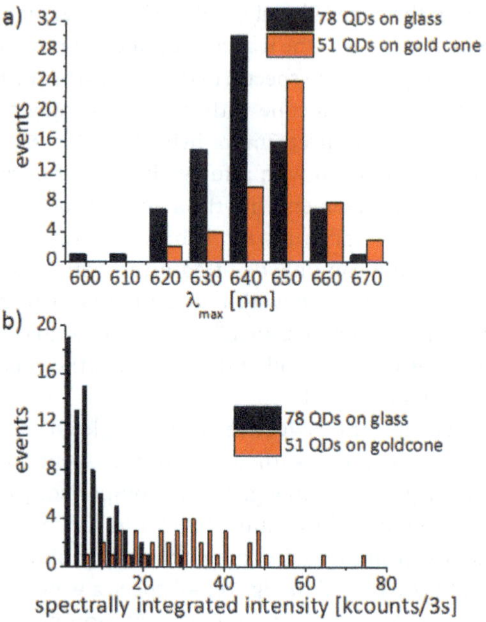

Fig. 8 (a) Spectral distribution of the intensity maxima and (b) the spectrally integrated intensity for single CdSe/ZnS-QDs dispersed on a glass cover slide (black) or attached to the tips of individual gold nanocones (red). The statistics are based on 51 QDs on gold cones and 78 QDs on glass cover slides, which all show typical single-molecule spectral behaviour, such as narrow PL-spectra fitted by Lorentzian line profiles and PL-intermittency as shown in Fig. 6 and 7. Wider spectra indicating that they may originate from more than one QD are not considered in the distributions.

C. Time-correlated single photon counting measurements

Intensity trajectories showing the emission intermittency of the spectrally integrated PL signal are often used in single-molecule/QD spectroscopy to prove the presence of a single quantum system. However, the blinking statistics of QDs can be very rich, involving multi-exciton decay[58–60] and charge trapping,[61,62] and are still the subject of an on-going scientific debate. The time trace shown in Fig. 9 is typical for a single QD attached to the tip of a gold cone and illustrates the blinking behaviour, which seems to persist in the hybrid systems we have investigated so far. Intensity trajectories were recorded with an APD (Perkin Elmer, SPCM-AQR-14) followed by single photon counting electronics. Here we show a time interval of 18 seconds with a time bin lasting 0.5 ms. For recording the images and spectra, the QDs were excited in the diffraction limited focus of an RPM with a wavelength of 488 nm and power of 1 μW. The intensity trajectory together with a zoom-in of one second show intensity jumps between a maximum level at about 160 counts/0.5 ms and a minimum intensity level at about 5 counts/0.5 ms, which is the background signal. These levels occur as maxima in the distributions plotted on the right of the respective time trace. An intermediate intensity level at 80 counts/0.5 ms can also be observed for some time periods. The longest dark intervals last up to about one second, while the shortest ones that can be resolved here are on the order of 1 ms. Even faster blinking behavior

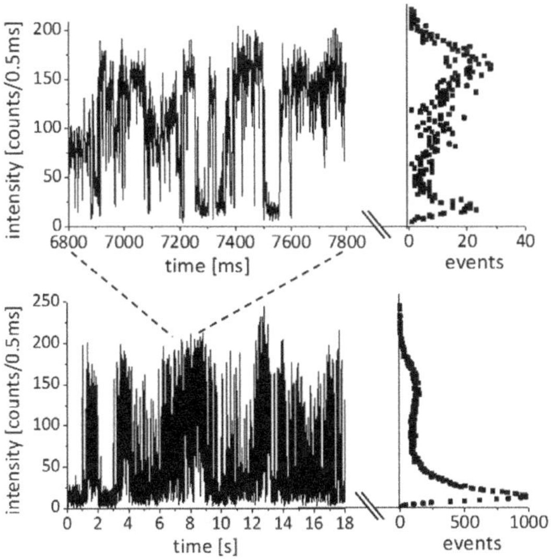

Fig. 9 Intensity trajectory of a single CdSe/ZnS-QD bound to the tip of a gold nanocone with a zoom-in of one second. The histograms on the right show how often the PL signal reaches a certain intensity value.

may be hidden in the time traces due to the time binning of 0.5 ms. The observed on–off blinking behavior is a strong argument that only one QD is active in this hybrid system.

In some cases, not shown here, we also observed repeated jumping between the background and two or even three intensity levels. This behavior could either be explained by two or three closely spaced QDs per tip, or by one single QD that can emit from several levels with different intrinsic emission quantum efficiencies $\Phi(t)$.[63]

To further investigate the influence of plasmon modes on the excited state of single QDs, we also performed fluorescence lifetime measurements. A pulsed supercontinuum laser (SuperK Extreme EXB-4, NKT Photonics, Denmark) operating at a repetition rate of 20 MHz was used as an excitation source. Using an acousto-optical tunable filter (AOTF, SuperK Select, NKT Photonics, Denmark), a laser wavelength of 488 nm was chosen in combination with a time-correlated single photon counting system (Picoquant, Berlin, Germany) with a spectrally integrating detector (PDM series, Micro Photon Devices, Italy) providing a nominal timing resolution <50 ps. The lifetime histograms were integrated until a reasonable intensity for the following fitting procedures (at least 10^3 counts for the peak channel) was collected. The raw data were fitted with a bi-exponential decay function convoluted with the instrument response function (IRF) and the intensity-weighted average fluorescence lifetime values were calculated according to:

$$\langle \tau \rangle = \frac{\sum_i A_i \cdot {\tau_i}^2}{\sum_i A_i \cdot \tau_i}.$$

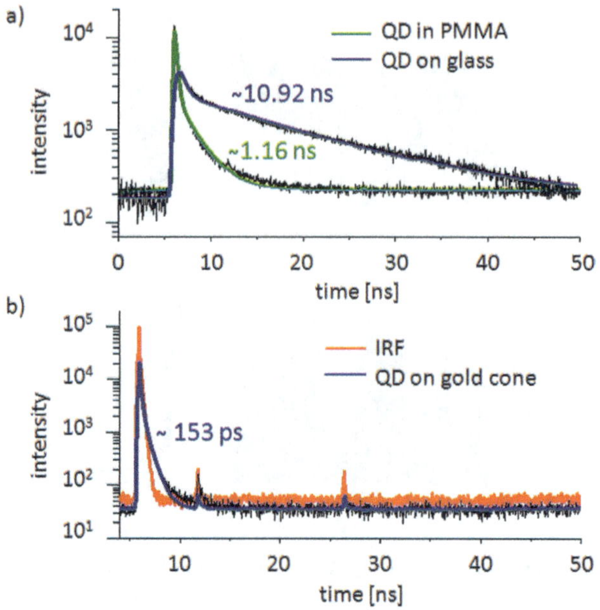

Fig. 10 Original decay data is presented in black. (a) PL intensity decay curves of two isolated single CdSe/ZnS-QDs deposited on a glass cover slide with a fluorescence lifetime of about 10.9 ns (blue) and one in PMMA with a fluorescence lifetime of about 1.16 ns (green). (b) For a QD chemically bound to the tip of a gold cone (blue), the decay is almost as short as the IRF (red) with a calculated lifetime of about 150 ps. The peaks at 11.8 ns and 26.4 ns originate from reflections of the excitation pulse in the optical system, which can arise in certain setup arrangements but only rarely interfere with the fitting procedures.

Fig. 10 shows typical fluorescence decay curves of QDs in different environments. The lifetime of an isolated CdSe/ZnS-QD on a glass cover slide can be described as a bi-exponential decay with an average lifetime on the order of 10 ns. In PMMA we observe a fluorescence lifetime around 1.6 ns and a multi-exponential decay, as has also been observed in other polymers.[63] Both values are in good agreement with the literature.[64,65] In contrast, the QD bound to the gold nanocone shows a significantly shorter lifetime. The measured 153 ps is close to the duration of the IRF and reflects an upper limit.

IV. Conclusions

Gold nanocones with a tunable plasmon resonance present an interesting and promising model for the analysis of optical processes in hybrid structures. The use of a hybrid system of QDs on a conical antenna may lead to an improved understanding of the interaction between QDs and plasmonic nanostructures, and the influence on the QD's emission. The gold cones were designed in such a way that they have a longitudinal plasmon resonance at the PL maximum of the QD. For excitation we used 488 nm radiation, which is off-resonance with the longitudinal mode of the gold cones. By comparing the PL spectra of 51 hybrid systems of QDs indicating single QD behavior, each on top of a conical gold antenna, we observe a red-shift of about 10 nm and in addition a PL intensity on average five times higher than for 78 single QDs of the same batch deposited on glass cover slides. The presence of single QDs was verified by recording intensity trajectories showing distinct on/off blinking. The dark intervals last from one millisecond, as limited by our detection electronics, to about one second. Our results indicate that the luminescence of single QDs and their decay rates can be enhanced by directly binding them to the tips of plasmonic nanocones. The current manuscript reflects work in progress. Further experiments, particularly concerning fluorescence lifetimes, fluorescence intermittency and emission polarization, will be performed.

Acknowledgements

We would like to thank E. Nadler for taking the SEM images of the gold cones. This work was performed in the context of the European COST Action MP1302 Nanospectroscopy. A. J. M. acknowledges partial financial support from DFG project ME 1600/5-3. Support by the Carl Zeiss Foundation is gratefully acknowledged by J. F.

References

1 J. M. Gerton, L. A. Wade, G. A. Lessard, Z. Ma and S. R. Quake, *Phys. Rev. Lett.*, 2004, **93**, 180801.
2 E. Shafran, B. D. Mangum and J. M. Gerton, *Phys. Rev. Lett.*, 2011, **107**, 037403.
3 A. Hartschuh, H. H. Qian, A. J. Meixner, N. Anderson and L. Novotny, *Nano Lett.*, 2005, **5**, 2310–2313.
4 P. Anger, P. Bharadwaj and L. Novotny, *Phys. Rev. Lett.*, 2006, **96**, 113002.
5 S. Kuhn, U. Hakanson, L. Rogobete and V. Sandoghdar, *Phys. Rev. Lett.*, 2006, **97**, 017402.
6 F. B. Urena, M. P. Kreuzer, S. Itzhakov, H. Rigneault, R. Quidant, D. Oron and J. Wenger, *Adv. Mater.*, 2012, **24**, OP314–OP320.
7 A. G. Curto, G. Volpe, T. H. Taminiau, M. P. Kreuzer, R. Quidant and N. F. van Hulst, *Science*, 2010, **329**, 930–933.
8 A. Kinkhabwala, Z. F. Yu, S. H. Fan, Y. Avlasevich, K. Mullen and W. E. Moerner, *Nat. Photonics*, 2009, **3**, 654–657.
9 K. Hennessy, A. Badolato, P. M. Petroff and E. Hu, *Photonics and Nanostructures*, 2004, **2**, 65–72.
10 C. J. Wang, L. Huang, B. A. Parviz and L. Y. Lin, *Nano Lett.*, 2006, **6**, 2549–2553.
11 D. L. Huffaker, G. Park, Z. Zou, O. B. Shchekin and D. G. Deppe, *Appl. Phys. Lett.*, 1998, **73**, 2564–2566.
12 I. Fushman, D. Englund, A. Faraon, N. Stoltz, P. Petroff and J. Vuckovic, *Science*, 2008, **320**, 769–772.
13 A. Mohammadi, V. Sandoghdar and M. Agio, *New J. Phys.*, 2008, **10**, 105015.
14 A. M. Kern, A. J. Meixner and O. J. F. Martin, *ACS Nano*, 2012, **6**, 9828–9836.
15 N. Engheta, *Science*, 2007, **317**, 1698–1702.
16 P. Viste, J. Plain, R. Jaffiol, A. Vial, P. M. Adam and P. Royer, *ACS Nano*, 2010, **4**, 759–764.
17 J. Do, R. Schreiber, A. A. Lutich, T. Liedl, J. Rodriguez-Fernandez and J. Feldmann, *Nano Lett.*, 2012, **12**, 5008–5013.

18 J. C. Garno, Y. Yang, N. A. Amro, S. Cruchon-Dupeyrat, S. Chen and G.-Y. Liu, *Nano Lett.*, 2003, **3**, 389–395.

19 T. van der Sar, E. C. Heeres, G. M. Dmochowski, G. de Lange, L. Robledo, T. H. Oosterkamp and R. Hanson, *Appl. Phys. Lett.*, 2009, **94**, 173104.

20 A. W. Schell, G. Kewes, T. Schröder, J. Wolters, T. Aichele and O. Benson, *Rev. Sci. Instrum.*, 2011, **82**, 073709.

21 J. Fulmes, R. Jäger, C. Schäfer, S. Jäger, D. A. Gollmer, A. Horrer, E. Nadler, D. Zhang, T. Chassé, A. J. Meixner, D. P. Kern and M. Fleischer, *Nanoscale*, 2015, 14691–14696.

22 D. Bera, L. Qian, T. K. Tseng and P. H. Holloway, *Materials*, 2010, **3**, 2260–2345.

23 C. T. Yuan, P. Yu, H. C. Ko, J. Huang and J. Tang, *ACS Nano*, 2009, **3**, 3051–3056.

24 A. Hatef, S. M. Sadeghi, E. Boulais and M. Meunier, *Nanotechnology*, 2013, **24**, 015502.

25 A. Horrer, C. Schäfer, K. Broch, D. A. Gollmer, J. Rogalski, J. Fulmes, D. Zhang, A. J. Meixner, F. Schreiber, D. P. Kern and M. Fleischer, *Small*, 2013, **9**, 3987–3992.

26 A. Mooradian, *Phys. Rev. Lett.*, 1969, **22**, 185–187.

27 M. R. Beversluis, A. Bouhelier and L. Novotny, *Phys. Rev. B: Condens. Matter Mater. Phys.*, 2003, **68**, 115433.

28 G. T. Boyd, Z. H. Yu and Y. R. Shen, *Phys. Rev. B: Condens. Matter Mater. Phys.*, 1986, **33**, 7923–7936.

29 A. Tcherniak, S. Dominguez-Medina, W. S. Chang, P. Swanglap, L. S. Slaughter, C. F. Landes and S. Link, *J. Phys. Chem. C*, 2011, **115**, 15938–15949.

30 M. Yorulmaz, S. Khatua, P. Zijlstra, A. Gaiduk and M. Orrit, *Nano Lett.*, 2012, **12**, 4385–4391.

31 H. Hu, H. Duan, J. K. Yang and Z. X. Shen, *ACS Nano*, 2012, **6**, 10147–10155.

32 E. Dulkeith, T. Niedereichholz, T. A. Klar, J. Feldmann, G. von Plessen, D. I. Gittins, K. S. Mayya and F. Caruso, *Phys. Rev. B: Condens. Matter Mater. Phys.*, 2004, **70**, 205424.

33 M. B. Mohamed, V. Volkov, S. Link and M. A. El-Sayed, *Chem. Phys. Lett.*, 2000, **317**, 517–523.

34 E. Kretschmann, *Z. Physik*, 1971, **241**, 313–324.

35 H. Raether, *Surface Plasmons on Smooth and Rough Surfaces and on Gratings*, Springer Verlag, 1988.

36 C. Schafer, D. A. Gollmer, A. Horrer, J. Fulmes, A. Weber-Bargioni, S. Cabrini, P. J. Schuck, D. P. Kern and M. Fleischer, *Nanoscale*, 2013, **5**, 7861–7866.

37 J. N. Farahani, D. W. Pohl, H. J. Eisler and B. Hecht, *Phys. Rev. Lett.*, 2005, **95**, 017402.

38 C. D. Geddes and J. R. Lakowicz, *J. Fluoresc.*, 2002, **12**, 121–129.

39 L. Novotny and B. Hecht, *Principles of nano-optics*, Cambridge University Press, Cambridge, New York, second edn, 2012.

40 A. M. Kern, D. Zhang, M. Brecht, A. I. Chizhik, A. V. Failla, F. Wackenhut and A. J. Meixner, *Chem. Soc. Rev.*, 2014, **43**, 1263–1286.

41 R. X. Bian, R. C. Dunn, X. S. Xie and P. T. Leung, *Phys. Rev. Lett.*, 1995, **75**, 4772–4775.

42 H. Fredriksson, Y. Alaverdyan, A. Dmitriev, C. Langhammer, D. S. Sutherland, M. Zaech and B. Kasemo, *Adv. Mater.*, 2007, **19**, 4297–4302.

43 M. Fleischer, F. Stade, A. Heeren, M. Haffner, K. Braun, C. Stanciu, R. Ehlich, J. K. H. Horber, A. J. Meixner and D. P. Kern, *Microelectron. Eng.*, 2009, **86**, 1219–1221.

44 M. Fleischer, D. Zhang, K. Braun, S. Jager, R. Ehlich, M. Haffner, C. Stanciu, J. K. H. Horber, A. J. Meixner and D. P. Kern, *Nanotechnology*, 2010, **21**, 065301.

45 A. G. Curto, T. H. Taminiau, G. Volpe, M. P. Kreuzer, R. Quidant and N. F. van Hulst, *Nat. Commun.*, 2013, **4**, 1750.

46 J. H. Song, T. Atay, S. F. Shi, H. Urabe and A. V. Nurmikko, *Nano Lett.*, 2005, **5**, 1557–1561.

47 F. Wackenhut, A. V. Failla, T. Züchner, M. Steiner and A. J. Meixner, *Appl. Phys. Lett.*, 2012, **100**, 263102.

48 T. Zuchner, A. V. Failla and A. J. Meixner, *Angew. Chem., Int. Ed.*, 2011, **50**, 5274–5293.

49 T. C. Isabell, P. E. Fischione, C. O'Keefe, M. U. Guruz and V. P. Dravid, *Microsc. Microanal.*, 1999, **5**, 126–135.

50 M. Antognozzi, A. Sentimenti and U. Valdre, *Microsc., Microanal., Microstruct.*, 1997, **8**, 355–368.

51 I. Campillo, J. M. Pitarke, A. Rubio, E. Zarate and P. M. Echenique, *Phys. Rev. Lett.*, 1999, **83**, 2230–2233.

52 J. Cao, Y. Gao, H. E. Elsayed-Ali, R. J. D. Miller and D. A. Mantell, *Phys. Rev. B: Condens. Matter Mater. Phys.*, 1998, **58**, 10948–10952.

53 T. V. Shahbazyan, *Nano Lett.*, 2013, **13**, 194–198.

54 S. Jäger, *Dissertation*, Tübingen, 2014.

55 A. I. Chizhik, A. M. Chizhik, D. Khoptyar, S. Bär and A. J. Meixner, *Nano Lett.*, 2011, **11**, 1131–1135.

56 J. Kim, J. Lee and K. Kyhm, *Appl. Phys. Lett.*, 2011, **99**, 213112.

57 Y. Ito, K. Matsuda and Y. Kanemitsu, *Phys. Rev. B: Condens. Matter Mater. Phys.*, 2007, **75**, 033309.

58 K. E. Knowles, E. A. McArthur and E. A. Weiss, *ACS Nano*, 2011, **5**, 2026–2035.

59 K. Sahu, H. Wu and M. A. Berg, *J. Phys. Chem. B*, 2013, **117**, 15257–15271.

60 R. Schmidt, C. Krasselt, C. Gohler and C. von Borczyskowski, *ACS Nano*, 2014, **8**, 3506–3521.

61 A. A. Cordones and S. R. Leone, *Chem. Soc. Rev.*, 2013, **42**, 3209–3221.

62 M. Califano and F. M. Gómez-Campos, *Nano Lett.*, 2013, **13**, 2047–2052.

63 G. Schlegel, J. Bohnenberger, I. Potapova and A. Mews, *Phys. Rev. Lett.*, 2002, **88**, 137401.

64 W. G. J. H. M. van Sark, P. L. T. M. Frederix, A. A. Bol, H. C. Gerritsen and A. Meijerink, *ChemPhysChem*, 2002, **3**, 871–879.

65 L. Zhu, S. Samudrala, N. Stelmakh and M. Vasilyev, *Opt. Express*, 2012, **20**, 3144–3151.

Faraday Discussions

PAPER

Enhanced Raman scattering from aromatic dithiols electrosprayed into plasmonic nanojunctions

Patrick Z. El-Khoury,*[a] Grant E. Johnson,[a] Irina V. Novikova,[b] Yu Gong,[a] Alan G. Joly,[a] James E. Evans,[b] Mikhail Zamkov,[c] Julia Laskin[a] and Wayne P. Hess*[a]

Received 6th April 2015, Accepted 1st June 2015
DOI: 10.1039/c5fd00036j

We describe surface enhanced Raman spectroscopy (SERS) experiments in which molecular coverage is systematically varied from 3.8×10^5 to 3.8×10^2 to 0.38 molecules per μm^2 using electrospray deposition of ethanolic 4,4′-dimercaptostilbene (DMS) solutions. The plasmonic SERS substrate used herein consists of a well-characterized 2-dimensional (2D) array of silver nanospheres (see El-Khoury *et al.*, *J. Chem. Phys.*, 2014, **141**, 214308), previously shown to feature uniform topography and plasmonic response, as well as intense SERS activity. When compared to their ensemble averaged analogues, the spatially and temporally averaged spectra of a single molecule exhibit several unique features including: (i) distinct relative intensities of the observable Raman-active vibrational states, (ii) more pronounced SERS backgrounds, and (iii) broader Raman lines indicative of faster vibrational dephasing. The first observation may be understood on the basis of an intuitive physical picture in which the removal of averaging over multiple molecules exposes the tensorial nature of Raman scattering. When an oriented single molecule gives rise to the recorded SERS spectra, the relative orientation of the molecule with respect to vector components of the local electric field determines the relative intensities of the observable vibrational states. Using a single molecule SERS framework, described herein, we derive a unique molecular orientation in which a single DMS molecule is isolated at a nanojunction formed between two silver nanospheres in the 2D array. The DMS molecule is found lying nearly flat with respect to the metal. The derived orientation of a single molecule at a plasmonic nanojunction is consistent with observations (ii) and (iii). In particular, a careful inspection of the temporal spectral variations along the recorded single molecule SERS time sequences reveals that the time averaged SERS backgrounds arise

[a]Physical Sciences Division, Pacific Northwest National Laboratory, Richland, WA 99352, USA. E-mail: patrick.elkhoury@pnnl.gov; wayne.hess@pnnl.gov
[b]Environmental and Molecular Sciences Laboratory, Pacific Northwest National Laboratory, Richland, WA 99352, USA
[c]Department of Physics and the Center for Photochemical Sciences, Bowling Green State University, Bowling Green, OH 43403, USA

from individual molecular events, marked by broadened SERS signatures. We assign the broadened spectra along the SERS time sequence – which sum up to a SERS background in the averaged spectra – to instances in which the π-framework of the DMS molecule is parallel to the metal at a classical plasmonic nanojunction. This also accounts for Raman line broadening as a result of fast vibrational dephasing, and driven by molecular reorientation at a plasmonic nanojunction. Furthermore, we report on the molecular orientation dependence of single molecule SERS enhancement factors. We find that in the case of a single DMS molecule isolated at a plasmonic nanojunction, molecular orientation may affect the derived single molecule SERS enhancement factor by up to 5 orders of magnitude. Taking both chemical effects as well as molecular orientation into account, we were able to estimate a single molecule enhancement factor of $\sim 10^{10}$ in our measurements.

1 Introduction

The dynamic coupling between molecular polarizabilities and electric fields leads to Raman scattering. Surface-enhanced Raman scattering (SERS)[1–4] is a variant of Raman scattering in which the driving fields are typically light waves trapped at the surface of metal nanostructures in the form of localized surface plasmons. The effect is particularly evident in assemblies of plasmonic nanostructures, where the plasmonic eigenmodes of individual particles hybridize to form new collective modes in the assembly.[5–7] Confined and enhanced local electric fields at plasmonic nanojunctions formed between interacting plasmonic particles rendered SERS particularly popular, ever since the possibility of single molecule detection sensitivity in such constructs was demonstrated.[8,9] The single/few molecular SERS regime is slowly becoming routinely accessible through the use of engineered plasmonic substrates, as highlighted in recent reviews.[10,11] In this contribution, we examine the information content in Raman scattering from aromatic dithiols, namely 4,4′-dimercaptostilbene (DMS) molecules, addressed at well-defined plasmonic nanojunctions.

Pioneering SERS works were based on the low-concentration approach to explore the single-molecule SERS realm.[8,9] Namely, the samples used therein featured one (or fewer) molecular scatterer per effective SERS probing volume, defined by a high numerical aperture microscope objective. The ultra-low concentration approach was challenged over the years, particularly for colloidal solutions, where several convoluted factors come into play in the interpretation of fluctuating Raman spectra.[10] Besides obvious considerations such as the number of plasmonic particles per unit volume, shape and polydispersity of the particles, and physisorption or chemisorption efficiency of the molecular reporters onto the particles,[10] the major challenge lies in our inability to precisely control or even directly characterize molecules and plasmonic constructs over the relevant length scales (sub- to a few- nm) in solution. Even in the case of solid or dried plasmonic substrates well-characterized using high-resolution microscopic techniques such as transmission electron microscopy (TEM), typical sample preparation methods introduce further challenges. Namely, SERS samples are conventionally prepared by (i) mixing plasmonic particles with dilute solutions of SERS analytes, (ii) drop/spin casting dilute analyte solutions onto solid/dried plasmonic substrates, or (iii) incubating plasmonic substrates in dilute analyte solutions and allowing time for adsorption/chemisorption to take place. In all (i)–(iii), the exact location and number of probed molecule(s) in the effective SERS sampling area vary for different plasmonic substrates/molecular SERS reporters and are overall ill-determined. For instance, factors such as preferable adsorption/chemisorption of analytes at specific sites of metal substrates are not easily discernible. In this regard, isolated engineered prototypical dumbbell antennae, defined by an aromatic dithiol at the junction formed between two plasmonic nanospheres and prepared through (i) are more telling.[12,13] Indeed, such constructs have been recently used to explore some of the underlying physics in single molecule SERS based on strong relative intensity and background fluctuations in Raman spectral sequences, *vide infra*.

Although much evidence supports it, the inference of single molecule sensitivity in ambient SERS is not trivial. The origin of intensity fluctuations in few/single molecule SERS in sequentially recorded SERS (and tip-enhanced Raman scattering, TERS) spectra is still a subject of debate, four decades after the first observation of SERS.[13,14] Even the original observation[9] of quantized probability distributions corresponding to 0, 1, 2, and 3 molecules in the SERS probing volume was later challenged on the basis of insufficient statistical sampling.[10] More recent works[13,15,16] have associated mode-specific intensity variations in non-resonant SERS and TERS with molecular reorientation at plasmonic nanojunctions. Namely, the removal of orientational averaging in the single molecule regime exposes the tensorial nature of Raman scattering. As such, relative intensity fluctuations may be used to track the orientation of single molecules relative to the local electric fields.[13,15,16] In this context, the use of non-resonant reporters in SERS and TERS turns into a comparative advantage, as technical (photobleaching, photochemistry, resonant heating, *etc.*) and theoretical challenges[14] arise in interpreting surface- and tip-enhanced *resonant* Raman scattering. In the following, we will provide a brief overview of our recent SERS and TERS studies, focused at understanding the origins of inelastic light scattering from aromatic dithiols at plasmonic nanojunctions. The brief summary justifies and leads into the strategy adopted herein to achieve and demonstrate single molecule detection sensitivity in Raman scattering from aromatic dithiols at plasmonic junctions.

In a series of recent reports from our group, we examined the information content in SERS and TERS point spectra[16–18] and images[19,20] recorded by taking advantage of light confinement at plasmonic junctions, all using DMS and a structurally related biphenyl-4,4′-dithiol (BPDT) system as prototypical molecular reporters. We distinguished between two regimes encountered in TERS of the two dithiols, where plasmonic junctions are formed between the metal AFM tip and a metal surface.[16–20] In contact mode TERS experiments, in which a plasmonic junction is formed between a metal AFM tip *in contact* with a metal surface coated with dithiols,[17,18] we identified optical signatures[13] of current-carrying charge-transfer plasmons. Indeed, a recent quantum analysis of plasmonic junctions[21] suggests that in the case of strongly coupled plasmonic particles (0 1 nm particle separation), conductive overlap is established. The onset of electron tunneling is accompanied by broadened Raman scattered signals in which only molecular parentage is apparent.[13,17,18] In contrast, *non-contact* mode TERS experiments in which the AFM probe oscillates atop the dithiol-coated metal substrate resulted in familiar tip-enhanced molecular line spectra.[16,19] Relative intensity fluctuations in sequences of *non-contact* mode TERS spectra were evident, and were associated with molecular reorientation at plasmonic nanojunctions.[16,19] Notably, the effect was recognizable in TERS point spectra[16] and also governed nanoscale chemical images recorded using the same technique.[19,20] Overall, the interplay between molecules and plasmons is evident in SERS and TERS experiments conducted both at the classical and quantum regimes of plasmons. Nonetheless, the transition to the quantum realm of current-carrying plasmons renders structural/chemical selectivity based on single/few molecule SERS and TERS rather difficult to obtain. This is a result of significant broadening of the DMS and BPDT spectra as a result of charge shuttling across the plasmonic gap.[13,17,18]

With the above consideration in mind, we recently designed a sensitive and well-characterized SERS substrate that exclusively operates in the classical regime of plasmons, where the enhancement of recognizable molecular line spectra is optimal.[22] It consists of a self-assembled 2D array of Ag nanospheres with an average particle diameter : inter-particle separation distance of 9 : 3.7 nm. The structures of the individual particles and their assemblies were characterized using TEM.[22] The plasmonic response of the nanoparticle network was probed using two-photon photoemission electron microscopy (TP-PEEM) and SERS.[22] TEM and TP-PEEM statistics revealed the structure and plasmonic response of the network to be fairly homogeneous on a length scale of tens of microns. This translated into a uniform SERS response from BPDT molecules adsorbed onto different sites of the nanoparticle network. Finite-difference time-domain (FDTD) simulations additionally revealed that plasmonic enhancement factors on the order of 10^6 are attainable at the nanogaps formed between the Ag nanospheres in the 2D array.[22] Combined with modest chemical enhancement factors attainable by aromatic dithiols at plasmonic nanojunctions,[12,13] this substrate is potentially capable of broadcasting the Raman signature of a single molecule.

In this work, we use the aforementioned SERS substrate.[22] For the above-mentioned reasons, we do not rely on mixing, spin/drop casting, or incubation to functionalize the plasmonic substrate with DMS molecules. Rather, we deposit DMS molecules over the 2D array of silver nanospheres using *ambient* electrospray deposition (ESD).[23–25] We find that ESD produces a uniform distribution of intact DMS molecules and allows for superior control over the number of molecules within the effective SERS probing area. This is ideal for this proof-of-principle study in which coverages ranging from a single molecule to molecular ensembles are systematically prepared and probed using SERS. At the smallest coverage tested herein (0.38 molecules per μm^2), we identify and analyze characteristic spectral signatures of SERS at the single molecule limit. Removal of orientational averaging exposes the tensorial nature of single molecule Raman scattering.[16–20,22] The measured relative intensities of the observable vibrational states at the single molecule limit are distinct from their analogues that were recorded from molecular ensembles. This feature allows us to identify an average orientation of a single molecule with respect to the local electric fields defined by a plane-polarized laser field incident to the 2D array of silver nanospheres. The derived 3D orientation of a single DMS molecule at a classical plasmonic nanojunction accounts for our observations of time-fluctuating SERS backgrounds along sequentially recorded SERS spectra. The backgrounds observed in the spatially and temporally averaged spectra recorded at classical plasmonic nanojunctions are associated with individual events: instances along SERS spectral sequences in which a DMS molecule is flat with respect to the metal. This only requires a 30° rotation around the long axis of a DMS molecule from its derived initial orientation. The derived atomistic picture also accounts for our observation of broadened Raman lines at lower molecular coverage. The observed SERS line broadening at the single molecule limit is associated with rapid dephasing of molecular vibrations when the π-framework of DMS strongly interacts with the metal surface. We end our analysis by providing an estimate of the recorded single molecule SERS enhancement factor both with and without taking molecular orientation into account. We find that for a DMS molecule isolated at a plasmonic nanojunction, molecular orientation may affect estimations of single molecule SERS enhancement factors by up to five orders of magnitude.

2 Experimental methods

2.1 Ambient ESD

A 0.1 mM solution of DMS (Sigma-Aldrich) in ethanol (Gold Shield, 200 proof) was successively diluted to 1 μM, 1 nM, and 1 pM aliquots. For ambient ESD experiments, samples were injected through a pulled fused silica capillary (~75 μm inner diameter) tip at a flow rate of 50 μL hour^{-1}. A 1 mL glass syringe (Hamilton) mounted on a microprocessor-controlled syringe pump (KD Scientific) was used to feed the electrospray capillary. Samples were prepared in increasing order of concentration to avoid cross-contamination. Ample amounts of pure solvent (and target sample solutions) were also used to clean the syringe and capillary between successive sample preparations. After 1 hour of deposition (50 μL total volume) using a positive ionization mode (4 kV), an area (>1 cm diameter) of adsorbed species was faintly visible at the surface of the electrically grounded SERS substrates, described below. As such, we estimate the upper limits of coverage produced by ESD to be 3.8×10^5, 3.8×10^2, and 0.38 molecules per μm^2 for the samples prepared using 1 μM, 1 nM, and 1 pM solutions, respectively.

2.2 SERS substrate

Control experiments were performed by either ESD or drop casting a 0.1 mM solution of DMS in ethanol onto a 15 nm polycrystalline Ag film evaporated onto a 0.1 mm-thick microscope glass slide by arc-discharge physical vapor deposition (Cressington 208HR, target: Ted Pella Inc., 99.99% purity). The film thickness was monitored *in situ* using a quartz crystal microbalance, and the surface roughness measured from freshly evaporated Ag films with AFM indicates a root-mean-square (RMS) height distribution of ~3 nm.[15–17]

Concentration-dependent SERS measurements were performed using a recently reported[22] and thoroughly characterized SERS substrate comprising a 2D array of Ag nanospheres atop a sputtered silver substrate, as described above. In the following, we provide a brief description of previous observations that are of particular relevance to this investigation. The reader is referred to the original work[22] for more specifics. The UV-Vis absorption profile of the synthesized colloidal Ag nanoparticles in chloroform exhibited a sharp plasmon resonance centered at 415 nm, indicative of a narrow size distribution of oleylamine-capped nanoparticles. TEM images corroborated the UV-Vis observables, revealing a relatively narrow size distribution of nearly spherical Ag nanoparticles, featuring an average diameter of 9 ± 2 nm. The recorded TEM images also revealed that drop casting the as-synthesized nanoparticle solution yielded patches of uniformly self-assembled 2D Ag nanoparticle networks. They are best described as hexagonal close packed distributions of Ag nanospheres with an average inter-particle separation distance of 3.7 nm in the network. The plasmonic properties of the assemblies were probed using TP-PEEM. Oversampled TP-PEEM image statistics revealed the plasmonic response of the network to be homogeneous, namely, 90% of the recorded enhancement factors were within a factor of $4\left(\left(\frac{E}{E_0}\right)^4\right)$ from the average. The homogeneity in structure and plasmonic response over a length scale of tens of microns gauged from TEM and TP-PEEM analysis translated into a uniform SERS response from BPDT molecules adsorbed onto different sites of the network from a 0.1 mM ethanolic solution. Finally, FDTD simulations revealed that plasmonic enhancement factors on the order of 10^6 are attainable at the nanogaps formed between the Ag nanospheres in the array following in-plane-polarized irradiation at 514 nm.

2.3 SERS experiments

SERS measurements were conducted using an inverted optical microscope (Axiovert 200, Zeiss). The incident 514 nm continuous wave laser light (Innova 300, Coherent) is attenuated using a variable neutral density filter wheel (150–300 μW μm^{-2} used throughout), reflected off a dichroic beamsplitter, and focused onto the sample surface using an oil-immersion objective (1.3 NA, 100×). The diameter of the diffraction-limited laser spot was measured to be 260 nm using non-contact mode TERS imaging of a DMS-coated thin Ag film.[19] As such, our effective SERS probing area was 0.053 μm^2. The polarization of the incident laser was controlled using a half wave plate, which allows polarization rotation in the sample plane. The polarization of the incident laser field was readjusted prior to recording each of the spectral sequences reported herein, such that the molecular signal at 1575 cm^{-1} was maximized. No polarizers were used to analyze the backscattered radiation, which was collected through the same objective, transmitted though the dichroic beamsplitter, and filtered through a long pass filter. The resulting light was detected using a liquid nitrogen-cooled charge coupled device (CCD) coupled to a spectrometer (Holespec f/1.8i, Kaiser Optical System). The effective instrument spectral resolution in the micro-Raman experiments was 8 cm^{-1}.

2.4 Theoretical framework

Calculations were performed using methodologies implemented in Gaussian 09,[26] and employed the B3LYP[27–30] exchange–correlation functional in conjunction with the 6-311^{++}G** (ref. 31 and 32) basis set. Unconstrained geometry optimization was performed to energy-minimize the structure of DMS in its ground electronic state. Frequency-dependent molecular polarizability derivative tensors were computed using the Raman optical activity[33,34] module of Gaussian 09 at an incident wavelength of 514 nm. Standard Euler-averaged Raman scattering activities were evaluated near the ground state minimum according to

$$S_n = 45\bar{\alpha}_n'^2 + 7\gamma_n'^2 \quad (1)$$

in which $\overline{\alpha}'_n$ and γ'^2_n are the spherical and anisotropic parts of the molecular polarizability derivative tensors given by

$$\overline{\alpha}'_n = \frac{1}{3}\left(\frac{\partial\alpha_{xx}}{\partial Q_n} + \frac{\partial\alpha_{yy}}{\partial Q_n} + \frac{\partial\alpha_{zz}}{\partial Q_n}\right) \tag{2}$$

and

$$\gamma'^2_n = \frac{1}{2}\left(\frac{\partial\alpha_{xx}}{\partial Q_n} - \frac{\partial\alpha_{yy}}{\partial Q_n}\right)^2 + \frac{1}{2}\left(\frac{\partial\alpha_{yy}}{\partial Q_n} - \frac{\partial\alpha_{zz}}{\partial Q_n}\right)^2 + \frac{1}{2}\left(\frac{\partial\alpha_{zz}}{\partial Q_n} - \frac{\partial\alpha_{xx}}{\partial Q_n}\right)^2 + 3\left(\frac{\partial\alpha_{xy}}{\partial Q_n}\right)^2 + 3\left(\frac{\partial\alpha_{yz}}{\partial Q_n}\right)^2 + 3\left(\frac{\partial\alpha_{zx}}{\partial Q_n}\right)^2 \tag{3}$$

where α_{ij} are diagonal ($i = j$) and off-diagonal ($i \neq j$) molecular polarizability derivative tensor elements and Q_n are the coordinates for the nth vibrational mode. To ease spectral assignments, we also performed a full anharmonic vibrational frequency analysis based on perturbation correction for the harmonic approximation.[35] The resulting theoretical Raman spectra are thus compared to their experimental analogues without applying any scaling factors. Throughout the text, the computed spectra are represented as a sum of Lorentzians individually broadened by 8 cm^{-1} to match our effective instrument resolution, and centered at the calculated anharmonic frequency shifts. To compare the computed Raman scattering activities with experimental intensities, a useful quantity is the differential Raman scattering cross section, $\frac{\text{d}\sigma}{\text{d}\Omega}$, defined in terms of S_n as

$$\frac{\text{d}\sigma}{\text{d}\Omega} = \frac{(2\pi)^4}{45}\frac{h{\omega_\text{s}}^4}{8\pi^2 c\omega_n}\frac{S_n}{1-\text{e}^{-\left(\frac{hc\omega_n}{k_\text{b}T}\right)}} \tag{4}$$

in which ω_s is the scattered frequency in cm^{-1}, ω_n is the frequency of the nth vibrational mode in cm^{-1}, and the calculated S_i is typically in $\text{Å}^4\ \text{amu}^{-1}$, governed by eqn (1)–(3).

In the realm of single/few molecules, orientational averaging according to eqn (1)–(3) is no longer appropriate.[13,15,16,19] In this case, the scattering tensor that governs Raman activity may be expressed as

$$S_n{}^2 = \sum_n \left|E_\text{s}^\text{L}\alpha'_n(\Omega)E_\text{i}^\text{L}\right|^2 \tag{5}$$

where $E^\text{L}_{\text{i,s}}$ are the enhanced incident and scattered local electric fields, α'_n is the molecular polarizability derivative tensor of the nth vibrational eigenstate, and $\Omega = \{\alpha,\beta,\gamma\}$ are the Euler angles which determine molecular orientation relative to the local fields. This formalism stresses that a single molecule assumes a unique orientation with respect to the local electric fields. As such, the observable relative intensities in single-molecule SERS may be used to track the orientation of a molecule in 3D where $E^\text{L}_{\text{i,o}}$ are known, *e.g.* at a hot spot formed between two interacting plasmonic particles.[12,13] Alternatively, if molecular orientation is known, the vector components of the local electric field may be inferred from the relative intensities in single molecule SERS spectra. If neither molecular orientation nor local electric fields are known, removal of orientational averaging according to eqn (5) yields a unique set of relative intensities for every orientation of a molecule with respect to the local fields. This framework was recently used to track reorientation dynamics of molecules at plasmonic junctions in SERS and TERS experiments.[13,16,19] Overall, it provides an intuitive framework that may be used to rationalize observables such as relative intensity fluctuations in the single molecule realm. Its limitations have been noted elsewhere,[36,37] and will be further stressed in context.

3 Results and discussion

3.1 Control experiments: sputtered silver film on glass

We begin our analysis by comparing SERS spectra recorded from two samples prepared either by ESD or drop casting a 0.1 mM solution of DMS in ethanol onto freshly sputtered Ag films on glass. The results are illustrated in Fig. 1, in which panels A (drop casted) and C (ESD) show several spectra recorded from randomly selected spots separated by tens of microns on the metal surface. We note that each of the spectra shown in the aforementioned two panels represents a time average of 60 consecutively collected spectra that were individually integrated for 10 s at one position. The spatially and temporally averaged spectra in the two cases are expanded and shown in panels B (drop casted) and D (ESD) of the same figure, along with their corresponding standard deviations. It is immediately noticeable that the percent deviation from the average spectrum is twice narrower for the sample prepared using ESD. The substrate coated using ESD features a more spatially uniform DMS distribution. Nonetheless, the

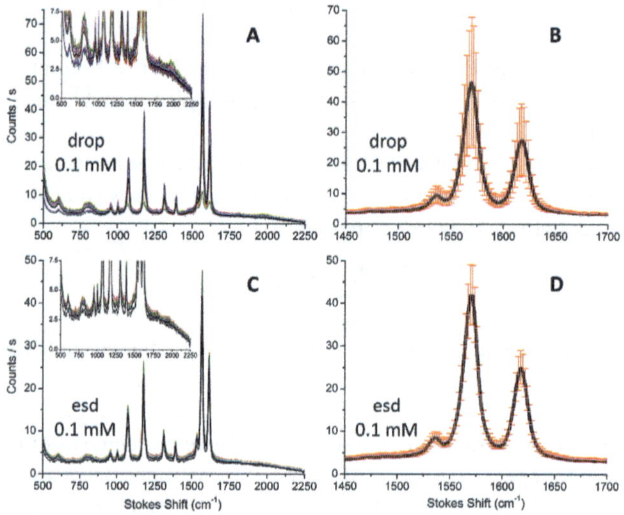

Fig. 1 Control experiments: SERS spectra recorded from samples prepared either by drop casting (A and B) or electrospray deposition (C and D) of a 0.1 mM solution of 4,4′-dimercaptostilbene in ethanol. Shown in panels A and C are several time-averaged spectra recorded from randomly selected spots on a sputtered silver substrate. Shown in panels B and D are the spatially and temporally averaged spectra for each case (in the 1450–1700 cm^{-1} region) along with the standard deviation of the various spectra shown in panels A and C, respectively.

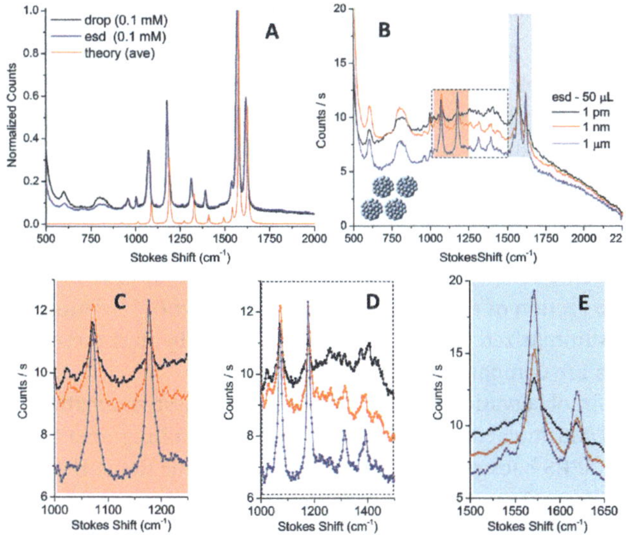

Fig. 2 (A) The spatially and temporally averaged SERS spectra shown in Fig. 1 (control experiments) are shown on the same plot, along with the orientationally averaged Raman spectrum of an isolated DMS molecule calculated at the B3LYP/6-311^{++}G** level of theory. Note that in the computed spectrum, the intensities are obtained from dynamic polarizability calculations ($\lambda_{ex} = 514$ nm) and the frequency shifts are obtained after explicit anharmonic frequency corrections including third and fourth derivatives, see text for more details. (B) Spatially and temporally averaged SERS spectra recorded from the 2D Ag nanoparticle array at DMS coverages of 3.8×10^5 (1 μM), 3.8×10^2 (1 nM) and 0.38 molecules per μm^2 (1 pM). Three spectral regions of interest are highlighted in panel B, and in turn expanded in panels C–E.

spectral features in the 500–1700 cm^{-1} region of the spectrum are very similar in the two cases. This is more obvious in Fig. 2A, where the two spatially and temporally averaged spectra are normalized and compared on the same plot. Apart from minor background differences in the 500–800 cm^{-1} region, no noticeable differences are observed between the two experimental spectra both in terms of frequency shifts and relative intensities. These control experiments reveal that ESD yields a uniform distribution of molecules for SERS analysis, all while preserving the chemical identity of the molecules. Moreover, given a total volume of 50 μL deposited in an area of ~0.79 cm^2 (1 cm diameter, lower limit), an upper limit of 3.8×10^7 molecules per μm^2 may be estimated. This value assumes (i) 100% transfer of molecules from the 0.1 mM DMS solution to the Ag substrate, and (ii) that all the molecules are packed in a circular cross-section with a diameter of 1 cm.

The theoretical orientationally averaged B3LYP/6-311^{++}G** Raman spectrum, computed according to eqn (1)–(3) is also shown in Fig. 2A, for comparison. We find that the level of theory used herein reproduces the corresponding experimental observables in which molecular ensembles are probed, both in terms of frequency shifts and relative intensities. The latter is an important observation to keep in mind in going through the ensuing theoretical analysis of the samples featuring increasingly sparse molecular distributions.

3.2 Ensemble *vs.* single molecule SERS: 2D nanosphere array

The second set of SERS spectra were recorded from three different samples prepared by ESD of 1 μM, 1 nM, and 1 pM solutions of DMS in ethanol. We estimate upper limits of 3.8×10^5, 3.8×10^2 and 0.38 molecules per μm^2 for the samples prepared using 1 μM, 1 nM, and 1 pM solutions, respectively. It is perhaps worth re-iterating that the effective SERS probing area is 0.053 μm^2, which is 19 times smaller than the 1 μm^2 unit area used in our estimation of molecular coverage on the surface of the substrate. The SERS substrate in this set of measurements consists of a 2D array of plasmonic silver nanospheres; see the methods sub-section of this work and our previous work[22] for more specifics. The results are summarized in Fig. 2B–E, where spatially and temporally averaged SERS spectra are compared on the same plot.

Three main observations are highlighted. First, systematically decreasing the number of DMS molecules in the effective SERS probing area leads to a more pronounced SERS background. The background spans the entire spectral region of interest (500–2250 cm^{-1}) and increases in magnitude as a function of decreasing molecular coverage. Second, the relative intensities of the major characteristic molecular Raman lines of DMS, expanded in panels C and E of Fig. 2, also exhibit systematic variations as a function of molecular coverage. This is noticeable in Fig. 2E, where the ratio of the predominant aromatic C=C stretching (1575 cm^{-1}) to the vinyl C=C stretching (1625 cm^{-1}) vibrations increases with decreasing surface coverage of DMS. A similar trend is observed when the relative scattering activities of the C–S stretching (1085 cm^{-1}) and aromatic C–H in-plane rocking (1186 cm^{-1}) vibrations are compared in Fig. 2C. Third, the widths of the observable Raman lines increase as a function of decreasing molecular coverage. The effect is most evident in the 1200–1500 cm^{-1} spectral region shown in Fig. 2D, but also noticeable in Fig. 2C and E. This particular observation is counter-intuitive at first sight, as inhomogeneous broadening would be expected to lead to the opposite trend. In fact, pictures in which contributions from individual molecules with slightly different resonances sum up to an overall broad line shape typical of the ensemble are often invoked in this context. We will illustrate why the aforementioned spectral analysis may lead to erroneous conclusions in the interpretation of single molecule *vs.* ensemble-averaged spectra, at least in our case of Raman scattering of aromatic dithiols at plasmonic nanojunctions. Overall, the key to understanding all three observations is in the analysis of the spatial and temporal variations in Raman scattering activities, both of which are masked in the spatially and temporally averaged spectra discussed thus far.

3.3 Spectral fluctuations in single molecule SERS

Time-averaged SERS spectra recorded at randomly selected spots on three substrates prepared by ESD of 1 μM (panel A), 1 nM (panel C), and 1 pM (panel E) DMS solutions in ethanol are presented in Fig. 3. Variations in the magnitudes of the recorded SERS backgrounds are most evident in panel E, *i.e.* in the single molecule regime. Less pronounced SERS background variations are still observable in the hundreds-of-molecules regime (panel C), but are negligible at higher molecular coverage (panel A). Our primary analysis of the temporal variations in SERS activity at select spots on each sample is also shown in the same figure

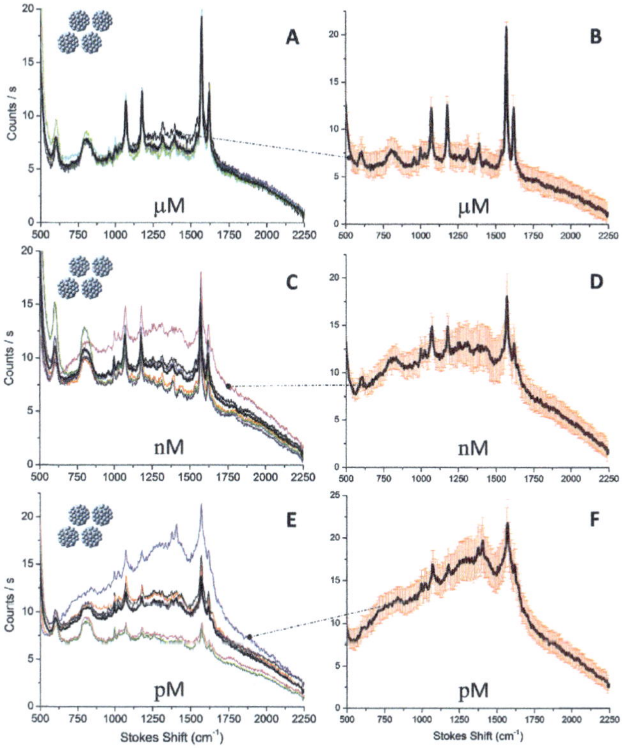

Fig. 3 Panels A, C, and E show time-averaged spectra recorded from randomly selected spots on the 2D Ag nanoparticle array at different DMS coverages. Each spectrum shown is an average of 60 spectra, each of which was time-integrated for 10 s. Panels B, D, and F show 3 temporally averaged spectra along with the standard deviations for the SERS time-sequences that were averaged to yield the selected spectra. The concentrations used in electrospray deposition of DMS solutions are denoted in the insets.

(panels B, D, and F). Namely, shown are time averaged spectra at one spot for each of the three DMS coverages along with their corresponding standard deviations derived from sequences of 60 consecutively recorded spectra, each of which was individually integrated for 10 s. The standard deviation error bars again show that ensemble averaged SERS spectra (panel B) exhibit much less-pronounced background fluctuations when compared to their analogues that were recorded from samples featuring lower molecular coverages of DMS (panels D and F). Before examining the nature of the fluctuating background further, we note that the relative intensities of the predominant Raman-allowed vibrational states are very similar when the time-averaged spectra recorded at different positions of the substrate and at the same molecular coverage are compared to one another. This is to be expected at high molecular coverage (panel A): the response is orientationally averaged. At low molecular coverage (panel E), this observation is indicative of similar local environments felt by the different (individual) molecules probed. Given a uniform SERS substrate, and subsequently a uniform plasmonic response,[22] this observation suggests that similar molecular orientations are assumed at the various plasmonic nanojunctions probed in this work.

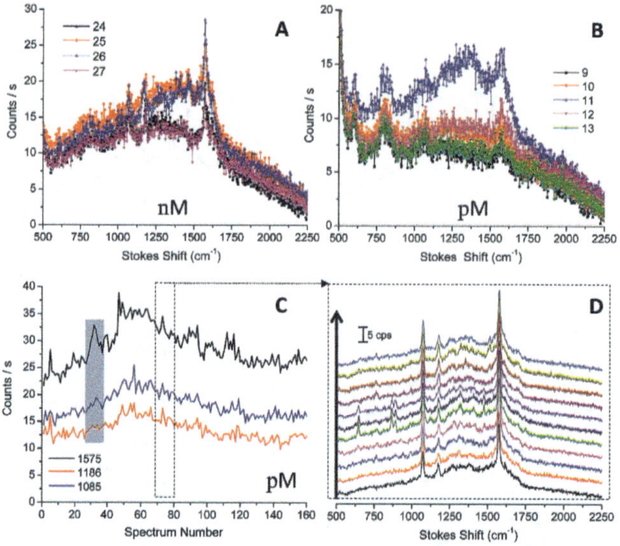

Fig. 4 Spectral variations along SERS time sequences at 3.8×10^2 DMS molecules per μm^2 (1 nM, panel A) and 0.38 DMS molecules per μm^2 (1 pM, panel B). Note that the spectra shown in panels A and B are selected sequentially recorded SERS spectra along a time-sequence, where each spectrum is individually time-integrated for 10 s. Panels C and D show another selected SERS spectral sequence. In panel C, we monitor the time evolution of SERS intensity at three vibrational resonances of DMS. Two highlighted regions in panel C are noted. The first (transparent blue rectangle) shows distinct relative intensity fluctuations along the SERS time sequence. The second (dashed rectangle, spectra 70–80) is expanded in panel D, which further illustrates temporal relative intensity fluctuations along the SERS time sequences.

To better illustrate the nature of the fluctuating backgrounds we now turn the reader's attention to Fig. 4, where representative spectra along the SERS time sequences are shown. Sequentially recorded spectra ($\Delta t = 10$ s) are plotted in panels A and B, where subtle background fluctuations are observable in the hundreds-of-molecules regime, and dominate the SERS response in the single molecule regime. As the background fluctuations in Fig. 4B mask more subtle relative intensity fluctuations, we selected another spectral sequence in which the SERS background is relatively stable over time for further analysis. In this particular time sequence, frequency slices taken at select molecular resonances are shown in panel C, and sequentially recorded spectra (10 s per spectrum) along the representative time sequence are shown in panel D. It is shown in panel C that the different molecular resonances exhibit distinct time evolution trends along the SERS trajectory. This is evident, for example, around spectrum 30 ($t = 300$ s, highlighted with a transparent light blue rectangle) where an intensity spike at 1575 cm^{-1} is not evident in the 1186 and 1085 cm^{-1} time traces. In panel D, the most obvious intensity fluctuations along the time sequence are observed in the 500–1000 cm^{-1} region of the spectrum, in which the appearance/disappearance of at least three distinct vibrational signatures may be readily discerned. More subtle relative intensity fluctuations are also observable in the 1250–1500 cm^{-1} region of the spectrum. Relative intensity fluctuations in SERS and TERS at plasmonic junctions in the single molecule regime have been recently associated with molecular reorientation dynamics according to eqn (5). In principle, the same analysis carried out in the previous works[13,15,16,19] may be performed to extract a unique relative orientation of the molecule with respect to local electric fields for each of the spectra along the time trajectory. We refrain from duplicating the prior analyses, to which the reader is referred. Rather, we will focus our analysis on identifying the most likely time-averaged molecular orientation of DMS throughout the entire SERS spectral sequence shown in Fig. 4C and D, to illustrate the concept.

3.4 Removal of averaging: orientation-specific single molecule SERS

In Fig. 5, we time-average the SERS sequences shown in Fig. 4C and D. Notice how the relative intensities of the observable vibrational states are distinct from the ensemble averaged SERS spectra shown in Fig. 1 and 2. Namely, (i) the vinyl C=C stretching vibration (1625 cm^{-1}) is suppressed, (ii) the C–S stretching vibration (1085 cm^{-1}) is more pronounced than the aromatic C–H in-plane rocking (1186 cm^{-1}), and (iii) the vibrational modes in the 1200–1500 cm^{-1} region are heavily broadened and largely buried in the underlying background. On the basis of (i) our SERS-active substrate preparation method of choice (ESD) which yields an upper limit of 0.02 molecules (on average) in our effective SERS probing area, and (ii) the distinct relative intensity patterns observed in the different spectra recorded at low molecular coverage, we can attribute the observed spectrum to the Raman signature of a single DMS molecule. More evidence in support of this assignment will follow. Using the molecular polarizability derivative tensors derived from B3LYP/6-311^{++}G** calculations, we identify a unique average orientation of a single molecule with respect to vector components of the local electric fields according to eqn (5). This is first done in the molecular frame by fixing the molecule and rotating the incident and scattered electric fields until a satisfactory match is found.[13,16] Projecting into the laboratory frame, defined in the inset of Fig. 5, and recognizing that the local electric fields are unidirectional (in the x-direction defined in Fig. 5), we obtain an average orientation in which the

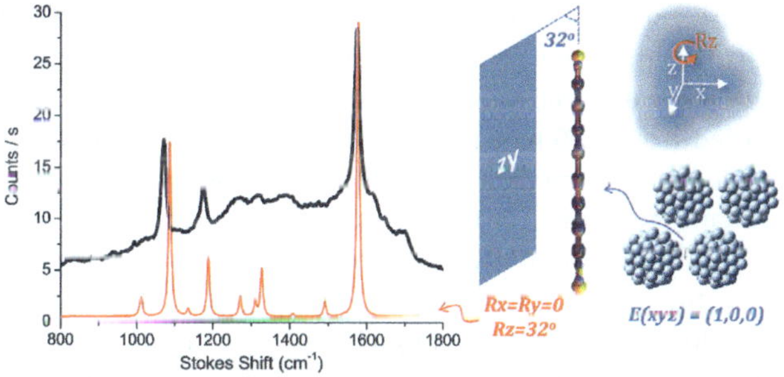

Fig. 5 The SERS sequence shown in panels C and D of Fig. 4 is time averaged (black trace) and compared to the Raman spectrum of an oriented single DMS molecule (red trace) simulated according to eqn (5) in the main text. The figure also shows the laboratory frame and schematic illustrations of the substrate and derived orientation of the molecule with respect to the local electric fields defined by our experimental geometry. The reader is referred to the main text for more details.

molecule is nearly flat with respect to the surface of the Ag spheres defining the nanojunction. Specifically, the long axis of the molecule is oriented along the z-axis of the laboratory frame, and the molecule's π-framework is rotated by 32° with respect to the zy plane of the same frame. That the field is unidirectional may be rationalized on the basis of our substrate and experimental geometry. Namely, given a laser source polarized in the sample plane (inverted optical microscopy setup), the inter-particle regions are the only regions in which the local electric fields are maximally enhanced such that the optical signature of a single molecule is detectable.[22] Note that assuming xz-plane polarized laser irradiation to excite the plasmonic junction defined by a molecule positioned between two particles aligned along xz (see the inset of Fig. 5) would yield the same relative orientation of the molecule in the corresponding plasmonic junction. As illustrated in Fig. 5, the simulated Raman spectrum of the oriented single molecule matches the intensity pattern observed experimentally reasonably well. Overall, we employ $3N - 6$ (the number of vibrational degrees of freedom) equations to solve for molecular orientation relative to the vector components of the incident/scattered local fields; the derived mathematical solution is rigorous. That said, there are limitations to the framework used herein,[13,36,37] mostly stemming from the approximations invoked in the level of theory employed to compute the molecular polarizability derivative tensors. Those include assuming (i) separable polarizabilities of the molecule and metallic systems, (ii) that the molecular polarizabilities which govern Raman scattering activities are invariant to molecular orientation with respect to the metal substrate, and (iii) truncation at the dipole limit[13] in the expansion of molecular polarizability. These approximations may prove inadequate in many cases, three of which have been noted in recent reports.[13,36,37] Nonetheless, the molecular orientation derived from our analysis accounts for many of the experimental observables described thus far, as discussed below.

3.5 Spectral fluctuations, background fluctuations, and line broadening

The main observations in SERS from 3.8×10^5, 3.8×10^2, and 0.38 DMS molecules per μm^2 adsorbed on a 2D Ag nanosphere assembly are now reiterated, and an expanded interpretation of our experimental observables is provided. The relative intensities of the observable vibrational eigenstates in the single molecule regime are distinct from their analogues that were recorded from the molecular ensembles. When compared to ensemble average measurement, hundreds of spectra recorded from different regions of the substrate containing only a few DMS molecules per probing area exhibit similar trends: (i) an increased ratio of the predominant aromatic C=C stretching (1575 cm^{-1}) to the vinyl C=C stretching (1625 cm^{-1}) vibrations, and (ii) an increased ratio of the C–S stretching (1085 cm^{-1}) to the aromatic C–H in-plane rocking (1186 cm^{-1}) vibrations. The recorded intensity patterns may be uniquely reproduced according to eqn (5). The picture painted through this analysis necessitates that the DMS molecule resides at a nanojunction defined by two silver nanoparticles in the array and is lying nearly flat at the plasmonic nanojunction.

Decreasing the number of molecules in the SERS probing area leads to a larger SERS background. These backgrounds are individual events captured along sequentially recorded SERS spectra, and representative sequentially recorded spectra are shown in Fig. 4A and B. Such events are most noticeable at a coverage of 0.38 molecules per μm^2, obscured but still observable at 3.8×10^2 molecules per μm^2, and mostly averaged out at higher molecular coverage. Individual background fluctuation events sum up to comprise the overall background observed in the temporally averaged spectra. We associate the background-like SERS spectra along the SERS sequences to instances in which the π-framework of the DMS molecule is parallel to the metal surface. Given the average molecular orientation derived from the relative intensities in the single molecule regime, this only requires a 32° rotation around the z axis of the laboratory frame, see Fig. 5. Similar broadened Raman signals have been previously observed in SERS from flat adsorbates, a classical example of which is the SERS spectrum of benzene on Au.[38] They are thought to arise from aromatic ring-metal surface electron density overlap.[39] We similarly associate the intermittent broadened molecular spectra with instances at which the π-framework of DMS is flat with respect to the metal. Rapid dephasing of the vibrational transitions induced by transient flat adsorption events would also explain our observation of broadened Raman lines, see Fig. 2C–E. A closer inspection of Fig. 5 also reveals that in the 1200–1500 cm^{-1} spectral region, at least 5 predicted vibrational transitions appear as broadened transitions atop the averaged background which spans the entire frequency range. In this context, it appears that homogeneous broadening governs the observable line shapes in junction plasmon-enhanced SERS in the single molecule limit. To comment on why some modes are more rapidly dephased than others would require a more extensive mode-specific theoretical analysis, a challenge we will take up in future work.

3.6 Single molecule SERS enhancement factor

Using the signal derived from the spectrum shown in Fig. 5, we may estimate a SERS enhancement factor (EF) according to[13,39]

$$I = \mathrm{EF}\ I_0 \eta_{\max} \frac{\partial\sigma}{\partial\Omega}\mathrm{d}\Omega = 390\ \mathrm{s}^{-1} \tag{6}$$

Based on a count rate of 390 s^{-1} obtained by integrating over the 1575 cm^{-1} band, an incident intensity of $I_0 = 7.76 \times 10^{22}$ photons cm^{-2} s^{-1} (300 $\mu W\ \mu m^{-2}$), an upper limit for detection efficiency $\eta_{\max} = 0.1$, a Raman scattering cross-section $\frac{\partial\sigma}{\partial\Omega} = 7.47 \times 10^{-28}\ cm^2\ sr^{-1}$ (calculated according to eqn (4) for an incident laser wavelength of 514 nm, $S_{1575\ cm^{-1}} = 11\ 818$ Å^4 amu^{-1}), and a solid angle $\mathrm{d}\Omega = \pi$, we derive an enhancement factor of 2.1×10^6 for the bright aromatic C=C stretching vibration at 1575 cm^{-1}. A plasmonic enhancement factor on the order of 10^6 is certainly attainable at the plasmonic nanojunctions within the 2D Ag nanosphere array SERS substrate used in this work.[22] The largest error in the current estimation arises from the computed $\frac{\partial\sigma}{\partial\Omega}$ value, which (i) does not account for molecular orientation, as $S_{1575\ cm^{-1}}$ is an orientationally averaged quantity (see eqn (1)–(3)), and (ii) is derived from an isolated gas-phase DMS molecule and hence does not account for static and dynamic changes in molecular polarizability as a result of chemisorption/adsorption onto a metallic surface. Although quantitative estimates of both (i) and (ii) are challenging, we will attempt to shed light on both effects by relying on our density functional theory calculations and single molecule framework and experiments.[15,16,19] Given the molecular orientation derived from our single molecule Raman simulation according to eqn (5), we find that the largest Raman scattering cross-section for the 1575 cm^{-1} mode is attainable at $E_i^L = E_s^L = (0,0,1)$ in the laboratory frame, see Fig. 5. This is expected, given the large polarizability associated with π-conjugated electrons aligned along the *z*-axis. When $E_i^L = E_s^L = (1,0,0)$, *i.e.* the same relative orientation used to produce the fit shown in Fig. 5, we find that the calculated EF is underestimated by a factor of $\sim 10^5$ where the rotationally averaged value of $S_{1575\ cm^{-1}}$ is used to calculate EF based on eqn (6). On the other hand, using an $S_{1575\ cm^{-1}}$ value derived from the isolated molecule yields an overestimated EF value, as static and dynamic polarizability changes to the molecule upon chemisorption are expected to render DMS a more efficient Raman scatterer. To account for chemical effects, we consider another model structure in which DMS is chemisorbed onto the vertex of a finite pyramidal Ag_{20} cluster and recalculate $S_{1575\ cm^{-1}}$.[18,19] Here, only a small factor of 30× increase in Raman scattering activity (30× decrease in EF) is obtained, consistent with a prior analysis of chemical enhancement factors in SERS of aromatic dithiols.[12] Taking both the molecular orientation dependence and chemical effects into account allows us to re-estimate the EF to be $\sim 10^{10}$, required to observe 390 counts per second from a single DMS molecule under the above-described experimental conditions. To the best of our knowledge, this is the first report in which molecular orientation is taken into account in estimating SERS enhancement factors.

4 Conclusions

We report SERS from a uniform 2D array of Ag nanospheres coated with increasingly sparse distributions of DMS molecules using ESD. Knowing the concentration and volume of the DMS solutions deposited into an area >1 cm in diameter, we are able to estimate an upper limit for the number of molecules in our effective SERS probing area. The lowest concentration used featured $\ll 1$ molecule per probed area. Systematically decreasing the molecular coverage of DMS resulted in spectral changes: (i) more pronounced SERS background signals at lower concentrations, (ii) increased widths of the observable vibrational lines at lower concentrations, and (iii) systematic changes in the relative intensities of the observable vibrational signatures of DMS. The backgrounds observed in the spatially and temporally averaged SERS spectra are associated with individual events that are particularly evident at pM concentrations, observable but less pronounced at nM concentrations, and less evident at higher molecular coverage. Namely, transient flattening of a DMS molecule at a plasmonic junction leads to a broadened response of molecular origin. Overall, these individual background fluctuations sum up to a SERS background in the time-averaged signals. This picture is also consistent with the observed Raman line broadening. In this regard, homogeneous broadening of a molecule interacting with its local environment governs the observable line shapes in junction plasmon-enhanced SERS from DMS molecules at classical plasmonic junctions. The observation of distinct relative intensities in the low *vs.* high coverage regimes is associated with removal of Euler averaging which exposes the tensorial nature of Raman scattering in the single molecule limit. Using a representative sequence recorded at the single molecule limit, we identify a time-averaged orientation of a single DMS molecule at a plasmonic junction formed between two Ag particles in the array. Taking the derived molecular orientation into account allows us to estimate a SERS enhancement factor of up to $\sim 10^{10}$ in our measurements. Overall, our analysis suggests that molecular orientation is a key factor that needs to be taken into consideration when estimating single molecule SERS enhancement factors. Herein, we find that in the case of a nearly flat-lying π-conjugated molecule, molecular reorientation may affect the derived SERS enhancement factor by about five orders of magnitude.

The inference of single molecule detection sensitivity in this work was not proven with a definite test, *e.g.* using the bi-analyte approach.[10] Nonetheless, given (i) our SERS-active substrate preparation method of choice (ESD) which yields an upper limit of 0.02 molecules in our effective SERS probing area, (ii) the above-mentioned spectral signatures that can be associated with the removal of orientational averaging, and (iii) a derived single molecule SERS enhancement factor of $\sim 10^{10}$, it is likely that all the spectra that were recorded at ultralow molecular coverage are those of single molecules. Our analysis also suggests that time-dependent orientational dynamics which affect the electronic structure of a molecular system chemisorbed/physisorbed onto metal substrates comprise another factor which needs to be considered in an effort to accurately account for single molecule SERS observables. We hope that this contribution fuels future theoretical and experimental works aimed at attaining a detailed understanding of specific molecule–metal interactions which govern surface-enhanced linear and nonlinear ultrasensitive spectroscopies.

Acknowledgements

WPH, JL, AGJ, YG, and GEJ acknowledge support from the US Department of Energy (DOE), Office of Science, Office of Basic Energy Sciences, Division of Chemical Sciences, Geosciences & Biosciences. PZE acknowledges support from the Laboratory Directed Research and Development Program (LDRD) through a Linus Pauling Fellowship at Pacific Northwest National Laboratory (PNNL), an allocation of computing time from the National Science Foundation (TG-CHE130003), and the use of the Extreme Science and Engineering Discovery Environment. A portion of the work was performed using EMSL, a national scientific user facility sponsored by the DOE's Office of Biological and Environmental Research and located at PNNL. This work also benefitted from PNNL Institutional Computing resources. PNNL is a multiprogram national laboratory operated for DOE by Battelle.

References

1 M. Fleischmann, P. J. Hendra and A. J. McQuillan, *Chem. Phys. Lett.*, 1974, **26**, 163.

2 D. L. Jeanmaire and R. P. Van Duyne, *J. Electroanal. Chem.*, 1977, **84**, 1.

3 M. Moskovits, *J. Chem. Phys.*, 1978, **69**, 4159.

4 A. Otto, *Surf. Sci.*, 1978, **75**, L392.

5 N. J. Halas, S. Lal, W.-S. Chang, S. Link and P. Nordlander, *Chem. Rev.*, 2011, **111**, 3913.

6 E. Hao and G. C. Schatz, *J. Chem. Phys.*, 2004, **120**, 357.
7 F. J. Garcıa de Abajo, *J. Phys. Chem. C*, 2008, **112**, 17983.
8 S. Nie and S. R. Emory, *Science*, 1997, **275**, 1102.
9 K. Kneipp, Y. Wang, H. Kneipp, L. T. Perelman, I. Itzkan, R. R. Dasari and M. S. Feld, *Phys. Rev. Lett.*, 1997, **78**, 1667.
10 E. C. Le Ru and P. G. Etchegoin, *Annu. Rev. Phys. Chem.*, 2012, **63**, 65.
11 B. Pettinger, P. Schambach, C. J. Villagomez and N. Scott, *Annu. Rev. Phys. Chem.*, 2012, **63**, 379.
12 M. Banik, A. Nag, P. Z. El-Khoury, A. Rodriguez-Perez, N. Guarrotxena, G. C. Bazan and V. A. Apkarian, *J. Phys. Chem. C*, 2012, **116**, 10415.
13 M. Banik, P. Z. El-Khoury, A. Nag, A. Rodriguez-Perez, N. Guarrottxena, G. C. Bazan and V. A. Apkarian, *ACS Nano*, 2012, **6**, 10343.
14 M. D. Sonntag, D. Chulhai, T. Seideman, L. Jensen and R. P. Van Duyne, *J. Am. Chem. Soc.*, 2013, **135**, 17187.
15 P. Z. El-Khoury and W. P. Hess, *Chem. Phys. Lett.*, 2013, **581**, 57.
16 P. Z. El-Khoury, D. Hu and W. P. Hess, *J. Phys. Chem. Lett.*, 2013, **4**, 3435.
17 P. Z. El-Khoury, D. Hu, V. A. Apkarian and W. P. Hess, *Nano Lett.*, 2013, **13**, 1858.
18 P. Z. El-Khoury and W. P. Hess, *Nano Lett.*, 2014, **14**, 4114.
19 P. Z. El-Khoury, T. W. Ueltschi, A. L. Mifflin, D. Hu and W. P. Hess, *J. Phys. Chem. C*, 2014, **118**, 27525.
20 P. Z. El-Khoury, Y. Gong, P. Abellan, B. W. Arey, A. G. Joly, D. Hu, J. E. Evans, N. D. Browning and W. P. Hess, *Nano Lett.*, 2015, **15**, 2385.
21 R. Esteban, A. G. Borisov, P. Nordlander and J. Aizpurua, *Nat. Commun.*, 2012, **3**, 1.
22 P. Z. El-Khoury, E. Khon, Y. Gong, A. G. Joly, P. Abellan, J. E. Evans, N. D. Browning, D. Hu, M. Zamkov and W. P. Hess, *J. Chem. Phys.*, 2014, **141**, 214308.
23 A. Li, Z. Baird, S. Bag, D. Sarkar, A. Prabhath, T. Pradeep and R. G. Cooks, *Angew. Chem., Int. Ed.*, 2014, **53**, 12528.
24 A. K. Badu-Tawiah, C. Wu and R. G. Cooks, *Anal. Chem.*, 2011, **83**, 2648.
25 K. D. D. Gunaratne, V. Prabhakaran, Y. M. Ibrahim, R. V. Norheim, G. E. Johnson and J. Laskin, *Annu. Rev. Anal. Chem.*, 2011, **4**, 83.
26 M. J. Frisch, G. W. Trucks and H. B. Schlegel, *et al.*, *GAUSSIAN 09, revision D.01*, Gaussian, Inc., Wallingford, CT, 2009.
27 A. D. Becke, *J. Chem. Phys.*, 1993, **98**, 5648.
28 C. Lee, W. Yang and R. G. Parr, *Phys. Rev. B: Condens. Matter Mater. Phys.*, 1988, **37**, 785.
29 S. H. Vosko, L. Wilk and M. Nusair, *Can. J. Phys.*, 1980, **58**, 1200.
30 P. J. Stephens, F. J. Devlin, C. F. Chabalowski and M. J. Frisch, *J. Phys. Chem.*, 1994, **98**, 11623.
31 R. Krishnan, J. S. Binkley, R. Seeger and J. A. Pople, *J. Chem. Phys.*, 1980, **72**, 650.
32 A. D. McLean and G. S. Chandler, *J. Chem. Phys.*, 1980, **72**, 5639.
33 K. Ruud, T. Helgaker and P. Bour, *J. Phys. Chem. A*, 2002, **106**, 7448.
34 O. Quinet and B. Champagne, *J. Chem. Phys.*, 2001, **115**, 6293.
35 V. Barone, *J. Chem. Phys.*, 2005, **122**, 014108.
36 P. Z. El-Khoury, E. J. Bylaska and W. P. Hess, *J. Chem. Phys.*, 2013, **139**, 174303.

37 P. Z. El-Khoury, K. Honkala and W. P. Hess, *J. Phys. Chem. A*, 2014, **118**, 8115.
38 Z. Gao, J. P. Davies and M. J. Weaver, *J. Phys. Chem.*, 1990, **94**, 6858.
39 P. Z. El-Khoury, S. J. Peppernick, D. Hu and W. P. Hess, *J. Phys. Chem. C*, 2013, **117**, 7260.

Faraday Discussions

PAPER

Imaging the behavior of molecules in biological systems: breaking the 3D speed barrier with 3D multi-resolution microscopy

Kevin Welsher† and Haw Yang*

Received 22nd May 2015, Accepted 30th June 2015
DOI: 10.1039/c5fd00090d

The overwhelming effort in the development of new microscopy methods has been focused on increasing the spatial and temporal resolution in all three dimensions to enable the measurement of the molecular scale phenomena at the heart of biological processes. However, there exists a significant speed barrier to existing 3D imaging methods, which is associated with the overhead required to image large volumes. This overhead can be overcome to provide nearly unlimited temporal precision by simply focusing on a single molecule or particle *via* real-time 3D single-particle tracking and the newly developed 3D Multi-resolution Microscopy (3D-MM). Here, we investigate the optical and mechanical limits of real-time 3D single-particle tracking in the context of other methods. In particular, we investigate the use of an optical cantilever for position sensitive detection, finding that this method yields system magnifications of over 3000×. We also investigate the ideal PID control parameters and their effect on the power spectrum of simulated trajectories. Taken together, these data suggest that the speed limit in real-time 3D single particle-tracking is a result of slow piezoelectric stage response as opposed to optical sensitivity or PID control.

Introduction

Imaging is an attempt to capture a static snapshot of the contents of a scene. When a scene is dynamically changing, several snapshots, or frames, are taken in rapid succession to capture the progression of a temporally evolving event. At the macroscale, our ability to capture dynamic events is limited in the spatial domain primarily by the size and number of pixels in our detector, while the temporal resolution is dictated by how frequently frames can be read out. With the development of high resolution and high speed cameras, nearly any macroscopic event

Department of Chemistry, Princeton University, New Jersey, USA. E-mail: hawyang@princeton.edu
† Current address: Department of Chemistry, Duke University, Durham, North Carolina 27708. Email: kevin.welsher@duke.edu.

can be captured, from the rupturing of a balloon to the exploding of a bullet from the barrel of a gun.

Imaging in live cell microscopy is aimed at capturing the state and dynamics of cellular behavior. At the heart of these cellular scale events are molecules, such as proteins, small molecule co-factors and nucleic acids, whose dynamics and interactions lead to all cellular and organism level function. At these microscales, we encounter many of the same obstacles as macroscale imaging, with a couple of added challenges. First, spatial resolution ceases to be limited by the pixel density of our detector, but rather by the diffractive nature of light, which limits the resolvability of objects to about half the wavelength light used, meaning that structures below ~200 nm cannot be resolved with light from the visible spectrum[1] and that we cannot rely on simple scattering or reflection of light to follow a specific molecule, but need specific labeling strategies so that only the molecule of interest yields a signal while others do not. As a result, live cell imaging has relied heavily on fluorescence labeling to follow specific molecular details, despite existing in a sea of other unlabeled molecules. Live cell microscopy also suffers from limited temporal resolution. The reliance on fluorescence, where a single molecule may only yield a few thousand photons per second, has led to longer integration times and slower "frame rates". At the same time, the narrow depth of field has commonly been overcome by stacking several images together at different focal planes, further reducing the temporal resolution. The combined lack of spatial, temporal and 3D resolution has long hampered our ability to observe molecular phenomena on their native length scale (down to <1 nm) and on a time scale relevant to molecular dynamics (sub-millisecond), while simultaneously placing these interactions and dynamics in their macromolecular biological context. The traditional methods of observing molecular biology in context have been optical imaging methods, starting with the most predominantly used and commercially available method: confocal microscopy.

Scanning methods

In confocal fluorescence microscopy, as originally implemented,[2–4] the sample was physically scanned through the detection volume to generate an image pixel by pixel. Current laser scanning microscopes (LSMs) implement resonant galvanic scanners, which can easily achieve video rate scans of a single optical section.[5] However, imaging with LSMs is still fundamentally limited by the amount of time the scanner spends on a particular pixel (the pixel dwell time). As the scan rate increases, the pixel dwell time drops. For example, when sampling a 512×512 image at 1 Hz, the pixel dwell time is about 4 μs, assuming zero time for mirror flyback. To achieve video rate (30 Hz), the dwell time becomes 133 ns, drastically reducing the signal to noise in each pixel. Today, LSMs based on resonant scanners can achieve modest signal-to-noise 3D volumes on the order of 0.1 volumes per second (VPS).[6] To overcome these short pixel dwell times, a parallelized approach was needed. This led to the implementation of Nipkow spinning disks.[7] These high speed confocal units have really pushed the limits of confocal microscopy, allowing rendering of 3D volumes at up to 1 VPS,[8] with near confocal performance.[9] While the spinning disk method allows faster imaging, it is a difficult method to implement, as the pinhole array is constructed for a particular objective (typically 100×, NA = 1.4), making switching to larger length scales difficult. Moreover, these time scales (≥1 second) are far too slow to monitor chemical dynamics at the cellular or subcellular level.

Due to the optical diffraction limit, the smallest resolvable features visible to confocal microscopes are on the order of 200 nm in the lateral dimensions and 600 nm in the axial dimensions.[1] Early attempts to increase the spatial resolution of these microscopes included the addition of an additional detection/excitation objective to increase the solid angle covered by the microscope and to effectively reduce the point-spread function (PSF) of the system. This "4Pi" microscopy[10] increased the axial resolution of a confocal microscope down to ~100 nm.[11] Other methods have sought to bypass the diffraction limit entirely. Nonlinear methods, such as stimulated emission depletion microscopy (STED)[12] and reversible saturable optical fluorescence transitions (RESOLFT),[13] use shaped excitation pulses to effectively turn off fluorescence emission in a certain area. By carefully shaping the laser pulse to define a spherical area with a node of zero intensity at the focal spot, the focal spot size can be effectively reduced in size due to the depletion of the surrounding emission. In RESOLFT, photoswitchable probes are set to their dark state everywhere except in a small dub-diffraction region around the focus. In STED, the depletion pulse promotes stimulated emission to the point that the excited state lifetime is zero and there is a very low chance of spontaneous emission. Both of these approaches have been extremely successful in surpassing the diffraction limit, with STED reaching isotropic focal spots with a resolution down to 30 nm, enabling imaging of mitochondrial cristae in live cells.[14]

While all of these point-scanning methods have made huge leaps in the ability to resolve biological features approaching the molecular scale, they carry certain limitations when it comes to placing these phenomena in the context of larger scale biological processes, such as in multicellular organisms. For point scanning methods, observing large scales means trading off for slower observation times, especially when the super-resolution voxels necessitate extremely small step sizes. For instance, high speed STED imaging has been achieved with frame rates up to 28 Hz, but only for very small imaging areas (2.5 μm × 1.8 μm).[15] This video rate temporal resolution is quickly lost when the observation area needs to be expanded. This has started to be addressed by using massively parallelized STED,[16] though this method still requires a few hundred camera frames per image, which precludes it from being used for high speed 3D imaging. Similar improvements have been made in RESOLFT, achieving a super-resolved image of a 120 μm × 100 μm field of view in less than 1 second.[17] Of course, these imaging rates will be greatly decreased as these methods are scaled to 3 dimensions.

Super-resolution localization methods

Other methods have relied on localization of single fluorophores. Despite the limitations of optical microscopes, the location of single fluorophores can be known with great precision. The image of a fluorophore, which is effectively a point source, is a Gaussian profile, which is determined by the PSF of the objective lens. The spread of the profile is determined by the diffraction limit, but the precision with which the center of the distribution can be measured is determined by the number measurements that are made on the molecule of interest. Given enough photons and perfect instrumentation, the location of a single molecule can be known down to nanometer resolution.[18,19] The difficulty with applying this method to real biological systems is the difficulty in locating a single molecule among many. Several methods have tried to solve this problem using blinking or photoswitchable probes. Sparse sets of these fluorophores are imaged in a sample repeatedly until the full image is reconstructed. These methods (stochastic optical reconstruction microscopy (STORM)[20] and fluorescence photoactivated localization microscopy (PALM[21] and fPALM[22])) have shown the ability to resolve phenomena down to the 10s of nm scale.

While these methods were initially implemented for investigating phenomena in two dimensions, they have been extended to three dimensions *via* a variety of different methods. One method is astigmatic imaging, which uses a cylindrical lens to deliberately skew the PSF such that the image of a single molecule changes as a function of defocus (STORM, 20–30 nm lateral, 60 nm axial).[23,24] Other methods have relied on the simultaneous imaging of two focal planes (biplane imaging, BP-PALM, 30 nm lateral, 75 nm axial)[25–27] or interferometry (iPALM, 20 nm lateral, 10 nm axial).[28] Perhaps the most successful approach implemented thus far is the double helix PSF (DH-PSF, 10 nm lateral, 10 nm axial),[29] which uses a spatial light modulator or phase mask to encode 3D information into the PSF of a single emitter. Unfortunately, all of these efforts to extend these single molecule approaches to three dimensions suffer from a limited range. While the astigmatism, biplane and double helix PSF methods have proven to be great methods for gathering high resolution axial localization data without requiring scanning optical slices, they come with the drawback that they only maintain their precision over 1–2 microns at the most,[30] severely limiting the extension to larger structures. Further work on the SLM or phase mask approach has extended this range to the 5–6 micron range.[31] However, as these PSFs become more complex, the molecules in each frame will need to be separated by greater and greater distances for individual spots to be identified. On the temporal side, these approaches require the observation of many switching events to isolate individual fluorophores, which initially limited their temporal resolution to 10 s or more per image. Recent advances, however, have shown improved temporal sampling, down to 1–2 seconds per image.[32,33] Ultimately, the advent of faster imaging sCMOS based cameras will improve this limited temporal sampling, which has already shown the potential to achieve 32 frames per second.[34]

Rapid volumetric imaging

Several methods have tried to solve the problem of imaging large 3D volumes with high temporal resolution. One of these methods is selective plane illumination microscopy (SPIM).[35] This method uses an excitation beam that is spread out by a cylindrical lens and delivered to the sample perpendicular to the collection objective, creating a plane of illumination, which allows for optical sectioning. Fast imaging of volumes of 400 × 400 × 200 μm with neuronal action potentials every 6 seconds have been achieved.[36] The axial resolution of this method is unfortunately limited by the spread of the excitation beam, on the order of 5 microns. This method has been improved by rapidly scanning a laser beam to create an illumination plane, allowing for more intense illumination and faster acquisition times, acquiring volumes of 1000 × 1000 × 600 μm in 60–90 seconds with 300 nm lateral and 1000 nm axial resolution.[37,38] More recently, Hillman and colleagues have taken a different approach, using a rotating polygonal mirror to scan a volume at high speed, showing the ability to achieve 3D imaging at up to 10 volumes per second.[39] However, these methods are still on the "cellular" scale, and may have great promise for studying organism level behavior, but will struggle to study the behavior of molecules.

Probably the most impressive rapid volumetric imaging achieved to date has been the so-called "Bessel-beam" approach. This approach overcomes the limited axial resolution of the SPIM approach by using a narrow excitation beam created using an annular transmission mask and a low magnification objective lens, which is perpendicular to the observation objective. This narrow excitation beam is then scanned through the sample to generate a 3D image. The implementation of this Bessel beam with two-photon excitation enables axial resolution down to 500 nm.[6] Impressively, entire cellular processes can be observed at rates approaching 1 volume per second.[40]

Possibly the most well established super-resolution method, which aims to combine the benefits of super-resolution with larger scale observations, is the structured illumination microscope (SIM).[41] By implementing excitation fields with frequencies near the spatial frequencies of the sample, a lower beat frequency, observable by a traditional microscope objective, can be created. This method has shown the ability to improve the lateral resolution down to ~110 nm and axial resolution down to ~350 nm.[42,43] 3D-SIM has been improved to achieve 0.25 VPS in single color and 0.11 VPS in two color imaging. These speeds have been far surpassed by a completely analog version, wherein the need to collect 10–100 digital images per optical section is eliminated, allowing 100 Hz imaging in two dimensions.[44]

Despite efforts to push for higher temporal and spatial resolution in microscopy, there is still a huge range of length and time scales that go unobserved. Fig. 1 shows a representative resolution/scale plot for the temporal and spatial capabilities of several of the methods above. The plot shows the resolution of each method, both spatial and temporal as well as the spatial "scale", or the spatial extent over which imaging is possible. For each method, the best spatial and temporal resolution for 3D measurements was chosen, with the spatial resolution taken from the worst of the lateral or axial resolution. The spatial scale was selected given the results for the highest frame rates achieved for each method. Here, we have omitted characterization of the temporal scale, as this is difficult to compare without comparing all methods under similar conditions to account for photobleaching, toxicity, *etc.* In the upper right corner of the plot, we find the traditional point scanning methods. Moving to the left (towards greater spatial resolution), we find the "super-resolution" methods. For instance, the single molecule methods, such as DH-PSF and iPALM, show remarkable spatial resolution, as illustrated by their circles on the left hand side of the plot. In the lower right hand side, we have the rapid volumetric imaging methods, which have achieved some improvement in their temporal resolution, but still have relatively limited spatial resolution.

When plotting the data in this way there is a noticeable lack of methods that approach the native spatial and temporal scales of molecular interactions (nanometer and microsecond, respectively). Upon inspection, the reason for this lack is obvious: imaging an entire scene carries far too much observational overhead to allow for the capturing of the sub-millisecond dynamics of a single molecule. This can be thought of as the inverse of the scaling problem observed in

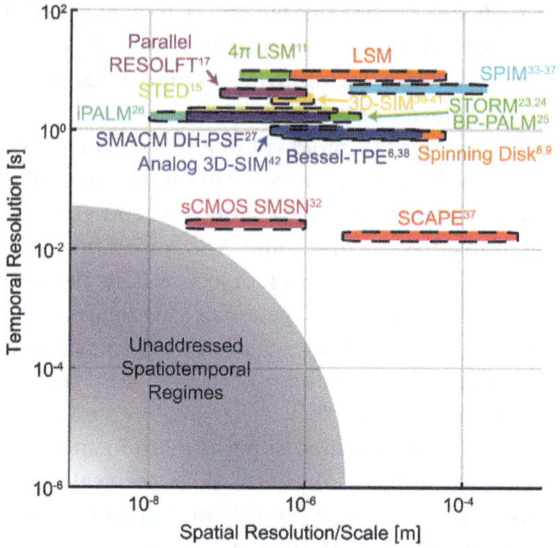

Fig. 1 Spatiotemporal resolution of common imaging methods. Each rectangle represents the spatial and temporal precision of a 3D imaging method. The left bound of each rectangle represents the spatial precision of each method. In the case of methods where there is anisotropy between the lateral and axial resolution (as is commonly the case), the worse of the two is plotted. The right edge of each rectangle represents the maximum spatial scale that each method can address. Again, the smaller of the lateral and axial scale is plotted. On the temporal axis, the bottom of each rectangle represents the fastest imaging rate for each method in 3D for the given spatial precision and scale. While the spatial scales along the x-axis have been fully covered, this has been done at the expense of the temporal precision, as demonstrated by the lack of methods that can reach time scales below 10 ms.

LSM or other point scanning methods. When increasing the spatial scale of our observation, by necessity, the temporal resolution will suffer from the increased data overhead at a given spatial resolution. Given this reasoning, it is easy to imagine that unlimited temporal sampling can be achieved in the limit of an extremely small spatial scale and short exposure time. Effectively, if we want to understand the dynamics of a molecule with the highest possible precision, we should focus solely on that molecule.

To address this shortcoming, we developed a new hybrid implementation where the high spatial and temporal resolution real-time 3D single-particle tracking[45–48] is combined with the large-scale contextual information of a conventional two-photon scanning fluorescence microscope. This method, called 3D Multi-resolution Microscopy (3D-MM) is a breakthrough for correlating fast subcellular dynamics in a larger cellular context.[49] 3D-MM is implemented through the spectral separation and data synchronization of two imaging modalities: a real-time 3D single-particle tracking module and a two-photon LSM. The real-time 3D single particle-tracking module locks quickly moving particles in the focus of the objective *via* a piezoelectric stage and an optical feedback loop running at 100 kHz. The repeated centroid measurement of the particle at a high rate results in spatial resolution down to 10–20 nm, placing it on the left hand side of Fig. 1. Further, the high sampling rate imparts a time resolution of 10 μs, a vast improvement over other methods. The reason for this improved temporal resolution is straightforward. Focusing solely on a single probe in the focal volume of the objective and disregarding contextual resolution leads to an extremely narrow spatial scale. The spatial scale is essentially equal to the spatial resolution. As a result, real-time 3D single particle tracking is able to beat super-resolution methods in terms of the combined spatial and temporal resolution, but is incapable of placing these high resolution measurements in a larger scale biological context. In simpler terms, the highest possible spatiotemporal precision can be achieved by discarding all observational overhead, with no measurement of the areas around the particle. Combining this real-time measurement of a single probe with the large scale context provided by the two-photon LSM yielded the first fully resolved, real-time 3D data of an extracellular nanoparticle approaching and landing on the cell surface. The power of 3D-MM to address multiple spatiotemporal scales is demonstrated in Fig. 2. Fluorescently labeled 100 nm nanoparticles labeled with the TAT peptide[50] were prepared as described in ref. 49 and allowed to bind to the surface of fluorescent protein labeled NIH-3T3 cells. The 3D-MM technique allows the simultaneous observation of the larger cellular environment *via* the two-photon LSM (Fig. 2a and b) while recording the real-time 3D position of the nanoparticle at 100 kHz. By tracing the particle's position, the isotropic precision (<15 nm in *XYZ*) carves out 3D sub-diffraction structures, such as hemispherical and cylindrical membrane protrusions, which are unresolvable by conventional imaging methods (Fig. 2c and d). Data like these could be revolutionary to the fields of virology and drug delivery.

3D-MM is a powerful technique that is able to bridge multiple spatiotemporal scales in a single method, which is a necessity for the evaluation of molecular dynamics in biological systems. Critical to the application of the 3D-MM technique is the capacity to follow single particles in real-time in three dimensions with high precision. The ability to follow diffusive particles is dependent on two factors. First, there must be an optical readout of the particle's position, which is sensitive to very small changes. Second, there needs to be a sufficiently fast response mechanism to keep the particle in the focus of the objective. By keeping the particle in the small focal volume, we can use single point detectors and achieve near unlimited temporal resolution, as shown in Fig. 3.

Optical response

While the optical feedback differs with each implementation,[45–49,51–53] each real-time 3D single-particle tracking method relies on getting the position sensitive detection of the particle to be tracked with high temporal resolution. The simplest way one can think of doing this is to magnify the image of the particle onto a position sensitive detector. The largest challenge here is achieving a large enough magnification such that the detector is sensitive to nanometer scale displacements of the particle in real-time. For a typical high magnification system (*e.g.* 100×), a 1 nm displacement of a particle will only result in a 100 nm displacement of the particle image given a standard tube lens configuration. If this signal is split onto different detectors, there may not be sufficient signal to noise to detect this difference. To achieve higher magnification, the most straightforward approach is the use a lens pair after the microscope tube lens in order to increase

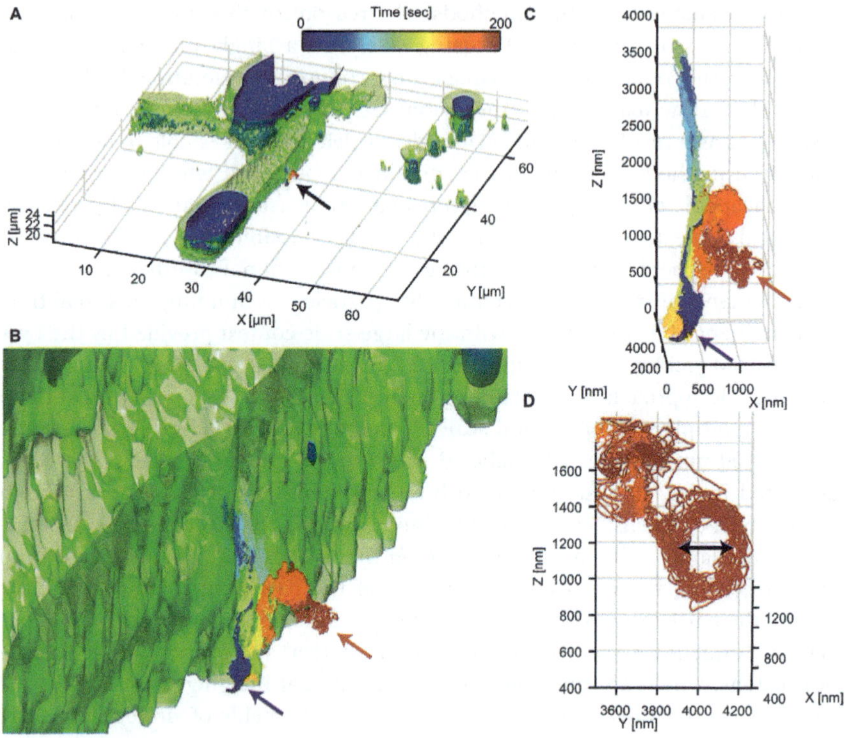

Fig. 2 3D-MM measurement of a membrane bound nanoparticle. (A) 100 nm TAT labeled fluorescent nanoparticles were tracked along the 3D cellular surface using the 3D-MM method. The nanoparticles were allowed to bind to NIH-3T3 cells expressing cytosolic GFP and CFP NLS, which served as labels for the 2P-LSM readout. The green surface represents the extent of the cytosol, while the blue surface shows the nucleus. The black arrow shows the location of the 3D nanoparticle trajectory within the 70 µm × 70 µm × 4 µm field of view. (B) A zoomed in image of the 3D-MM trajectory in (A), showing the particle diffusing along the membrane. As the particle diffuses, it traces out 3D structures, such as hemispherical (blue arrow) and cylindrical (red arrow) protrusions. (C) Edge-on view of the particle diffusing along the membrane surface, with multiple membrane features resolved by the particle's trajectory. (D) A zoomed in section of a cylindrical protrusion of ~250 nm in diameter carved out by the diffusing nanoparticle. Note that the high precision of the real-time 3D tracking is converted into a high-resolution structure by integrating the oversampled position over a long time.

the total system magnification. For example, a combination of a 75 mm and a 750 mm lens pair following the tube lens would result in a further 10× magnification, bringing the total system magnification to 1000×. However, this results in a detection mechanism that would change greatly with defocus of the target particle. As the goal is 3D particle tracking, we do not want our ability to track in *X* or *Y* to be affected by the ability to track in *Z*. The elegant solution to this problem is the optical cantilever.[47,48]

The optical cantilever delivers nanometer resolution to real-time 3D single-particle tracking by optically converting nanometer linear displacements of the target into angular variations, which at large distances become discernable shifts in the projected images. The mechanism behind the optical cantilever is similar to that of the lens pair magnification suggested above, without the final long focal length lens (above: 750 mm). Instead of focusing the particle's image onto the position sensitive detector, the infinity space projection of the particle's image is split, in this case using a prism mirror, onto two different detectors. The principle is shown in the cartoon inset of Fig. 4. As the particle is displaced in the object plane by a distance Δ, the angle with which the light exits the objective back aperture is deflected by angle θ, given by the relation:

$$\tan(\theta) = \left(\frac{\Delta}{f_{\text{obj}}}\right) \cdot \left(\frac{f_{\text{tube}}}{f_{\text{cantilever}}}\right) \tag{1}$$

where f_{obj}, f_{tube} and $f_{\text{cantilever}}$ are the focal lengths of the object lens, tube lens and optical cantilever lens, respectively. The offset, d, of the infinity space output of the optical cantilever is then given by:

$$\tan(\theta) = \frac{d}{L} \tag{2}$$

where L is the length of the optical cantilever, which is the distance between the cantilever lens and the prism mirror. Substituting for $\tan(\theta)$ yields:

$$d = \left(\frac{\Delta}{f_{\text{obj}}}\right) \cdot \left(\frac{f_{\text{tube}}}{f_{\text{cantilever}}}\right) \cdot L. \tag{3}$$

The ratio of L to $f_{\text{cantilever}}$ can be thought of as an effective magnification in addition to the magnification from the ratio of the tube lens to the objective lens focal length. The offset of the infinity space beam at the prism mirror is linearly

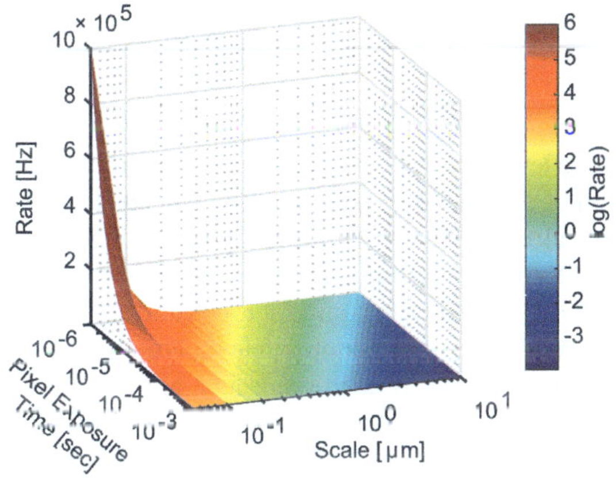

Fig. 3 Imaging rate *versus* spatial scale. For all the imaging methods, there is a trade-off between the maximum spatial scale that can be addressed and the imaging or frame rate. This is most obvious in the case of point scanning methods, such as LSM or STED. The plot shows the relationship between the maximum spatial scale addressed, the pixel dwell time and the overall imaging rates. When imaging volumes on the 10s of micron scale, the highest achievable rates are on the order of 1 volume per second (VPS), even with a pixel dwell time of 1 μs. As the spatial scale and pixel dwell time are reduced, the imaging rate increases exponentially. This implies that to achieve the highest possible measurement rates, it is necessary to minimize both the pixel integration time and the spatial scale addressed, which is exactly what real-time 3D single-particle tracking achieves.

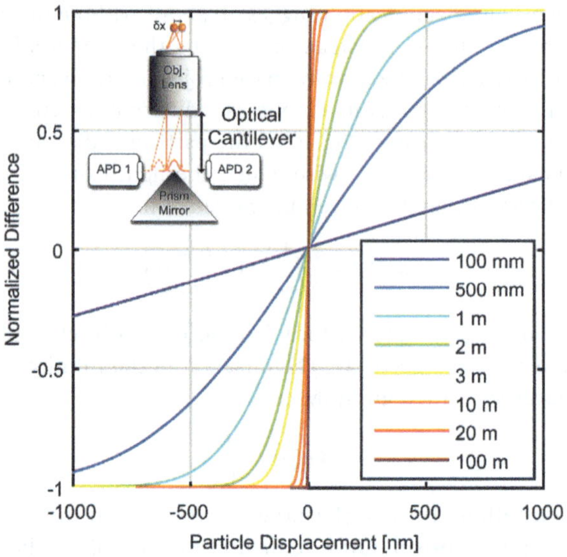

Fig. 4 The optical cantilever. The cartoon inset demonstrates the mechanism of the optical cantilever. As a particle is displaced in the object plane, the output light collected by the objective exits the back focal plane at an angle determined by the focal length of the lens. This infinity space light is then projected onto a prism mirror, which splits the light onto two detectors. The angular deviation of the beam leads to displacement of the center of the beam at the prism mirror and this displacement is linearly proportional to the distance between the back focal plane of the lens and the prism mirror, which is referred to as the optical cantilever. The plot shows the simulated response of the normalized difference of two detectors as a function of particle displacement in the object plane for different optical cantilever lengths. For short lengths (100 mm), the response is linear over a long range, but has a low sensitivity to particle displacement. For long lengths (100 m), the response becomes very steep, but is only linear over about 8 nm. The ideal response occurs for cantilever lengths on the order of a few meters.

proportional to the length of the optical cantilever, meaning that, to increase the optical resolution of the tracking system, one only needs to increase the distance between the optical cantilever lens and the position sensitive detector. As an illustrative example, the recent implementation of 3D-MM used a 100× Zeiss Plan Apo (NA = 1.4, f_{obj} = 1.65 mm), a 200 mm tube lens and a 75 mm optical cantilever lens to track 100 nm fluorescent nanoparticles.[49] For a particle displacement of 1 nm, the simple magnification of the objective and tube lens (Mag ≈ 121×) results in an image offset of 121 nm. However, given an optical cantilever length of 2 meters, this 1 nm displacement becomes a 3.2 μm image offset, yielding an effective magnification of 3200×.

The effect of the optical cantilever on the simulated optical response of the position sensitive detectors is shown in Fig. 4. To model the optical system response, a point source of 500 nm wavelength light (FWHM ≈ 268 nm) is imaged onto the prism mirror *via* the optical cantilever. The point source is imaged onto the intermediate image plane *via* the tube lens, which is then projected into infinity space by the optical cantilever lens. It should be noted that the optical cantilever lens serves two purposes here. First, as described above, it magnifies the angular displacement of the parallel bundle of rays exiting the microscope. Second, and almost as significant, the optical cantilever lens serves to shrink the bundle of rays. This means that for a given displacement at the prism mirror, the optical response will be even greater due to the smaller size of the infinity space Gaussian beam, which in this case has FWHM ≈ 1 mm if divergence is neglected. The process variable (PV) that is used for real-time tracking can be calculated by splitting this infinity space beam onto two perfectly integrating detectors as a function of particle displacement in the object plane. Fig. 4 shows the resulting calculation of this error function for a range of different optical cantilever lengths, ranging from 100 mm to 100 m. While cantilever lengths of 10 m or greater are experimentally untenable, they serve as a guide to the ultimate limit of this approach. At short cantilever lengths (100 mm), the optical error function is almost perfectly linear over at least 2 μm in the object space, but the position readout is shallow, with a slope of $\sim 3 \times 10^{-4}$ nm^{-1}. At the other extreme, an optical cantilever of 100 m would yield a control slope of ~ 0.2 nm^{-1}, but would only be linear over a distance of about 8 nm. In practice, it is found that an optical cantilever length of about 2 m gives the best results for real-time 3D single-particle tracking. Using this model, this distance should yield a control slope of 4.3×10^{-3} nm^{-1} over a range of about half a micron.

Despite the simplicity of the model used above to predict the optical response as a function of the optical cantilever length, it agrees quite well with experimentally measured values. To compare the predicted control to the real system,

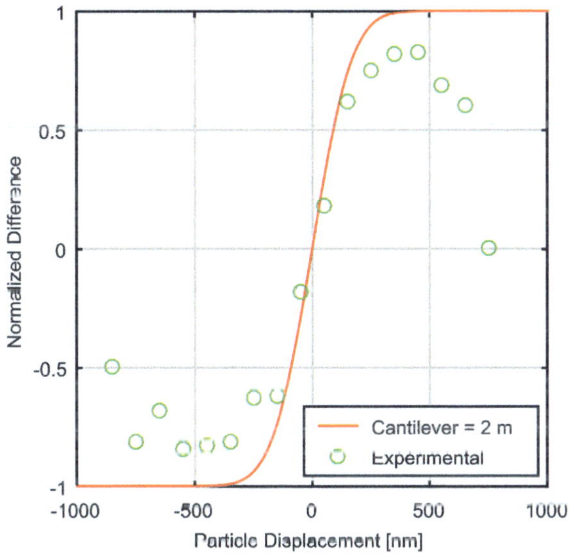

Fig. 5 Calculated *versus* experimental response curves of a 2 m optical cantilever. The experimental response of a 2 m optical cantilever was measured by scanning a 110 nm fluorescent nanoparticle through the focal volume of a 100× objective lens (f_{obj} = 1.65 mm). Following the optical cantilever (f_{tube} = 200 mm, $f_{cantilever}$ = 75 mm, L = 2 m), the emission light is split by a prism mirror onto two single-photon counting APDs. The normalized difference in the intensities of the two detectors are calculated as a function of the particle position. The calculated (red line) and measured (green circles) responses of the system are in remarkably good agreement.

110 nm fluorescent beads (Tetraspeck, Life Technologies) were scanned by a piezoelectric stage (MadCity Labs Nano-LPQ). The beads were excited with a 642 nm diode laser (Coherent Cube) operating at ~5 μW in the focus of a 100× Zeiss Plan Apo objective. The fluorescence was collected through the same objective and focused by a 200 mm achromatic doublet tube lens (Thorlabs) onto the intermediate image plane. The intermediate image was projected at infinity by a 75 mm achromatic doublet. After a distance of ~2 m, the collimated output from the 75 mm cantilever lens was split by a prism mirror. The resulting split images were focused by 45 mm achromats (Thorlabs) onto two single photon counting APDs (Excelitas SPCM-AQRH-13). The particle was scanned in 100 nm steps and the fluorescence was collected by both APDs for a period of 10 ms. The process variable (PV) is calculated as the difference in the intensity of the two detectors divided by their sum:

$$\mathrm{PV} = \frac{I_1 - I_2}{I_1 + I_2}. \tag{4}$$

The result of the particle scan is shown in Fig. 5 (green circles), along with the predicted control for a 2 m optical cantilever (red line). The two curves show remarkably good agreement near the focus, with nearly identical control slopes. The main differences between the two curves likely arise from the difference

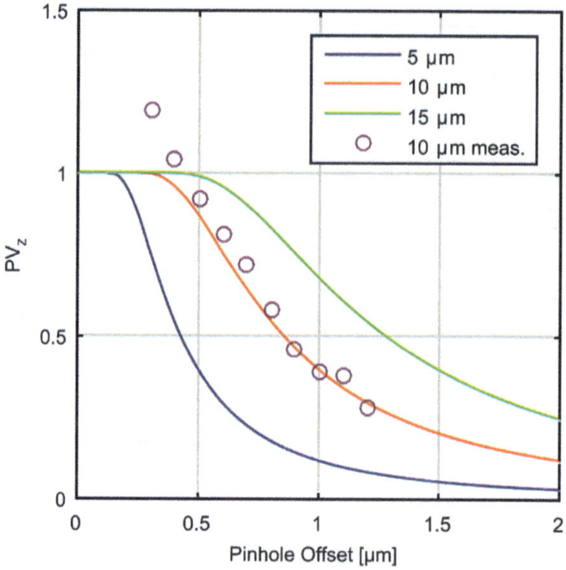

Fig. 6 The offset pinhole enables high resolution Z tracking. A short focal length lens ($f = 2.9$ mm) is used to image the target particle onto a pinhole. The plot shows the relative light intensity collected by a detector immediately after the pinhole, normalized to the light collected for XY tracking. The size of the pinhole and offset from the XY focal volume affects both the steepness and range of control allowed, with a 10 μm pinhole providing a compromise between the two factors. The purple circles show the results of a fluorescent nanoparticle scanned through the tracking volume and detected with the 10 μm offset pinhole, showing good agreement with the predicted behavior.

between the actual experimental setup and the simplified calculation. First of all, the experiment relies on real detectors with small (~180 μm) active areas. As a result, the detector may not collect all the photons as the particle displaces further and further from the prism mirror center. This, along with the presence of a non-zero background (dark counts and background scattering), preclude the experimental process variable from reaching ±1 at the extremes of particle displacement. Secondly, the calculated slope seems to slightly outperform the experimental realization. This is likely due to not accounting for beam divergence over the length of the optical cantilever. Even in the case of a small divergence, the infinity beam will become wider at the prism mirror than in the divergence-free case, resulting in a slightly reduced control slope.

In addition to lateral position sensing, a change in the axial position of the particle must also be measured in real-time with high precision. This can be achieved by imaging the particle onto an offset confocal pinhole with a short focal length lens (here $f = 2.9$ mm). As in typical confocal microscopy, objects which lie above or below the optical section defined by the pinhole will yield a lower signal. Displacing the confocal pinhole away from the focus of the tracked particle allows a nearly linear index into the particle's position. Fig. 6 shows the predicted optical tracking response for 5, 10 and 15 micron pinholes at various displacements from the objective's focal volume. For all three pinholes tested, there is a flat response at small displacements (<0.5 μm in the object space), followed by a rapid drop in the Z process variable, which is the Z-detector readout (I_z) after normalization to the XY detectors, as discussed above for the optical cantilever ($I_x + I_y$). While the 5 μm pinhole yields the sharpest control slope, it also restricts light collection to the highest degree. The 15 μm pinhole has a higher throughput, but shallower control. The 10 μm pinhole shows a good intermediate response. For comparison, the simulated 10 μm pinhole response is compared to an experimentally measured scan of a fluorescent particle, as described above. In this case, the particle is scanned through the objective's focal volume in the Z-direction.

Feedback control

The optical response is only one factor that governs the performance of real-time 3D single particle tracking microscopies. Detecting deviations in the particle's position at high speeds are only half the battle. The other half of the challenge is active feedback control to maintain the particle in the focal volume of the objective, which is what ultimately allows extremely high sampling rates (allowed due to the small spatial scale, see Fig. 3). Generally, this feedback control is applied using optical position sensitive detection, proportional–integral–derivative (PID) control[54] and high speed, 3D piezoelectric stages. The process variable from the optical readout is converted to an analog voltage, as dictated by the PID control parameters, and fed to the stage to move the sample and maintain the tracked particle in the focal volume of the objective lens.

Both the PID control parameters and the response speed and accuracy of the piezoelectric stage contribute to the ultimate limits of real-time 3D single-particle tracking. In general PID control, the control applied to maintain a certain setpoint in a system is governed by the following equation:

$$C_i = K_P \cdot \text{err}_i + K_I \cdot \sum_{n=0}^{i} \text{err}_n + K_D \cdot (\text{PV}_i - \text{PV}_{i-1}) \tag{5}$$

where C_i is the control signal to be applied, and K_P, K_I and K_D are the proportional, integral and derivative control weights, respectively. K_P operates purely on the current state of the system, K_I operates on the integrated state of the system over time and K_D operates on the recent change of the system. The system error, err, is defined as follows:

$$\text{err}_i = \text{SP} - \text{PV}_i. \tag{6}$$

SP is the setpoint that is maintained by the control loop and PV is the current state of the system. In the case of the optical cantilever detection described above, the goal is to keep the particle's image centered with respect to the prism mirror. This occurs when both detectors have the same reading and the error function will have a value of zero (as in Fig. 4 and 5). This means that to maintain the particle in the center of the focus of the objective lens, SP is set to 0. As the particle is displaced away from the center, PV will take on a nonzero value, which will be used as the input for the PID control.

While there are a number of approximate methods for predicting the values of the control weights, they are routinely optimized based on the response of the system to the control. In this case, the control weights are chosen based on which yield the longest trajectories on the most quickly diffusing particles. Here, we

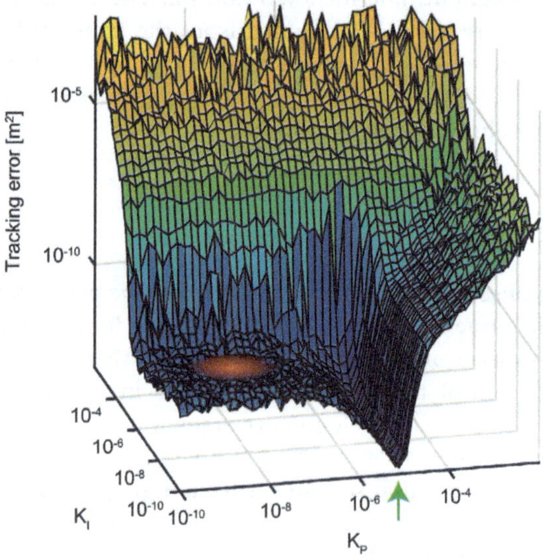

Fig. 7 PID tuning for the optical tracking response. The effect of different values of K_P and K_I were simulated by tracking a 60 nm nanoparticle undergoing a Brownian random walk in one dimension. The optical response was taken from the data in Fig. 4, while the stage response was modeled as a 500 Hz low pass filter. The plot shows that the empirically derived values (red region) lie in a local minimum, which is relatively flat. The plot also reveals a global minimum in the tracking error at a critical value of K_P that is relatively independent of K_I (green arrow).

develop a simple methodology for predicting the values of the control weights and show the results for one dimension. First, 1D trajectories of 60 nm diffusing particles are generated with 10 μs step sizes. Then, a simulated control algorithm is applied, which uses the empirically measured optical response function for the 2 m optical cantilever detection described above. The output of the simulated control is run through a 500 Hz low pass filter, which simulates the behavior of the piezoelectric stage. This low-pass filter is a good approximation for the piezoelectric stage for two reasons. First, the fastest piezoelectric stages have small step response times on the order of a few milliseconds. Second, these stages typically have resonant frequencies on the order of 1 kHz when unloaded and down to 500 Hz when loaded. Near these frequencies, the stage position oscillates for even low amplitude signals. To avoid this, control signals are either designed specifically to avoid high frequencies, or a low-pass filter is deliberately inserted into the system to block them.

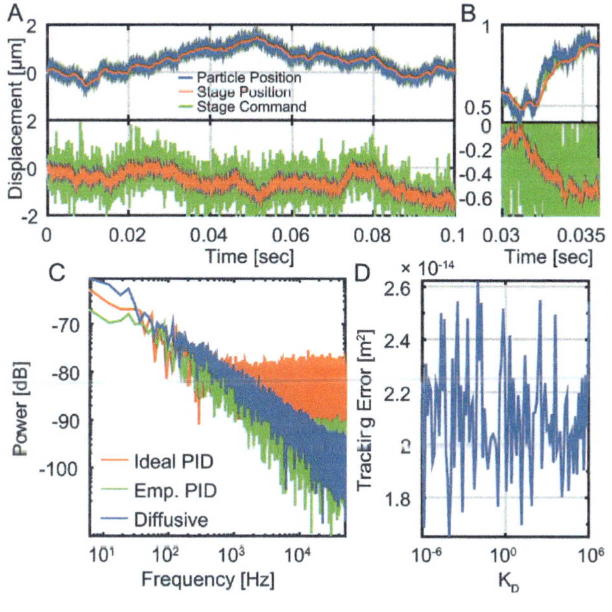

Fig. 8 Simulated 1D tracking for empirically derived and ideal PID parameters. (A) Tracking was simulated for the empirically derived (top) as well as "ideal" (bottom) PID parameters, as found in Fig. 5. For the empirical parameters, the tracked stage position (red) is smoothed relative to the actual particle position (blue), while for the ideal parameters, the tracked stage position almost identically matches the particle position. (B) A zoomed in section of the trajectories shown in (A). The empirical tracking (top, red curve) is smoothed and delayed relative to the particle position (top, blue curve). In the case of the ideal parameters (bottom), the stage control (green) is driven with high frequency and high amplitude to match the particle's position. (C) Power spectrum of the simulated particle position (blue), empirical PID stage commands (green) and ideal PID stage commands (red). The empirical PID parameters closely follow the frequency response of the particle itself, avoiding high frequencies, which lead to an overshoot and ringing in the piezoelectric stage. The ideal parameters show large amplitude high frequency components above the piezoelectric stage resonant frequency (~1 kHz). (D) Evaluation of the effect of K_D on the simulated trajectories, showing that no value has a significant effect on the tracking error.

This simulated tracking system was then used to pinpoint the values of the best control parameters. Different values of K_P and K_I were applied to 10 trajectories lasting 1 ms each (100 data points). The integrated squared error between the simulated stage and particle positions was integrated over all ten trajectories and used as an evaluation of the tested parameters. Fig. 7 shows the results of testing trajectories for 50 logarithmically spaced values of K_P and K_I from 10^{-10} to 10^{-2}, with K_D being held at zero. As can be clearly seen from the plot, high values of both K_P and K_I lead to tracking instability. Interestingly, the empirically found best parameters (shown as the red spot on the plot in Fig. 7) can be seen as a local minimum of the control surface. Within this local minimum, there is relatively little effect of the value of K_P or K_I on the tracking error. It is also apparent that these empirically derived values lie significantly far from the optimal control parameters, which appear as a "ravine" in the control space (as indicated by the green arrow in Fig. 7). These values produce simulated tracking errors, which are 10^2 times smaller than in the case of the empirical parameters. This could lead to the tracking of much smaller and faster molecules.

To further investigate the behavior of the simulated tracking system under these different control conditions, Fig. 8a shows a simulated trajectory for the empirically determined parameters (top) and the ideal parameters (bottom). The empirically determined parameters issue stage commands (green), which closely match the actual particle position (blue). The resulting stage position (red) appears as a smoothed and slightly delayed signal, which tracks the particle position. This differs greatly from the newly calculated ideal PID parameters,

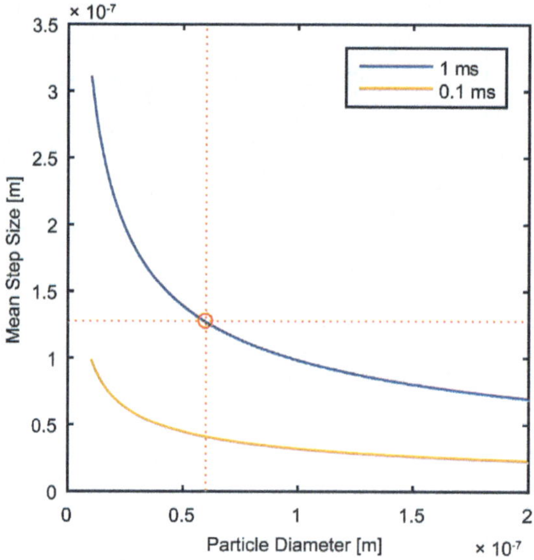

Fig. 9 Mean step size *versus* response time. The blue curve shows the mean diffusive step size expected at 1 ms delay for Brownian diffusion. The red circle highlights that for a 60 nm particle this step size is ~127 nm, on the order of the radius of the PSF of a 100× objective lens. The yellow curve shows that if the response time of the system can be reduced to 0.1 ms, the expected step size can be maintained below this cut-off for all particle diameters, even down to the scale of single molecules.

which drive the stage command with high amplitude such that the stage position almost exactly matches the particle position. The power spectrum of the different control methods shown in Fig. 8c gives some insight into the reason why the ideal parameters are not found to be the same as the empirical parameters. Shown in blue is the power spectrum of the diffusive particle, which shows a distinctively Brownian character. The empirical parameters (green) yield a power spectrum that closely matches the particle spectrum, with a small increase in frequencies above 10 kHz. The ideal parameters (red), however, show much higher amplitude signals in the regime beyond 1 kHz. If these parameters are applied to a real stage, which does not behave strictly as a low pass filter, there will be significant ringing and overshoot, leading to poor tracking or even damage to the stage. As such, these calculations would suggest that the response of real-time tracking systems is more dominated by the piezoelectric stage than the optical detection modalities.

The one parameter that is frequently neglected in the experimental case is K_D, as, typically, no noticeable difference in tracking is observed for any value of K_D. To see if any value of K_D can be used to improve the speed and accuracy in tracking, we simulated the tracking response for 100 values of K_D, logarithmically spaced between 10^{-6} and 10^6. As shown in Fig. 8d, there is no effect of K_D on the tracking performance (negative values were also tested). This is not a surprising result, as a diffusing particle is highly random and the next step in the process will have almost no relation to the previous step. Further, derivative control is notoriously susceptible to noise, which would overwhelm any possible benefit that derivative control might add to this system.

Conclusion and outlook

In the overarching goal of changing real-time 3D single-particle tracking to real-time 3D single-molecule tracking, it is necessary that fast and sensitive detection and feedback mechanisms are found. From the above, it is clear that the limiting factor in real-time 3D single-particle tracking is the limited temporal and frequency response of 3D piezoelectric stages. As an illustrative example, Fig. 9 shows the mean step size for particles with varying hydrodynamic diameters. The blue curve shows that a 60 nm particle makes a mean step size of ~127 nm in 1 ms. In other words, in this time window, a particle can go from the center of the focal volume to the edge between measurements. This is significant because this is the typical response time of a piezoelectric stage and the empirically determined fastest particle that can be tracked by the optical cantilever method has a radius of 60 nm. This strongly suggests that the response time of the stage limits tracking of faster particles. If faster response mechanisms allow an order of magnitude faster response (yellow curve, 0.1 ms step size), this could pave the way for following the behavior of single molecules in aqueous environments, allowing the study of single molecule dynamics in nearly any environment.

Acknowledgements

This work was supported by the US Department of Energy and Princeton University.

References

1 A. Diaspro, *Confocal and Two-Photon Microscopy: Foundations, Applications and Advances*, Wiley-Liss Inc, New York, 2002.

2 M. Minsky, Memoir On Inventing The Confocal Scanning Microscope, *Scanning*, 1988, **10**, 128–138.

3 G. J. Brakenhoff, P. Blom, P. J. Barends and C. Bakker, High-Resolution Confocal Scanning Light-Microscopy, *Ultramicroscopy*, 1979, **4**, 115.

4 G. J. Brakenhoff, H. T. M. Vandervoort, E. A. Vanspronsen, W. A. M. Linnemans and N. Nanninga, 3-Dimensional Chromatin Distribution In Neuro-Blastoma Nuclei Shown By Confocal Scanning Laser Microscopy, *Nature*, 1985, **317**, 748–749.

5 M. J. Sanderson and I. Parker, in *Methods Enzymol.*, ed. I. Parker and G. Marriott, Academic Press, 2003, vol. 360, pp. 447–481.

6 T. A. Planchon, L. Gao, D. E. Milkie, M. W. Davidson, J. A. Galbraith, C. G. Galbraith and E. Betzig, Rapid three-dimensional isotropic imaging of living cells using Bessel beam plane illumination, *Nat. Methods*, 2011, **8**, 417–423.

7 M. Petráň, M. Hadravský, M. D. Egger and R. Galambos, Tandem-Scanning Reflected-Light Microscope, *J. Opt. Soc. Am.*, 1968, **58**, 661–664.

8 R. G. McAllister, D. R. Sisan and J. S. Urbach, Design and optimization of a high-speed, high-sensitivity, spinning disk confocal microscopy system, *J. Biomed. Opt.*, 2008, **13**, 054058.

9 E. Wang, C. M. Babbey and K. W. Dunn, Performance comparison between the high-speed Yokogawa spinning disc confocal system and single-point scanning confocal systems, *J. Microsc.*, 2005, **218**, 148–159.

10 S. Hell and E. H. K. Stelzer, Properties Of A 4pi Confocal Fluorescence Microscope, *J. Opt. Soc. Am. A*, 1992, **9**, 2159–2166.

11 S. Hell and E. H. K. Stelzer, Fundamental Improvement Of Resolution With A 4pi-Confocal Fluorescence Microscope Using 2-Photon Excitation, *Opt. Commun.*, 1992, **93**, 277–282.

12 T. A. Klar, S. Jakobs, M. Dyba, A. Egner and S. W. Hell, Fluorescence microscopy with diffraction resolution barrier broken by stimulated emission, *Proc. Natl. Acad. Sci. U. S. A.*, 2000, **97**, 8206–8210.

13 M. A. Schwentker, H. Bock, M. Hofmann, S. Jakobs, J. Bewersdorf, C. Eggeling and S. W. Hell, Wide-field subdiffraction RESOLFT microscopy using fluorescent protein photoswitching, *Microsc. Res. Tech.*, 2007, **70**, 269–280.

14 R. Schmidt, C. A. Wurm, A. Punge, A. Egner, S. Jakobs and S. W. Hell, Mitochondrial Cristae Revealed with Focused Light, *Nano Lett.*, 2009, **9**, 2508–2510.

15 V. Westphal, S. O. Rizzoli, M. A. Lauterbach, D. Kamin, R. Jahn and S. W. Hell, Video-rate far-field optical nanoscopy dissects synaptic vesicle movement, *Science*, 2008, **320**, 246–249.

16 B. Yang, F. Przybilla, M. Mestre, J. B. Trebbia and B. Lounis, Large parallelization of STED nanoscopy using optical lattices, *Opt. Express*, 2014, **22**, 5581–5589.

17 A. Chmyrov, J. Keller, T. Grotjohann, M. Ratz, E. d'Este, S. Jakobs, C. Eggeling and S. W. Hell, Nanoscopy with more than 100 000 'doughnuts', *Nat. Methods*, 2013, **10**, 737–740.

18 R. E. Thompson, D. R. Larson and W. W. Webb, Precise Nanometer Localization Analysis for Individual Fluorescent Probes, *Biophys. J.*, 2002, **82**, 2775–2783.

19 A. Yildiz, J. N. Forkey, S. A. McKinney, T. Ha, Y. E. Goldman and P. R. Selvin, Myosin V walks hand-over-hand: Single fluorophore imaging with 1.5-nm localization, *Science*, 2003, **300**, 2061–2065.

20 M. J. Rust, M. Bates and X. Zhuang, Sub-diffraction-limit imaging by stochastic optical reconstruction microscopy (STORM), *Nat. Methods*, 2006, **3**, 793–796.

21 E. Betzig, G. H. Patterson, R. Sougrat, O. W. Lindwasser, S. Olenych, J. S. Bonifacino, M. W. Davidson, J. Lippincott-Schwartz and H. F. Hess, Imaging Intracellular Fluorescent Proteins at Nanometer Resolution, *Science*, 2006, **313**, 1642–1645.

22 S. T. Hess, T. P. K. Girirajan and M. D. Mason, Ultra-high resolution imaging by fluorescence photoactivation localization microscopy, *Biophys. J.*, 2006, **91**, 4258–4272.

23 B. Huang, W. Wang, M. Bates and X. Zhuang, Three-Dimensional Super-Resolution Imaging by Stochastic Optical Reconstruction Microscopy, *Science*, 2008, **319**, 810–813.

24 S. A. Jones, S. H. Shim, J. He and X. W. Zhuang, Fast, three-dimensional super-resolution imaging of live cells, *Nat. Methods*, 2011, **8**, 499–505.

25 M. F. Juette, T. J. Gould, M. D. Lessard, M. J. Mlodzianoski, B. S. Nagpure, B. T. Bennett, S. T. Hess and J. Bewersdorf, Three-dimensional sub-100 nm resolution fluorescence microscopy of thick samples, *Nat. Methods*, 2008, **5**, 527–529.

26 E. Toprak, H. Balci, B. H. Blehm and P. R. Selvin, Three-dimensional particle tracking via bifocal imaging, *Nano Lett.*, 2007, **7**, 2043–2045.

27 P. Prabhat, Z. Gan, J. Chao, S. Ram, C. Vaccaro, S. Gibbons, R. J. Ober and E. S. Ward, Elucidation of intracellular recycling pathways leading to exocytosis of the Fc receptor, FcRn, by using multifocal plane microscopy, *Proc. Natl. Acad. Sci. U. S. A.*, 2007, **104**, 5889–5894.

28 G. Shtengel, J. A. Galbraith, C. G. Galbraith, J. Lippincott-Schwartz, J. M. Gillette, S. Manley, R. Sougrat, C. M. Waterman, P. Kanchanawong, M. W. Davidson, R. D. Fetter and H. F. Hess, Interferometric fluorescent super-resolution microscopy resolves 3D cellular ultrastructure, *Proc. Natl. Acad. Sci. U. S. A.*, 2009, **106**, 3125–3130.

29 M. A. Thompson, M. D. Lew, M. Badieirostami and W. E. Moerner, Localizing and Tracking Single Nanoscale Emitters in Three Dimensions with High Spatiotemporal Resolution Using a Double-Helix Point Spread Function, *Nano Lett.*, 2010, **10**, 211–218.

30 M. Badieirostami, M. D. Lew, M. A. Thompson and W. E. Moerner, Three-dimensional localization precision of the double-helix point spread function *versus* astigmatism and biplane, *Appl. Phys. Lett.*, 2010, **97**, 161103.

31 Y. Shechtman, L. E. Weiss, A. S. Backer, S. J. Sahl and W. E. Moerner, Precise 3D scan-free multiple-particle tracking over large axial ranges with Tetrapod point spread functions, *Nano Lett.*, 2015, **15**, 4194–4199.

32 L. Zhu, W. Zhang, D. Elnatan and B. Huang, Faster STORM using compressed sensing, *Nat. Methods*, 2012, **9**, 721–723.

33 S.-H. Shim, C. Xia, G. Zhong, H. P. Babcock, J. C. Vaughan, B. Huang, X. Wang, C. Xu, G.-Q. Bi and X. Zhuang, Super-resolution fluorescence imaging of organelles in live cells with photoswitchable membrane probes, *Proc. Natl. Acad. Sci. U. S. A.*, 2012, **109**, 13978–13983.

34 F. Huang, T. M. P. Hartwich, F. E. Rivera-Molina, Y. Lin, W. C. Duim, J. J. Long, P. D. Uchil, J. R. Myers, M. A. Baird, W. Mothes, M. W. Davidson, D. Toomre and J. Bewersdorf, Video-rate nanoscopy using sCMOS camera-specific single-molecule localization algorithms, *Nat. Methods*, 2013, **10**, 653–658.

35 J. Huisken, J. Swoger, F. Del Bene, J. Wittbrodt and E. H. K. Stelzer, Optical sectioning deep inside live embryos by selective plane illumination microscopy, *Science*, 2004, **305**, 1007–1009.

36 T. F. Holekamp, D. Turaga and T. E. Holy, Fast three-dimensional fluorescence imaging of activity in neural populations by objective-coupled planar illumination microscopy, *Neuron*, 2008, **57**, 661–672.

37 P. J. Keller, A. D. Schmidt, J. Wittbrodt and E. H. K. Stelzer, Reconstruction of Zebrafish Early Embryonic Development by Scanned Light Sheet Microscopy, *Science*, 2008, **322**, 1065–1069.

38 P. J. Keller, A. D. Schmidt, A. Santella, K. Khairy, Z. Bao, J. Wittbrodt and E. H. K. Stelzer, Fast, high-contrast imaging of animal development with scanned light sheet-based structured-illumination microscopy, *Nat. Methods*, 2010, **7**, 637–642.

39 M. B. Bouchard, V. Voleti, C. S. Mendes, C. Lacefield, W. B. Grueber, R. S. Mann, R. M. Bruno and E. M. C. Hillman, Swept confocally-aligned planar excitation (SCAPE) microscopy for high-speed volumetric imaging of behaving organisms, *Nat. Photonics*, 2015, **9**, 113–119.

40 L. Gao, L. Shao, C. D. Higgins, J. S. Poulton, M. Peifer, M. W. Davidson, X. Wu, B. Goldstein and E. Betzig, Noninvasive Imaging beyond the Diffraction Limit of 3D Dynamics in Thickly Fluorescent Specimens, *Cell*, 2012, **151**, 1370–1385.

41 M. G. L. Gustafsson, Surpassing the lateral resolution limit by a factor of two using structured illumination microscopy, *J. Microsc.*, 2000, **198**, 82–87.

42 L. Schermelleh, P. M. Carlton, S. Haase, L. Shao, L. Winoto, P. Kner, B. Burke, M. C. Cardoso, D. A. Agard, M. G. L. Gustafsson, H. Leonhardt and J. W. Sedat, Subdiffraction multicolor imaging of the nuclear periphery with 3D structured illumination microscopy, *Science*, 2008, **320**, 1332–1336.

43 L. Shao, P. Kner, E. H. Rego and M. G. L. Gustafsson, Super-resolution 3D microscopy of live whole cells using structured illumination, *Nat. Methods*, 2011, **8**, 1044.

44 A. G. York, P. Chandris, D. D. Nogare, J. Head, P. Wawrzusin, R. S. Fischer, A. Chitnis and H. Shroff, Instant super-resolution imaging in live cells and embryos *via* analog image processing, *Nat. Methods*, 2013, **10**, 1122–1126.

45 G. A. Lessard, P. M. Goodwin and J. H. Werner, Three-dimensional tracking of individual quantum dots, *Appl. Phys. Lett.*, 2007, **91**, 224106.

46 K. McHale, A. J. Berglund and H. Mabuchi, Quantum Dot Photon Statistics Measured by Three-Dimensional Particle Tracking, *Nano Lett.*, 2007, **7**, 3535–3539.

47 H. Cang, C. S. Xu, D. Montiel and H. Yang, Guiding a confocal microscope by single fluorescent nanoparticles, *Opt. Lett.*, 2007, **32**, 2729–2731.

48 H. Cang, C. M. Wong, C. S. Xu, A. H. Rizvi and H. Yang, Confocal three dimensional tracking of a single nanoparticle with concurrent spectroscopic readouts, *Appl. Phys. Lett.*, 2006, **88**, 223901.

49 K. Welsher and H. Yang, Multi-resolution 3D visualization of the early stages of cellular uptake of peptide-coated nanoparticles, *Nat. Nanotechnol.*, 2014, **9**, 198–203.

50 S. Console, C. Marty, C. García-Echeverría, R. Schwendener and K. Ballmer-Hofer, Antennapedia and HIV Transactivator of Transcription (TAT) "Protein Transduction Domains" Promote Endocytosis of High Molecular Weight Cargo upon Binding to Cell Surface Glycosaminoglycans, *J. Biol. Chem.*, 2003, **278**, 35109–35114.

51 N. P. Wells, G. A. Lessard, P. M. Goodwin, M. E. Phipps, P. J. Cutler, D. S. Lidke, B. S. Wilson and J. H. Werner, Time-Resolved Three-Dimensional Molecular Tracking in Live Cells, *Nano Lett.*, 2010, **10**, 4732–4737.

52 J. J. Han, C. Kiss, A. R. M. Bradbury and J. H. Werner, Time-Resolved, Confocal Single-Molecule Tracking of Individual Organic Dyes and Fluorescent Proteins in Three Dimensions, *ACS Nano*, 2012, **6**, 8922–8932.

53 Y. Katayama, O. Burkacky, M. Meyer, C. Bräuchle, E. Gratton and D. C. Lamb, Real-Time Nanomicroscopy *via* Three-Dimensional Single-Particle Tracking, *ChemPhysChem*, 2009, **10**, 2458–2464.

54 A. Kiam Heong, G. Chong and L. Yun, PID control system analysis, design, and technology, *IEEE Trans. Control Syst. Technol.*, 2005, **13**, 559–576.

Faraday Discussions

PAPER

Size dependent efficiency of photophoretic swimmers†

Andreas P. Bregulla and Frank Cichos*

Received 18th June 2015, Accepted 30th June 2015
DOI: 10.1039/c5fd00111k

We investigate experimentally the efficiency of self-propelled photophoretic swimmers based on metal-coated polymer particles of different sizes. The metal hemisphere absorbs the incident laser power and converts its energy into heat, which dissipates into the environment. A phoretic surface flow arises from the temperature gradient along the particle surface and drives the particle parallel to its symmetry axis. Scaling the particle size from micro to nanometers, the efficiency of converting optical power into motion is expected to rise with the reciprocal size for ideal swimmers. However, due to the finite size of the metal cap, the efficiency of a real swimmer reveals a maximum depending sensitively on the details of the metal cap shape. We compare the experimental results to numerical simulations.

1 Introduction

The field of artificial micro-swimmers has recently attracted much interest. Over the last decade, numerous types of such artificial swimmers driven by self-thermophoretic,[1–3] self-diffusiophoretic[4–6] or self-electrophoretic[7,8] surface flows have been proposed. The diversity of the driving mechanism with the simplicity of the design and controllability of their motion make them a valuable model system to study processes far from equilibrium. When such self-propelled objects are starting to interact at higher densities, coherent collective motion is observed in which the swimmers align, and form flocks, swarms or other complex patterns.[6,9–12]

Most of these self-propelled particles are driven by an osmotic pressure difference across the particles surface caused by symmetry breaking of *e.g.* temperature or concentration profiles along the particle surface realized by an asymmetric material composition. As a result of the osmotic pressure difference,

Molecular Nanophotonics Group, Institute of Experimental Physics I, University of Leipzig, 04103 Leipzig, Germany. E-mail: cichos@physik.uni-leipzig.de

† Electronic supplementary information (ESI) available: Additional experimental details on the sample preparation and the size dependent velocity; derivation of the mean squared displacement and velocity correlation. See DOI: 10.1039/c5fd00111k

a phoretic surface flow occurs, which leads, in combination with hydrodynamic boundary conditions, to a directed motion of the Janus particle.

For this propulsion they rely on externally supplied energy drawn, for example, from electromagnetic waves or chemical fuel. The efficiency of this conversion process can be evaluated by the ratio of the output power and the input power, $\varepsilon = P_{out}/P_{in}$. The output power involves the sum of all viscous losses causing the directed motion and thus requires a detailed knowledge of the created phoretic flow field. While this flow field has been determined theoretically,[13–15] no efficiency calculations have been based on it so far. It is known, however, that the efficiency of most artificial swimmers is very low with values of typically less than $\varepsilon = 10^{-7}$.[16]

A hydrodynamic upper bound of the efficiency was derived by Sabass and Seifert.[17] They predicted that the efficiency of a surface driven swimmer increases as its size decreases. For diffusiophoretic Janus type particles (driven by concentration gradients caused by catalytic reactions), theoretical studies on the size dependent efficiency[18] support this dependence but reveal a deviation for small particle sizes. Measurements of their size dependent velocity[19] exist but without considering the efficiency of the propulsion and their dependence on size.

Here we focus on self-propelled thermophoretic swimmers. We investigate their size dependent phoretic velocity to access the scaling of the relative efficiency experimentally and theoretically.

The efficiency of phoretic swimmers has been addressed by calculating the power required to pull a colloidal particle of the same radius R as the swimmer in a liquid of viscosity η (see eqn (1)) with a constant velocity equal to the phoretic velocity V_{th} of the swimmer.[20] The power required to drag this particle is then related to the Stokes' friction F_{st} of this probe colloid and the swimmer's velocity:

$$\varepsilon = \frac{F_{st} V_{th}}{P_{in}} \propto \frac{R {V_{th}}^2}{P_{in}}. \tag{1}$$

The right side of eqn (1) is useful to study the dependence of the efficiency on the size of the swimmer, as for the example carried out theoretically by Sabass,[17] while a quantitative estimate of the efficiency ($\varepsilon = 7.6 \times 10^{-15}$) yields a similar order of magnitude as reported before.[16] The input power P_{in} in eqn (1) relates, in general, to the energy picked up by the particle and involves different processes for different types of swimmers. In the case of catalytic swimmers, this input power is difficult to define. Sabass and Seifert therefore consider a sum of the power lost in viscous dissipation and the propulsion power as the input power. In the present case of photophoretic swimmers, a thin metal hemisphere on a transparent polystyrene sphere is absorbing a fraction P_{abs} of the incident laser light power P_0. The absorbed electromagnetic energy is transformed rapidly into heat, which is released into the environment. After a short time period, a stationary temperature gradient arises at the surface of the particle, which induces a phoretic flow[13] that drives the particle along its symmetry axis. The input power P_{in} is therefore well defined and corresponds to the absorbed electromagnetic power P_{abs}. This is different from the definition of Sabass as it includes the power required to maintain a temperature field around the hemisphere, even if this would not generate motion.

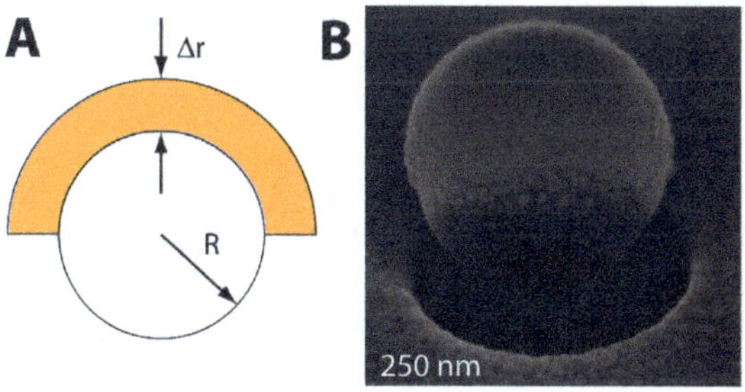

Fig. 1 (A) Definition of the relevant parameters for the numerical calculations. The particle has a radius R and a thermal conductivity κ_{PS}, the gold cap has a thickness Δr and a thermal conductivity κ_{Au}, and the environment has a thermal conductivity κ_{water}. (B) Electron microscopy image of a $R = 0.25$ μm Janus particle.

A study of the size dependence of the swimmer efficiency thus boils down to the size dependence of the absorbed power P_{abs} and phoretic velocity V_{th}.

The absorbed power is related to the incident power P_0 of the laser by the absorption cross section $\sigma_{abs}(R)$ and the area A illuminated by the laser: $P_{abs}(R) = \sigma_{abs}(R)P_0/A$.

Thus the size dependence is hidden in the absorption cross section of the metal cap.

The phoretic velocity on the other side can be expressed as the product of the thermodiffusion coefficient D_T and the surface temperature gradient $\nabla T_{\parallel}$ (eqn (2)):

$$\vec{V}_{th} = -D_T \vec{\nabla}_{\parallel} T \tag{2}$$

Both quantities may depend on size even though Piazza *et al.* find a size independent thermodiffusion coefficient D_T for colloidal particles.[21] Inserting these dependencies, the efficiency reads:

$$\varepsilon \propto \frac{R V_{th}^2}{P_{abs}} \propto \frac{R\left(D_T \nabla_{\parallel} T\right)^2}{\sigma_{abs}}. \tag{3}$$

In the present work, we will therefore explore the size dependence of the photophoretic swimming efficiency by measuring the dependence of the phoretic velocity and the optical properties on the particle radius, ranging from $R = 100$ nm up to 625 nm. Throughout all of these studies, the gold cap thickness $\Delta r = 50$ nm of the Janus particle is kept constant (see Fig. 1 for details). As the surface temperature profile is difficult to access experimentally, we rely on finite element simulations (FEM) for this part.

2 Experimental

The motion of Janus swimmers of $R = 100$ nm to $R = 625$ nm has been studied by dark field microscopy. A heating laser with a wavelength of $\lambda = 532$ nm is focused onto the sample region. Images are recorded at an inverse frame rate of $\Delta t = 15$ ms and analyzed by a single particle tracking algorithm, as described in ref. 22.

The single particle tracking yields the position and the orientation of the swimmer. To ensure a constant heating laser intensity at the particle position during a measurement, two different feedback based measurement schemes are employed. The first scheme is photon nudging,[22,23] where the particle position and orientation are captured in real time and the propulsion is switched according to the particle orientation. This method allows us to trap and hold the particle close to a target position of constant laser intensity for a practically unlimited amount of time. The propulsion periods within this scheme can be analyzed by projecting the displacement of the particle in adjacent frames onto the orientation (angle ϕ_i in the lab frame) of the particle. The thermophoretic velocity V_{th} is then the mean displacement divided by the inverse frame rate Δt:

$$V_{\mathrm{th}} = \frac{1}{\Delta t}\left\langle \left(\vec{x}_{i+1} - \vec{x}_i\right)\begin{pmatrix}\cos\phi_i\\ \sin\phi_i\end{pmatrix}\right\rangle_i \tag{4}$$

where $\langle\ldots\rangle_i$ denotes the ensemble average.

Below a particle radius of 250 nm, the orientation of the particle cannot be determined from the particle images (see Fig. 2c). To ensure constant illumination intensity for these measurements, the sample was moved by a x–y-piezo stage to keep the particle in the center of the illumination. The thermophoretic velocity can then be determined by either the velocity autocorrelation VAC(δ) of the particle (eqn (5)) or the long time limit of the mean squared displacement. The velocity autocorrelation may be used if the inverse frame rate is significantly smaller than the rotation time of the particle. Here the rotation time and the velocity can be obtained separately. Using an inverse frame rate of $\Delta t = 5$ ms is still sufficient to study the $R = 160$ nm swimmers ($\tau_{\mathrm{R}} = 11.2$ ms).

$$\mathrm{VAC}(\delta) = \langle \vec{V}(t)\cdot\vec{V}(t+\delta)\rangle_i = {V_{\mathrm{th}}}^2 \exp(-\delta/\tau_{\mathrm{R}}) \tag{5}$$

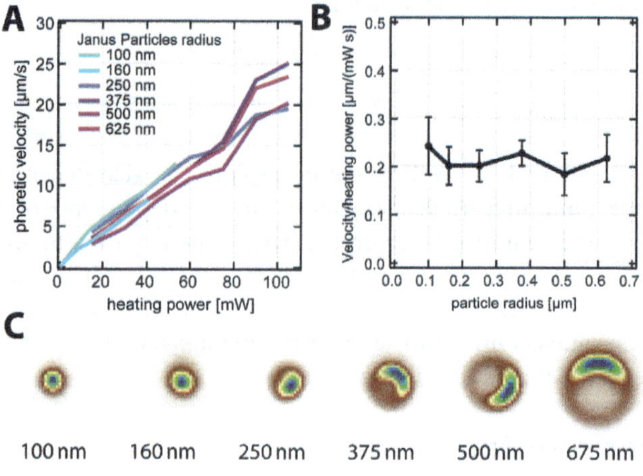

Fig. 2 (A) Measured thermophoretic velocity as a function of the heating power for different particle radii ranging from 100 nm up to 625 nm. The thickness of the gold cap was 50 nm for all sizes. (B) Average slope for the power dependence of the velocity as a function of the particle radius. (C) Typical dark field images of the Janus particles of different sizes in false color representation.

In the case of the $R = 100$ nm swimmers, we have employed the long time limit of the mean squared displacement. At short times, the mean squared displacement shows a ballistic behavior ($\mathrm{MSD}(\delta) \propto \delta^2$), which becomes linear in the time lag δ for long times, since the rotational Brownian motion randomizes the propulsion direction. The effective diffusion coefficient in the long time limit D_{eff} is then related to the diffusion coefficient D of the Janus swimmer without propulsion plus the square of the propulsion velocity V_{th} times the rotational diffusion time τ_{R} (eqn (6)).

$$D_{\mathrm{eff}} = D + {V_{\mathrm{th}}}^2 \tau_{\mathrm{R}}/2 \tag{6}$$

A full derivation of the mean squared displacement[24–27] is summarized in the ESI† together with the obtained mean squared displacements and velocity autocorrelations. To obtain the phoretic velocity, we have used the Debye–Stokes–Einstein predictions for a $R = 125$ nm swimmer (particle radius plus half of the cap thickness).

3 Size dependence of the photophoretic velocity

With the help of the above described methods, we have determined the phoretic velocity of $R = 100$ nm to $R = 625$ nm Janus particles as a function of the incident laser power P_0. As displayed in Fig. 2A, the thermophoretic velocity scales linearly for all particle sizes with the heating power. Plotting the average slope of the velocity increment with the heating power further reveals that the photophoretic velocity is independent of the particle size over the studied size range (Fig. 2B). Since the phoretic velocity is the product of the thermodiffusion coefficient D_{T} and the temperature gradient along the particle surface $\nabla_{\|}T$, either both are independent of the particle size or their size dependencies exactly cancel each other out. A more detailed insight into this relationship may thus be obtained by considering the temperature gradient (Section 3.2), which depends on the absorbed electromagnetic power defining the temperature rise at the gold cap and the shape of the particle.

3.1 Size dependence of the absorbed power

The size dependence of the absorbed power $P_{\mathrm{abs}}(R)$ is encoded in the absorption cross section $\sigma_{\mathrm{abs}}(R)$ of the metal cap by $P_{\mathrm{abs}} = \sigma_{\mathrm{abs}} I$, where $I = P_0/A$, which is the incident intensity of the heating laser. We here assume that the metal cap is a homogeneous absorber with no particular resonances (*i.e.* plasmon resonances). The absorbed power then depends on the volume that the incident electromagnetic wave can penetrate. The total volume of the metal cap is:

$$V_{\mathrm{cap}} = \frac{2}{3}\pi\left(3\Delta r R^2 + 3\Delta r^2 R + \Delta r^3\right). \tag{7}$$

For metals, electromagnetic waves may only penetrate up to a certain depth, known as the skin-depth δ. If the cap thickness is larger than this skin-depth, $\Delta r > \delta$ (for gold and a wavelength of 532 nm, the skin-depth is $\delta \approx 45$ nm (ref. 28)), then the effective absorbing volume can be approximated by replacing Δr by the skin-depth δ in eqn (7). Thus, considering a constant cap thickness Δr or an effective absorption within a layer given by the skin-depth δ, the absorbing volume scales in the leading order of Δr with R^2. If the particle radius R becomes comparable to the cap thickness and tends towards zero, then the two latter terms in eqn (7) become important, where the last term resembles the volume of the cap for a particle radius of $R = 0$. The experimental situation for different particle radii and a constant gold cap thickness leads to $P_{abs} \propto R^2$ also for the case when $\Delta r \approx \delta$.

A different scaling behavior appears if the cap thickness is constant relative to the particle radius, *i.e.* $\Delta r/R = f$, where f is a constant. In this case, the absorbing volume, the absorption cross section, and the absorbed power scale with R^2 as long as $R > \delta/f$. If the cap thickness decreases with the particle radius and reaches values below the skin-depth (*i.e.* $R < \delta/f$), the absorbing volume scales with R^3. This case, however, is not realized in the experiment.

To access the scaling of the absorption cross section experimentally, we consider the size dependence of the scattering cross section and conclude on the scaling of the absorption cross section. For this purpose, we have extracted the integrated scattering intensity from the dark field images as a function of time. The use of two different particle sizes in one sample allows us to determine the relative scattering intensity.

Two example time traces of the scattering intensity for particles of $R = 500$ nm and $R = 375$ nm are shown in Fig. 3A. The strong fluctuations in the data are the result of the rotational motion of the particle. In Fig. 3B, the average relative scattering cross section normalized to the value for the $R = 250$ nm particle is shown in a double logarithmic plot. Particles above a radius of 250 nm reveal a slope of two. Such a scaling for the absorption and scattering cross-section is expected for larger spherical gold particles, as displayed in Fig. 5 in the ESI.† We therefore conclude from the scaling of the relative scattering cross section that the absorption cross section depends on the square of the radius and therefore on the surface, as expected. Smaller particles ($R = 100$ nm and $R = 160$ nm) deviate from this scaling, most likely due to a changing effective thickness of the gold layer. Based on these results for the relative scattering cross section, we assume that the absorbed power scales as $P_{abs} \propto R^2$ at a constant incident intensity $I = P_0/A$.

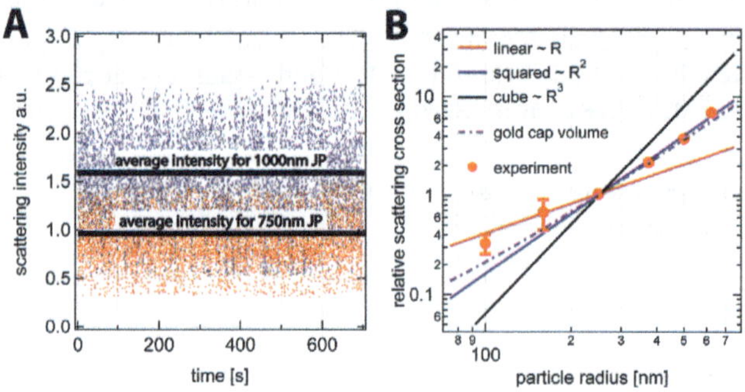

Fig. 3 (A) Example measurements of the scattering intensity *vs.* time for 2 different particle sizes. The strong fluctuations are a result of the rotation of the particle relative to the dark field illumination. (B) The relative average scattering cross section (red dots) for different particle sizes.

3.2 Size dependence of the temperature profile

The second parameter, which determines the photophoretic velocity of the particle, is the surface temperature gradient. To obtain the temperature profile for a given geometry, the steady state heat eqn (8) with the thermal conductivity κ and the heat source density Q_V has been solved numerically with the help of FEM simulations (COMSOL 4.2 Multiphysics). For the Janus particles, analytical results in the limit of a thin cap are also presented. The heat source density Q_V is the ratio between the absorbed power P_{abs} (see Section 3.1) and the volume V_{cap} of the cap.

$$\kappa \nabla^2 T = -Q_V \tag{8}$$

We have studied the temperature profile for a constant cap thickness $\Delta r = 50$ nm but with varying particle radii. The used parameters for the numerical calculations are reported in the ESI.† The results for the temperature profile along the particle surface for different particle radii but a constant gold cap thickness are displayed in Fig. 4A. The temperature gradient along the surface $\nabla T_{\parallel}$ is then proportional to the temperature difference and the detailed shape of the temperature profile. Since the temperature difference between the gold and the polymer side scales with the absorbed laser power, we may express the temperature gradient as the product of the absorbed laser power and a geometry factor $h(R)$, which contains all the details of the shape of the temperature profile. Both parameters depend on the particle radius R:

$$\nabla T_{\parallel} \propto \frac{\Delta T_{max} - \Delta T_{ps}}{R} \propto P_{abs}(R)h(R). \tag{9}$$

Assuming a constant cap thickness of $\Delta r = 50$ nm, the geometry factor scales as depicted in Fig. 4B. The results of the numerical solution of eqn (8) show that above a particle radius of $R = 100$ nm, $h(R)$ drops with R^{-2}. This corresponds to the thin cap limit ($\Delta r \ll R$), which has been obtained before in ref. 22. For smaller particle radii ($R < 100$ nm), the geometry factor $h(R)$ levels off, since the cap is no longer smaller than the particle.

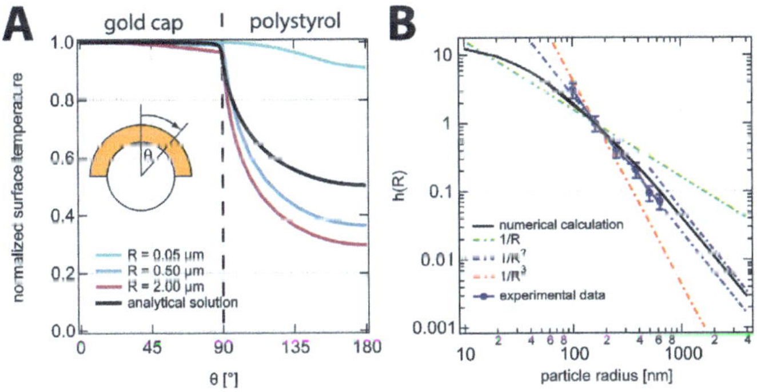

Fig. 4 (A) Representative example of the temperature profile along the particle surface. (B) Dependence of the temperature gradient normalized to the scaling of the absorbancy $\sigma_{abs}(R)$ of the particle for numerical calculations (black line) in comparison to the experimental data, assuming a quadratic scaling of $\sigma_{abs} \sim R^2$ (blue dots).

The parameter $h(R)$ can be extracted from the experimentally determined phoretic velocity by rewriting eqn (2) and (9) into:

$$h_{\text{exp}}(R) = \frac{V_{\text{th}}}{D_{\text{T}} P_{\text{abs}}}. \quad (10)$$

Since the measured phoretic velocity is size independent and D_{T} is assumed to be size independent as well, the experimentally obtained geometry factor scales with the inverse of the absorption cross section for all studied radii and therefore with R^{-2}. The difference between the numerical calculations and the experimental data is within the obtained error of the phoretic velocity but shows slightly different scaling, especially for the smallest particles. This deviation is attributed to the ideal geometry of the gold cap assumed within the numerical calculations, which is different from the shape depicted in the electron microscopy images (see Fig. 1 and ESI†).

4 Size dependent propulsion efficiency

The above obtained results may now be combined to yield an expression for the efficiency of the self-thermophoretic swimmer. Similar to the temperature gradient, we may express the efficiency using eqn (1) and (9) as the product of the absorbed power P_{abs} and a size dependent geometry factor $k(R) = Rh(R)^2$.

$$\varepsilon \propto \frac{R(\nabla_{\parallel} T)^2}{\sigma_{\text{abs}}} = P_{\text{abs}} Rh(R)^2 = P_{\text{abs}} k(R) \quad (11)$$

Thus, two main parameters influence the efficiency and its size dependence. The first is a geometry factor $k(R)$, which tells us how an optically induced and fixed temperature difference is converted into a temperature gradient. The second parameter P_{abs} describes the ability of the swimmer to harvest energy from the incident laser power.

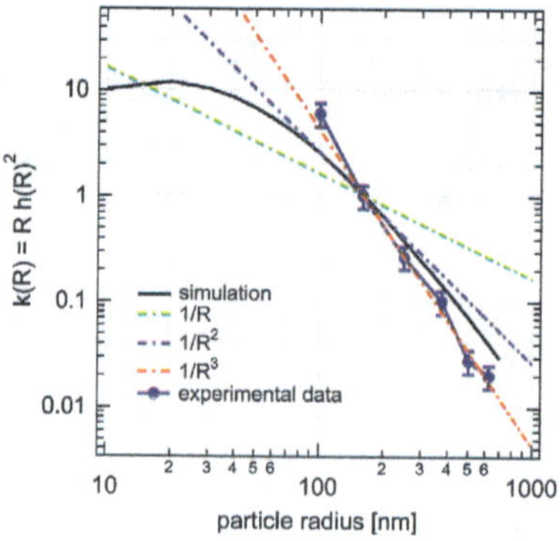

Fig. 5 The intrinsic scaling of the efficiency by the geometry summarized in $k(R)$ as a function of the particle radius, displayed in a double logarithmic plot. The numerical estimate of an idealistic cap shape (black solid line) is compared to the experimentally obtained values (blue dots). The dashed line acts as a guide to the eye.

The geometry factor $k(R)$ scales with R^{-3} for large particles and shows a maximum at sizes around 20 nm, as visible in Fig. 5 (black curve). Even though the experimentally studied swimmers are not in the size range to reveal this maximum, numerical calculations show that the position of the maximum is sensitive to the thickness and shape of the gold cap. For very small particle sizes, the Janus swimmer turns into a passive gold half-sphere, which still absorbs light but does not generate a directed motion due to the missing temperature gradient along the particle surface. Thus, the efficiency has to reveal a maximum. The experimentally studied swimmer sizes nicely follow the expected R^{-3}.

The power absorbed by the particle also directly affects the efficiency scaling, meaning that on one hand, at a fixed particle radius R, the swimming efficiency increases linearly with the heating power, which is equivalent to a linear increase in the temperature difference across the self-thermophoretic Janus swimmer. This result is in contrast to the result in ref. 17, where the efficiency is independent of the energy stored in the system. This difference arises from the fact that our definition of the input power includes the energy that is required to maintain the whole temperature field around the Janus swimmer even though the radial temperature gradients are not contributing to the propulsion of the swimmer. Sabass and Seifert, however, only consider the viscous losses by the particle motion and in the boundary layer.[17] Further, the power dependence reported here suggests, at first glance, no upper bound for the efficiency, whereas other swimmer efficiencies are bounded *e.g.* by the reaction rates for catalytic swimmers.[18,20,29] Similar effects may also be expected as, in general, a saturation of the absorption may occur at very high laser powers. However, this is typically close to the damage threshold of the materials. On the other hand, at a fixed incident heating laser intensity, it is concluded from the experiments that the absorbed power scales with the swimmer radius squared, R^2. The product of the absorbed power and the geometry factor thus reveal, in the limit of large swimmers, a $1/R$ dependence, which agrees well with earlier theoretical predictions.[17]

The deviation of the experimental results from the numerical data can be related to three aspects. First, the simple geometry which was assumed for the numerical calculation might not represent the real Janus particles accurately enough. Second, as discussed above, the scaling of the absorption cross section of the real swimmer is probably more complex. And third, the assumption that the thermodiffusion coefficient is size independent has only been shown for passive plastic particles in an external temperature profile and has to be verified for self propelled photophoretic swimmers.

5 Conclusion

In conclusion, the size dependent efficiency of photophoretic Janus particles has been studied experimentally and with the help of numerical calculations. The efficiency of the studied system can be separated into a product of their capability to build up a surface temperature gradient summarized by a geometry factor and the ability to harvest the externally supplied energy as quantified by the absorbed power. Numerical calculations show a scaling of the efficiency with the reciprocal size for large particle sizes $R \gg \Delta r$, which is also confirmed by experimental data, including the size dependence of the propulsion velocity and the scaling of the scattering cross-section for self-thermophoretic Janus swimmers with a radius between $R = 100$ nm and $R = 625$ nm. For small particle sizes, the calculations reveal that the detailed shape of the metal hemisphere becomes important, which leads to a maximum in the efficiency of a real system. In contrast to other considerations of the efficiency, our results show a linear dependence of the efficiency on the absorbed power of the particle, which is due to the fact that a large amount of energy is spent in maintaining a temperature field around the particle, which is not causing a directed motion of the particle.

Acknowledgements

The authors acknowledge financial support for this research from SFB TRR 102 and Sächsische Forschergruppe FOR 877.

References

1 H.-R. Jiang, N. Yoshinaga and M. Sano, *Phys. Rev. Lett.*, 2010, **105**, 268302.
2 I. Buttinoni, G. Volpe, F. Kümmel, G. Volpe and C. Bechinger, *J. Phys.: Condens. Matter*, 2012, **24**, 284129.
3 L. Baraban, R. Streubel, D. Makarov, L. Han, D. Karnaushenko, O. G. Schmidt and G. Cuniberti, *ACS Nano*, 2013, **7**, 1360–1367.
4 J. R. Howse, R. A. L. Jones, A. J. Ryan, T. Gough, R. Vafabakhsh and R. Golestanian, *Phys. Rev. Lett.*, 2007, **99**, 048102.
5 J. G. Gibbs and Y.-P. Zhao, *Appl. Phys. Lett.*, 2009, **94**, 163104.
6 J. Palacci, S. Sacanna, A. P. Steinberg, D. J. Pine and P. M. Chaikin, *Science*, 2013, **339**, 936–940.
7 W. F. Paxton, K. C. Kistler, C. C. Olmeda, A. Sen, S. K. S. Angelo, Y. Cao, T. E. Mallouk, P. E. Lammert and V. H. Crespi, *J. Am. Chem. Soc.*, 2004, **126**, 13424–13431.
8 M. Pumera, *Nanoscale*, 2010, **2**, 1643–1649.
9 I. Buttinoni, J. Bialké, F. Kümmel, H. Löwen, C. Bechinger and T. Speck, *Phys. Rev. Lett.*, 2013, **110**, 238301.
10 I. Theurkauff, C. Cottin-Bizonne, J. Palacci, C. Ybert and L. Bocquet, *Phys. Rev. Lett.*, 2012, **108**, 268303.
11 J. A. Cohen and R. Golestanian, *Phys. Rev. Lett.*, 2014, **112**, 068302.
12 J. Bialke, T. Speck and H. Löwen, *J. Non-Cryst. Solids*, 2015, **407**, 367–375.
13 T. Bickel, A. Majee and A. Würger, *Phys. Rev. E: Stat., Nonlinear, Soft Matter Phys.*, 2013, **88**, 012301.
14 M. Yang and M. Ripoll, *Soft Matter*, 2013, **9**, 4661–4671.
15 F. M. Weinert and D. Braun, *Phys. Rev. Lett.*, 2008, **101**, 168301.
16 W. Wang, T.-Y. Chiang, D. Velegol and T. E. Mallouk, *J. Am. Chem. Soc.*, 2013, **135**, 10557–10565.
17 B. Sabass and U. Seifert, *Phys. Rev. Lett.*, 2010, **105**, 218103.
18 B. Sabass and U. Seifert, *J. Chem. Phys.*, 2012, **136**, 064508.
19 S. Ebbens, M.-H. Tu, J. R. Howse and R. Golestanian, *Phys. Rev. E: Stat., Nonlinear, Soft Matter Phys.*, 2012, **85**, 020401.
20 M. J. Lighthill, *Commun. Pure Appl. Math.*, 1952, **5**, 109–118.
21 M. Braibanti, D. Vigolo and R. Piazza, *Phys. Rev. Lett.*, 2008, **100**, 108303.
22 A. P. Bregulla, H. Yang and F. Cichos, *ACS Nano*, 2014, **8**, 6542–6550.

23 B. Qian, D. Montiel, A. Bregulla, F. Cichos and H. Yang, *Chem. Sci.*, 2013, **4**, 1420–1429.
24 M. Schienbein and H. Gruler, *Bull. Math. Biol.*, 1993, **55**, 585–608.
25 W. Alt and G. Hoffmann, in *Lecture Notes in Biomathematics. Biological Motion*, Heidelberg Springer Verlag, 1990, pp. 254–268.
26 H. Risken, *The Fokker–Planck Equation*, Heidelberg Springer Verlag, 1984.
27 H. Qian, M. Sheetz and E. Elson, *Biophys. J.*, 1991, **60**, 910–921.
28 R. L. Olmon, B. Slovick, T. W. Johnson, D. Shelton, S.-H. Oh, G. D. Boreman and M. B. Raschke, *Phys. Rev. B: Condens. Matter Mater. Phys.*, 2012, **86**, 235147.
29 S. Childress, *J. Fluid Mech.*, 2012, **705**, 77–97.

Faraday Discussions

PAPER

Single molecule study of non-specific binding kinetics of LacI in mammalian cells

Laura Caccianini,[a] Davide Normanno,[bcde] Ignacio Izeddin†*[a] and Maxime Dahan*[a]

Received 23rd June 2015, Accepted 22nd July 2015
DOI: 10.1039/c5fd00112a

Many key cellular processes are controlled by the association of DNA-binding proteins (DBPs) to specific sites. The kinetics of the search process leading to the binding of DBPs to their target locus are largely determined by transient interactions with non-cognate DNA. Using single-molecule microscopy, we studied the dynamics and non-specific binding to DNA of the Lac repressor (LacI) in the environment of mammalian nuclei. We measured the distribution of the LacI–DNA binding times at non-cognate sites and determined the mean residence time to be $\tau_{1D} = 182$ ms. This non-specific interaction time, measured in the context of an exogenous system such as that of human U2OS cells, is remarkably different compared to that reported for the LacI in its native environment in *E. coli* (<5 ms). Such a striking difference (more than 30 fold) suggests that the genome, its organization, and the nuclear environment of mammalian cells play important roles on the dynamics of DBPs and their non-specific DNA interactions. Furthermore, we found that the distribution of off-target binding times follows a power law, similar to what was reported for TetR in U2OS cells. We argue that a possible molecular origin of such a power law distribution of residence times is the large variability of non-cognate sequences found in the mammalian nucleus by the diffusing DBPs.

1 Introduction

Biological processes such as transcription, replication or repair, and their regulation are controlled by the association of DBPs to specific sites. Understanding

[a]*Laboratoire Physico-Chimie Curie, Institut Curie, PSL Research University, CNRS UMR168, Université Pierre et Marie Curie-Paris 6, 11 rue Pierre et Marie Curie, 75005 Paris, France. E-mail: maxime.dahan@curie.fr*

[b]*Centre de Recherche en Cancérologie de Marseille, CNRS UMR7258, 27 boulevard Leï Roure, 13273 Marseilles, France*

[c]*INSERM U1068, 13009 Marseilles, France*

[d]*Institut Paoli-Calmettes, 13009 Marseilles, France*

[e]*Aix-Marseille Université, 13284 Marseilles, France*

† Currently at: Institut Langevin, ESPCI ParisTech, PSL Research University, 1 rue Jussieu, 75005 Paris, France. E-mail: ignacio.izeddin@espci.fr

the mechanisms by which DBPs find their target sequence—typically on the order of ten base pairs (bp) among a genome composed of millions to billions of bp—is thus critical.[1] The current paradigm for the target search process involves a combination of three-dimensional diffusion in the nucleoplasm and one-dimensional sliding on DNA (see review by Mirny *et al.*).[2] This facilitated diffusion (FD) model was first proposed[3] to explain the (apparently[4]) faster-than-diffusion DNA association rate of LacI proteins reported in the seminal work of Riggs and colleagues.[5] A key element in the FD model is that the non-specific interaction of DBPs with DNA largely controls the search rate. Therefore, it is an important goal to properly determine the non-specific binding times (referred hereafter as τ_{1D}).

In past years, single molecule (SM) assays[6–8] have started complementing conventional approaches such as the recovery after photobleaching or fluorescence correlation spectroscopy (see review by Mueller *et al.*[9]) for the study of nuclear dynamics and organization. In particular, SM experiments provide direct ways of measuring non-specific binding kinetics in live cells. The first SM measurements of transcription factor dynamics in living cells were those reported by Elf and collaborators for LacI in *E. coli*.[10] They could infer a rapid non-specific association to DNA, with $\tau_{1D} < 5$ ms, and later probe the FD model.[11] More recently, SM measurements have been extended to mammalian systems. Compared to *E. coli*, DBPs in mammalian cells must find their specific DNA sequence among a much larger pool of competing non-specific sites. Experiments in mammalian cells have now been performed on a few different systems, including p53 in human lung carcinoma H1299 cells,[12] Sox2/Oct4 in mouse embryonic stem cells,[13] glucocorticoid receptors in human breast cancer MCF-7 cells,[14] and TetR in human osteosarcoma U2OS cells.[15] Interestingly, in all these cases, the mean non-specific interaction times appear to be a few seconds, several orders of magnitude longer than the values reported for LacI in *E. coli*.[10] This thus raises the question of whether these markedly different non-specific binding kinetics are due to the intrinsic properties of the DBPs that were studied, the difference in DNA organization between bacteria and mammalian cells, or the experimental techniques that were used. In the present study, we address this question by measuring the diffusion dynamics and the non-specific binding properties of LacI in U2OS cells in order to directly compare our results with those obtained for the same protein but in bacteria.

2 Results

To address the non-specific LacI–DNA interactions in mammalian cells, we chose an engineered cell line containing repeated insertions of the *lac*O binding sequence. More precisely, we used the human osteosarcoma cell line U2OS 2-6-3, which contains 200 insertions of an artificial gene, each including 256 *lac*O repeats.[16] Since the human genome does not contain any *lac*O sequence, the sites of the specific and non-specific interactions between LacI molecules and DNA could be unambiguously separated with this cellular system.

We transfected U2OS 2-6-3 cells with two different plasmids: NLS-LacI-GFP and NLS-LacI-HaloTag. After overnight transfection, we stained the cells with HaloTag TMR ligand (at high concentration). The NLS-LacI-GFP plasmid served as a visual aid to unambiguously locate the target locus. Fluorescence images of the transfected nuclei in both GFP and TMR channels displayed a homogenous and diffuse population of LacI molecules and bright fluorescent loci, corresponding to an accumulation of LacI recruited at the target sequence (Fig. 1).

2.1 Single-molecule dynamics of LacI in mammalian cells

To investigate the LacI dynamics at the single molecule level, we used a low concentration of HaloTag TMR ligand (25 pM) to label the cells, which allowed us to work in a diluted regime, suited for single-particle-tracking (SPT) experiments. We acquired SM fluorescence images of LacI-HaloTag TMR ligand proteins under continuous illumination with a 561 nm laser (at ~1 kW cm^{-2}) with an EMCCD camera at a frame rate of 197 Hz. Under these imaging conditions, the signal from a single dye was sufficient to overcome the background coming from out-of-focus dyes and from endogeneous autofluorescence. We next localized the protein's position in each frame (with an accuracy of ~20–40 nm) and computed a set of individual trajectories away from the specific target locus, which was identified in the GFP channel (Fig. 1). The length of the trajectories is limited either by axial diffusion or by photobleaching of the dye. We performed experiments in 6 cells accumulating a total of 1525 individual traces longer than 6 frames, with a median trace length of 29 frames (145 ms). We then computed the time-averaged mean square displacement (MSD) of each individual track. The MSD curves of a sample with 10 individual traces are plotted in Fig. 2a and reveal a wide range of diffusion behaviors of LacI in the nucleus of mammalian cells. To quantify the diffusion dynamics, we calculated the instantaneous diffusion coefficient, D_{Inst}, from the linear fit of the initial points (2nd to 5th) of each MSD curve. We found that the distribution of instantaneous diffusion coefficients spanned over three

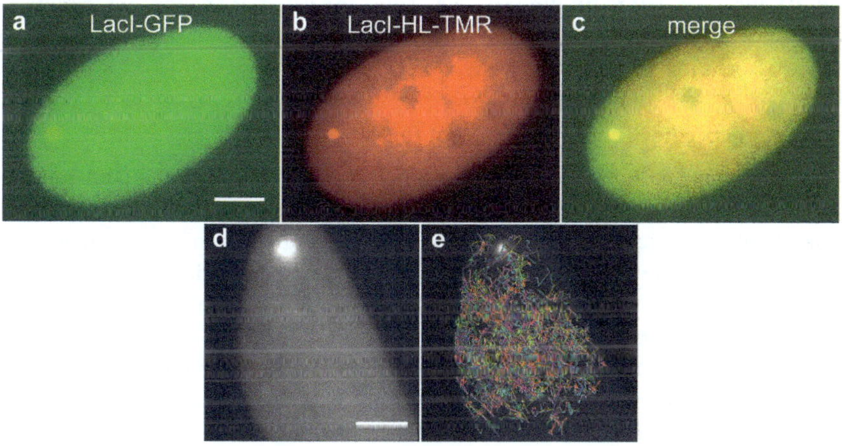

Fig. 1 (a–c) Fluorescence images of a U2OS 2 6 3 cell nucleus co-transfected with NLS-LacI-GFP (green) and NLS-LacI-HaloTag labeled with HaloTag TMR ligand at high concentration (1 nM) (red). The accumulation of proteins on the bright spot indicates the position of the binding locus (target site). (d and e) Fluorescence image of a cell transfected with NLS-LacI-GFP (d) and NLS-LacI-HaloTag stained with a low concentration (25 pM) of HaloTag TMR ligand for the SM experiments. In panel (e), the overlaid trajectories resulting from our single-particle detection and raw-tracking analysis are shown. Note that for the traces at the specific binding site and elsewhere in the nucleoplasm, only the latter was considered for our non-specific binding analysis. Scale bar = 5 μm.

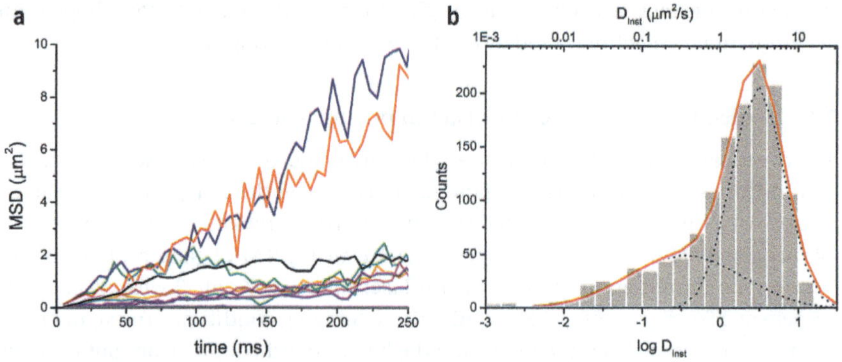

Fig. 2 (a) Mean square displacement (MSD) curves of 10 individual traces of LacI-HL-TMR molecules, exemplifying the broad distribution of LacI diffusivity. (b) Histogram of the instantaneous diffusion coefficients, D_{Inst}, from individual MSD curves, extending over three orders of magnitude (1525 trajectories, 6 cells).

orders of magnitude, from 0.01 μm^2 s^{-1} to more than 10 μm^2 s^{-1} (Fig. 2b). The distribution was approximately bimodal with the first peak centered around 5 μm^2 s^{-1}, which we attributed to proteins diffusing in the nucleoplasm, and the second one centered around 0.3 μm^2 s^{-1}, corresponding to proteins interacting (at least transiently) with DNA. The fraction of the traces displaying an instantaneous diffusion coefficient, D_{Inst}, smaller than 0.1 μm^2 s^{-1}, which we considered the benchmark for an immobile protein—being the diffusivity of chromatin-bound markers as H2B histones of the same order[7,14]—was equal to $f \sim 8\%$. This fraction f can be expressed as: $\tau_{1D}/(\tau_{1D} + \tau_{3D})$, where τ_{3D} is the average time spent in 3D diffusion before binding to DNA. Thus, the value of f yields an estimate of the ratio $\tau_{3D}/\tau_{1D} \sim 11.5$. Note that $1/\tau_{3D}$ and $1/\tau_{1D}$ correspond to the association and dissociation rates, respectively.

2.2 Non-specific binding time of LacI to DNA

We next aimed to more carefully examine the role of non-specific DNA interactions on the dynamics of LacI molecules and estimate τ_{1D}. To this end, we followed an approach recently developed for the analysis of the non-specific interactions of the Tet repressor in mammalian cells.[15] In brief, we analyzed the changes in the instantaneous diffusion coefficient along each individual trajectory by computing D_{Inst} in a running window of 40 ms. We identified events in which the molecule reversely switched from fast to much slower diffusion (Fig. 3a), likely corresponding to a transition from 3D diffusion to non-specific binding to DNA. The distribution of the interaction times was approximately exponential with a decay time of 37 ms (Fig. 3b), a value which is ~8 times larger than the non-specific binding times of LacI in *E. coli* ($\tau < 5$ ms)[10] but smaller than other transcription factors reported in eukaryotes, like p53 (~1.7 s)[12] and Sox2 (~0.8 s).[8] Nevertheless, we also observed long binding events, exceeding several seconds, which were difficult to capture due to the photobleaching of the fluorophore. In order to examine them and circumvent photobleaching, we performed time-lapse experiments in which 5 ms exposure images were interlapsed with periods of 50 and 500 ms. We then considered the proteins that remained

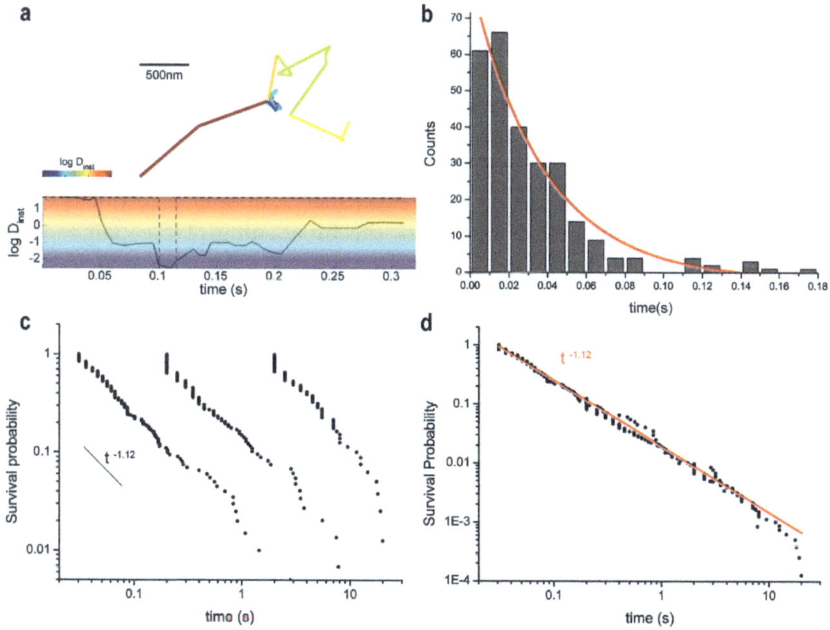

Fig. 3 (a) Example of a molecule undergoing a transition from the diffusive state to non-specific DNA binding and its instantaneous diffusion coefficient, D_{Inst}, calculated over a running window of 40 ms. (b) Distribution of binding times retrieved with the running window analysis, measured with a 95% confidence that the events represent real deviation from diffusive behavior. The exponential fit with a characteristic decay time of $\tau = 37$ ms is shown in red. (c) Survival probability of the non-specific binding events of LacI molecules for (from left to right) continuous imaging (197 Hz) and time-lapse (20 and 2 Hz) experiments. The black line is a visual guide corresponding to t^{γ} with $\gamma = -1.12$. (d) Global survival probability curve resulting from rescaling of the three temporal series. The red line represents the best fit to the data with a power law t^{γ} function, with $\gamma = -1.12$.

localized within a region of half a pixel (80 nm) and calculated their survival probability, SP(τ), defined as the probability to stay bound for a time longer than τ (Fig. 3c). The same analysis was also performed for the case of continuous imaging experiments and yielded a power law distribution (t^{γ}) of the non-specific binding survival probability, with $\gamma = -1.12$, all the more evident in the normalized curves over all the time regimes under investigation (Fig. 3d). The numerical integration of the global survival probability curve provided us with an estimation of the mean non-specific binding time $\tau_{1D} = 182$ ms, from which we could deduce $\tau_{3D} = 2.1$ s.

3 Discussion

Our SM experiments provide quantitative information on the diffusion dynamics of LacI proteins in the nucleus of human cells. In particular, with this exogenous system we gained insight into the role of non-specific interactions with non-cognate DNA sequences and their impact on nuclear protein's diffusion. Compared to the results obtained in *E. coli*, we note several striking differences.

First, the mean non-specific binding time (182 ms) in U2OS is about 30 times larger than the upper bound (5 ms) estimated in *E. coli*.[10] While we do not think that this explains the discrepancy between the results, it is worth mentioning a technical difference between our experiments and those in bacteria. In *E. coli*, the non-specific binding time was not obtained, as in our case, by analyzing individual trajectories due to the limited brightness and photostability of the fluorescent proteins used to tag LacI in those experiments. Instead, an estimate of τ_{1D} was indirectly determined by measuring the broadening of the point-spread function (PSF) as a function of the illumination time. A width of the PSF exceeding the diffraction limit was interpreted as evidence that the protein had detached during the recording time.

In our experiments, we were able to recover the full distribution of non-specific binding times, and not only the mean value. The observed distribution is broad and scales as a power law. The longest binding events exceeded a few seconds and were not found for the LacI in bacteria. Interestingly, a power law distribution has also been found for the SP of non-specific binding times of TetR in U2OS 2-6-3 cells,[15] but with a different exponent (-0.7 *vs.* -1.12 for the LacI). The molecular origin of these broad distributions is not well understood yet. A possible cause is the large variability of non-specific sequences encountered by the diffusing proteins in the nucleus of mammalian cells, going from random sequences to quasi-specific sites, leading to heterogeneity in the binding affinities. Notably, it can be shown that from the cumulative sum of an exponential distribution of DNA-binding energies, a power law distribution of binding times can emerge.[17,18] This scenario is in contrast to bacteria, where the genomic DNA is much shorter than in eukaryotes and the number of non-cognate sequences, which might act as (transient) decoy sites, is much smaller.

The fraction f of time spent non-specifically bound to DNA is higher in bacteria ($\sim$87% in *E. coli* *vs.* $\sim$8% in U2OS 2-6-3 cells). This is equivalent to saying that τ_{3D} is much shorter in bacteria, less than 1 ms compared to $\sim$2 s in our measurements. This, in turn, means that the association rate, $1/\tau_{3D}$, is much higher in bacteria. To evaluate the theoretical diffusion-limited association rate k, it is necessary to estimate the concentration c of the DNA (expressed in terms of base-pairs). Assuming a volume of 1 μm^3 for *E. coli* and 500 μm^3 for a human cell nucleus and given the respective size of the genome (4.6 millions bp and 3 billions bp), c is in fact comparable in both cases and on the order of 10 mM. Thus, we expect $k = Dca$ (where $D \sim 5\ \mu m^2\ s^{-1}$, which is the diffusion coefficient, and a is taken to be equal to 1 nm) to be on the order of $10^5\ s^{-1}$. Even though neither of the measurements (in *E. coli* or in human cells) reached the diffusion limit, the discrepancy with the measurement of τ_{3D} in U2OS is especially striking. This might be due to the fact that genomic DNA is globally more accessible in prokaryotes compared to in higher eukaryotes, where DNA is compacted within nucleosomes and can be covered by many other proteins, specifically or non-specifically bound and competing for the same accessible sites. Also, the bound and unbound times for different DBPs in mammalian cells are closer to what we observed for LacI proteins in the same cellular system compared to in bacteria (Table 1). This points to the potential role of the nuclear environment of mammalian cells in dictating the diffusion dynamics of DBPs and their interactions with DNA.

Table 1 Comparison between the 3D diffusion and the non-specific binding times reported for different transcription factors (TFs) in several cellular lines from different organisms

Protein	Cell	τ_{1D}	τ_{3D}	Ref.
LacI	U2OS	182 ms	2.1 s	This study
LacI	*E. coli*	<5 ms	<1 ms	Elf *et al.*[10]
TetR	U2OS	2 s	6 s	Normanno *et al.*[15]
p53	H1299	1.7 s	1.8 s	Mazza *et al.*[12]
GR	MCF/U2OS	1.5 s	—	Gebhardt *et al.*[14]
Sox2	Mouse ES	0.8 s	3.3 s	Chen *et al.*[8]

4 Conclusion

Our experiments illustrate how kinetic rates can now be retrieved using single molecule imaging of DBPs in live cells. They also point to open questions about the role of chromatin organization on the specific or non-specific association rates with DNA. Unquestionably, together with powerful genome-editing tools[19] and advanced imaging techniques,[20,21] single molecule techniques are expected to play an important role in our understanding of nuclear dynamics and organization.

References

1 D. Normanno, M. Dahan and X. Darzacq, *Biochim. Biophys. Acta, Gene Regul. Mech.*, 2012, **1819**, 482–493.

2 L. Mirny, M. Slutsky, Z. Wunderlich, A. Tafvizi, J. Leith and A. Kosmrlj, *J. Phys. A: Math. Theor.*, 2009, **42**, 434013.

3 O. G. Berg, R. B. Winter and P. H. von Hippel, *Biochemistry*, 1981, **20**, 6929–6948.

4 S. E. Halford, *Biochem. Soc. Trans.*, 2009, **37**, 343.

5 A. Riggs, S. Bourgcois and M. Cohn, *J. Mol. Biol.*, 1970, **53**, 401–417.

6 I. I. Cisse, I. Izeddin, S. Z. Causse, L. Boudarene, A. Senecal, L. Muresan, C. Dugast-Darzacq, B. Hajj, M. Dahan and X. Darzacq, *Science*, 2013, **341**, 664–667.

7 I. Izeddin, V. Récamier, L. Bosanac, I. I. Cissé, L. Boudarene, C. Dugast-Darzacq, F. Proux, O. Bénichou, R. Voituriez, O. Bensaude, M. Dahan and X. Darzacq, *eLife*, 2014, **3**, DOI: 10.7554/eLife.02230.

8 J. Chen, Z. Zhang, L. Li, B.-C. Chen, A. Revyakin, B. Hajj, W. Legant, M. Dahan, T. Lionnet, E. Betzig, R. Tjian and Z. Liu, *Cell*, 2014, **156**, 1274–1285.

9 F. Mueller, T. J. Stasevich, D. Mazza and J. G. McNally, *Crit. Rev. Biochem. Mol. Biol.*, 2013, **48**, 492–514.

10 J. Elf, G. W. Li and X. S. Xie, *Science*, 2007, **316**, 1191–1194.

11 P. Hammar, P. Leroy, A. Mahmutovic, E. G. Marklund, O. G. Berg and J. Elf, *Science*, 2012, **336**, 1595–1598.

12 D. Mazza, A. Abernathy, N. Golob, T. Morisaki and J. G. McNally, *Nucleic Acids Res.*, 2012, **40**, e119.

13 B. Chen, L. A. Gilbert, B. A. Cimini, J. Schnitzbauer, W. Zhang, G.-W. Li, J. Park, E. H. Blackburn, J. S. Weissman, L. S. Qi and B. Huang, *Cell*, 2013, **155**, 1479–1491.

14 J. C. M. Gebhardt, D. M. Suter, R. Roy, Z. W. Zhao, A. R. Chapman, S. Basu, T. Maniatis and X. S. Xie, *Nat. Methods*, 2013, **10**, 421–426.

15 D. Normanno, L. Boudarene, C. Dugast-Darzacq, J. Chen, C. Richter, F. Proux, O. Bénichou, R. Voituriez, X. Darzacq and M. Dahan, *Nat. Commun.*, 2015, **6**, 7357.

16 S. M. Janicki, T. Tsukamoto, S. E. Salghetti, W. P. Tansey, R. Sachidanandam, K. V. Prasanth, T. Ried, Y. Shav-Tal, E. Bertrand, R. H. Singer and D. L. Spector, *Cell*, 2004, **116**, 683–698.

17 J. P. Bouchaud and A. Georges, *Phys. Rep.*, 1990, **195**, 127–293.

18 R. Metzler and J. Klafter, *The random walk's guide to anomalous diffusion*, 2000.

19 J. D. Sander and J. K. Joung, *Nat. Biotechnol.*, 2014, 1–9.

20 S. Abrahamsson, J. Chen, B. Hajj, S. Stallinga, A. Y. Katsov, J. Wisniewski, G. Mizuguchi, P. Soule, F. Mueller, C. D. Darzacq, X. Darzacq, C. Wu, C. I. Bargmann, D. A. Agard, M. Dahan and M. G. L. Gustafsson, *Nat. Methods*, 2013, **10**, 60–63.

21 B. Hajj, M. El Beheiry, I. Izeddin, X. Darzacq and M. Dahan, *Phys. Chem. Chem. Phys.*, 2014, **16**, 16340–16348.

Faraday Discussions

PAPER

Millisecond single-molecule localization microscopy combined with convolution analysis and automated image segmentation to determine protein concentrations in complexly structured, functional cells, one cell at a time

Adam J. M. Wollman and Mark C. Leake

Received 12th May 2015, Accepted 6th July 2015
DOI: 10.1039/c5fd00077g

We present a single-molecule tool called the CoPro (concentration of proteins) method that uses millisecond imaging with convolution analysis, automated image segmentation and super-resolution localization microscopy to generate robust estimates for protein concentration in different compartments of single living cells, validated using realistic simulations of complex multiple compartment cell types. We demonstrate its utility experimentally on model *Escherichia coli* bacteria and *Saccharomyces cerevisiae* budding yeast cells, and use it to address the biological question of how signals are transduced in cells. Cells in all domains of life dynamically sense their environment through signal transduction mechanisms, many involving gene regulation. The glucose sensing mechanism of *S. cerevisiae* is a model system for studying gene regulatory signal transduction. It uses the multi-copy expression inhibitor of the GAL gene family, Mig1, to repress unwanted genes in the presence of elevated extracellular glucose concentrations. We fluorescently labelled Mig1 molecules with green fluorescent protein (GFP) *via* chromosomal integration at physiological expression levels in living *S. cerevisiae* cells, in addition to the RNA polymerase protein Nrd1 with the fluorescent protein reporter mCherry. Using CoPro we make quantitative estimates of Mig1 and Nrd1 protein concentrations in the cytoplasm and nucleus compartments on a cell-by-cell basis under physiological conditions. These estimates indicate a ~4-fold shift towards higher values in the concentration of diffusive Mig1 in the nucleus if the external glucose concentration is raised, whereas equivalent levels in the cytoplasm shift to smaller values with a relative change an order of magnitude smaller. This compares with Nrd1 which is not involved directly in glucose sensing, and which is almost exclusively localized in the nucleus under high and low external glucose levels. CoPro facilitates time-resolved quantification of protein concentrations in single functional cells, and enables the distributions of concentrations across a cell population

University of York, Physics and Biology, UK. E-mail: mark.leake@york.ac.uk

to be measured. This could be useful in investigating several cellular processes that are mediated by proteins, especially where changes in protein concentration in a single cell in response to changes in the extracellular chemical environment are subtle and rapid and may be smaller than the variability across a cell population.

1. Introduction

A number of different methods already exist for quantifying protein copy number in cells, mainly involving bulk ensemble average biochemical techniques of cell lysates, including tag-affinity quantification and mass spectrometry,[1] but more recently using fluorescence measurements on living cells. Affinity methods (reviewed here[2]) involve tagging a protein of interest and quantifying it using western blots or ELISA, and have achieved zeptomolar (*i.e.* 100s of molecules per cell) resolution.[3] The entire yeast proteome has been mapped using affinity methods.[4] Mass spectrometry combined with chromatography can generate peptide spectra which can be identified using peptide databases[5] and has been used to map the proteome in the malaria parasite, *Plasmodium falciparum*.[6] Fluorescence detection has been used with flow cytometry to measure the copy numbers of >4000 GFP tagged proteins in *S. cerevisiae*.[7] Comparing the cytometry results with western blots, proteins with >8000 copies per cell were precisely quantified. However, the sensitivity of detection fell off steeply with lower copy numbers, and at copy numbers equivalent to 2000–4000 molecules per cell this sensitivity dropped to ~50%. Fluorescence detection using microscopy methods have enabled more precise quantification – for example a YFP-tagged, chromosomally integrated protein library was used by Taniguchi *et al.* with live cells investigated using fluorescence microscopy combined with automated microfluidics to study the *E. coli* proteome.[8]

In our study here, we have developed a fluorescence microscopy single-molecule tool we call the CoPro (concentration of proteins) method to generate robust estimates of protein concentrations inside different subcellular compartments of living cells. To characterize our method we used robust simulations of realistic complex cell shapes and compartments corresponding to typical rod-shaped prokaryotic bacteria and spherical eukaryotic yeast cells. We then applied our approach experimentally using *E. coli* bacterial cells containing protein labelled with fluorescently labelled replisome component DnaQ to compare against the earlier method of Taniguchi *et al.*, used for their *E. coli* proteome studies. Following this we ultimately applied the method experimentally to a more complex budding yeast cell system for which we could control the protein concentrations in different subcellular compartments by precise manipulation of the extracellular environment. The glucose sensing pathway of budding yeast was an excellent choice in this regard, offering a complex cellular system which has multiple subcellular compartments and with the protein concentrations of key regulators of the signal transduction response known to vary in response to extracellular glucose concentration from earlier biochemical and standard epifluorescence microscopy studies.

All cells dynamically sense their environment through signal transduction mechanisms, and in the majority of these mechanisms gene regulation is involved in responding to environmental changes (a notable exception being chemotaxis in many bacteria such as *E. coli* which utilize a protein-only regulatory network in the cell). These gene regulatory mechanisms rely on cascades of protein–protein interactions which transmit signals from sensory elements to responsive elements within each cell. The glucose sensing mechanism in budding yeast, *Saccharomyces cerevisiae*, is a model system for the study of signal transduction which involves gene regulation. The multi-copy inhibitor (a protein called Mig1) of the GAL gene expression protein is an essential transcription factor in this mechanism. Mig1 is a Cys^2–His^2 zinc finger DNA binding protein[9] which binds several glucose-repressed promoters.[10–13] In the presence of elevated concentration levels of extracellular glucose it is poorly phosphorylated and predominantly located in the nucleus[14,15] where it recruits a repression complex to the DNA.[16] If extracellular glucose concentrations levels are depleted, Mig1 is phosphorylated by the sucrose non-fermenting protein (Snf1),[17–19] resulting in a redistribution of the mean localization of Mig1 into the cytoplasm.[14,20,21] Thus, Mig1 concentration levels in the cell nucleus and cytoplasm serve as a readout of glucose signal transduction in budding yeast.

Standard epifluorescence microscopy has been used previously to quantify the ratio of mean fluorescence pixel intensity of GFP-tagged Mig1 in the cytoplasm with the nucleus in live budding yeast cells.[15] Bendrioua *et al.* used microfluidics assays to observe changes in this ratio of mean intensity values in real-time in response to glucose changes. They observed a rapid (<1 min) response of cells to extracellular glucose concentration changes in a history-dependent manner dependent on cells' previous exposure to glucose, indicative of cellular sensory adaptation. However, using these methods they were not able to quantify the absolute number of Mig1 molecules in the nucleus or the cytoplasm but relied instead only on the fluorescence intensity as an uncalibrated readout of Mig1 concentration. Here, we have used our home-built single-molecule narrowfield microscope, in combination with automated super-resolution localization microscopy, image segmentation and convolution analysis, to directly quantify the concentration of Mig1 in the nucleus and cytoplasm of live yeast cells at high and low glucose concentrations.

In narrowfield microscopy, the area of a collimated epifluorescence laser excitation field in the focal plane is reduced compared to conventional epifluorescence by an order of magnitude to have a typical width of 5–10 μm which can be concentrated to illuminate just a single cell, similar to slimfield which has a comparably sized beam waist in the focal plane but is generated using a focused laser beam in the sample.[22] Both narrowfield and slimfield illumination generate high local excitation intensity fields of typically a few kW cm^{-2}, which enable single fluorescent protein detection above camera readout noise at millisecond sampling times, which is required to track the movement of individual molecules diffusing in relatively low viscosity cellular environments, such as the cell cytoplasm.[23] We have previously used millisecond imaging to quantify the concentration of bacterial replisome protein components in single *E. coli* bacterial cells.[24] In our study here, we extend these methods to quantify the concentration of fluorescently labelled molecules in more complexly structured yeast cells. These cells have a total volume an order of magnitude greater and have a significantly higher level of native autofluorescence. They also contain complex subcellular compartments, potentially with different concentrations in different compartments.

A narrowfield fluorescence image of a cell compartment containing fluorescently-labelled components is comprised of four principal intensity components: (1) a background signal from camera noise; (2) a cellular autofluorescence background from naturally fluorescent molecules which are native to the cell; (3) foreground spots of fluorescence of varying degrees of brightness corresponding to one or more fluorophores which have been introduced in the cell to label specific molecular components – fluorophores which are colocalized in space to within less than the optical resolution limit of 200–300 nm are detected as being part of the same 'fluorescent spot' (for example, a molecular complex of effective diameter of a few nm may be composed of N repeating subunits of the same protein, and if this protein subunit is labelled with a single specific fluorophore, the brightness of the fluorescence spot that we observe in the far-field diffraction-limited regime of our microscope is N times the brightness of a single fluorophore) within the microscope depth of field; (4) a background pool of, typically, rapidly diffusing, fluorescently-labelled molecules that are not detected as distinct spots of fluorescence.

Each of these separate intensity components must be quantified to obtain the total concentration of fluorescently-labelled molecules. The autofluorescence background and camera noise were characterized by narrowfield microscopy images of wild-type parental cells (*i.e.* cells containing no fluorescent protein labelling). The mean intensity of the autofluorescence was subtracted from the fluorescence signal to obtain the fluorescent protein foreground signal. The number of molecules in distinct 'spots' or 'foci' of fluorescence, which have a mean effective diameter of a few hundred nm consistent with the measured point spread function (PSF) width of our microscope, was measured using our bespoke super-resolution localization microscope[25,26] which could objectively track automatically detected candidate fluorescent spots over time using robust probabilistic criteria. The brightness of these spots can be compared against a separate calibration obtained for the brightness due to a single fluorophore (for example, one can measure step-like intensity changes due to photobleaching of single fluorescent protein molecules obtained in separate *in vitro* experiments in which single fluorescent protein molecules have been immobilized *via* specific chemical conjugation to a microscope coverslip surface).

2. Materials and methods

2.1 Strains and growth conditions

We used a DnaQ-GFP *E. coli* strain, adapted from DnaQ-YPet[24] by PCR amplification of eGFP with A206K mutation[27] for monomerization, and KAN-I using primers with 50 nt overhangs homologous to the insertion region. *E. coli* strains were grown in 56 salts minimal media supplemented with 4% glucose and grown overnight at 37 °C. We used the MATa MIG1-GFP-HIS3 NRD1-mCherry-hphNT1METLYS *S. cerevisiae* strains in the BY4741 background.[15] Yeast strains were grown in YNB minimal media supplemented with 4% glucose overnight at 30 °C.

2.2 Purification of fluorescent proteins

His-tagged mCherry and GFP genes were amplified by PCR and cloned into PET vectors. These were transformed into BL21 PLysS *E. coli* and grown to saturation overnight in 2l 2YT media supplemented with 20 mg mL^{-1} kanamycin. Overnight, cultures were induced with 100 μg mL^{-1} IPTG and allowed to express protein for 5 hours at 20 °C. The induced culture was pelleted by centrifugation, resuspended in lysis buffer (10% glycerol, 50 mM Tris–HCl, 150 mM NaCl, 10 mM imidazole, 1 mM DTT, pH 8) and lysed by sonication. The lysate was cleaned by centrifugation and the supernatant applied to pre-packed NTA columns (His Gravitrap, GE Healthcare) equilibrated with lysis buffer at 4 °C. The column was washed with lysis buffer before the protein was eluted with lysis buffer supplemented with 500 mM imidazole. 1 mL fractions were collected and run on a denaturing gel. Fractions containing the protein were pooled and dialysed overnight in storage buffer (50% glycerol, 50 mM Tris–HCl, 150 mM NaCl, 10 mM imidazole, pH 8). The protein was then aliquoted and flash frozen in liquid nitrogen to be stored at −80 °C.

2.3 Narrowfield microscope

A bespoke inverted fluorescence microscope was constructed using a Zeiss microscope body with a 100× TIRF 1.49 NA Olympus oil immersion objective lens and a *xyz* nanopositioning stage (Nanodrive, Mad City Labs). Fluorescence excitation used 50 mW Obis 488 nm and 561 nm lasers. A dual pass GFP/mCherry dichroic with 25 nm transmission windows centred on 525 nm and 625 nm was used underneath the objective lens turret. The beam was expanded 0.5× and 1× for imaging bacteria and yeast cells respectively with a series of lenses on selectable flipper mounts, to generate an excitation field of intensity ~6 W cm^{-2}. Beam intensity profiles were measured directly by raster scanning in the focal plane while imaging a sample of fluorescent beads. A high speed camera (iXon DV860-BI, Andor Technology, UK) was used to image at typically 5 ms per frame (this rapid sampling speed was required to image diffusing molecules in the cell) with the magnification set at ~80 nm per pixel. The camera CCD was split between a GFP and mCherry channel using a bespoke colour splitter consisting of a dichroic centred at pass wavelength 560 nm and emission filters with 25 nm bandwidths centred at 525 nm and 594 nm. The microscope was controlled using our in-house bespoke LabVIEW (National Instruments) software.

2.4 *In vitro* microscopy

In vitro experiments were performed in a simple tunnel slide flow-chamber constructed from strips of double-sided tape creating a channel on a standard glass microscope slide and covered with a plasma-cleaned BK7 glass coverslip, creating a chamber 5–10 μL in volume, using a protocol adapted from earlier studies.[28,29] In brief, the PSF was measured using 20 nm diameter fluorescein beads (Invitrogen) diluted by a factor of 1000 in PBS and one volume injected into the flow chamber. Beads were left to sediment onto the coverslip for 10 min with the chamber inverted before excess beads were washed out with 10 flow-chamber volumes of PBS. *In vitro* fluorescent proteins, GFP and mCherry were imaged by flowing one flow-chamber volume of 1 μg mL^{-1} anti-GFP or anti-DsRed antibodies respectively. After 5 min incubation at RT in the inverted flow-chamber excess antibody was washed away with 10 flow-cell volumes of PBS. One flow-cell volume of 1 μg mL^{-1} fluorescent protein was then injected, incubated for 5 min and washed. To focus on the coverslip surface in brightfield, 1000-fold dilution of 300 nm diameter polystyrene beads (Invitrogen) was added to the slide, incubated and washed.

2.5 *In vivo* microscopy

Budding yeast and *E. coli* cells were imaged on agarose pads.[24] In brief, gene frames (Life Technologies) were stuck to a glass microscope slide to form a well and 500 μL YNB or 56 salts, for yeast or *E. coli*, plus 1% agarose pipetted into the well. The pad was left to dry at room temperature before 5 μL overnight yeast or *E. coli* culture was pipetted in 6–10 droplets onto the pad. This was covered with a plasma-cleaned glass coverslip and imaged immediately. The overnight yeast culture was used as high (4%) glucose concentration and 4% glucose included in the agarose pad. For low glucose concentrations, the overnight culture was spun down, washed and resuspended in YNB with no glucose. Imaging consisted of finding a cell in brightfield mode, recording a stack of 10 brightfield images before taking stacks of 100–1000 frames of fluorescent images for each fluorescent channel separately.

2.6 Image analysis

The diffusive pool fluorescence (due to particles which diffused too rapidly, and/or which were too dim, and/or which were too close together to be detected as a distinct fluorescent spot) was modelled as a 3D convolution integral of the normalized PSF, P, of our imaging system, over the whole cell. Fig. 1 illustrates the model diagrammatically in the case of a simple spherical cell. Each camera detector pixel of physical area ΔA has an equivalent area dA in the conjugate image plane of the sample mapped in the focal plane of the microscope, such that $\mathrm{d}A = \Delta A/M$ where M is the total magnification between the camera and the sample. The measured intensity, I', in a conjugate pixel area dA is the sum of the foreground fluorophore intensity I plus autofluorescence (I_a) plus detector noise

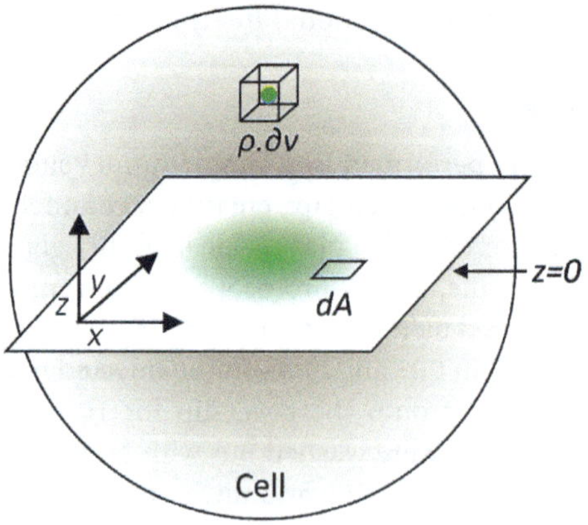

Fig. 1 Diagram illustrating the convolution integral over a single cell, in this simple example, assumed to be spherical. The intensity at an area element, dA, in the microscope's focal plane is the sum of the PSFs of all of the fluorophores in the cell or the integral of the PSF multiplied by the concentration, ρ, over the cell volume, v. The focal plane is marked as $z = 0$.

(I_d). I is the is the sum of the contributions from all of the non-autofluorescence fluorophores in the whole of the cell, such that:

$$I(x_0, y_0, z_0)\mathrm{d}A = \sum_{i=1}^{\text{all cell voxels}} E\rho I_s P(x_i - x_0, y_i - y_0, z_i - z_0). \quad (1)$$

Here I_s is the characteristic integrated intensity of a single fluorophore and ρ is the fluorophore density in units of molecules per voxel (*i.e.* a pixel volume unit). E is a function representing the change in the laser profile excitation intensity over the cell. In a uniform excitation field $E = 1$. For narrowfield microscopy the excitation intensity is uniform in z but has a 2D Gaussian profile in the lateral xy plane parallel to the microscope focal plane. In a non-saturating regime for photon emission flux of a given fluorophore, the brightness of that fluorophore, assuming simple single-photon excitation, is proportional to the local excitation intensity,[24] thus:

$$E(x, y, z) = \exp\left[-\left(\frac{x^2 + y^2}{2\sigma_{xy}{}^2}\right)\right]. \quad (2)$$

Here σ_{xy} is the Gaussian width of the excitation field in the focal plane, which we configure to be 3 μm and 6 μm in our microscope for the *E. coli* and budding yeast configurations respectively. In slimfield there is a z-dependence also with a Gaussian sigma width which is ~2.5 times that of the σ_{xy} value, but in narrowfield it is independent of z, which made narrowfield a more ideal choice of illumination here for larger cells such as yeast. Thus:

$$I(x_0, y_0, z_0)\mathrm{d}A = \sum_{i=1}^{\text{all cell voxels}} \rho I_s \exp\left[-\left(\frac{x^2 + y^2}{2\sigma_{xy}}\right)\right] P(x_i - x_0, y_i - y_0, z_i - z_0). \quad (3)$$

Defining $C(x_0,y_0,z_0)$ as the numerical convolution integral (also containing the Gaussian excitation field) over the specific cell being imaged, and assuming the fluorophore density averaged over time is uniform in space in any given subcellular cell compartment, we can calculate the fluorophore density from each pixel in the image as:

$$\rho = \frac{I(x_0, y_0, z_0)\mathrm{d}A}{C(x_0, y_0, z_0)I_s} = \frac{I'(x_0, y_0, z_0) - (I_a + I_d)}{C(x_0, y_0, z_0)I_s}. \quad (4)$$

Here, the mean value of the total background noise ($I_a + I_d$) can be calculated from images of cells which do not contain any foreign fluorophores. To calculate the mean value of the fluorophore density in the diffusive pool of fluorophores in a given cell compartment we can average over all ρ estimates corresponding to all pixels inside that compartment whose boundaries have been determined by automated image segmentation and whose pixels are not associated with distinctly detected spots of fluorescence due to tracked assemblages of fluorophores (*e.g.* a molecular complex). This generates a robust estimate for fluorophore density in the diffusive pool which is unique not just to a specific single cell, but also provides a rough estimate to a specific single compartment within that cell. This rough estimate is useful in providing a simple preliminary and computationally non-intensive quantitation because it assumes no prior knowledge of subcellular structures within the cell.

In a more general case of multiple cellular compartments of different sizes, locations and shapes we can extend this model, assuming that the time-averaged pool concentration within each given intracellular compartment (for example, the nucleus, the cytosol compartment, and many other types of cell organelle) can be characterized by a mean value subject to small fluctuations across the extent of the compartment. However we do not assume that the mean values within different cellular compartments are necessarily equal.

Thus, in the case of two such cellular compartments containing the cytoplasm and the nucleus we have:

$$I(x_0, y_0, z_0)\mathrm{d}A = I_\mathrm{s} \exp\left[-\left(\frac{x^2+y^2}{2\sigma_{xy}}\right)\right] \times \left(\begin{array}{l} \displaystyle\sum_{i=1}^{\text{cytoplasm voxels}} \rho_\mathrm{c} P(x_i - x_0, y_i - y_0, z_i - z_0) \\ + \displaystyle\sum_{i=1}^{\text{nucleus voxels}} \rho_\mathrm{n} P(x_i - x_0, y_i - y_0, z_i - z_0) \end{array}\right). \quad (5)$$

Here, ρ_c and ρ_n refer to the mean cytoplasmic and nuclear concentrations respectively. By analysing all pixel data from both compartments we can then use least-squares regression analysis in Matlab to estimate mean values for ρ_c and ρ_n. This can be generalised to any number of intracellular compartments:

$$I(x_0, y_0, z_0)\mathrm{d}A = I_\mathrm{s} \exp\left[-\left(\frac{x^2+y^2}{2\sigma_{xy}}\right)\right] \times \sum_{j=1}^{\substack{\text{number of}\\ \text{compartments}}} \sum_{i=1}^{\substack{\text{compartment}\\ \text{voxels}}} \rho_j P(x_i - x_0, y_i - y_0, z_i - z_0) \quad (6)$$

where ρ_j is the mean concentration of the jth compartment. This has an advantage of taking into account the full contribution of all compartments to each observed pixel intensity in the image. These values can then be used to generate a modified convolution integral I' in eqn (4), thus allowing the pixel-by-pixel variation of ρ to be estimated. An important feature of this general model is that each separate compartment does not necessarily have to be modelled by an ideal geometrical shape (such as a sphere, for example) but can be any enclosed 3D volume provided its boundaries are well-defined to allow numerical integration in eqn (6).

2.7 Analytical PSF

The 3D PSF can be approximated as a series of 2D Fourier transforms of the pupil function at different z slices:[30]

$$P(x, y, z) = \int_{k_x}\int_{k_y} F(k_x, k_y, z)\exp\left(j(k_x x + k_y y)\right)\mathrm{d}k_x \mathrm{d}k_y \quad (7)$$

where k_x and k_y are the 2D Fourier coordinates and $F(k_x,k_y,z)$ is the pupil function which describes the field distribution in the pupil of the objective lens; when $\sin\theta_\mathrm{i} = \mathrm{NA}/n_\mathrm{i}$, F is given by:[31]

$$F(k_x,k_y,z) = A(\theta_\mathrm{i})\exp(jk_0\varphi(\theta_\mathrm{i},\theta_\mathrm{s},z)). \quad (8)$$

Otherwise, $F = 0$. The angles $\theta_{i,s}$ can be defined in Fourier coordinates (see Fig. 2):

$$\sin\theta_{i,s} = \frac{k_x^{\,2} + k_y^{\,2}}{k_{i,s}}. \tag{9}$$

Here $k_{i,s}$ is the wavenumber, $k_{i,s} = 2\pi n_{i,s}/\lambda$, of the emitted light (we approximate the wavelength to the peak emission wavelength, $\lambda = 515$ nm for GFP and 610 nm for mCherry) through the immersion medium (refractive index $n_i = 1.515$), specimen (refractive index $n_s = 1.33$) or k_o is the wavenumber through a vacuum and NA is the numerical aperture of the objective lens (in our study here, NA = 1.49). $A(\theta_i)$ is the apodization function for an emitting point source such as a fluorophore:[32]

$$A(\theta_i) = (\cos\theta_i)^{-\frac{1}{2}} \tag{10}$$

The phase function $\varphi(\theta_i, \theta_s, z)$ describes the optical path difference of a wavefront exiting the pupil compared to a reference wavefront and is the sum of a defocus term φ_d and an aberration term φ_a. The defocus term is approximated by:[30]

$$\varphi_d \approx n_i z(1 - \cos\theta_i) \tag{11}$$

and the aberration term, caused by differences in the immersion and sample media can be derived from geometric optics as:

$$\varphi_a = d(n_s\cos\theta_s - n_i\cos\theta_i) \tag{12}$$

where d is the distance from the front surface of the objective lens to the focal plane (~400 μm in our case).

3. Results and discussion

3.1 Estimating the PSF

To perform the convolution analysis, the PSF of the microscope was first experimentally estimated. Here, we measured the PSF of the our microscope using narrowfield illumination by acquiring experimental fluorescence images of 20 nm

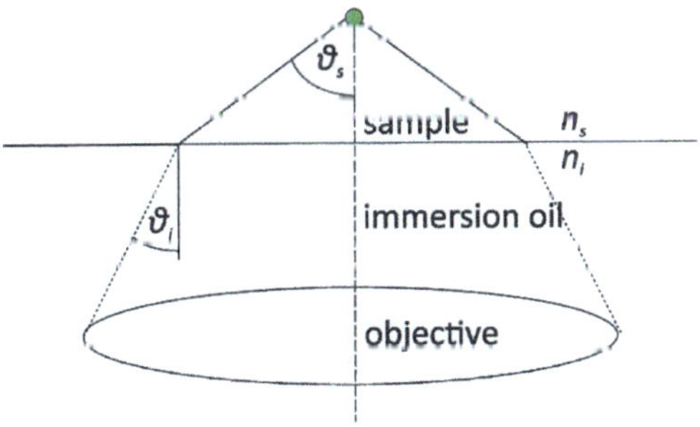

Fig. 2 Schematic of light emitted by a point source collected by an objective lens. A ray of fluorescence emission light is traced from the source (green) at an angle θ_s in the sample media (*i.e.* water-based minimal media) with refractive index n_s, refracted at the interface of the immersion oil to an angle θ_i in the immersion oil media with refractive index n_i.

diameter green fluorescent beads (Invitrogen) immobilized on the glass coverslip surface of our microscope flow-chamber.[33] The beads were imaged at 100 nm intervals in z which was controlled by an automated piezo nanostage (Nanodrive, Mad City Labs) controlled by our own bespoke software (LabVIEW, National Instruments) over a z-range ± 1 μm above/below the focal plane. Time-series images of six different beads using the same imaging conditions as for live cell microscopy were averaged pixel-by-pixel to form the experimental PSF (Fig. 3), which was then background-corrected by subtracting the mean local background and then normalized by dividing by the total summed pixel intensities of the detected foreground spot image. At absolute values of z beyond ~1 μm from the focal plane, the weak defocused intensity fluorescence signal from single beads was difficult to resolve above camera noise. However, by modelling the experimental PSF data obtained over the ± 1 μm z-range using an analytical PSF formulation we could then extend the range of the PSF determination to cover the full extent of single yeast cells whose diameter is typically ~5 μm, but can be as high as ~10 μm at certain stages in the cell cycle for budding yeast cells.

For this analytical approximation we used a Matlab implementation[34] of the Stokseth method[30] (Fig. 3), outlined in the Materials and methods section 2.7. This models the PSF as the Fourier transform of the pupil function which describes the field distribution in the pupil of the objective lens for given wavelengths of light.

The analytical function and experimental data were found to be in good agreement to the experimental PSF, with an associated mean chi-squared value of ~65 when measured on a pixel-by-pixel basis across each PSF image equivalent to a probability confidence interval P.

3.2 Fluorescent proteins *in vitro*

To test the single-molecule detection capabilities of the narrowfield microscope and to measure the characteristic intensity of single fluorophores, we imaged purified GFP (Clontech eGFP, with the addition an A206K mutation to inhibit GFP

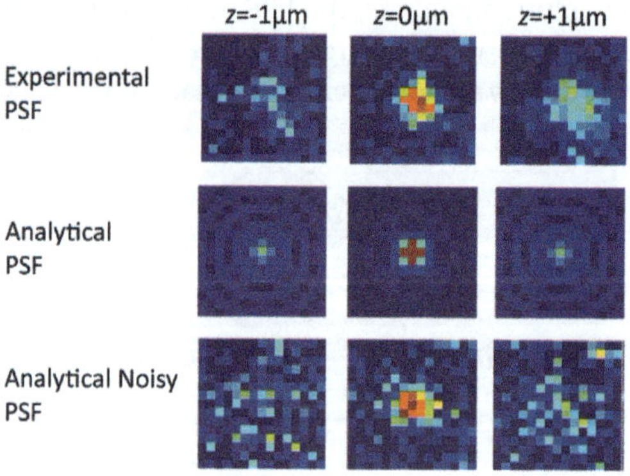

Fig. 3 The measured, experimental PSF is shown (upper panel) over a 2 μm range in z centred on the focal plane ($z = 0$), alongside the analytical PSF (middle panel) and the analytical PSF convolved with localisation error and noise for qualitative comparison (lower panel). The chi-squared value in comparing the analytical PSF model to the experimental over the z range −1 to +1 μm was 65 equating to a goodness of fit of probability confidence $p < 0.001$.

dimerization[27]) and mCherry (Clontech) immobilized on the coverslip surface *via* antibody conjugation. Representative images of GFP and mCherry are shown in the left-hand panels of Fig. 4a and b. Rapid photobleaching occurred within only a few image frames so the first frame is shown. Bright spots were tracked over time using millisecond sampled images, analysed using our custom Matlab software.[25,26] The software objectively identifies candidate bright spots by a combination of pixel intensity thresholding and image transformation. The threshold is set using the pixel intensity histogram as the full width half maximum of the peak in the histogram which corresponds to background pixels. A series of erosion and dilation is applied to the thresholded image to remove individual bright pixels due to noise. A final erosion step then leaves a single pixel at each candidate spot co-ordinate. The intensity centroid and characteristic intensity, defined as the sum of the pixel intensities inside a 5 pixel radius region of interest around the spot minus the local background[35] and corrected for non-uniformity in the excitation field are determined by iterative Gaussian masking[36] which resulted in a mean localization precision of ~40 nm and ~55 nm for GFP and mCherry molecules, respectively. Localisation precision was measured as the standard deviation of the spot centroid over time. Spots are accepted as real if their signal-to-noise ratio is above a threshold which was pre-determined from simulated data, using a realistic noise distribution. Initially this threshold was set generously, equivalent to a level of 35% false positive detection probability per image frame, however additional tracking criteria for subsequent image frames were applied (spots are only ultimately accepted if they constitute a track or trajectory, meaning: (i) they last for at least 3 consecutive image frames; (ii) the intensity centroid displacement between spots in consecutive images is 5 pixels

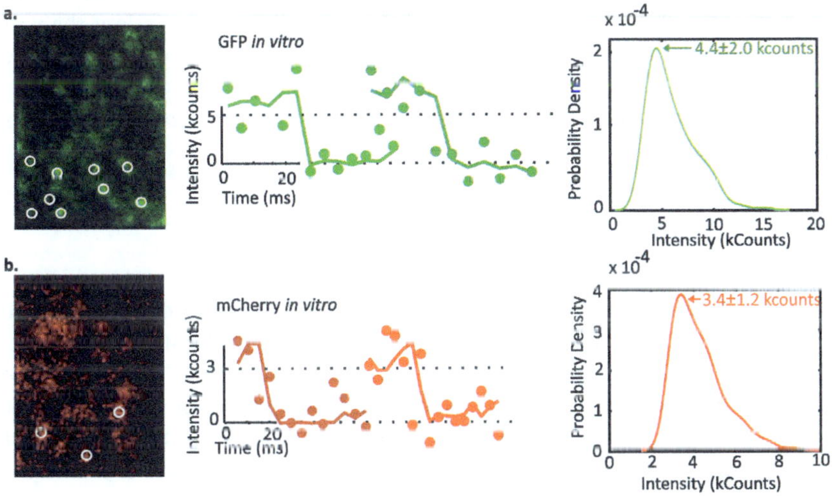

Fig. 4 *In vitro* single-molecule fluorescent protein characterization. Surface immobilisation assay for (a) GFP, and (b) mCherry, showing typical fluorescence images (left panels, white indicating example autodetected spots from our bespoke localisation and tracking software); typical measured photobleach traces (middle panels); and probability distributions of single-molecule intensity values value (right panels) with peak and half width at half maximum (HWHM) error indicated.

(one spot region of interest) or less; (iii), the width and integrated intensity of a spot in an image frame in a given track is within 50% of that measured in the previous image frame). These additional tracking acceptance criteria reduced the likelihood of false positive detection to <1%.

Spots are linked into trajectories based on their proximity to neighbouring spots in subsequent image frames, their integrated intensity and their estimated size based on a 2D unconstrained Gaussian fit to their experimental PSF intensity profile. The middle panels in Fig. 4 show intensity as a function of time of GFP and mCherry spots which were 'overtracked' (*i.e.* where the integrated intensity continued to be measured at the intensity centroid to visually indicate the level of local background intensity in the absence of the spot), after a spot had photobleached to zero mean integrated intensity. These traces are overlaid using a bespoke Chung–Kennedy edge-preserving filter[37,38] and show a step-like drop to mean zero intensity, indicative of a molecular signature for single fluorescent proteins. Kernel density estimations (KDEs) were used to obtain the distribution of all integrated spot intensity measurements of GFP and mCherry images.[39] This involves a 1D convolution of the spot integrated intensity data with a Gaussian kernel of unitary integrated area (*i.e.* equivalent to a total of just one data point) and an optimised bandwidth of ~600 counts which was determined objectively from the software, and are shown on the right panels of Fig. 4. This approach results in significant objectification of the displayed distribution in comparison to standard histogram methods.

The distributions peak at ~4400 counts and ~3400 counts on our EMCCD detector for GFP and mCherry respectively which we use as best estimates for the characteristic brightness values. We use these estimates as opposed to mean values from the distributions which are biased towards marginally higher values than the peak values due to a tail on the distribution comprised in part from a minority of detected spots which include two or more individual fluorescent protein molecules whose separation of the coverslip surface is less than the optical resolution limit. Step-wise photobleaching data measured from live budding yeast cells indicated similar levels of brightness to within ~10% consistent with earlier *in vivo* single-molecule studies of *E. coli* bacteria using fluorescent proteins.[24,28,29]

3.3 Localisation precision

Since localisation microscopy is required in the estimation of diffusive pools of pixels in a given compartment we sought to characterize the localisation precision of our narrowfield illumination single-molecule microscope over a range of different fluorescent spot contrast values. As a model for bright fluorescent spots we captured time series of surface-immobilised 20 nm green fluorescent beads (Invitrogen) which had been exposed to varying pre-bleach laser exposures. By varying the bleach times we were able to generate images of spots over a range of effective spot signal-to-noise ratios. Spots were tracked as before and their localisation precision determined as the standard deviation in intensity centroid position over time. Localisation precision against the signal-to-noise ratio is plotted in Fig. 5 in blue. The data was fitted using the Thompson equation[36] for localisation precision of 2D data, given by:

$$\left\langle (\Delta x)^2 \right\rangle = \frac{s^2 + a^2/12}{N} + \frac{8\pi s^4 b^2}{a^2 N^2}. \tag{13}$$

Here, s is the sigma width of the PSF (170 nm for green fluorescent beads), a is the conjugate equivalent pixel size in the sample focal plane (80 nm), N is the number of photons collected belonging to a spot and b is the number of background photons collected. The intensity, I, collected from a given detected spot of fluorescence is equal to the number of photons emitted from the spot multiplied by a constant G which incorporates camera gain and total photon collection efficiency between the sample and the camera detector. The signal-to-noise ratio was defined as the integrated spot intensity divided by the standard deviation of the local background of the spot (calculated from pixels which are within a square 17 × 17 pixel array centred on the intensity centroid of the spot but excluding pixels that are contained within the central 5 pixel radius circular region of interest which comprises the spot integration area, equivalent to ~8 counts for the fluorescent bead data, constant for each set of imaging conditions) multiplied by the area of the spot (80 pixels). The Thompson equation was fitted to the spot intensity data to generate the localisation precision as a function of the signal-to-noise ratio for the fluorescent nanobead data, shown in black in Fig. 5 with 90% confidence intervals as dotted lines, using an optimised collection constant of $G = 0.1$ and background photon count, $b = 5$ photons. The localisation precision of GFP and mCherry is also shown in Fig. 5 in green and red respectively (note, since mCherry has a larger PSF width than the green fluorescent beads of the GFP there is a marginal deviation from the fit extrapolated from the green bead data).

3.4 Concentration measurements in *E. coli*

As proof-of-principle we tested the CoPro method on the single cytoplasmic cellular compartment of model *E. coli* bacteria using a cell strain consisting of the

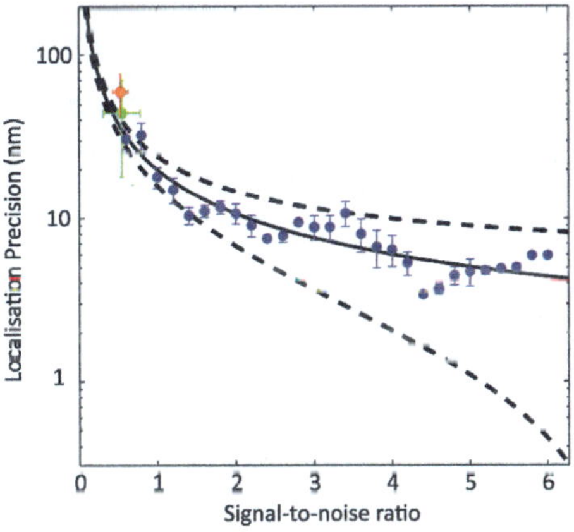

Fig. 5 Mean localisation precision which we measured against the signal-to-noise ratio for fluorescent beads (blue), GFP (green) and mCherry (red) with error bars showing the standard deviation on a semilog plot, each datapoint sampled from a set of $n = 5$–10 beads. The black line is the fitted Thompson model for localisation precision (using parameters $b = 5$, $G = 0.1$) with 90% confidence bounds for fit (black dotted lines). The mean mCherry signal is smaller than GFP by ~26%, but also its PSF width is larger by ~10% resulting in a slight deviation from the black fit as predicted by the Thompson *et al.* model.[36]

replisome protein DnaQ fused to GFP using chromosomal integration. This was essentially identical to a cell strain developed earlier, but which instead used the yellow YPet fluorescent protein as the fluorophore tag. This cell strain's DnaQ protein copy number per cell had previously been estimated using both quantitative western blots and a different convolution method using slimfield illumination which used different experimentally derived PSF estimates.[24] We took narrowfield fluorescence images of the DnaQ-GFP tagged cell strain and quantified the cellular concentration of DnaQ.

Fig. 6 shows images of a representative DnaQ-GFP cell and quantification of DnaQ concentration. A brightfield image of a cell is shown in grey in Fig. 6a, with the segmented outline of the cell which was obtained from the raw cell fluorescence image using an automatically varying threshold based on the pixel intensity distribution shown overlaid in orange. This raw cell boundary image segmentation, in the case of *E. coli* cells, could be modelled as a 'sausage' shape (shown overlaid in white on the raw fluorescence image shown in green in Fig. 6b). The segmentation threshold was configured to correspond to low intensity autofluorescence which was observed to be delocalised in the cytoplasm, and so was not restricted to the spatial distribution of GFP-tagged material in the cytoplasm compartment. The centroid, orientation and major and minor axis lengths of the cell area were used to define the unique fitted 'sausage' function around each cell, which consisted of a rectangle capped by a half-circle at either end,[40] which was an accurate 2D projection of the 3D *E. coli* cell shape of a cylinder capped by two hemispheres.[41] Distinct bright detected spots in the fluorescence image were

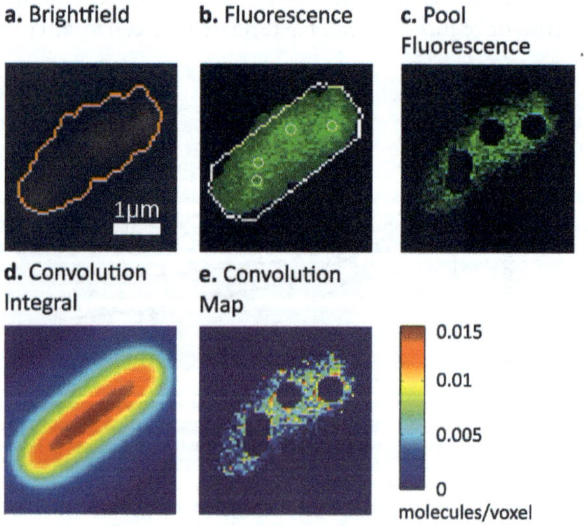

Fig. 6 (a) Brightfield image (grey) of a typical single live *E. coli* cell, with the segmented cell outline obtained from the fluorescence data (as outlined in the main text) shown overlaid (orange). (b) GFP fluorescence image (green) with found spots marked (white circles); the white line is the unique fitted 'sausage' function for that cell. (c) Pool fluorescent pixels with detected distinct spots now excluded from the image. (d) Projection image of PSF corresponding to the *E. coli* cell integrated over its cell volume. (e) Protein concentration map of the cell (which is the image of (c) divided by the image of (d) pixel-by-pixel).

found using the same methods as for the *in vitro* data, and are shown overlaid in Fig. 6b as white circles.

The fluorescence pixels associated just with the diffusive pool are shown in Fig. 6c and were determined by taking only pixels within the segmented cell area and removing pixels in a 5 pixel radius around the bright spot centroids (*i.e.* corresponding to the same region of interest area used for the integrated spot intensity determination). The number of DnaQ molecules in bright spots was determined by dividing the total intensity in the 5 pixel radius around the spot by the intensity of a single *in vitro* GFP and was typically <10 molecules. The mean autofluorescence background and camera detector noise background intensity values were also subtracted from all pool pixel intensity values. The convolution integral C, shown as a heat map in Fig. 6d, was obtained by integrating the analytical PSF over the sausage function which defined the cell boundaries. A map of the diffusive pool concentration was obtained by dividing the pool pixel intensity values by the corresponding convolution integral pixel values, on a pixel-by-pixel basis, and by the characteristic peak integrated intensity for a single molecule GFP obtained from the *in vitro* surface-immobilized assay. The pixel map for this calculation is shown in Fig. 6e in units of molecules per voxel.

The KDE of cell DnaQ protein concentration sampled from a population of only seven cells is shown in Fig. 7, showing that a probability distribution for a cell population can be generated with this method using relatively low numbers of cells. The overall effective DnaQ concentration in the entire compartment was defined as the mean DnaQ protein concentration in the diffusive pool plus the total amount of DnaQ detected in any distinct fluorescent spots divided by the effective volume of the cellular compartment, and was used to determine the total copy number of DnaQ molecules in each individual cell by multiplying the estimate of the overall effective DnaQ concentration by the calculated volume given by the cell's unique sausage function. The mean copy number of DnaQ ± standard deviation was calculated to be 350 ± 120 molecules which agrees well with the copy number of DnaQ-YPet, measured previously using the YPet fluorescent protein and a different convolution method, of 270 ± 160 molecules.[24] The distribution of DnaQ concentration per cell is broad, reflecting cell-to-cell

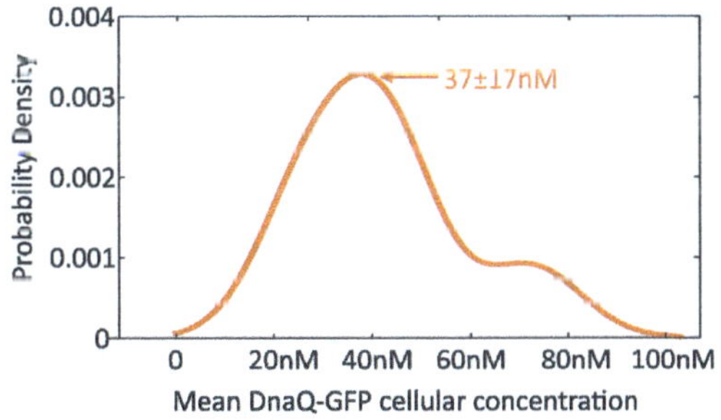

Fig. 7 Example distribution of mean cellular DnaQ concentration which illustrates that even with only a few cells in a population (here, $n = 7$ *E. coli* bacteria cells) we can reconstruct a sample probability distribution; mean and HWHM error are indicated. The relative scale on the probability density axis is set to ensure normalization conditions (*i.e.* the area under the curve is exactly unity).

variation – this information is lost when quantifying protein concentrations in cells using traditional bulk ensemble biochemical assays. These variations may be caused by each cell being at a different phase in the cell cycle but may also reflect that each cell is an individual and, to fully understand cellular behaviour, this individuality must be characterized.

3.5 Concentration measurements of simulated structured cells

To extend our method to more complexly structured cells, we first tested it on realistic simulated images from different cell types containing multiple cellular compartments. Fig. 8a shows two simulated budding yeast images. The left panels indicate a simulated nuclear concentration of Mig1 which is double the cytoplasmic concentration. Fig. 8a right panels are a simulated image in which the nuclear concentration is zero with all the simulated Mig1 molecules localized uniformly to the cytoplasm. The cell periphery is labelled in orange and the nuclear periphery in cyan. The cell and the nucleus were modelled as convolution integrals of spheres with radii of 32 and 15 pixels respectively containing uniform fluorophore concentrations. Realistic noise was added in the lower panels by adding normally distributed random pixels with a standard deviation ~20% peak cytoplasm intensity. Images were simulated over a range of nuclear concentrations from zero to five times a constant cytoplasmic value.

Each pixel intensity value is the sum of the convolution integral of each separate compartment in the cell (eqn (5)) so the concentration in each compartment was determined by solving a set of simultaneous equations for each pixel value using our bespoke Matlab-coded software. Equations were solved by linear least squares regression analysis with the only constraint that concentration cannot be negative. Fig. 8b shows the measured cytoplasmic concentration as determined by our CoPro algorithm on the simulated data against the simulated nuclear concentration, with error bars given by the standard deviation over five

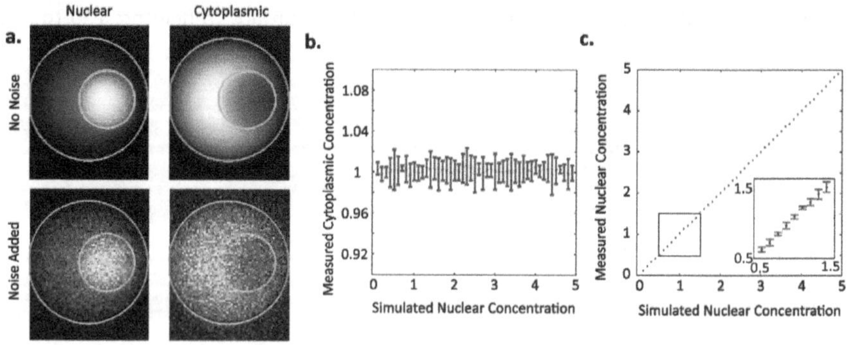

Fig. 8 (a) Simulated images of a spherical yeast cell with a spherical nucleus, showing (left panels) nuclear concentration set to be twice the cytoplasmic concentration, and (right panels) zero concentration in the nucleus with all Mig1 localized in the cytoplasm. No noise (upper panels) and realistic added noise (lower panels) are shown. (b) Measured cytoplasmic concentration for cells (n = 300 cells, made up from 5 repeats at each of 60 different protein concentrations) with varying simulated nuclear concentrations; standard deviation error bars are indicated. (c) Measured nuclear concentration as a function of simulated nuclear concentration, with (inset) a zoom-in of the simulated data and the fit.

repeated simulations. Fig. 8c shows the same for the nucleus with an insert to show an expanded section of the data and the fit. The linear fits of Fig. 8b and c are associated with chi-squared values equivalent to a probability confidence interval *P*.

To demonstrate the generality of our method we also simulated a different cell type with multiple different sized components and different locations, shown without and with noise in Fig. 9a left and right respectively. The cell is rod-shaped, like an *E. coli* bacterium, and contains three spherical compartments of different volumes and internal protein concentrations. Although not modelled on any particular system, this cell resembles, for example, fluorescently labelled carboxysomes in cyanobacteria. The mean concentration in each compartment across five repeated simulations was again calculated by solving the simultaneous equations from four separate convolution integrals at each pixel and is plotted against the simulated concentration in Fig. 9b. The measured concentrations all agree very well with the simulated concentrations, demonstrating this method is also robust in this example of non-spherical cells containing multiple different cellular compartments.

3.6 Concentration measurements in budding yeast

We then applied these methods to quantify experimentally the concentration of Mig1 protein molecules in budding yeast cells at high and low glucose conditions, and also to estimate the concentration of the RNA polymerase protein Nrd1 as a control, since Nrd1 is not directly involved in the glucose sensing pathway. Fig. 10 shows images of a representative dual-label Mig1-GFP:Nrd1-mCherry cell with the corresponding quantification of Mig1 concentration. Fig. 10a shows a brightfield non-fluorescence image of the cell with the nuclear membrane and cell membrane boundaries overlaid in cyan and orange respectively. The algorithm used for image segmentation was similar to that employed for *E. coli* above, based on the fluorescence image of Mig1 (shown in Fig. 10b) for the cell boundary and

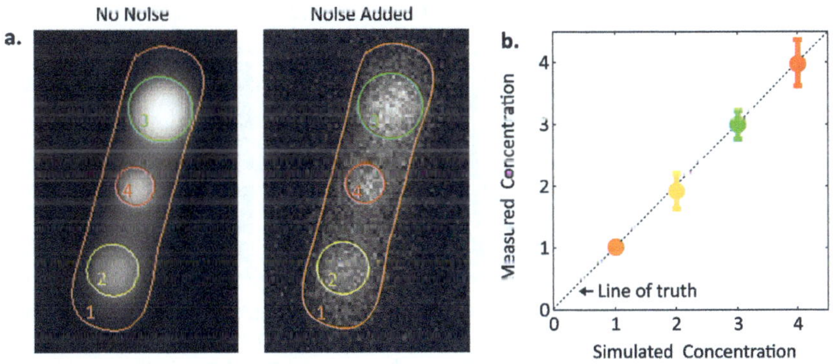

Fig. 9 Simulated bacterial 'sausage-shaped' cell (region 1) with three differently sized spherical organelle compartments and different concentrations (regions 2–4). No noise (left panel) and realistic added noise (right panel) are shown. (b) Mean and standard deviation concentrations of these different regions measured using the CoPro method plotted against the real simulated values, with the 'line of truth' added (black dashed line).

Nrd1 (shown in Fig. 10c) for the nucleus, as before configuring the threshold levels to be sensitive to delocalised cellular autofluorescence.

Bright spots were tracked as before, and shown overlaid in white in Fig. 10b. Due to much higher levels of cellular GFP fluorescence and autofluorescence, spots are harder to display in the image, so a zoomed-in cut-out on the figure shows a typical fluorescent spot with the intensity display levels adjusted appropriately. The nuclear and cytoplasmic concentrations were determined separately. The distinctly detected spots of fluorescence were removed from the fluorescence images and background correction applied. The concentrations in the nucleus and cytoplasm were determined as for the simulated images, modelling the cell and nucleus as spheres with radii determined from the segmentation. Fig. 10d shows a spatial map of the pool concentration fluctuation obtained by dividing the background-corrected fluorescence image by the sum of the convolution integrals in the cytoplasm and nucleus multiplied by their concentration value and the characteristic GFP intensity. Each pixel in the fluctuation map represents the percentage difference from the compartment mean.

The KDE distribution of Mig1 concentration in the nucleus and cytoplasm at high and low glucose is shown in Fig. 11 left and centre, respectively. At high levels of extracellular glucose concentration, the concentration of Mig1 is much higher (roughly by a factor of 4) in the nucleus than in the cytoplasm. At low extracellular glucose concentrations the concentration of Mig1 is more similar in the nucleus and cytoplasm but still elevated in the nucleus by ~30%. As a control, we measured the concentration of the Nrd1 protein at high and low extracellular glucose concentrations. Nrd1 was found to be almost exclusively localized in the nucleus, with the peak in the Nrd1 concentration distribution indicating a copy number of ~2000 molecules. The distributions for the Nrd1 nuclear concentration are shown in the right panel of Fig. 11 and were shown by Student *t*-tests to be independent of glucose concentration.

The mean copy number of Mig1 molecules in the nucleus and cytoplasm is shown in Table 1, indicating mean and standard deviation values. Using bulk

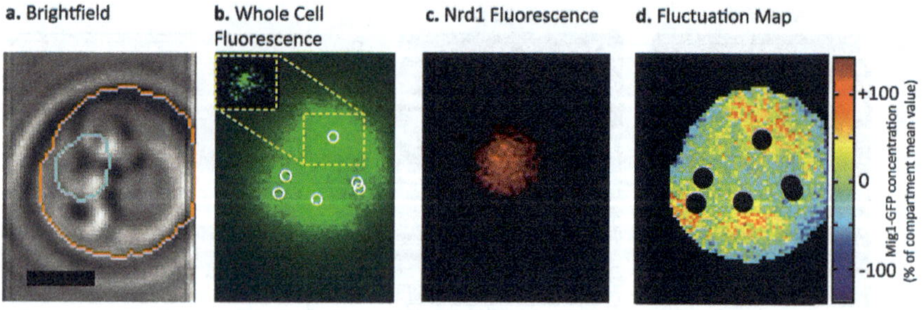

Fig. 10 (a) Brightfield image of a cell with automated segmentation outline of the nucleus (cyan) and cell outer boundary (orange); 2 μm scale bar indicated. (b) GFP fluorescence image, reporting the localization of Mig1, with found spots (white circles) marked and cutaway (yellow) with adjusted intensity levels to indicate the position of the underlying detected spot. (c) mCherry fluorescence image, reporting the localization of Nrd1 in the nucleus. (d) Spatial distribution map of Mig1 concentration, indicating the fluctuation from the mean compartment concentration value with respect to position across the cell image.

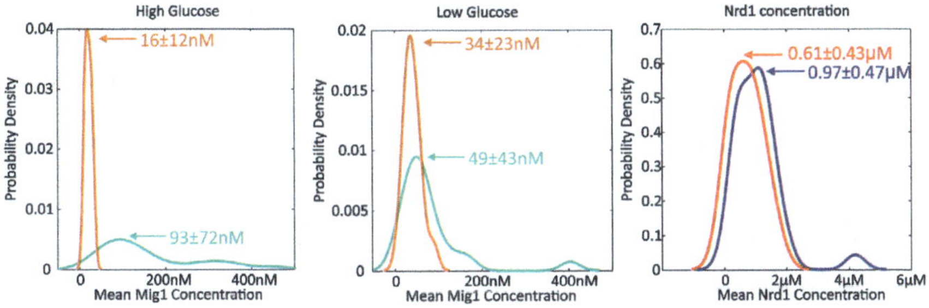

Fig. 11 Distribution of cytoplasmic (orange) and nuclear (cyan) Mig1 protein concentrations of $n = 25-30$ cells at (a) high and (b) low glucose and distribution, and equivalent for (c) Nrd1 concentration (high glucose in blue and low glucose in red) with the peak values ± HWHM indicated.

ensemble average affinity methods, the total copy number of Mig1 in the whole cell was estimated to be ~830 molecules per cell (ref. 4) which agrees well with our results – the authors' conservative assessment of error on this copy number estimate was ~100%, which illustrates one of the key advantages of our single-molecule method. The Mig1 concentration is higher in the nucleus than the cytoplasm at low glucose concentrations (~30%), although by much less than at high glucose concentrations (~400%). This suggests that some Mig1 molecules interact in the nucleus even at low levels of extracellular glucose concentration.

4. Conclusions

The CoPro method for determining the protein concentration in live cells was first tested using the DnaQ protein concentration in *E. coli* as a previously studied system, and found to produce similar results to within experimental error. In applying CoPro to budding yeast cells we were able to quantify changes in protein concentration in cellular compartments in response to controlled environmental changes; here, by changing the extracellular glucose concentration and then using CoPro to monitor the protein concentration of the protein Mig1 which performs a biological role as a response regulator in the glucose sensing pathway in yeast. By measuring the concentration of Mig1 in both the nuclear and cytoplasmic compartments in each budding yeast cell, cell-by-cell, we were able to obtain distributions across cell populations, enabling observation of subtle concentration shifts. These results show promise for the investigation of future

Table 1 Copy number of Mig1 molecules, rounded to the nearest molecule, in the nucleus and cytoplasm

	Mean number of Mig1 molecules per cell	Standard deviation (molecules per cell)
High glucose cytoplasm	542	200
High glucose nucleus	249	88
Low glucose cytoplasm	1070	400
Low glucose nucleus	141	57

biological systems which may exhibit relatively small changes in the concentration of protein in a particular cellular compartment for a given cell, which may be smaller than the variability across the whole cell population and thus hidden where traditional ensemble average approaches were used in the assessment of protein copy numbers in cells. These distributions of protein concentration also render probabilistic information for the number of a specific protein type in a cell, and how they are distributed spatially between different regions of the cell, which is invaluable information that can be correlated back to stochastic models for gene expression activity.

Importantly, the CoPro method is entirely general in regards to the shape and size of different cellular compartments. Although the case of budding yeast cells involves ostensibly spherical nuclei and cells, the algorithm only requires that the 3D volume of each cellular compartment is well-defined. In the case of hypothetical, asymmetrical cellular compartments which are difficult to model as ideal geometrical shapes, for example, the appropriate numerical integrations can still be performed provided sensible physical estimates of the compartment shapes and relative orientations can be made. These, for example, could be enabled using a separate imaging technique in separate experiments, such as transmission electron microscopy of thin sample sections, or utilizing confocal microscopy to obtain *z*-sections though the cell. The time resolution of typical confocal microscopy is not high enough to follow molecular motions inside living cells, however it can be used potentially to determine the complex 3D shapes of appropriate cellular compartments in conjunction with CoPro which then has the rapid time resolution to map out spatial fluctuations in protein concentration in real time on each separate image frame obtained from millisecond single-molecule microscopy.

Although fluorescence microscopy has been used for quantitative bacterial proteomics previously, both from our own earlier work[24] and those of others,[8] here we have demonstrated a new method which can be applied to larger, more complexly structured eukaryotic cells to quantify protein concentration distributions at a subcellular level. Budding yeast cells have a well-defined approximately spherical structure, allowing for simple convolution integrals but using numerical integration and 3D microscopy techniques such as light sheet microscopy, more complex eukaryotic cells could in principle be investigated. Although the cellular locations used here were the nucleus and cytoplasm, these methods could similarly be applied to other cellular organelles, including mitochondria or vacuoles. Our method also utilises high-speed narrowfield microscopy and can thus be readily combined with single particle tracking to fully quantify protein dynamics in living cells.

In our study we have used a bespoke narrowfield laser illumination technique which enables millisecond fluorescence microscopy at a single-molecule precise detection sensitivity level. This rapid sampling rate is comparable to the mobility time scale of single proteins in low viscosity cellular environments. For example, a typical fluorescently-labelled protein in the cell cytoplasm has an apparent diffusion coefficient equivalent to a few $\mu m^2 s^{-1}$, implying that it will diffuse its own point spread function width of the associated fluorescent 'spot' image after just a few milliseconds of observation. This therefore sets a benchmark for the maximum permitted camera exposure time for a single image frame, as above this level such a fluorescent spot appears significantly blurred in a typical image frame and so will fail to be detected as a distinct spot.

Diffusion, however, is a stochastic process; therefore, some single protein molecules may still diffuse greater distances on some given individual image frames and so will fail to be detected, depending also on whether their nearest-neighbour mean separation is less than the optical resolution limit of *ca.* 200–300 nm. Molecular complexes containing more protein molecule subunits have a higher molecular weight and are likely to have a larger effective Stokes radius and thus a lower diffusion coefficient, in addition to their associated fluorescent spots being brighter. Therefore, utilizing CoPro with rapid millisecond imaging enables *separate* experimental quantification between the effective compartment concentrations of proteins present in distinct molecular complexes and those present in lower stoichiometry states that diffuse more rapidly. However, the algorithms of the CoPro method will still work with less rapid imaging rates than those we use here, if these are not technically feasible on a given fluorescence microscope setup; the resultant analytical output will simply indicate fewer, or potentially no, distinctly detected fluorescent spots, but rather output an increased proportion of protein in the diffusive pool in a given compartment. This reduces some aspects of biological insight in regards to lacking the capability to infer the presence or not of distinct molecular complexes in the protein population, but still results in robust quantitative estimates for the total effective numbers of protein monomer units within a cell compartment. A reduced imaging rate also reduces the capacity for *time-resolved* measurements of protein concentration in living cells; instead, slow imaging generates *steady-state* information, but which still has utility in being quantitative on a cell-by-cell and compartment-by-compartment basis.

Our method here utilizes the *in vitro* estimate for the single-molecule brightness of GFP and mCherry. We measured this as being within ~10% of the equivalent *in vivo* brightness in budding yeast cells, consistent with earlier stoichiometry studies of *E. coli* molecular complexes using step-wise photobleaching analysis of fluorescent proteins.[31,30,39] However, it may be possible that in some specialized cellular compartments there is a significantly different pH to the rest of the cell, a good example of which might be lysosomes. In such a case large differences in pH may manifest as a more significant difference in the brightness of single fluorescent proteins in that compartment compared to the rest of the cell. In this circumstance the CoPro method could still be utilized with the modification of a different equivalent brightness value for separate compartments, which could quantified using similar step-wise photobleaching procedures outlined here but pooling statistics into separate distinct compartments on the basis of automated image segmentation. Although we use fluorescent proteins as reporter labels the CoPro method can generalise to other fluorescent labels; these may ultimately be selected to have less sensitivity to changes in local cellular pH, and indeed may also be brighter than fluorescent proteins and have an improved associated localisation precision due to a greater signal-to-noise ratio as evidenced by Fig. 5.

Potential issues of using other fluorophores tags beyond fluorescent proteins however include possible homo-FRET/quenching effects. In our study here we could detect no significant correlation between the stoichiometry of tracked fluorescent spots (of Mig-GFP) and the size of the single-molecule photobleach step of the fluorescent protein label (here of GFP), consistent with the earlier single-molecule fluorescent protein stoichiometry studies alluded to previously. This is indicative of an absence of any measurable homo-FRET or quenching effect. The Förster radius of a fluorescent protein FRET pair is in the range 4–5 nm, whereas the closest two fluorescent protein molecules can physically get to each other is a comparable distance due to the steric hindrance from their beta barrel structure. This indicates that non-radiative energy transitions due to the interaction of electrons in molecular orbitals, whether due to hetero- or homo-FRET, have a relatively small associated signal – the paucity of published single-molecule FRET studies using fluorescent protein FRET pairs lies in testament to this. However, hypothetical quenching may of course be measurable from smaller dyes were they to be used, and so these effects may need to be characterized in order to minimize associated errors on stoichiometry and protein concentration measurements.

The methods outlined here illustrate not only how light microscopy has evolved from a qualitative observational tool into a highly quantitative instrument,[23] but also how bespoke tools from physics can be developed to characterize properties of the living component of soft matter at the single-molecule length scale[42] not just at a controlled, reductionist *in vitro* level[43] but also to gain molecular-level insight into the physiologically relevant context of single functioning, living cells.[44–49] Combining this automated CoPro method with multicolour single-molecule real-time fluorescence imaging[50,51] may also enable quantitative estimation of dynamic protein concentration changes of multiple interacting proteins in live cells, which is an appealing route towards investigating native biochemistry, one molecule at a time.

Acknowledgements

We thank Sviatlana Shashkova and Stefan Hohmann (University of Gothenburg, Sweden) for donation of yeast cell strains and assistance with yeast cell culturing. We thank Aisha Syeda and Peter McGlynn (University of York, UK) for assistance with adapting the DnaQ-YPet strain to DnaQ-GFP. With thank Holly Hathrell (University of York, UK) for assistance with characterizing the sensitivity of our spot detection algorithms. MCL is assisted by a Royal Society URF and research funds from the Biological Physical Sciences Institute (BPSI) of the University of York, UK.

References

1 P. Picotti, B. Bodenmiller, L. N. Mueller, B. Domon and R. Aebersold, *Cell*, 2009, **138**, 795–806.
2 M. Uhlén, *Biotechniques*, 2008, **44**, 649–654.
3 M. Pawlak, E. Schick, M. A. Bopp, M. J. Schneider, P. Oroszlan and M. Ehrat, *Proteomics*, 2002, **2**, 383–393.
4 S. Ghaemmaghami, W.-K. Huh, K. Bower, R. W. Howson, A. Belle, N. Dephoure, E. K. O'Shea and J. S. Weissman, *Nature*, 2003, **425**, 737–741.
5 R. Aebersold and M. Mann, *Nature*, 2003, **422**, 198–207.
6 L. Florens, M. P. Washburn, J. D. Raine, R. M. Anthony, M. Grainger, J. D. Haynes, J. K. Moch, N. Muster, J. B. Sacci, D. L. Tabb, A. A. Witney,

D. Wolters, Y. Wu, M. J. Gardner, A. A. Holder, R. E. Sinden, J. R. Yates and D. J. Carucci, *Nature*, 2002, **419**, 520–526.
7 J. R. S. Newman, S. Ghaemmaghami, J. Ihmels, D. K. Breslow, M. Noble, J. L. DeRisi and J. S. Weissman, *Nature*, 2006, **441**, 840–846.
8 Y. Taniguchi, P. J. Choi, G.-W. Li, H. Chen, M. Babu, J. Hearn, A. Emili and X. S. Xie, *Science*, 2010, **329**, 533–538.
9 M. Lundin, J. O. Nehlin and H. Ronne, *Mol. Cell. Biol.*, 1994, **14**, 1979–1985.
10 J. O. Nehlin, M. Carlberg and H. Ronne, *EMBO J.*, 1991, **10**, 3373–3377.
11 C. J. L. Klein, L. Olsson and J. Nielsen, *Microbiology*, 1998, **144**, 13–24.
12 R. Ghillebert, E. Swinnen, J. Wen, L. Vandesteene, M. Ramon, K. Norga, F. Rolland and J. Winderickx, *FEBS J.*, 2011, **278**, 3978–3990.
13 J. R. Broach, *Genetics*, 2012, **192**, 73–105.
14 M. J. De Vit, J. A. Waddle and M. Johnston, *Mol. Biol. Cell*, 1997, **8**, 1603–1618.
15 L. Bendrioua, M. Smedh, J. Almquist, M. Cvijovic, M. Jirstrand, M. Goksör, C. B. Adiels and S. Hohmann, *J. Biol. Chem.*, 2014, **289**, 12863–12875.
16 M. A. Treitel and M. Carlson, *Proc. Natl. Acad. Sci. U. S. A.*, 1995, **92**, 3132–3136.
17 F. C. Smith, S. P. Davies, W. A. Wilson, D. Carling and D. G. Hardie, *FEBS Lett.*, 1999, **453**, 219–223.
18 J. Ostling, M. Carlberg and H. Ronne, *Mol. Cell. Biol.*, 1996, **16**, 753–761.
19 J. Ostling and H. Ronne, *Eur. J. Biochem.*, 1998, **252**, 162–168.
20 E. Frolova, *Nucleic Acids Res.*, 1999, **27**, 1350–1358.
21 M. J. DeVit and M. Johnston, *Curr. Biol.*, 1999, **9**, 1231–1241.
22 M. Plank, G. H. Wadhams and M. C. Leake, *Integr. Biol.*, 2009, **1**, 602–612.
23 A. J. M. Wollman, R. Nudd, E. G. Hedlund and M. C. Leake, *Open Biol.*, 2015, **5**, 150019.
24 R. Reyes-Lamothe, D. J. Sherratt and M. C. Leake, *Science*, 2010, **328**, 498–501.
25 A. J. M. Wollman, H. Miller, Z. Zhou and M. C. Leake, *Biochem. Soc. Trans.*, 2015, **43**, 139–145.
26 H. Miller, Z. Zhaokun, A. J. M. Wollman and M. C. Leake, *Methods*, 2015, DOI: 10.1016/j.ymeth.2015.01.010.
27 D. Landgraf, B. Okumus, P. Chien, T. A. Baker and J. Paulsson, *Nat. Methods*, 2012, **9**, 480–482.
28 M. C. Leake, J. H. Chandler, G. H. Wadhams, F. Bai, R. M. Berry and J. P. Armitage, *Nature*, 2006, **443**, 355–358.
29 M. C. Leake, N. P. Greene, R. M. Godun, T. Granjon, G. Buchanan, S. Chen, R. M. Berry, T. Palmer and B. C. Berks, *Proc. Natl. Acad. Sci. U. S. A.*, 2008, **105**, 15376–15381.
30 P. A. Stokseth, *J. Opt. Soc. Am.*, 1969, **59**, 1314.
31 H. H. Hopkins, *Proc. R. Soc. A*, 1955, **231**, 91–103.
32 B. M. Hanser, M. G. L. Gustafsson, D. A. Agard and J. W. Sedat, *Opt. Lett.*, 2003, **28**, 801.
33 C.-J. Lo, M. C. Leake and R. M. Berry, *Biophys. J.*, 2006, **90**, 357–365.
34 P. Pankajakshan, L. Blanc-Feraud, J.-C. Olivo-Marin and J. Zerubia, in *2008 5th IEEE International Symposium on Biomedical Imaging: From Nano to Macro*, IEEE, 2008, pp. 740–743
35 Q. Xue and M. C. Leake, *Proc. - 2009 IEEE Int. Symp. Biomed. Imaging From Nano to Macro, ISBI 2009*, 2009, pp. 1158–1161.
36 R. E. Thompson, D. R. Larson and W. W. Webb, *Biophys. J.*, 2002, **82**, 2775–2783.

37 M. C. Leake, D. Wilson, M. Gautel and R. M. Simmons, *Biophys. J.*, 2004, **87**, 1112–1135.
38 M. C. Leake, D. Wilson, B. Bullard and R. M. Simmons, *FEBS Lett.*, 2003, **535**, 55–60.
39 M. C. Leake, *Phys. Chem. Chem. Phys.*, 2014, **16**, 12635–12647.
40 Q. Xue, N. S. Jones and M. C. Leake, *2010 7th IEEE Int. Symp. Biomed. Imaging From Nano to Macro, ISBI 2010-Proc.*, 2010, pp. 161–164.
41 C.-J. Lo, M. C. Leake, T. Pilizota and R. M. Berry, *Biophys. J.*, 2007, **93**, 294–302.
42 Q. Xue, O. Harriman and M. C. Leake, *J. Phys.: Conf. Ser.*, 2011, **286**, 012001.
43 M. C. Leake, A. Grützner, M. Krüger and W. A. Linke, *J. Struct. Biol.*, 2006, **155**, 263–272.
44 T. Lenn, M. C. Leake and C. W. Mullineaux, *Biochem. Soc. Trans.*, 2008, **36**, 1032–1036.
45 T. Lenn and M. C. Leake, *Open Biol.*, 2012, **2**, 120090.
46 M. C. Leake, *Commun. Integr. Biol.*, 2010, **3**, 415–418.
47 S.-W. Chiu and M. C. Leake, *Int. J. Mol. Sci.*, 2011, **12**, 2518–2542.
48 W. A. Linke and M. C. Leake, *Phys. Med. Biol.*, 2004, **49**, 3613–3627.
49 M. C. Leake, *Philos. Trans. R. Soc. London, Ser. B*, 2013, **368**, 20120248.
50 I. Llorente-Garcia, T. Lenn, H. Erhardt, O. L. Harriman, L.-N. Liu, A. Robson, S.-W. Chiu, S. Matthews, N. J. Willis, C. D. Bray, S.-H. Lee, J. Y. Shin, C. Bustamante, J. Liphardt, T. Friedrich, C. W. Mullineaux and M. C. Leake, *Biochim. Biophys. Acta*, 2014, **1837**, 811–824.
51 A. Badrinarayanan, R. Reyes-Lamothe, S. Uphoff, M. C. Leake and D. J. Sherratt, *Science*, 2012, **338**, 528–531.

Faraday Discussions

PAPER

A study of SeqA subcellular localization in *Escherichia coli* using photo-activated localization microscopy†

Jacek T. Mika,‡*[a] Aster Vanhecke,‡[a] Peter Dedecker,[a] Toon Swings,[b] Jeroen Vangindertael,[a] Bram Van den Bergh,[b] Jan Michiels[b] and Johan Hofkens[a]

Received 4th May 2015, Accepted 10th July 2015
DOI: 10.1039/c5fd00058k

Escherichia coli (*E. coli*) cells replicate their genome once per cell cycle to pass on genetic information to the daughter cells. The SeqA protein binds the origin of replication, *oriC*, after DNA replication initiation and sequesters it from new initiations in order to prevent overinitiation. Conventional fluorescence microscopy studies of SeqA localization in bacterial cells have shown that the protein is localized to discrete foci. In this study we have used photo-activated localization microscopy (PALM) to determine the localization of SeqA molecules, tagged with fluorescent proteins, with a localization precision of 20–30 nm with the aim to visualize the SeqA subcellular structures in more detail than previously possible. SeqA–PAmCherry was imaged in wild type *E. coli*, expressed from plasmid or genetically engineered into the bacterial genome, replacing the native *seqA* gene. Unsynchronized cells as well as cells with a synchronized cell cycle were imaged at various time points, in order to investigate the evolution of SeqA localization during the cell cycle. We found that SeqA indeed localized into discrete foci but these were not the only subcellular localizations of the protein. A significant amount of SeqA–PAmCherry molecules was localized outside the foci and in a fraction of cells we saw patterns indicating localization at the membrane. Using quantitative PALM, we counted protein copy numbers per cell, protein copy numbers per focus, the numbers of foci per cell and the sizes of the SeqA clusters. The data showed broad cell-to-cell variation and we did not observe a correlation between SeqA–PAmCherry protein numbers and the cell cycle under the experimental conditions of this study. The numbers of SeqA–PAmCherry molecules per focus as well as the foci sizes also showed broad distributions indicating that the foci are likely not characterized by a fixed number of molecules. We also imaged an *E. coli* strain devoid of the dam methylase (*Δdam*) and

[a]Department of Chemistry, KU Leuven, Celestijnenlaan 200F, Box 2404, 3001 Heverlee, Belgium. E-mail: jacek.mika@chem.kuleuven.be
[b]Centre of Microbial and Plant Genetics (CMPG), KU Leuven, Kasteelpark Arenberg 20, Box 2460, 3001 Leuven, Belgium

† Electronic supplementary information (ESI) available. See DOI: 10.1039/c5fd00058k

‡ The authors contributed equally.

observed that SeqA–PAmCherry no longer formed foci, and was dispersed throughout the cell and localized to the plasma membrane more readily. We discuss our results in the context of the limitations of the technique.

Introduction

SeqA is an *Escherichia coli* (*E. coli*) DNA-binding protein, that is involved in regulation of DNA replication initiation.[1,2] During cell division *E. coli* replicates its DNA and there are mechanisms ensuring that each daughter cell receives one copy of the genome after cell division. In *E. coli* the DNA is usually methylated at GATC sites by the dam methylase. Newly replicated GATC sequences stay hemimethylated for a while until the dam enzyme gets the opportunity to methylate them. *E. coli* is able to recognize and inactivate new origins of replication because they are hemimethylated.[3] In a process termed DNA sequestration, SeqA binds the new, freshly replicated origin of chromosomal replication (*oriC*), specifically at the hemimethylated GATC sequences that are highly abundant in the *oriC* region, whereby it prevents DnaA from starting replication initiation.[1,2,4]

The subcellular localization of SeqA in *E. coli* is particular. It has been shown with immunofluorescence microscopy and GFP-tagging that the protein forms discrete foci inside *E. coli* that are predominantly localized on the bacterial nucleoid.[5–7] The number of foci per cell varies between cells and also depends on growth condition (amount of chromosomes), but in general these studies found between 1 and 4 foci per cell under slow growth conditions.[6,8,9] The nature of the foci is dynamic: as the cell grows a single focus divides into two foci that migrate in opposite directions within the cell.[5] With conventional fluorescence microscopy approaches, these foci appear as single spots and it is possible that additional detail of these structures is obscured by the diffraction limit.

The size of bacterial cells (typically in the orders of micrometers) renders them difficult to study with conventional light microscopy. Due to the diffraction limit of visible light ($\pm$250 nm) only limited detail of the cellular substructure is revealed. In the last ten years, several new approaches have been developed that aim at increasing the resolution of fluorescence light microscopy.[10] They are often referred to as superresolution microscopy techniques and give researchers a tool to look at structures of living cells with nanometer scale resolution.[11–15] Superresolution microscopy has already contributed to an improved understanding of bacterial subcellular organization. To name a few examples, photo-activated localization microscopy PALM,[11,12] has been used to visualize the chemotactic protein network,[16] the spatial organization of the bacterial nucleoid and nucleoid associated proteins[17–19] or to uncover features of the cytoskeletal structures formed by FtsZ and MreB[20,21] proteins. The improved localization precision of this imaging technique, capable of imaging single proteins, revealed organizational features such as clustering and measured the sizes of structures that are below the diffraction limit on the tens of nanometers scale.[22] Albeit with limitations originating from the nature of the fluorescent proteins (photophysics and maturation) PALM and other single molecule fluorescence microscopy techniques have been used to count or estimate protein copy numbers in *E. coli* cells.[23–26]

Inspired by the improved localization precision of PALM, we have employed this technique to visualize the subcellular localization of SeqA in *E. coli* with a spatial resolution that was previously not available to conventional fluorescence microscopy. To this end we have created *E. coli* strains where the native SeqA protein was replaced with a variant of the protein fused to fluorescent proteins (FPs: eYFP, mEos3.2 and PAmCherry) to investigate the protein at close to physiological expression levels. We have also performed PALM on cells overexpressing the SeqA-FP fusion from a plasmid in wild type cells and in cells devoid of the dam methylase. Using the stringent response, we have synchronized the cell cycle of the *E. coli* cells expressing SeqA–PAmCherry from the genome. Next, we looked into changes in subcellular localization of this protein and counted the protein copy numbers in the cells grown in rich medium at 37 °C and 30 °C using PALM. We found that while typically the SeqA protein localized to foci as reported previously, it was not the sole subcellular localization of the protein. A significant fraction of the SeqA–PAmCherry proteins localized outside of the foci and in some cells the protein was more enriched at the cell membrane or near the cell periphery. Additionally, we observed cells that did not display defined, large foci. In an attempt to answer the question whether SeqA had a fixed number of molecules per focus and a defined size we have counted the number of SeqA–PAmCherry molecules per focus and determined the sizes of SeqA clusters. The number of molecules displayed a broad distribution as did the sizes of the foci which led us to favor the possibility that SeqA–PAmCherry does not display a fixed number of molecules per focus. We discuss the visualization of SeqA subcellular localization with PALM and the protein number quantification attempt in light of the limitations of the technique.

Experimental procedures

Strains and growth

All microscopy experiments were performed on *E. coli* K-12 MG1655 and related strains. For experiments where SeqA tagged with a fluorescent protein was expressed from the genome, the native *seqA* gene was fused at its 3′ end to fluorescent protein genes, PAmCherry,[27] eYFP[21] or mEos3.2,[28] separated with a linker (ASPPPGRSR) and including a kanamycin resistance marker. Recombineering[29] was used to replace the native SeqA with the fusion genes (Table 1). The genomic knock-ins were validated with PCR and subsequently with DNA sequencing. In experiments where the SeqA–PAmCherry protein was overexpressed from a plasmid *E. coli* MG1655 or BW25113 Δdam were transformed with pBAD–SeqA–PAmCherry (this study).

Cells were grown in LB medium (10 g l^{-1} Bacto Tryptone (Becton Dickinson), 5 g l^{-1} Yeast extract (Becton Dickinson) and 10 g l^{-1} NaCl (Merck)) with vigorous

Table 1 Strains used in this study

Strain	Features	Source
MG1655	Wild type	Blattner and co-workers[30]
MG1655 *seqA–PAmCherry*	MG1655 *seqA–PAmCherry:km*^{R}	This study
MG1655 *seqA–eYFP*	MG1655 *seqA–eYFP:km*^{R}	This study
MG1655 *seqA–mEos3.2*	MG1655 *seqA–mEos3.2:km*^{R}	This study
BW25113 *Δdam*	BW25113 *Δdam* Keio collection JW3350	The Keio collection[31]

shaking in Erlenmeyer flasks at 37 °C or at 30 °C (expression from the genome in synchronized cells experiments) until mid-log phase ($OD_{600} = 0.3$–0.5) and harvested for microscopy, unless stated otherwise. For plasmid overexpression experiments cells were induced with 0.002% (w/vol) arabinose for 2–3 hours at 37 °C. Where necessary the cultures were supplemented with antibiotics: kanamycin (40 μg ml^{-1}) for the genomic knock-ins and ampicillin (100 μg ml^{-1}) for the strains harbouring pBAD–SeqA–PAmCherry.

Synchronization of cells

Where indicated, the cells expressing SeqA–PAmCherry from the genome were synchronized in their cell cycle using the stringent response.[32,33] The synchronization was performed as described by Ferullo and co-workers.[32] Briefly, *E. coli* cultures were grown to an $OD_{600} = 0.2$–0.3. Subsequently, 8 mM of the synchronizing agent serine hydroxamate (SHX, Sigma) was added to the cultures. After 90 min, when the cells have arrested their cell cycle and finished ongoing DNA replication rounds, SHX was removed from the cultures by centrifugation (5000*g*, 5 min). The cells were introduced into fresh LB medium and restarted growth, upon which a majority of the cells resumed their cell cycle in a synchronous fashion.

Sample preparation for PALM microscopy

1 ml of culture was harvested after growth. To decrease the fluorescence background from the LB medium, cells were washed with NaPGCl buffer (NaPGCl = 95 mM sodium phosphate, pH 7.0, 50 mM glucose plus 125 mM sodium chloride), isoosmotic to the LB medium. The cells were then fixed with a solution of 4% paraformaldehyde in NaPGCl buffer, similarly to what was described by Greenfield and coworkers.[16] To diminish cell motion the cells were spotted onto poly-L-lysine coated coverslips (poly-L-lysine 0.1% solution, Sigma-Aldrich) and imaging was performed.

PALM acquisition

PALM microscopy was carried out on a home-built set up based on an epifluorescence inverted microscope (IX83, Olympus, Japan) with a 60× TIRF oil immersion objective lens (APON 60XOTIRF, NA 1.49, Olympus, Japan). Images were captured with an electron-multiplying CCD camera (ImagEM C9100-13, Hamamatsu Photonics, Hamamatsu, Japan) and recorded using Hokawo Imaging software V2.5 (Hamamatsu Photonics, Hamamatsu, Japan). For mEos3.2 and PAmCherry a 561 nm laser was used for excitation (constant illumination) and a 405 nm laser was used for photoconversion. The output power of the 405 nm laser was adjusted for each experiment based on the observed switching rate in order to obtain a low emitter density (*i.e.*, less than one emitter per diffraction limited area in an individual cell in each frame) by use of neutral density filters. Fluorescence emission was collected with a 572 nm long pass and a 590/40 band pass filters. For eYFP a constant illumination of the 514 nm laser was used and the fluorescence emission was collected using a 530 nm long pass filter. The lasers used are from Coherent (Coherent, CA, USA) and the optical filters are from Chroma (Chroma technology, VT, USA). Samples were acquired with 100 ms exposure times. The duration of the PALM experiments was typically 10 000 to 15 000 frames. For the data sets with synchronized cells, where molecule counting was performed, the acquisition was continued until no more or hardly any photoactivation events could be detected upon visual inspection.

PALM analysis

The PALM raw data was analyzed using the Localizer software[34] in combination with Igor Pro (Wavemetrics, OR, USA). In the first step, each frame was analyzed to localize single emitters. Briefly, individual acquired images were segmented into emissive spots and noise, using the GLRT algorithm as described by Sergé and co-workers.[35] Subsequently, the exact position of the emitter was estimated by fitting a 2D Gaussian distribution to the measured brightness profiles, using the Levenberg–Marquardt least squares algorithm. The coordinates of the maximum of the fitted 2D Gaussian distribution were then stored as the position of the emitter for that frame.

Correction for multiple counting of identical emitters

To correct for multiple counting of the same fluorophores emitting over multiple frames or emitters reappearing after several frames of absence (*e.g.*, due to blinking or emitting too few photons to be recognized by the fitting algorithm) we used an algorithm implemented in Localizer (termed "consolidation of identical emitters"). This algorithm groups localized emitters that were observed in close vicinity in time and space. Two parameters were required by this algorithm: the maximum allowed distance between two positions and the maximum allowed dark time (in frames) to be considered as coming from the same emitter. In order to determine the optimal parameters for this algorithm the approach of Annibale and co-workers[36] was followed. The "consolidation of identical emitters" algorithm was repeated for different allowed jump distances and different allowed dark times. The total number of estimated emitters was plotted *versus* the maximum allowed dark times (Fig. S1 and Table S1†). The larger the maximum allowed dark time, the lower the number of estimated emitters. This decrease was composed of two regimes. The first one was the contribution of single emitters that remained dark/undetected for some frames in between two localization events, as a result of blinking or not being fitted by the algorithm. The second regime was the contribution of two different emitters that appeared close together in space and in time. The first regime should be steep and exponential, the second regime quasi linear. Both regimes were fitted using a procedure written in the Igor Pro analysis software based on the approach of Annibale and colleagues.[36] Based on this fitting the optimal maximum allowed jump distance and the maximum allowed dark time values were obtained for each data set and consolidation of identical emitters was performed (Fig. S1 and Table S1†). The result of this analysis was also used to estimate the average localization precision of 21–25 nm for PAmCherry, depending on the dataset. The procedure did not take into account incomplete maturation efficiency nor the efficiency of photoactivation.[27,37]

Calculating the sizes of SeqA foci

To obtain the sizes of individual SeqA–FP foci the L-clustering analysis was performed.[18,38] In each analyzed cell large clusters that resemble the typical SeqA foci[5,6,8,39] were selected upon visual inspection and a region of interest was selected around the area of the cluster/focus. Subsequently, L-clustering analysis was performed at this region and a distribution curve was obtained. The maximum of this curve was taken as the representative size of the cluster (SeqA focus). In the case where the L-clustering curve showed two clearly defined maxima, the maximum corresponding to the smaller cluster size was chosen. The rationale for that is that due to dynamic nature of the SeqA foci, a distribution with the two maxima is likely to represent a dividing SeqA focus. The small, often solitary proteins localized outside of the big typical SeqA foci were not analyzed.

Results

PALM imaging reveals more information about SeqA structures in *Escherichia coli*

In conventional fluorescence microscopy the diffraction limit of light makes it difficult to resolve features of imaged objects that are smaller than 250 nm. It was shown extensively that SeqA forms discrete foci that appear as small, single spots in conventional fluorescence microscopy images.[5–9] Driven by the increased localization precision of molecules enabled by PALM, we anticipated to find more detail of the SeqA foci properties using this technique. In Fig. 1 conventional fluorescence microscopy is compared with PALM imaging in two *E. coli* cells expressing SeqA-PAmCherry from the genome. The conventional fluorescence microscopy images (Fig. 1A2 and B2) were reconstructed from PALM data, where the localized molecules were rendered as diffraction limited spots (~250 nm). In the PALM images (Fig. 1A3 and B3) the single localized SeqA–PAmCherry molecules are rendered as 25 nm spots according to their average localization precision. In Fig. 1 it can be observed that the PALM images contain more detail than the conventional fluorescence microscopy images. Moreover, while some of the SeqA–PAmCherry foci in conventional fluorescence microscopy images appear as single objects/spots, in PALM reconstructions it can actually be observed that they are composed of two foci or possess additional features.

Upon visual inspection of the PALM images acquired in this study it was found that the structures formed by SeqA–PAmCherry in *E. coli* cells are heterogeneous (Fig. 2). As anticipated, the SeqA molecules did arrange into foci (Fig. 2, yellow arrows), as reported previously,[5,6,8,39,40] however the foci were irregular in size and shape (compare Fig. 2A and C with Fig. 2B) and it was also possible to observe cells without clearly defined (large) SeqA foci (Fig. 2H, see also section *Quantitative characterization of the amount of SeqA-PAmCherry molecules in Escherichia coli cells*). The SeqA–PAmCherry molecules were not solely confined to (large) foci and in a vast majority of cells it was possible to find molecules outside of the (large) foci often as solitary proteins (see Fig. 2A and D, magenta arrows for examples). In a small subset of the data, we counted the total amount of proteins per cell and the amount of proteins that were localized in (large) foci. The foci were identified upon visual inspection arbitrarily as structures/clusters clearly resembling the foci reported in previous studies.[5,6,8] The amount of proteins located within the (large) foci varied largely between cells (Fig. S2†) with on average only 30–34% of molecules located in the large foci. The purpose of this measurement is not to give absolute numbers of molecules that are within or outside the SeqA structures (as this may vary upon what one defines as a SeqA focus as well as on the maturation and detection efficiency of the fluorescent tags), but to provide example that SeqA molecules observed outside the large foci are abundant. Qualitatively, similar characteristics of SeqA foci were also observed for *E. coli* cells expressing

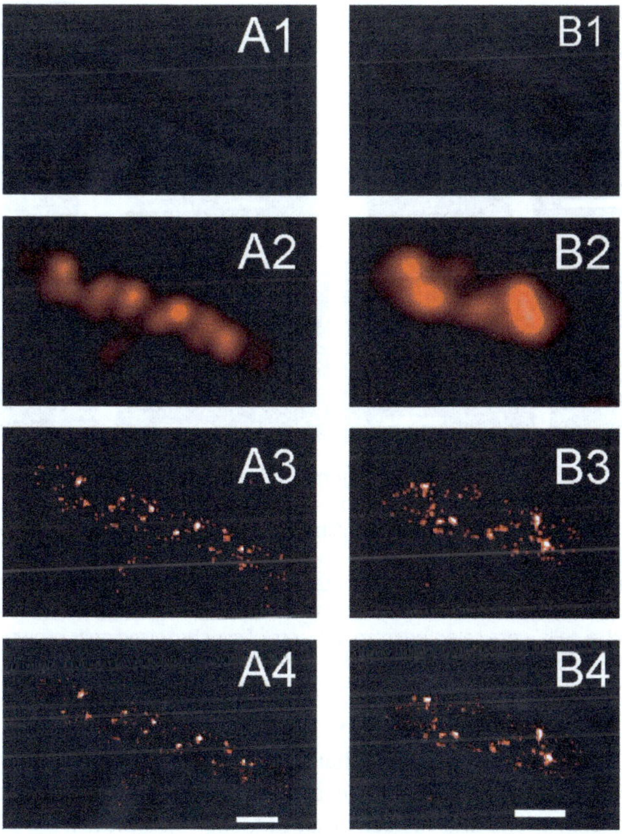

Fig. 1 PALM imaging shows more detail of the SeqA foci in *Escherichia coli* than conventional fluorescence microscopy. Examples of *E. coli* cells expressing SeqA–PAmCherry from the genome. (A1 and B1) Transmission images show the outline of the cells. (A2 and B2) Conventional fluorescence microscopy image of SeqA–PAmCherry. The image is reconstructed from the PALM where molecules are rendered as diffraction limited spots. (A3 and B3) PALM reconstructions of the SeqA–PAmCherry where each molecule is rendered as a 25 nm spot according to the average localization precision based on PALM fitting. (A4 and B4) Overlay of the transmittance image with the PALM reconstruction. Scale bar 1 μm

SeqA–eYFP genomically (Fig. S3†). While it is possible that a portion of the SeqA–fluorescent protein fusion (SeqA–FP) molecules located outside of the foci are artifacts resulting from PALM reconstructions (*e.g.*, sample impurities that were wrongly recognized during PALM reconstructions as SeqA–FP foci), we believe this localization is genuine. First, the small spots (putatively single molecule entities) outside of the foci were observed in most cells. Second, the fraction of molecules outside of the foci was high (Fig. S2†). And third, the single molecule-like localizations (small spots) were much more enriched inside the imaged cells compared to the area outside of the cells, which implies that these spots are not solely due to sample impurity (see below and Fig. S4†). It remains possible that a portion of the small spots are free PAmCherry molecules that were cleaved off from the SeqA–PAmCherry fusion protein.

As a test for false-positives, we have performed PALM reconstructions of wild type *E. coli* cells (MG1655) not expressing any FPs but imaged in the same way as

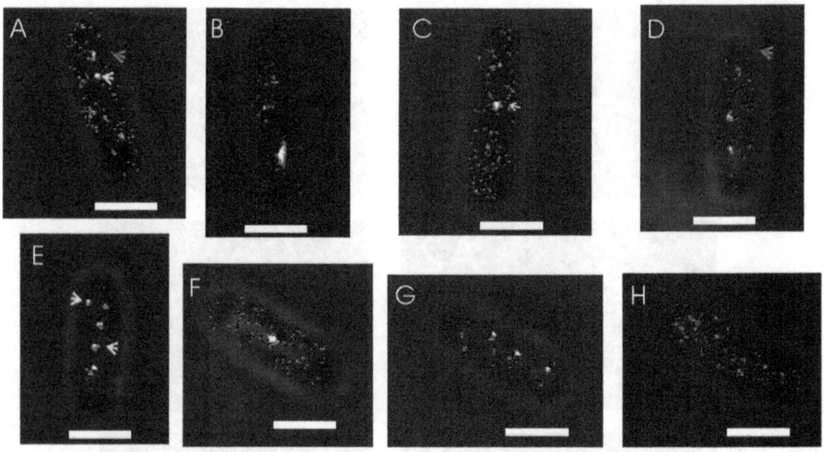

Fig. 2 PALM imaging of SeqA in *Escherichia coli* cells shows the molecules localized to heterogeneous structures. (A–H) Examples of *E. coli* cells expressing SeqA–PAmCherry from the genome. Typically SeqA–PAmCherry molecules are localized into (large) foci (examples indicated with yellow arrows). Some molecules are not confined to the clusters and are present as smaller, individual spots outside of the foci (examples indicated with magenta arrows). (H) Example of a cell where no defined (large) foci are observed. Representative cells were selected from datasets of synchronized cells grown at 37 °C and 30 °C in LB. Scale bar 1 μm.

the PAmCherry datasets. We found that the fraction of false-positive localizations in PALM reconstructions is negligible, as in the 120 imaged cells, 86 cells did not show any localizations and the cells that did, showed typically 1–5 localizations within the cell boundaries (Fig. S4†) giving on average 1.5 localizations per cell.

Some SeqA–PAmCherry molecules seem to localize at the membrane

Another detail of SeqA localization inside *E. coli* cells is that in some cells a fraction of the SeqA–PAmCherry molecules appeared to be localized or locally enriched at the membrane (Fig. 3, yellow arrows). This phenomenon was rare and upon visual inspection of the datasets for synchronized cells it was estimated to occur in about 12–18% of the cells (43 out of 336 cells, from 18 independent datasets, grown at 37 °C and 77 out of 419 cells, from 20 independent datasets, grown at 30 °C). It was also observed to occur in *E. coli* wild type strain and in the *Δdam* strain, both overexpressing SeqA–PAmCherry from the plasmid (pBAD–SeqA–PAmCherry, Fig. 9 and 10, respectively). The phenomenon was not observed in *E. coli* cells genomically expressing SeqA–YFP (Fig. S3†). A possible explanation for why the SeqA–eYFP molecules were not observed at the bacterial membrane might be the photophysical characteristics of eYFP.[21] It is expected that only a small fraction of the eYFP molecules are actually detected during PALM acquisition. Therefore the membrane localization might be missed due to not enough eYFP molecules being detected. To the best of our knowledge the membrane localization of SeqA has not been reported previously by fluorescence microscopy but there are reports indicating SeqA or SeqA-interacting proteins are present in membrane fractions in cellular fractionation experiments.[1,41–43]

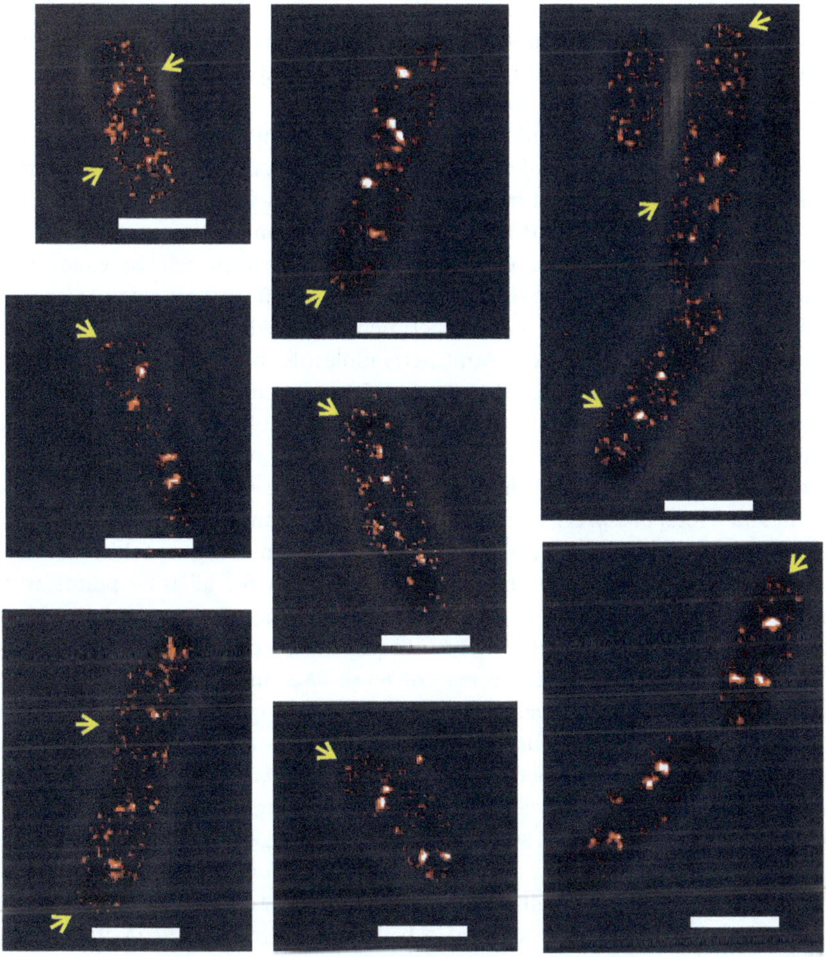

Fig. 3 PALM images of *Escherichia coli* cells where a fraction of the SeqA molecules localize to the membrane. Examples of *E. coli* cells in which a fraction of the genomically expressed SeqA–PAmCherry molecules were observed at the membrane (indicated by yellow arrows). Representative cells were selected from datasets of synchronized cells grown at 37 °C and 30 °C in LB. Different dataset were visually inspected for the membrane localization and it was found that in 43 out of 336 cells (18 independent datasets) grown at 37 °C and 77 out of 419 cells (20 independent datasets) grown at 30 °C a fraction of SeqA–PAmCherry molecules were localized at the membrane or near the cell periphery.

Numbers of SeqA-PAmCherry molecules in *Escherichia coli* cells

PALM was used previously to count the number of molecules in living bacterial cells.[23,44] Quantification of the number of molecules detected with PALM needs to be corrected for multiple counting of identical emitters that might result from the presence of the same emitter in subsequent frames, blinking and failure to recognize the emitter in subsequent frames by the PALM fitting algorithm. We corrected for this using the procedure outlined in the *Correction for multiple counting of identical emitters* section in the Experimental procedures. To facilitate the molecule counting and correction for multiple counting of identical emitters, it was crucial that the fluorophores did not move. For this aim cells were fixed with paraformaldehyde. Fixation with paraformaldehyde was performed for all the microscopy experiments in this study.

E. coli cells grown in liquid media are not synchronized in their cell cycle. We performed synchronization of cells using the stringent response.[19,32,33] To capture the evolution of SeqA foci during the cell cycle, cells were grown at 37 °C and 30 °C, synchronized with serine hydroxamate (SHX) and subsequently, after the synchronizing agent was removed and cells resumed growth, cells were harvested and prepared for microscopy (including fixation) at different time points (see the *Synchronization of cells* section in Experimental procedures for details). Fig. 4A and C show the results of SeqA–PAmCherry molecule counting in individual cells at different time-points after the synchronizing agent was removed and cells resumed growth. The amount of SeqA–PAmCherry molecules per cell showed a big spread between 17 and 302 molecules at 37 °C and between 15 and 195 molecules at 30 °C. Due to limited statistics (at 37 °C $n = 10$–24 cells at each time point and a total of 102 cells; at 30 °C $n = 11$–38 cells and a total of 182 cells) it is hard to draw conclusions to whether there are trends in changes of the amount of molecules as a function of the cell cycle. The data for all time points were combined (Fig. 4B and D) and there was an average of 97 ± 59.7 (standard deviation) and a median value of 74.5 SeqA–PAmCherry molecules per cell for cells grown at 37 °C and an average of 65 ± 33.5 and a median value of 58 molecules at 30 °C. To compensate for the fact that longer cells might have more molecules per cell the data for each cell was divided by the cell's length to obtain the number of molecules per 1 μm of cell. Similarly, the data showed a large spread (Fig. 4E and F) with cells grown at 37 °C having between 7 and 73 SeqA–PAmCherry molecules per 1 μm of cell and an average of 25 ± 13.3 molecules (median 21 molecules); cells at 30 °C (Fig. 4G and H) having between 5 and 62 molecules per 1 μm of the cell with an average of 21 ± 10.7 molecules (median 18 molecules). Cells grown at 30 °C also had on average fewer SeqA–PAmCherry molecules per cell than cells grown at 37 °C after the normalization for cell length. This observation might be rationalized by true differences in cell physiology resulting from growth at different temperatures but it cannot be excluded that at the lower temperature (30 °C) the maturation efficiency of PAmCherry in live cells is lower than for cells grown at 37 °C, which would affect the number of detected molecules.

Numbers of SeqA–PAmCherry foci in *Escherichia coli* cells

Since SeqA is known to organize into defined foci, upon visual inspection of the PALM images we determined the number of foci per cell in synchronized cells grown at 37 °C and 30 °C (see above for which structures were considered as large foci). Most cells showed between 1 and 4 foci per cell (Fig. 5). There was a significant fraction of cells that did not show defined (large) SeqA foci: 26% (20 out of 76 cells) and 29% (50 out of 171 cells) at 37 °C and 30 °C, respectively (Fig. 5, bottom panels). There were cells that showed more than 5 foci per cell with some cells having up to 12 foci. The larger foci numbers were much more prevalent for cells grown at 37 °C (13 out of 76 cells) than for cells grown at 30 °C (5 out of 171 cells).

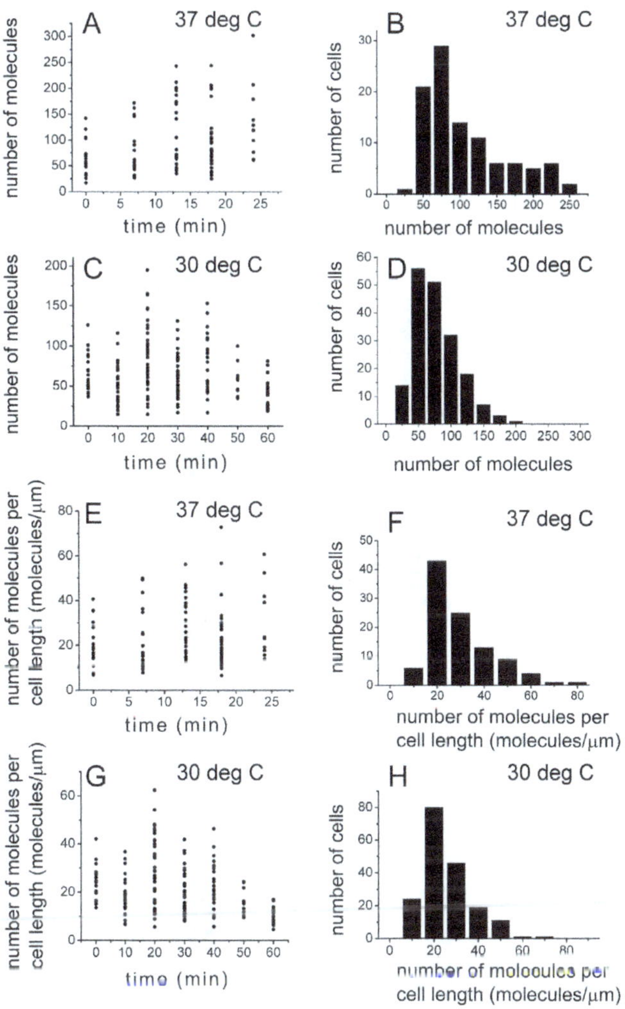

Fig. 4 Quantification of the number of SeqA–PAmCherry molecules in *Escherichia coli*. *E. coli* cells were grown in LB at 37 °C and at 30 °C and synchronized in the cell cycle (see main text for details). At various time points after the growth was resumed, the number of localized SeqA–PAmCherry molecules was calculated in each cell and corrected for multiple counting of the same emitters using a method described by Annibale[36] (see main text for details). (A and C) Number of molecules in each cell at a given time point after resumption of cell growth at 37 °C and at 30 °C, respectively. (B and D) Histograms of combined data from all time points in A and C. (E and G) The number of molecules per cell data (from A and C) was divided for each cell by the corresponding cell length, to correct for the possibility of larger cells having more molecules. (F and H) Histograms of combined data from all time points in E and G. (A, C, E and G) Each point represents a single cell. At 37 °C $n = 10–24$ cells at each time point and a total of 102 cells; at 30 °C $n = 11–38$ cells and a total of 182 cells.

Numbers of molecules per SeqA focus

We also sought out to count the number of SeqA–PAmCherry molecules per focus, to test whether the SeqA foci have a defined/fixed molecule number in each focus. The numbers of molecules per focus were characterized by a broad distribution

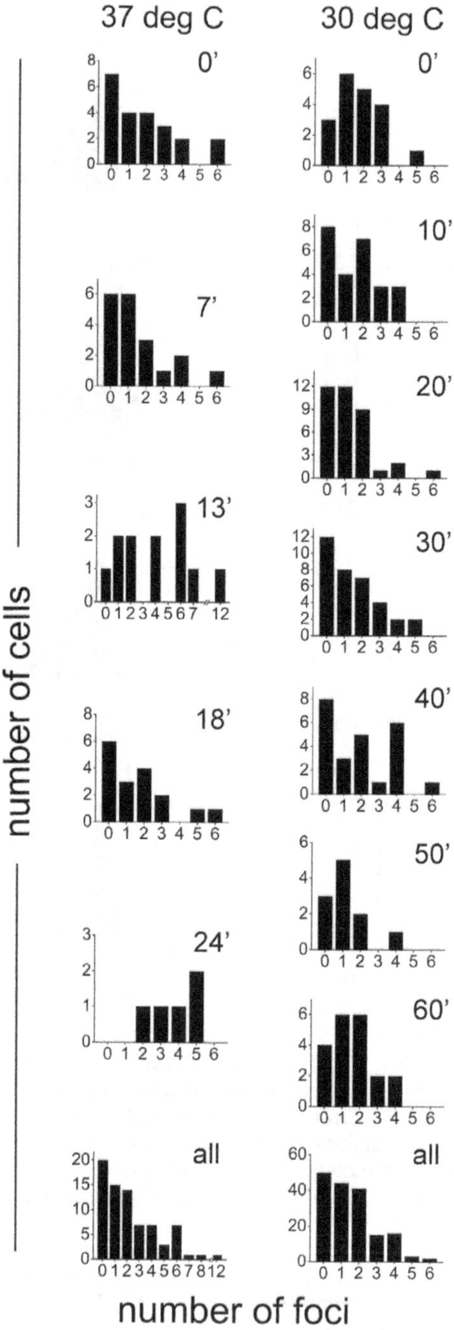

Fig. 5 Quantification of numbers of SeqA–PAmCherry foci in *Escherichia coli*. *E. coli* was grown and synchronized for the cell cycle (see main text for details) in LB medium at 37 °C and at 30 °C and the number of SeqA–PAmCherry foci was calculated in each cell after resumption of cell growth. Individual panels show the data for each given time point and temperature (left panels 37 °C and right panels at 30 °C). The bottom panels combine the data from all time points at each temperature. At 37 °C $n = 5$–22 cells at each time point and a total of 76 cells; at 30 °C $n = 11$–37 cells for each time point and a total of 171 cells.

for the different temperatures and at all time points tested (Fig. 6), with cells grown at 37 °C displaying foci with up to 55 SeqA–PAmCherry molecules per focus. At 30 °C the focus with highest amount of SeqA–PAmCherry counted 36 molecules (Fig. 6). Under the limited statistical sampling of this study (at 37 °C $n = 20–52$ foci at a given time point and a total of 172 foci; at 30 °C $n = 12–54$ foci at a given time point and a total of 248 foci) the numbers of molecules per focus at different time points had mostly similar distributions with the exception of the dataset at 24 min at 37 °C (Fig. 6). For the combined data from all the time points on average the numbers of SeqA–PAmCherry molecules per focus were comparable for cells grown at 37 °C and 30 °C, with most cells having between 10 and 15 molecules per focus, with cells grown at 30 °C having a median of 9 SeqA–PAmCherry molecules per focus and an average of 10.5 ± 6 molecules and at 37 °C having a median of 12 SeqA–PAmCherry molecules per focus and an average of 14.5 ± 9.6 molecules (Fig. 6 and 7A). A question now was whether the SeqA foci had a defined number of molecules per focus. We have tried to model the number of SeqA–PAmCherry molecules per focus in cells grown at 30 °C (to this end we combined data for all time points from cells grown at 30 °C obtaining a data set with the highest number of foci) with a binomial distribution assuming a fixed number of SeqA–PAmCherry per focus and assuming different PAmCherry detection efficiencies (*i.e.* 10%, 45% and 60%). In Fig. 7B it can be observed that within a data set of this size (248 foci) and with the distribution of molecules observed in the data from the experiment (gray bars) none of the binomial distributions fit well to the experimental data, as the latter ones have a broader distribution, implying that SeqA–PAmCherry does not have a fixed number of molecules per focus.

Quantification of the size of SeqA foci

To measure the sizes of the SeqA-FP foci, the PALM images were analyzed using the L-clustering algorithm (see *Experimental procedures*). We inspected the sizes of the foci in synchronized *E. coli* cells expressing SeqA–PAmCherry genomically at different time points after the cells had resumed growth. In Fig. 8 it can be observed that the SeqA foci are clearly smaller than the diffraction limit and that the sizes display a distribution between 30 and 130 nm with typical foci sizes in the 60–80 nm range. On average the clusters measured 76.9 ± 33 nm (median = 72.8 nm) for the combined data from cells grown at 37 °C (87 cells and 315 individual clusters, Fig. 8 and Table S2†) and 66.6 ± 30.2 nm (median = 62 nm) at 30 °C (174 cells, 456 individual clusters, Fig. 8 and Table S3†). In general we did not observe large changes in foci size at different timepoints after resumption of cell growth, with the exception of the timepoints at 18 min and 24 min at 37 °C, where the average foci size was 95.2 ± 44 and 90.1 ± 38.7 nm, respectively (Fig. 8 and Table S2†). These small changes are however close to the localization precision of PALM in our experimental set-up (21–25 nm). The foci sizes obtained for SeqA tagged with different FPs (PAmCherry, eYFP and mEos3.2) were very similar (Fig. S5A and Table S4†). The foci sizes obtained with data that were corrected for multiple counting of identical emitters were similar to the non-corrected ones (Fig. S5B and Table S5†), as were the foci sizes in synchronized cells compared to non-synchronized cells (Fig. S5C and Table S5†); the small differences were well within the localization precision.

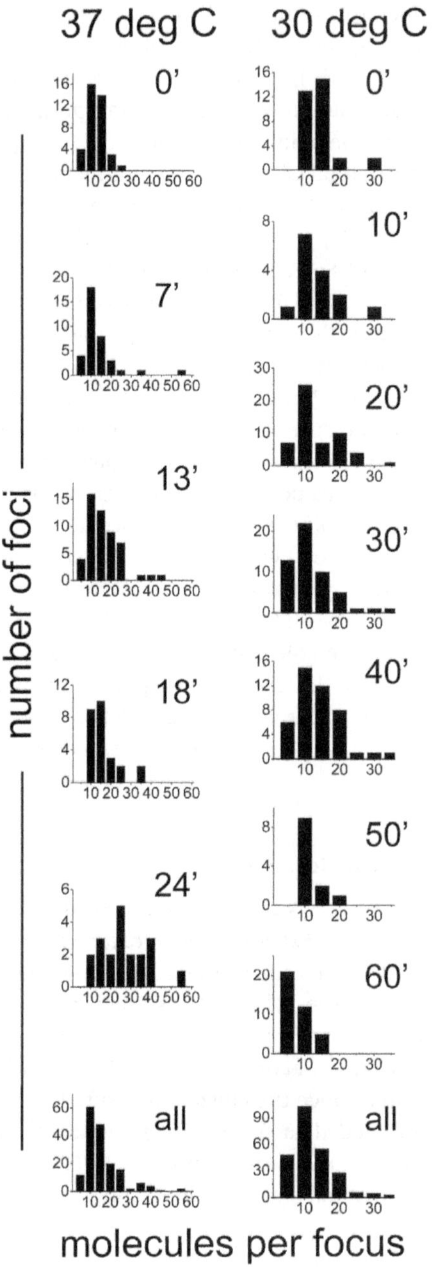

Fig. 6 Quantification of the numbers of molecules of SeqA–PAmCherry per focus in *Escherichia coli*. *E. coli* cells were grown in LB at 37 °C and at 30 °C and synchronized in the cell cycle (see main text for details). At various time points after the growth was resumed, the number of localized SeqA–PAmCherry molecules for each focus was calculated in each cell and corrected for multiple counting of the same emitters using a method described by Annibale[36] (see main text for details). Individual panels show the data for each given time point and temperature (left panels 37 °C and right panels at 30 °C). The bottom panels combine the data from all time points at each temperature. At 37 °C $n = 20–52$ foci at a given time point and a total of 172 foci; at 30 °C $n = 12–54$ foci at a given time point and a total of 248 foci.

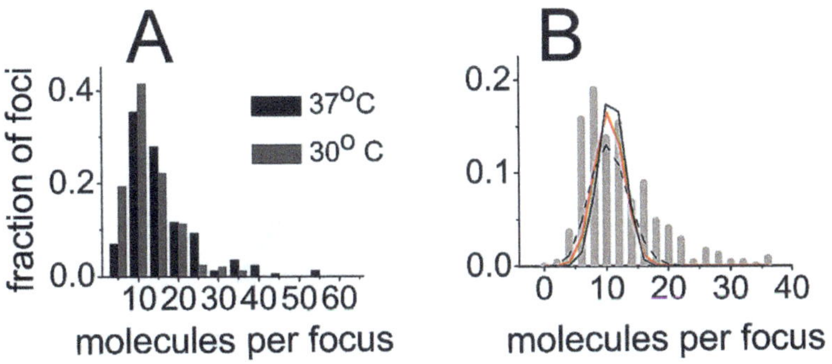

Fig. 7 SeqA foci do not have a fixed number of molecules per focus. (A) Histogram of combined numbers of SeqA–PAmCherry molecules per focus from all time points (as in Fig. 6) for all cells at 37 °C and at 30 °C (black bars and gray bars, respectively), displayed with 5 molecule bins. (B) Histogram of combined numbers of SeqA–PAmCherry molecules per focus from all time points at 30 °C plotted with a higher frequency of bins (gray bars, 2 molecules bins) than in A. The lines show a fit of binomial distribution of molecule detection events at different molecule detection efficiencies (60% black line, 45% red line and 10% black dotted line), assuming that the SeqA–PAmCherry foci have a fixed number of molecules per focus. The actual data (gray bars) have a broader distribution than the fitted lines implying that SeqA–PAmCherry does not have a fixed number of molecules per focus.

PALM imaging of SeqA–PAmCherry overexpressed from a plasmid

To compare SeqA localization in *E. coli* under conditions that resemble more closely the physiological expression levels of the protein (the SeqA–PAmCherry knock-in strain) with cells where SeqA–PAmCherry was overexpressed from a plasmid, we performed PALM imaging of *E. coli* harboring pBAD–SeqA–PAmCherry. The degree of overexpression varied between experiments and upon visual inspection it appeared that on average the cells overexpressing SeqA–PAmCherry had significantly more molecules, but the molecule numbers were not counted. As anticipated, we observed that SeqA–PAmCherry formed discrete foci (Fig. 9, magenta arrows) and in some cells the SeqA–PAmCherry molecules were also found at the membrane (Fig. 9, yellow arrows). The main difference in SeqA subcellular localization between the genomically expressed SeqA–PAmCherry (the knock-in strain) and the protein overexpressed from a plasmid (the strain harboring pBAD–SeqA–PAmCherry), was that in the latter case some cells showed regions of the cell with a high density of SeqA–PAmCherry molecules that appeared significantly larger than the SeqA foci. The non-uniform distribution of the molecules throughout the cell together with the fact that the proteins were localized mostly in the central region of the cell suggest that the SeqA–PAmCherry were residing at the nucleoid of the cells (Fig. 9, cyan arrows).

PALM imaging of SeqA–PAmCherry in *Δdam* cells

The SeqA protein preferentially binds hemimethylated DNA.[1,2] The *E. coli* DNA methyltransferase dam methylates newly synthesized DNA influencing the SeqA binding to DNA.[45] We sought to investigate the subcellular localization of SeqA in cells lacking the dam methylase with PALM. The *E. coli* Δdam strain harboring pBAD–SeqA–PAmCherry was imaged. In the majority of the cells, the overexpressed SeqA–PAmCherry was dispersed throughout the cell instead of localizing

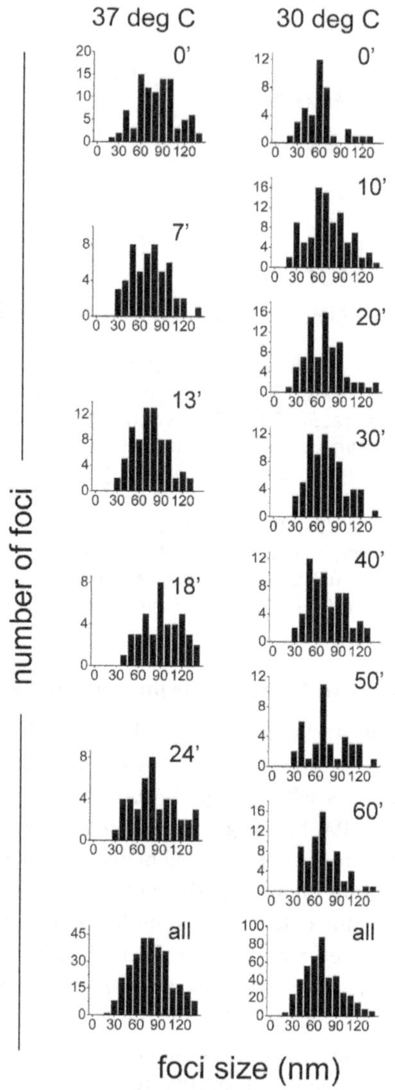

Fig. 8 Quantification of the sizes of SeqA–PAmCherry foci in *Escherichia coli*. *E. coli* cells were grown in LB at 37 °C and at 30 °C and synchronized in the cell cycle. At various time points after the growth was resumed, the sizes of SeqA–PAmCherry foci were calculated using L-clustering analysis (see main text for details) and displayed as histograms. Individual panels show the data for each given time point and temperature (left panels 37 °C and right panels at 30 °C). The bottom panels combine the data from all time points at each temperature. At 37 °C $n = 13–29$ cells and 51–95 foci at a given time point and a total of 87 cells and 315 foci; at 30 °C $n = 13–31$ cells and 38–93 foci at a given time point and a total of 174 cell and 456 foci.

into foci (Fig. 10). Upon visual inspection of PALM images only 5 out of 32 cells imaged showed structures that could be interpreted as foci. It was possible to observe that in some cells the SeqA protein seemed to localize to the membrane/cell periphery (Fig. 10, yellow arrows). In a previous study of SeqA localization in *E. coli Δdam* cells, where the wild type SeqA was imaged with immunofluorescence employing diffraction-limited conventional fluorescence microscopy, it was

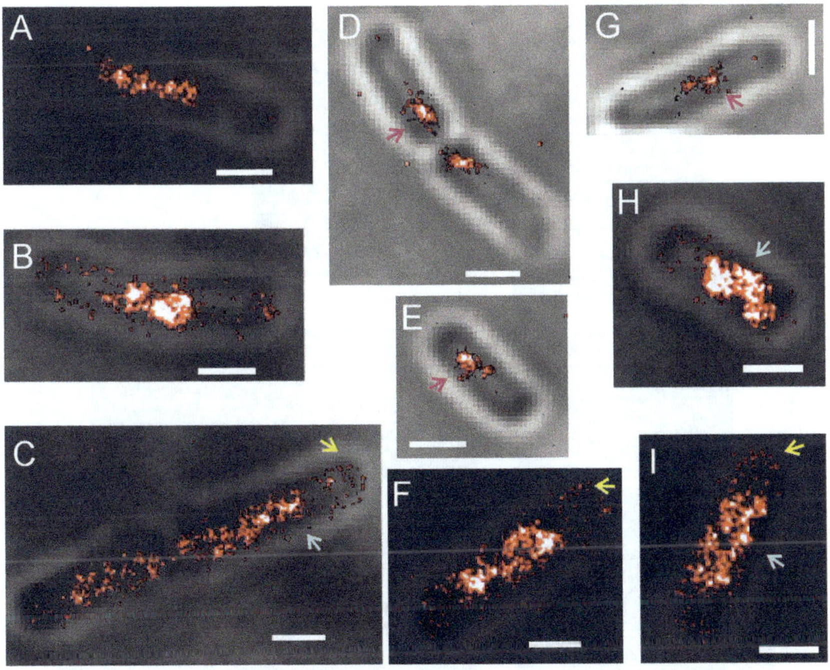

Fig. 9 PALM reconstructions of *Escherichia coli* cells overexpressing SeqA–PAmCherry from a plasmid. Examples of *E. coli* cells with pBAD–SeqA–PAmCherry. Overlay of transmittance image with the PALM reconstruction. The degree of SeqA–PAmCherry overexpression varied between experiments as did the number of localized molecules per cell. The SeqA–PAmCherry localization in the cell was heterogeneous, with some cells displaying SeqA foci (D, E, G, magenta arrows), some cells with SeqA localized at the membrane/cell periphery (C, F, I, yellow arrows) and some cells having a high concentration of SeqA molecules around the central region of the cell, that probably corresponds to the *E. coli* nucleoid; the latter structures are bigger than the typical SeqA foci (B, C, H, F, I, cyan arrows). Scale bar 1 μm.

also found that SeqA no longer formed discrete foci and it was concluded that the protein is dispersed throughout the nucleoid[6] or throughout the cell.[5] Upon PALM imaging we observed, that while the SeqA–PAmCherry is not homogeneously distributed throughout the cell, it is not solely confined to the nucleoid (central region of the cell) as there are cells with a clearly non-nucleoid like localization (Fig. 10).

Validation of the functionality of the SeqA–PAmCherry genomic knock-ins

We compared the growth rates of the wild-type strain (MG1655) expressing the native SeqA with those of the MG1655 *SeqA–PAmCherry* strain and we found that the genomic knock-in had no adverse effects on growth characteristics in LB under different temperatures tested (Fig. S6†). Western blots of the strains used in this study to perform microscopy show that qualitatively the expression level of SeqA in wild type cells is similar to the SeqA–FP levels in the genomic knock-ins (Fig. S8†). This is in good agreement with a previous report where it was demonstrated that a genomic knock-in of SeqA–eYFP has similar expression levels to those of SeqA in wild-type *E. coli* cells.[39]

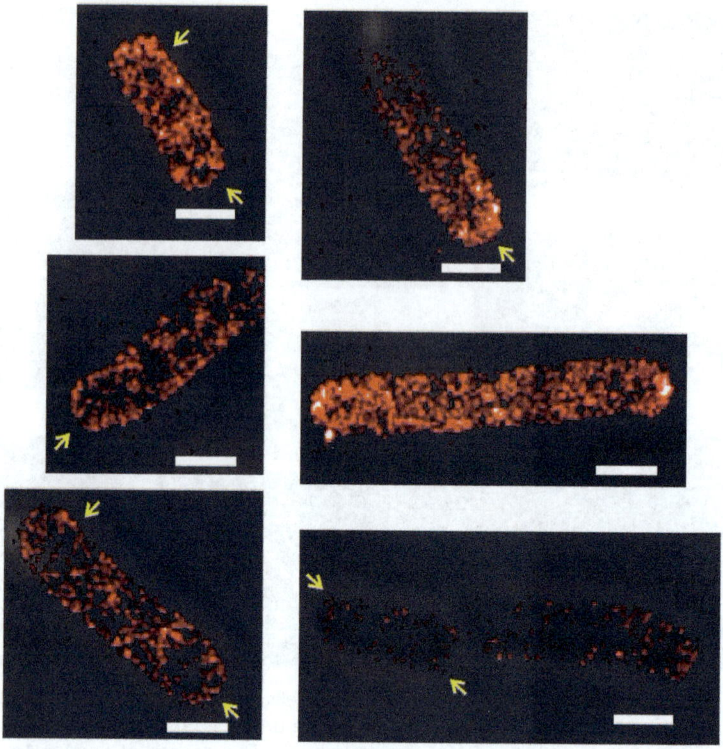

Fig. 10 PALM reconstructions of *Escherichia coli Δdam* cells overexpressing SeqA–PAmCherry from a plasmid. Examples of *E. coli Δdam* cells with pBAD–SeqA–PAmCherry. Overlay of transmittance image with the PALM reconstruction. The SeqA–PAmCherry molecules no longer form foci as prevalent subcellular localization and are localized throughout the cell with some cells displaying membrane/cell periphery localization (yellow arrows). Scale bar 1 μm.

The biological functionality of the SeqA–FP fusions has been characterized extensively before.[5,39] It has been shown that both the SeqA–GFP expressed from plasmid in the presence of native SeqA as well as SeqA–YFP expressed from the genome and replacing the native SeqA retain the ability to sequester the origins of replication, indicating that the fusion of SeqA to FPs does not compromise its biological function. Based on the similarity of the 3D structure between PAmCherry and eYFP (both are β-barrels of approximately similar molecular dimensions), the good qualitative agreement between the data obtained using the SeqA–eYFP and SeqA–PAmCherry genomic knock-ins (compare Fig. S3† with main text Fig. 1–3 and 8) and the fact that in the absence of dam methylase the SeqA–PAmCherry no longer organized into foci (Fig. 10) we conclude that there are strong indications to assume that the SeqA–PAmCherry can functionally replace the SeqA protein.

Discussion

SeqA is not solely localized into foci in *Escherichia coli*

The aim of this study was to obtain a more detailed view of SeqA subcellular localization, by employing a super-resolution microscopy technique – PALM. As shown previously SeqA localized into discrete foci. In some cases the improved localization precision of PALM allowed us to observe that the SeqA–PAmCherry structures that appeared as single discrete foci in diffraction limited microscopy were actually two foci in close vicinity (Fig. 1). A similar observation was reported by Helgesen and co-workers, who found that some foci that appear as single entities with conventional fluorescence microscopy are in fact composed of two clusters of SeqA molecules when reconstructed with super-resolution microscopy, using dSTORM imaging and SeqA labeled with fluorescent antibodies.[46]

The number of these foci per cell varies in most cells between 1 and 4 foci per cell with some cells showing up to 12 foci (Fig. 5). The high foci number per cell is consistent with other reports, where *E. coli* growing rapidly in rich media showed up to 12 foci per cell, with typically between 4 and 8 foci.[9] Yet, it was also possible to observe cells that did not display defined foci and under the experimental conditions of this study they made up to 26–29% of all the cells inspected. Cells that do not display SeqA foci have been observed previously with microscopy of SeqA–GFP fusions.[8,40] For cells grown in minimal media (slow growth conditions) the cells without SeqA foci composed between 20 and 45% of the imaged cells. However, these studies do not report cells devoid of SeqA foci when *E. coli* is grown in rich media (LB or similar). In our study the cells were grown in a rich medium (LB, fast growth conditions) and we still observed a fraction of cells without foci in the case of the SeqA–PAmCherry expressed from the genome in cells synchronized using SHX. Intriguingly, in the non-synchronized cells expressing SeqA–eYFP from the genome we did not observe high fractions of cells without foci (Fig. S3†). This might imply that the fraction of cells without foci observed in the synchronized cells could result from changes in cell physiology and nucleoid arrangement upon SHX treatment or that they reflect cells where there is no ongoing replication. We do not know whether the stringent response influences the SeqA propensity to arrange into foci. But it has been demonstrated that the stringent response is dependent on the presence of SeqA in cells, that during this response there is limited replication at *oriC* and that nucleoids of *E. coli* cells are decondensed.[33] We monitored cell growth after removal of the synchronizing agent (SHX) and we did observe resumption of growth, which indicates the end of the stringent response and release from the cell arrest (Fig. S7†). However, these are bulk measurements (optical density of the cultures) and it is possible that in a fraction of cells the cell cycle arrest is not relieved. We observed that foci were not the sole subcellular localization of SeqA–FP in *E. coli* as visualized by PALM. A fraction of SeqA–FP in cells expressing the fusion from the genome was localized as solitary or non-clustered proteins outside of the well defined large foci (Fig. 2 and S3†). As mentioned in the results section it is possible that the number of these molecules that are localized with PALM reconstructions outside of the foci might be due to the imperfections of the experimental set up used in this study (*e.g.*, sample impurities wrongly recognized by the PALM algorithm or free FPs that were cleaved off from the SeqA-FP fusion). If one assumes however that these localizations are real, these molecules might reflect SeqA molecules traveling to or between the existing foci or the beginnings of new SeqA foci. Since these are mostly single molecule localizations (solitary proteins or non-clustered protein oligomers) it is likely they were not visible in studies employing conventional fluorescence microscopy. In cells overexpressing SeqA–PAmCherry from the plasmid, where the protein expression level was much higher than in the genomically expressing cells, SeqA formed structures that were much bigger than the typical SeqA foci (Fig. 9), implying that it is possible to disturb the SeqA localization into foci.

Membrane localization of SeqA

In a fraction of cells (12–18% in cells expressing SeqA–PAmCherry from the genome, depending on growth temperature) we observed that some SeqA–PAmCherry molecules seemed to be localized at the membrane or the cell periphery (Fig. 3). SeqA is a soluble protein[47,48] and it has no domains that are expected to interact directly with the membrane. To the best of our knowledge membrane localization of SeqA with fluorescence microscopy has not been reported previously. There is, however, a great deal of reports of SeqA localization at the membrane coming from cell fractionation experiments. SeqA activity, measured as the binding of hemimethylated *oriC* DNA, was identified in membrane fractions[41,49] and the presence of SeqA in membrane fractions was further confirmed with an anti-SeqA antibody.[42,43] Shakibai and co-workers[42] showed that another protein in membrane fractions was auxiliary for *oriC* binding; this protein was named SeqB. D'Alençon and co-workers demonstrated that membrane fractions isolated from synchronized cells had highest SeqA activity during initiation of replication,[43] higher than membrane fractions from exponentially growing (non-synchronized) wild type cells. Moreover, Slater and co-workers report that SeqA is present but not enriched in membrane fractions as compared to the total cellular protein level of SeqA.[1] This could rationalize why we observe SeqA–PAmCherry at the membrane only in a fraction of cells and only for a small total amount of all molecule localizations with PALM in a given cell. In exponentially growing (non-synchronized) SeqA–eYFP expressing from the genome we did not observe SeqA to clearly localize at the membrane. This phenomenon might be partially explained by the specific photophysics of eYFP,[21] where only a small fraction of the total eYFP proteins are imaged (as mentioned in the results section) and some of the localizations might be missed by PALM. It is possible that the membrane localization of SeqA was not detected by conventional fluorescence microscopy previously as the much higher fluorescence intensity of the SeqA–FP foci might have overshadowed the membrane localized SeqA–FPs. Yet, in a single molecule localization microscopy technique such as PALM, it was possible to detect this phenomenon. This membrane localization was more common and more pronounced in cells overexpressing SeqA–PAmCherry from plasmid (Fig. 9), and even more so in the *Δdam* strain (Fig. 10). We speculate that as DNA binding sites get saturated or completely disappear, more SeqA molecules would be available to interact with SeqB or other putative interaction sites at the membrane, which makes the membrane localization easier to observe.

Quantification of the numbers of SeqA–PAmCherry molecules in cells

We have used PALM to count the numbers of SeqA–PAmCherry molecules in *E. coli* where the protein was expressed from the genome in synchronized cells. Within the statistically limited amount of cells probed in this study at each given time point (10–38 cells) we did not observe clear trends to changes in the numbers of SeqA copy numbers per cell as a function of cell cycle (Fig. 4). In general there was a big spread of the number of proteins per cell (15 to 300 molecules per cell) with on average 97 SeqA–PAmCherry molecules per cell for cells grown at 37 °C and an average of 65 molecules for cells grown at 30 °C. The smaller amount of molecules counted at 30 °C might reflect true differences in cell physiology or could result from a lower maturation efficiency of PAmCherry at this temperature. By quantifying the expression level of SeqA with western blotting using the anti-SeqA antibody Slater and co-workers report that wild type *E. coli* grown in minimal media has on average a 1000 SeqA molecules per cell (between 847 and 1435 SeqA molecules per cell depending on the strain and western blotting technique), which is about ten times more than was determined for the SeqA–PAmCherry fusion used in this study.[1] PAmCherry has been used previously for counting molecule numbers in microorganisms[23,50] and it was concluded that the numbers obtained with this probe are likely underestimated due to incomplete fluorophore maturation, fixation and premature photobleaching. Additionally Durisic and co-workers measured that the (*in vivo*) photoactivation efficiency of PAmCherry is only as little as 45% in *Xenopus* oocytes,[37] which will also contribute to an underestimation of the absolute molecule numbers. The photoactivation efficiency of PAmCherry is similar to those of other fluorescent proteins routinely used for PALM, which range between 39% (PA-GFP) and 61% (mEos2).[37] Yet, a clear advantage of PA–mCherry over other fluorophores used for molecule counting with PALM is that it blinks less[23,37] making the estimates of the amounts of unique emitters more reliable (Fig. S1†). Indeed, in this study we have corrected the numbers of localized molecules by employing the "consolidate identical emitters" procedure (see *Quantitative characterization of the amount of SeqA–PAmCherry molecules in Escherichia coli cells* in the Results section). We aimed to visualize all or nearly all PAmCherry molecules present in the samples as the data acquisition was continued until no more new photoactivation events could be detected upon visual inspection. Another possible source of error in the estimation of SeqA molecule numbers in cells, is the change of expression levels of the protein resulting from fusing it to the FPs. We have performed Western blots of wild type cells expressing SeqA and FP tagged expressing trains (eYFP, mEos3.2 and PAmCherry) and qualitatively we do not observe large changes in the expression level upon FP tagging as compared to the physiological level of SeqA (Fig. S8†). This is consistent with a previous report where it was shown that the genomically expressed SeqA–eYFP has a similar expression level to SeqA in wild-type cells.[39] We therefore conclude that the likely underestimated SeqA molecule numbers are not due to changes in expression level as a result of tagging the protein with FPs.

Broad distributions of SeqA molecule numbers per focus and foci sizes

Since SeqA forms discrete foci we sought to quantify the amount of SeqA–PAmCherry molecules per focus (Fig. 6). The number of molecules per focus was not observed to be dependent on the cell cycle, in the conditions of our cell cycle synchronization experiment. We observed a broad distribution of numbers of molecules per focus with on average 10–15 molecules per focus (for the largest dataset at 30 °C the average was 10.5 molecules per focus). Note that due to the properties of PAmCherry these numbers might be underestimated, as explained above. The modeling of the experimental distribution with a binomial distribution assuming a fixed amount of SeqA–PAmCherry molecules per focus and different fixed detection efficiencies as the only error source, did not fit well to the experimental data, which would be an indication that the foci do not have a fixed amount of SeqA–PAmCherry molecules per focus. However, peak widening may occur if there is more than one source of error.

In previous studies employing conventional fluorescence microscopy SeqA foci appeared as single spots and no further detail of these structures was resolved. With the improved localization precision of PALM microscopy we were able to show that these structures are indeed smaller than the diffraction limit (Fig. 1 and 8). By sizing the SeqA foci using the L-clustering analysis, we have shown that these structures have a distribution of sizes between 30 and 130 nm with most SeqA foci ranging between 60 and 80 nm. In a recent study employing dSTORM, Helgesen and co-workers[46] report the width of SeqA immunolabelled with Alexa647 to be 34 nm. The difference in reported foci sizes can partly be explained by the difference in localization precision or the sampling method employed (width of spots in the Helgesen study *vs.* L-clustering analysis in this study). The broad distribution of sizes of SeqA foci between different cells is similar to the distribution of numbers of molecules per focus, yet again implying that the foci do not have a fixed size.

It has been proposed that the SeqA complexes are dynamic in a threadmill-like fashion, *i.e.* growing in the leading end as SeqA molecules bind to newly replicated GATC sites and shrinking in the trailing end as dam methylates the older GATC sites.[2] While the exact mechanism of SeqA dynamics (assembly/disassembly and migration) remains to be fully elucidated, our claim that the SeqA foci do not have fixed numbers of molecules per focus, supports the threadmilling model or a gradual growth. The growing of SeqA structures (*e.g.*, resulting from the threadmilling) would rationalize the broad distributions of numbers of SeqA molecules per focus and the distribution of SeqA cluster sizes we observe.

Conclusions

In this study we have used super-resolution microscopy to study the subcellular localization of the *E. coli* DNA binding protein SeqA. This protein forms discrete foci in cells that have features sized below the diffraction limit. PALM enabled us to observe that some foci that can appear as single structures in conventional fluorescence microscopy are in fact composed of two foci. We have observed that not all of the SeqA-PAmCherry molecules are confined to foci, with some solitary molecules outside of the foci structures and in a fraction of cells some protein molecules localized to the membrane/cell periphery. While, to the best of our knowledge, the membrane localization of SeqA has not been demonstrated before with fluorescence microscopy, there were indications of this subcellular localization from cell fractionation studies. We thus believe the improved sensitivity of the technique used allowed us to observe details that were previously unavailable to diffraction-limited microscopy.

We also performed quantitative analysis of the numbers of molecules in cells that were synchronized in their cell cycle using the stringent response process. We found that the numbers of SeqA–PAmCherry molecules showed a broad distribution between cells with on average 50–100 molecules per cell, but no correlation between this number and the stage of cell cycle was found. In order to answer the question on whether the structures formed by SeqA have a fixed amount of molecules per focus, we have quantified the number of SeqA-PAmCherry molecules per focus and similarly found a broad distribution with a single focus containing on average 10–15 molecules. Furthermore, the SeqA foci sizes were also characterized by a distribution with most foci in the 60–80 nm range. We are inclined to believe that the SeqA foci have a dynamic nature and do not have a fixed sized or amount of molecules per focus.

The possible sources of error in this study might be of two origins. First, the cell physiology (and thus the subcellular localization of SeqA) could have been altered as a result of treatment with the cell cycle synchronization agent serine hydroxamate. Second, the PALM technique, especially its quantitative aspect, has still some limitations and room for improvement. The molecule number counting data reported here are most likely underestimated as a result of the rather low detection efficiency (resulting from photo-activation efficiency) of PAmCherry.

Nevertheless, despite the possible uncertainties, PALM contributed to understanding new biology as shown in this study and by others (for examples see the review of Gahlmann and Moerner[22]). Quantitative biology is an emerging field that aims to assign actual numbers to the key cellular parameters, such as protein copy numbers, stoichiometries of macromolecular complexes, reaction rates and diffusion coefficients *in vivo*.[51,52] To understand how proteins function in cells and to understand their regulatory mechanisms one needs to analyze molecules in their natural, cellular environment and not under idealized test tube conditions. Super-resolution imaging is undergoing constant development and new, brighter, monomeric, better-folding fluorescent proteins with better photophysical characteristics are made,[15] which will provide better tools to study biological processes quantitatively in living cells.

Acknowledgements

We would like to thank Prof. Kirsten Skarstad and Anne Wahl for the kind gift of the SeqA antibody and suggestions on western blotting. J. T. M. would like to thank Geert van den Bogaart and Wim Vandenberg for helpful discussions. J. T. M. would like to acknowledge funding from NWO-Rubicon (project number 825.12.026) and the Marie Curie Actions—Intra-European Fellowship (IEF, project number 329018). J. M., T. S. and B. B would like to acknowledge funding from KU Leuven Research Council (PF/10/010; IDO/13/008) and IAP-BELSPO, FWO (G.0B25.15, G.0471.12).

References

1 S. Slater, *et al.*, *E. coli* SeqA protein binds *oriC* in two different methyl-modulated reactions appropriate to its roles in DNA replication initiation and origin sequestration, *Cell*, 1995, **82**, 927–936.

2 T. Waldminghaus and K. Skarstad, The *Escherichia coli* SeqA protein, *Plasmid*, 2009, **61**, 141–150.

3 D. W. Russell and N. D. Zinder, Hemimethylation prevents DNA replication in *E. coli.*, *Cell*, 1987, **50**, 1071–1079.

4 M. Lu, J. L. Campbell, E. Boye and N. Kleckner, SeqA: a negative modulator of replication initiation in *E. coli.*, *Cell*, 1994, **77**, 413–426.

5 T. Onogi, H. Niki, M. Yamazoe and S. Hiraga, The assembly and migration of SeqA–Gfp fusion in living cells of *Escherichia coli.*, *Mol. Microbiol.*, 1999, **31**, 1775–1782.

6 S. Hiraga, C. Ichinose, H. Niki and M. Yamazoe, Cell cycle-dependent duplication and bidirectional migration of SeqA-associated DNA-protein complexes in *E. coli.*, *Mol. Cell*, 1998, **1**, 381–387.

7 S. Hiraga, C. Ichinose, T. Onogi, H. Niki and M. Yamazoe, Bidirectional migration of SeqA-bound hemimethylated DNA clusters and pairing of *oriC* copies in *Escherichia coli.*, *Genes Cells*, 2000, **5**, 327–341.

8 T. Brendler, J. Sawitzke, K. Sergueev and S. Austin, A case for sliding SeqA tracts at anchored replication forks during *Escherichia coli* chromosome replication and segregation, *EMBO J.*, 2000, **19**, 6249–6258.

9 Morigen, I. Odsbu and K. Skarstad, Growth rate dependent numbers of SeqA structures organize the multiple replication forks in rapidly growing *Escherichia coli.*, *Genes Cells*, 2009, **14**, 643–657.

10 B. Huang, M. Bates and X. Zhuang, Super-resolution fluorescence microscopy, *Annu. Rev. Biochem.*, 2009, **78**, 993–1016.

11 E. Betzig, *et al.*, Imaging intracellular fluorescent proteins at nanometer resolution, *Science*, 2006, **313**, 1642–1645.

12 S. T. Hess, T. P. K. Girirajan and M. D. Mason, Ultra-high resolution imaging by fluorescence photoactivation localization microscopy, *Biophys. J.*, 2006, **91**, 4258–4272.

13 M. J. Rust, M. Bates and X. Zhuang, Sub-diffraction-limit imaging by stochastic optical reconstruction microscopy (STORM), *Nat. Methods*, 2006, **3**, 793–795.

14 T. A. Klar, S. Jakobs, M. Dyba, A. Egner and S. W. Hell, Fluorescence microscopy with diffraction resolution barrier broken by stimulated emission, *Proc. Natl. Acad. Sci. U. S. A.*, 2000, **97**, 8206–8210.

15 W. Vandenberg, M. Leutenegger, T. Lasser, J. Hofkens and P. Dedecker, Diffraction-unlimited imaging: from pretty pictures to hard numbers, *Cell Tissue Res.*, 2015, **360**, 151–178.

16 D. Greenfield, *et al.*, Self-organization of the *Escherichia coli* chemotaxis network imaged with super-resolution light microscopy, *PLoS Biol.*, 2009, **7**, e1000137.

17 W. Wang, G.-W. Li, C. Chen, X. S. Xie and X. Zhuang, Chromosome organization by a nucleoid-associated protein in live bacteria, *Science*, 2011, **333**, 1445–1449.

18 S. F. Lee, M. A. Thompson, M. A. Schwartz, L. Shapiro and W. E. Moerner, Super-resolution imaging of the nucleoid-associated protein HU in *Caulobacter crescentus*, *Biophys. J.*, 2011, **100**, L31–L33.

19 C. Spahn, U. Endesfelder and M. Heilemann, Super-resolution imaging of *Escherichia coli* nucleoids reveals highly structured and asymmetric segregation during fast growth, *J. Struct. Biol.*, 2014, **185**, 243–249.

20 G. Fu, *et al.*, *In vivo* structure of the *E. coli* FtsZ-ring revealed by photoactivated localization microscopy (PALM), *PLoS One*, 2010, **5**, e12682.

21 J. S. Biteen, *et al.*, Super-resolutionimaging in live *Caulobacter crescentus* cells using photoswitchable EYFP, *Nat. Methods*, 2008, **5**, 947–949.

22 A. Gahlmann and W. E. Moerner, Exploring bacterial cell biology with single-molecule tracking and super-resolution imaging, *Nat. Rev. Microbiol.*, 2014, **12**, 9–22.

23 U. Endesfelder, *et al.*, Multiscale spatial organization of RNA polymerase in *Escherichia coli*, *Biophys. J.*, 2013, **105**, 172–181.

24 A. Badrinarayanan, R. Reyes-Lamothe, S. Uphoff, M. C. Leake and D. J. Sherratt, *In vivo* architecture and action of bacterial structural maintenance of chromosome proteins, *Science*, 2012, **338**, 528–531.

25 R. Reyes-Lamothe, D. J. Sherratt and M. C. Leake, Stoichiometry and architecture of active DNA replication machinery in *Escherichia coli*, *Science*, 2010, **328**, 498–501.

26 S. Bakshi, A. Siryaporn, M. Goulian and J. C. Weisshaar, Superresolution imaging of ribosomes and RNA polymerase in live *Escherichia coli* cells, *Mol. Microbiol.*, 2012, **85**, 21–38.

27 F. V. Subach, *et al.*, Photoactivatable mCherry for high-resolution two-color fluorescence microscopy, *Nat. Methods*, 2009, **6**, 153–159.

28 M. Zhang, *et al.*, Rational design of true monomeric and bright photoactivatable fluorescent proteins, *Nat. Methods*, 2012, **9**, 727–729.

29 K. A. Datsenko and B. L. Wanner, One-step inactivation of chromosomal genes in *Escherichia coli* K-12 using PCR products, *Proc. Natl. Acad. Sci. U. S. A.*, 2000, **97**, 6640–6645.

30 F. R. Blattner, *et al.*, The complete genome sequence of *Escherichia coli* K-12, *Science*, 1997, **277**, 1453–1462.

31 T. Baba, *et al.*, Construction of *Escherichia coli* K-12 in-frame, single-gene knockout mutants: the Keio collection, *Mol. Syst. Biol.*, 2006, **2**, 2006.0008.

32 D. J. Ferullo, D. L. Cooper, H. R. Moore and S. T. Lovett, Cell cycle synchronization of *Escherichia coli* using the stringent response, with fluorescence labeling assays for DNA content and replication, *Methods*, 2009, **48**, 8–13.

33 D. J. Ferullo and S. T. Lovett, The stringent response and cell cycle arrest in *Escherichia coli*, *PLoS Genet.*, 2008, **4**, e1000300.

34 P. Dedecker, S. Duwé, R. K. Neely and J. Zhang, Localizer: fast, accurate, open-source, and modular software package for superresolution microscopy, *J. Biomed. Opt.*, 2012, **17**, 126008.

35 A. Sergé, N. Bertaux, H. Rigneault and D. Marguet, Dynamic multiple-target tracing to probe spatiotemporal cartography of cell membranes, *Nat. Methods*, 2008, **5**, 687–694.

36 P. Annibale, S. Vanni, M. Scarselli, U. Rothlisberger and A. Radenovic, Quantitative photo activated localization microscopy: unraveling the effects of photoblinking, *PLoS One*, 2011, **6**, e22678.

37 N. Durisic, L. Laparra-Cuervo, A. Sandoval-Álvarez, J. S. Borbely and M. Lakadamyali, Single-molecule evaluation of fluorescent protein photoactivation efficiency using an *in vivo* nanotemplate, *Nat. Methods*, 2014, **11**, 156–162.

38 M. A. Kiskowski, J. F. Hancock and A. K. Kenworthy, On the use of Ripley's K function and its derivatives to analyze domain size, *Biophys. J.*, 2009, **97**, 1095–1103.

39 S. Fossum-Raunehaug, E. Helgesen, C. Stokke and K. Skarstad, *Escherichia coli* SeqA structures relocalize abruptly upon termination of origin sequestration during multifork DNA replication, *PLoS One*, 2014, **9**, e110575.

40 S. Adachi, T. Fukushima and S. Hiraga, Dynamic events of sister chromosomes in the cell cycle of *Escherichia coli*, *Genes Cells*, 2008, **13**, 181–197.

41 G. B. Ogden, M. J. Pratt and M. Schaechter, The replicative origin of the *E. coli* chromosome binds to cell membranes only when hemimethylated, *Cell*, 1988, **54**, 127–135.

42 N. Shakibai, *et al.*, High-affinity binding of hemimethylated *oriC* by *Escherichia coli* membranesis mediated by a multiprotein system that includes SeqA and a newly identified factor, SeqB, *Proc. Natl. Acad. Sci. U. S. A.*, 1998, **95**, 11117–11121.

43 E. D'Alençon, A. Taghbalout, R. Kern and M. Kohiyama, Replication cycle dependent association of SeqA to the outer membrane fraction of *E. coli.*, *Biochimie*, 1999, **81**, 841–846.

44 S.-H. Lee, J. Y. Shin, A. Lee and C. Bustamante, Counting single photoactivatable fluorescent molecules by photoactivated localization microscopy (PALM), *Proc. Natl. Acad. Sci. U. S. A.*, 2012, **109**, 17436–17441.

45 T. Waldminghaus, C. Weigel and K. Skarstad, Replication fork movement and methylation govern SeqA binding to the *Escherichia coli* chromosome, *Nucleic Acids Res.*, 2012, **40**, 5465–5476.

46 E. Helgesen, S. Fossum-Raunehaug, F. Sætre, K. O. Schink and K. Skarstad, Dynamic *Escherichia coli* SeqA complexes organize the newly replicated DNA at a considerable distance from the replisome, *Nucleic Acids Res.*, 2015, **43**, 2730–2743.

47 A. Guarné, Q. Zhao, R. Ghirlando and W. Yang, Insights into negative modulation of *E. coli* replication initiation from the structure of SeqA-hemimethylated DNA complex, *Nat. Struct. Biol.*, 2002, **9**, 839–843.

48 A. Guarné, *et al.*, Crystal structure of a SeqA–N filament: implications for DNA replication and chromosome organization, *EMBO J.*, 2005, **24**, 1502–1511.

49 A. Chakraborti, S. Gunji, N. Shakibai, J. Cubeddu and L. Rothfield, Characterization of the *Escherichia coli* membrane domain responsible for binding *oriC* DNA, *J. Bacteriol.*, 1992, **174**, 7202–7206.

50 D. Lando, *et al.*, Quantitative single-molecule microscopy reveals that CENP-A(Cnp1) deposition occurs during G2 in fission yeast, *Open Biol.*, 2012, **2**, 120078.

51 L. M. Gierasch and A. Gershenson, Post-reductionist protein science, or putting Humpty Dumpty back together again, *Nat. Chem. Biol.*, 2009, **5**, 774–777.

52 U. Moran, R. Phillips and R. Milo, SnapShot: key numbers in biology, *Cell*, 2010, **141**, 1262–1262e1.

Faraday Discussions

DISCUSSIONS

Plasmonics, Tracking and Manipulating, and Living Cells: general discussion

Thorben Cordes, William Moerner, Michel Orrit, Sergey Sekatskii, Sanli Faez, Paola Borri, Himangshu Prabal Goswami, Alex Clark, Patrick El-Khoury, Sandra Mayr, Jacek Mika, Guowei Lyu, Daniel Cross, Francisco Balzarotti, Wolfgang Langbein, Vahid Sandoghdar, Jens Michaelis, Arindam Chowdhury, Alfred J. Meixner, Niek van Hulst, Brahim Lounis, Fernando Stefani, Frank Cichos, Maxime Dahan, Lukas Novotny, Mark Leake and Haw Yang, FRSC

DOI: 10.1039/c5fd90093j

Michel Orrit opened a discussion of the paper by Paola Borri: You showed an interpretation of the spectral response of gold nanoparticles in terms of shift and broadening of the surface plasmon band. This empirical description of the response is model-independent. Did your data fit with only shift and broadening of the plasmon band?

When we tried a similar approach to describe the pump–probe signals of single gold nanospheres, we found that we had to introduce not only a shift and a broadening of the band, but also a change in integrated strength ("oscillator strength"). As experimental and calculated spectra of gold nanospheres in different solvents show, the oscillator strength of the plasmon is not always conserved, as the dielectric medium can significantly contribute to the polarization.

Paola Borri answered: We performed a "microscopic" model rather than an empirical lineshape fit. The model, described in detail in our paper,[1] contains broadening, shift and changes in the overall strength of the plasmon resonance by calculating the pump-induced change of the particle polarisability. This polarisability is calculated from the dielectric constant of gold and the surrounding medium. The dielectric constant of gold is a function of the electron and lattice temperatures, and these change as a consequence of the transfer of energy from the optically absorbed pump pulse to the surface plasmons which rapidly dephase into single electron excitations. The temporal evolution of the electron and lattice temperatures and the thermalisation with the dielectric environment are calculated using the well known two-temperature model. In short, the strength of the particle polarisability is fully embedded in our model and is not a separate 'fit' parameter.

1 Masia *et al.*, *Phys. Rev. B*, 2012, **85**, 235403.

Niek van Hulst commented: It is very nice how you can determine the spectral phase of plasmonic nanoparticles from the non-linear four-wave mixing (FWM) experiments, and even more nice that you can determine it for the coupled particles.

It seems the phase flips over 2π when going through resonance, unlike the expected π for a harmonic oscillator through resonance. Is this due to the fact that you excite non-linear, and quadratic dependence make the 2π? You showed some distance dependence in the presentation; as you get closer and closer the phase, it starts to deviate from π. Can you explain this?

Paola Borri replied: It is indeed interesting to observe that the measured FWM phase is very sensitive to the interparticle distance in dimers. We think this can have great potential for sensing measurements using dimers as 'plasmon rulers'. Our qualitative understanding of this behavior is shown schematically in Fig. 6 in our Faraday Discussion manuscript (DOI: 10.1039/c5fd00079c). Briefly, what we measure is the phase of the ratio between the pump-induced change in the nanoparticle dielectric function and the dielectric function itself. Because we are making this ratio, the phase evolution across the plasmon resonance is not that of a simple oscillator. Fig. 6 shows this ratio for a single spherical particle and indicates that if the pump and probe wavelength in the FWM experiment is blue-shifted relative to the plasmon resonance, the phase increases from π toward 2π. When a dimer is forming and we keep the wavelength in our FWM experiment constant, *via* the distance dependence red-shift of the coupled plasmon mode we are effectively creating this situation, *i.e.*, the wavelength in the experiment is blue-shifted relative to the longitudinal plasmon resonance of the dimer, and thus we expect the phase to increase toward 2π.

Patrick El-Khoury remarked: Some of the electron micrographs that were presented featured touching or nearly touching plasmonic nanoparticles. A recent quantum analysis of plasmonic junctions suggests that in the case of strongly coupled plasmonic nanoparticles (0–1 nm particle separation), conductive overlap is established.[1,2] It would be interesting to use your technique to examine not only localized, but also quantum plasmon resonances.

1 K. J. Savage *et al.*, *Nature*, 2012, **491**, 574.
2 S. F. Tan *et al.*, *Science*, 2014, **343**(6178), 1496.

Niek van Hulst asked: You did show absolute values for cross-sections of the nanoparticles. How did you manage to get the absolute values?

Paola Borri responded: For extinction cross-section measurements this can be obtained directly from integrating the wide-field differential transmission image around the nanoparticle area, as shown in our paper.[1] For scattering cross-section measurements the quantitative analysis is indeed more complicated. Here, it was achieved using nanodiamonds as scattering-only particles and by comparing the extinction cross-section with the measured dark-field scattering to determine a calibration parameter with which we convert dark-field scattering in cross-section units.

1 L. M. Payne *et al. Appl. Phys. Lett.* 2013, **102**, 131107.

Sergey Sekatskii opened a discussion of the paper by Alfred J. Meixner: The presented data showing the shortening of the quantum dot (QD) luminescence time when close to a gold nanocone are quite convincing. Still, in principle, this shortening can be caused either by an increase of the optical absorption/emission cross-section, or just by strong metallic quenching while corresponding cross-sections do not change a lot. Are there unambiguous evidences that here we definitely have the first case?

Alfred J. Meixner answered: This is a very good question. As is well known with gold nanostructures, there is always a competition between different effects, including increased absorption/emission on one side and non-radiative relaxation or quenching on the other side. Regarding the luminescence lifetime, of course both effects, an enhanced emission rate and enhanced quenching, can contribute to reduced luminescence lifetimes. However, in our experiments we observe that the QDs on the cones are on average five times brighter than those on glass (see Fig. 8b in DOI: 10.1039/c5fd00074b). This is an unambiguous sign that quenching is not the dominating effect in this configuration.

Brahim Lounis queried: When a quantum emitter is placed near a metallic structure it is often difficult to distinguish the radiative enhancement from the non-radiative processes. Can you comment on the possibility of measuring the radiative and non-radiative decay rates of individual hybrids through a controlled modification of the spontaneous emission rate by changing the environment index of refraction (as carried out by Brokmann *et al.*)?[1]

1 X. Brokmann, L. Coolen, M. Dahan, and J. P. Hermier, *Phys. Rev. Lett.*, 2004, **93**, 107403.

Alfred J. Meixner responded: This is a very good suggestion and a similar approach should also work for our samples for determining the radiative and non-radiative rates of QDs on gold cones independently. The idea would be to vary the optical environment on the dimensions of a wavelength in order to alter or further increase the radiative rate *e.g.* by covering the sample with a layer of material with a large refractive index. While changing these materials would vary the radiative rates, the local chemical environment being responsible for the non-radiative rate could be kept constant by firstly covering the cones with the QDs by a thin protective layer which would be kept constant.

Alex Clark asked: When your quantum dots are placed on the nanocones, is the spatial emission pattern of the dots adjusted? This could cause a change in how many photons you see through your microscope, and therefore contribute in some way to the data and skew the analysis of lifetime adjustments.

Alfred J. Meixner answered: The emission pattern of a QD on a cone can certainly be different from the emission pattern of a QD on a glass surface, and so are the lifetimes. Also, we definitely see more photons from QDs on cones than from QDs on glass as shown in Fig. 8 (DOI: c5fd00074b). In both cases we use the same inverted microscope and we do not expect that the collection efficiency of our high numerical aperture (NA) system has an influence on the luminescence lifetime.

Niek van Hulst queried: You did show some intriguing data on quantum dots coupling to the gold nanocones, recording both fluorescence lifetime and intensity. With decreasing distances the signal goes up, while lifetime goes down. The quantum dots have a high quantum efficiency, therefore I assume the enhancement must be mostly excitation enhancement. Is this correct? Do you know the actual quantum efficiency of these quantum dots? Similarly I assume the decreasing lifetime should be mainly radiative, rather than non-radiative, as otherwise the signal would collapse. Can you comment on that interplay?

Alfred J. Meixner replied: Thank you for this good question. The fluorescence quantum yield of our free quantum dots is nominally 20–40%, hence, as you say, it is well known that the close proximity of a metal nanostructure such as a cone can increase both the radiative and non-radiative rates. The fact that QDs are significantly brighter on the cones and have a significantly shorter lifetime than those on a glass substrate suggests that both the radiative rate and the excitation rate are enhanced. The non-radiative rate is not dominating, since as you say, the photoluminescence (PL) signal should collapse.

Wolfgang Langbein asked: How do you explain the lifetime decrease by a factor of 10 when going from QDs on glass to QDs in PMMA? This indicates a significant change in the non-radiative processes, for example by surface states or charging.

Alfred J. Meixner replied: It is well known that the fluorescence lifetime of a quantum system decreases with increasing refractive index of the surrounding medium. For a QD, the situation can be more complicated since the interaction with the surroundings can lead to an increase or change in both the radiative and non-radiative rates. For recent developments on this topic, refer to Ref. 1 and references therein.

1 A. Pillonnet *et al.*, *Opt. Exp.*, 2012, **20**(3), 3200–3208.

Paola Borri asked: Have you looked at the dependence of the coupling on the thickness of the quantum dot shell? Do you buy or fabricate these colloidal quantum dots?

Alfred J. Meixner replied: So far we have used only one type of commercially available QDs, as noted in the paper, but future experiments with different kinds of QDs and shell thicknesses are planned.

Thorben Cordes queried: How accurately can you place the quantum dots on the top of your nanostructure? How does this influence your results? Does the overall efficiency of deposition limit statistics?

Alfred J. Meixner replied: These are good points. 1) The fabrication process is designed in such a way that the quantum dots are only deposited near the very tips of the cones, which have a radius on the order of 10 nm. The accuracy of the placement additionally depends on how much of the embedding layer is removed during preparation. The process currently aims at exposing the top ~10 nm of the gold for QD attachment. The vertical position of the QD relative to the tip surface is determined by the linker molecules and QD shell thickness. The influence of the different parameters is described in detail in our recent paper published in Nanoscale.[1]

2) At this point we cannot yet conclusively answer the question of how the positioning influences the results. This is due to the fact that the maximum coupling effect is expected to be in the region of the highest near-field intensity which should be at the cone apex if the excitation light is polarized along the cone axis as shown in Figure 3a–c (see DOI: 10.1039/c5fd00074b). However, the statistical variation in the exact nanoscale shape and the sharpness of the individual nanocones leads to a distribution of coupling efficiencies. It is also known from the literature that the competition between intensity enhancement and quenching sensitively depends on the distance between the Au surface and the QD, hence, in work in progress, the distance to the surface will be varied to gain further information.

3) Since we aim at having on average only one QD per cone, and since the deposition remains a statistical process, many cones actually have no quantum dots on the top (~30%). As a consequence we have to look at a large number of QDs to obtain decent statistics, and to identify cones with single QDs. Nevertheless, as we can prepare arrays of cones and couple QDs to about 70% of them at the same time, our method is quite suitable to obtain good statistics in comparison to measurements involving scanning near-field tips.

1 J. Fulmes *et al.*, *Nanoscale*, 2015, 7, 14691–14696.

Himangshu Prabal Goswami enquired: The intermittent fluorescence behaviour of quantum dots is shown to obey an inverse power law behavior in time.[1] Recently this power law distribution in on-times was shown to be invalid for quantum dots on silicon oxides[2]. What happens to this power law distribution in the blinking times when you couple the quantum dots to gold nanocones? Also, does the plasmon lifetime affect the quantum dot blinking lifetime?

1 M. Kuno *et al.*, *J. Chem. Phys.*, 2000, **112**, 3117.
2 R. Schmidt *et al.*, *ACS Nano*, 2014, **8**(4), 3506–3521.

Alfred J. Meixner replied: Thank you for the comments on the power dependence of the blinking. We are aware of this debate but have not yet investigated the power dependence of the blinking statistics for QDs on cones. From the intensity trajectories measured so far, the off-times for QDs on cones appear to be significantly shorter than for QDs on glass, but the statistics still need to be quantitatively evaluated.

Vahid Sandoghdar remarked: These are very difficult experiments because one has to control the orientation of the emission and absorption dipoles and the exact relative position of the emitter and the antenna. QDs are known to have very fluctuating quantum efficiencies, so it is difficult to know what you have in each individual measurement unless you have characterized that very dot before coupling to the antenna. Also, separating the radiative and non radiative decay rates is a challenge.

Alfred J. Meixner replied: We agree that the presented experiments are challenging. However, we would like to comment on the individual issues: 1) With respect to the orientation of the absorption and emission dipole moments, in a recent work we have investigated the absorption polarization in the focal fields of a high NA objective lens formed by radially and azimuthally polarized laser beams where we can precisely control the orientation of the electric field component of the light in three dimensions. We found that CdSe/ZnS QDs can be excited isotropically.[1] Here the excitation field is polarized along the cone's axis and hence the QD should be able to efficiently absorb along the cone axis. In one of our next experiments we will check this behavior by exciting with azimuthally polarized light. Recently Gerton *et al.* have shown for similar CdSe/ZnS QDs that their emission orients itself along the axis of a sharp tip approaching the QD.[2] 2) We agree that QDs in contrast to atoms used in quantum optics have a very complicated emission behavior leading to a distribution of emission spectra, quantum efficiencies, *etc.* Also, the cones are not identical and contribute to a distribution of the experimental results. Even if the same QD was characterized before binding, the QD's optical properties could change during an experiment independent of the coupling given by the electromagnetic boundary conditions. Hence, any individual measurement on a single quantum system is only of limited value and some statistics are needed to observe trends. One advantage in our configuration is that extensive statistics can be gained from the hybrid structures that allow us to extract "typical" average behavior, in addition to providing the ability to compare each individual hybrid structure to the statistical ensemble.

1 A. I. Chizhik *et al.*, *Nano Lett.*, 2013, **13**, 194–198.
2 A. Ghimire *et al.*, *Sci. Rep.*, 2014, **4**, 6456.

Guowei Lyu communicated: (a) How would you confirm that only one (not two or three) quantum dot was attached to a gold nanocone? (b) When calculating the surface enhancement factor, how do you remove the signal caused by the metal?

Alfred J. Meixner replied: (a) In single molecule spectroscopy several spectral features which cannot be observed in an ensemble are accepted as criteria that the PL signal originates from a single molecule, or in our case, a single quantum emitter. These include (1) repeated on-off blinking in time trajectories, (2) abrupt turning into an off-state or abrupt photobleaching, (3) characteristic line shapes, which would be hidden in an inhomogeneously broadened spectrum, (4) the appearance as a diffraction limited spot in a raster-scanned image, (5) photon anti-bunching, *etc.* When the PL signal recorded from a cone with one or several QDs fulfils several of these criteria, it is very likely that it originates from one single QD. Additional proof could be gained from correlation measurements. (b) In the microscope configuration that was used, the gold luminescence from bare cones is too weak to be seen in the spectra or time traces of cones with QDs. The optical background signal in the off-periods of the time traces results from residual resistance and can be easily distinguished from the enhanced fluorescence of the QDs as shown in Fig. 7a–e (DOI: 10.1039/c5fd00074b).

William Moerner opened a discussion of the paper by Patrick El-Khoury: I am happy to see your description of the variety of SERS signals from dithiols on your carefully fabricated SERS-active substrates. These are similar to the Raman spectrum fluctuations we reported for thiols on bowtie nanoantennas.[1] Your report of huge transient Raman background fluctuations is fascinating. As a second comment, while you are nicely using electrospray to controllably add molecules to your surfaces, I would urge you and others in the community to use these substrates for analysis of molecules in solution which come down to the surface and sample the regions of high enhancements. This can apply not only to Raman signals but also to enhanced fluorescence emission from low-efficiency emitters. Clear bursts of fluorescence from single molecules can be seen, for example, for solution-phase molecules which sample the gap of a bowtie nanoantenna, as illustrated by Kinkhabwala *et al.*[2]

1 D. P. Fromm *et al.*, *J. Chem. Phys.*, 2006, **124**, 061101.
2 A. A. Kinkhabwala *et al.*, *Chem. Phys.*, 2012, **406**, 3–8.

Alfred J. Meixner commented: You often see a luminescence background with gold. Different background spectra and spectral fluctuations in surface enhanced Raman spectra are typical for low concentrations of molecules. These are single events and are a sign for achieving the single molecule limit.[1]

1 T. Vosgröne *et al.*, *ChemPhysChem*, 2005, **6**, 154–163.

Patrick El-Khoury responded: Indeed. Our on-going works aim at attaining a rigorous understanding of these optical signatures of single molecules (both background and relative intensity fluctuations of the observable vibrational eigenstates). There's a lot more to single molecule Raman spectroscopy than just chemical fingerprinting.

Guowei Lyu communicated: (a) What is the reason for various SERS backgrounds when the molecular coverage changes? (b) How do you determine the amount of molecules in the detecting area, and how do you avoid the influence of molecular aggregation?

Patrick El-Khoury responded: Key to understanding (a), is recognizing that the SERS backgrounds observed in this study are time-averaged individual molecular events (see figure 3, panels A and B of figure 4, and the corresponding discussion in the main text of DOI: 10.1039/c5fd00036j). When figures 4A and 4B are compared, it can be noticed that background fluctuations are more noticeable at the lowest molecular coverage reported herein (pm, 0.38 DMS molecules per square micron, see figure 4B), less pronounced at intermediate molecular coverage (nm, 3.8 x 10^2 DMS molecules per square micron, see Figure 4A), and largely suppressed at a higher molecular coverage (see Figure 3). These backgrounds are signatures of signal molecules that are naturally more pronounced at ultralow molecular coverage, and averaged over at higher molecular coverage.

Dr. Lyu is referred to section 2.1 of the manuscript (DOI: 10.1039/c5fd00036j) in response to the first section of question (b). Moreover, in addition to (i) our SERS-active substrate preparation method of choice (ESD) which yields an upper limit of 0.02 molecules in our effective SERS probing area, and (ii) our above-discussed observation of fluctuating SERS backgrounds that are associated with individual molecular events, the observation of distinct relative intensities of the vibrational Eigenstates in the single molecule realm precludes molecular aggregation. Note that spectra recorded from several randomly selected spots on the 2D

Ag nanoparticle array were considered in our analysis. These are reproducible observations.

Wolfgang Langbein opened a general discussion of the papers by Dr Borri, Dr Meixner and Dr El-Khoury: For plasmonic effects, the shape and crystallinity of the structure is of utmost importance. In many publications, plasmonic effects involving nanometer scale distances are attributed to specific mechanisms,[1,2] without knowledge of the 3D surface structure and crystallinity on this scale. Furthermore, theoretical models are typically not taking into account the surface roughness or atomic structure. As a result, the conclusions concerning the physical effects and mechanisms are questionable.

1 K. J. Savage *et al.*, *Nature*, 2012, **491**, 574–577.
2 C. Ciraci *et al.*, *Science*, 2012, **337**, 1072–1074.

Patrick El-Khoury responded: This is exactly why an all experimental approach to the problem is invaluable. I would also add that beyond topographic characterization is of utmost importance in this context. One needs independent access to the local electric fields over the length scales of relevance (*e.g.* using MP-PEEM, EELS).

Niek van Hulst also responded: Indeed, at the nanometer and Angstrom scale distances, the exact atomic structure (crystalline or not) is crucial, as seen in many SERS/TERS experiments. Yet at larger distances, beyond 5 nm, the atomic structure appears less critical: excitation and emission enhancement follow well the behaviour predicted based on bulk metallic/dielectric constants, even for polycrystalline plasmonic antennas. Thus, the interpretations of physical effects and mechanisms, on a spatial scale of 5 nm and larger, are quite consistent between theory and experiment.

Lukas Novotny opened a general discussion of the paper by Haw Yang: 1) If you used a beam with a dark centre (bottle beam) would you only get counts when the target leaves the center? Is this something you looked into and how does the signal to noise compare? 2) What kind of collection of numerical aperture did you use? Are you collecting all the photons and then splitting them 3 or 5 ways? 3) What fraction of the total amount of scattered photons is received by a single detector?

Haw Yang said in reply: 1) Yes, the shell-like beam shape could enable a different type of tracking algorithm. With imaging, for example, whenever the tracked particle moves away from the center of the microscope objective focus and hits the light shell, it will emit one photon. The detection event of such a photon could be used to move the sample to counter the movement of the target particle. In principle, this will allow a much faster feedback mechanism for weaker light emitters because the counter-movement steering could be done on a photon-by-photon basis. In our earlier studies we have indeed tried using a doughnut-shaped beam. Unfortunately, at that time, we were not able to create a stable hollow beam across all different kinds of media (in buffer and inside live cells). In our hands, the current avalanche photodiode (APD)-based beam splitting approach appears to be the most stable; therefore, it has been the default configuration for all the setups in our laboratory. 2) The microscope objective used for the 3D multi-resolution imaging was 100X, NA = 1.4 because in that experiment we detected photo luminescence from quantum dots instead of scattered light. For dark-field scattering-based experiments, the NA will be ~0.7–0.8. For the 3D multi-resolution experiment, 10% of the collected photons were directed to an electron multiplying charge coupled device (EMCCD) camera for monitoring and the remaining 90% were split 5 ways going into 5 different single-photon counting APDs (4 for *xy* sensing and 1 for *z* sensing). Typically, a total of 200,000 to 500,000 photons per second combining the photon counts from all 5 detectors will be needed for a stable tracking. 3) To achieve a stable *z* response, 45% of the total collected photons were directed to the *z*-axis APD. The other 45% of the collected photons were evenly divided between the 4 *xy* APDs. The final 10% of the collected photons were detected by an EMCCD camera for monitoring. We thank Professor Novotny for the probing questions and for the opportunity to clarify some of the fine points.

Niek van Hulst enquired: I do appreciate the advantage of 3D single-particle tracking to follow a particle in real-time and thus gain time resolution. Yet, comparing your method with conventional wide field particle tracking, you do lose the parallelism. Can you comment on what 3D single-particle tracking really gains compared to wide field imaging which currently achieves sub-millisecond time resolution with 20–50 nm spatial accuracy over a full field?

Haw Yang replied: The 3D multi-resolution imaging is beyond simple 3D single-particle tracking; it also delivers synchronously acquired 3D contextual insights using two-photon laser-scanning microscopy. That said, there exist of course a variety of different methods for tracking individual particles or molecules. Each type of method has its own advantages and disadvantages. A researcher would consider the nature of a scientific problem and decide which method would be most appropriate for solving that particular problem. It is difficult to be convinced that a single method can study all classes of problems. From the methodology perspective, a comparative study would have to be carefully carried out in order to answer this question fully. Short of such a study, one could perhaps discuss this question from an academic angle, say, phototoxicity. From the viewpoint of studying biological samples such as living cells, the 3D multi-resolution approach should, in principle, incur minimal phototoxicity to the live cell under study. The drawback of this approach is that it is high in content but low in throughput. On the other hand, for the wide-field imaging approach to produce 20–50 nm spatial accuracy in 3D and sub-millisecond time resolution for the full field, the photon flux must be sufficiently high in order for all the particles to achieve that kind of spatial localization accuracy. This is because the localization accuracy depends on the number of photons emanated from the single-particle light emitter. Since it is wide-field illumination, those areas where there are no particles of interest are also intensely illuminated; as such, one could imagine that the potential to incur photo-damage would be high. The advantage of a wide-field approach is that it is high throughput. In our opinion, all these different methods are complementary rather than competing; the community benefits from having such a wide array of tools so that many different kinds of problems can be studied. We thank Professor van Hulst for the comment and insightful question.

Sanli Faez asked: It is suggested that by reconstruction of the particle track, an image of the micro-environment can be reconstructed. Although the tracking method is very fast, image reconstruction just considering a simple random walk will be slow. Is it possible to "scan" the particle (for example by applying an external electric field) so that it spans a larger and more regular area (volume in 3D), over a certain period of time, than is covered by pure Brownian motion of the particle?

Haw Yang answered: We thank Professor Faez for this forward-looking question. Yes, absolutely. In fact, we have been actively pursuing this vision using the "photon nudging" idea that we have developed.[1] There will be some interesting technical challenges, as one can imagine, but we felt confident that it could be achieved in the near future.

1 B. Qian *et al.*, *Chem. Sci.*, 2013, **4**, 1420–1429.

Vahid Sandoghdar queried: The spatial and temporal precision and resolution in single particle tracking are tightly entangled with each other. Is it really possible for you to reach nm localisation and microsecond time resolution with small labels? Is there an advantage in your scheme, where you split the signal to several detectors in terms of photon budget and signal-to-noise ratio? You mentioned that you had also considered the idea of using a bottle beam. I would think that in that case you would know nothing about the particle unless it begins to leave the region of confinement. In other words, most of the time your particle is in the dark, giving very few photons, and therefore poor localization information. I guess this would work for general course tracking but not for high-end localization.

Haw Yang responded: The quoted localization precision was from an immobile nanoparticle, as it is in most other research groups reporting localization precision. While the sampling time resolution remains the same when interrogating a fast-moving nanoparticle, the localization precision for a dynamical phenomenon, indeed, will depend on numerous factors, including the brightness of the probe (photon flux on the detector), the response time of the system (in particular for the piezo stage), and the response time of the field-programmable gated array board, to name only a few. While we do not yet have a first-principle theory that will allow us to articulate the fundamental resolution limits, we did lay out an empirical way of characterizing dynamical localization precision in an earlier work.[1] In practice, we believe, it will depend on the specific type of application.

To answer the second question, one must consider alternatives in relative terms. The closest arrangement to our current set up would be a quadrant photodiode. Indeed, in our first paper,[2] we used a custom-made quadrant avalanche photodiode (APD) to sense the *xy* displacement. Unfortunately, at that time, the sensitivity of that quadrant APD was quite low such that it was not possible for us to follow low light-emitting particles such as a single quantum dot.

Furthermore, for APDs, the gap between the sensing elements must be sufficiently large to avoid arcing between the APD sensing elements. In addition to losing photons when they hit the gap, the width of the gap is also an important factor in the overall *xy* displacement sensitivity. On the other hand, using the prism splitting configuration and highest sensitivity single-element single-photon counting APD, we were able to overcome these shortcomings seen for a quadrant APD. Of course, it is entirely possible that better, more efficient means will become available in the near future.

For the last question/comment: You are absolutely right. Furthermore, one of the main goals for us to implement in this way is to be able to do time-dependent spectroscopy on the fast 3D moving nanoprobe, or any analytes attached to it. Presumably it would be very difficult, if not impossible, to do that using a doughnut-shaped beam for tracking.

We thank Professor Sandoghdar for the insightful questions and an opportunity to clarify a few points.

1 C. S. Xu *et al.*, *J. Phys. Chem. C*, 2007, **111**, 32–35.
2 H. Cang, *et al.*, *Appl. Phys. Lett.*, 2006, **88**, 223901.

Wolfgang Langbein commented: You are claiming to "break the 3D speed barrier with 3D multi-resolution microscopy", but what you do is single particle tracking of a large gold nanoparticle and rather slow volumetric imaging around the particle. The imaging speed is still limited by the slowest imaging process, so I do not see how a 3D speed barrier is being broken.

Haw Yang responded: It would appear that this comment is directed toward our first publication in 2006,[1] which was indeed a work reporting real-time 3D tracking of single gold nanoparticles of 250 nm. The 3D multi-resolution imaging, however, is conceptually different from conventional microscopy. The idea was first put forward in 2007 in a proof-of-principle work.[2] In that work, we have clearly laid out the central problem; that is, indiscriminate scanning is inefficient if one is only interested in a small but fast-moving object and its vicinity. The idea we put forward in the 2007 paper[2] was to let a fast-moving nanoprobe guide an imaging system. At that time, however, we used a simple 2D camera for imaging and hence did not have 3D sectioning capability for the lower-resolution contextual part of the image. It was not until recently that we were able to implement a full 3D-sectioning two-photon laser-scanning microscopy (2PLSM) slaved by the fast-moving nanoprobe.[3] The concept is fundamentally different from the conventional, indiscriminate scanning. In that work, we intentionally constructed a luminescent nanoprobe that mimicked the HIV-1 viral particle (~110 nm in diameter, with peptide coating derived from HIV-1). In this way, the vicinity of the virus-like particle was better sampled by the 2PLSM compared to regions farther away from the dynamically changing region of interest (the virus-like particle). The 3D multi-resolution approach has allowed us to uncover many interesting phenomena and images in full 3D when a virus-like particle approaches a live cell. Those phenomena, we believe, would not have been possible to capture using any of the conventional forms of microscopy.

We thank Dr Langbein for this opportunity to make the points clearer.

1 H. Cang *et al.*, *Appl. Phys. Lett.*, 2006, **88**, 223901.
2 H. Cang *et al.*, *Opt. Lett.*, 2007, **32**, 2729–2731.
3 K. Welsher *et al.*, *Nat. Nanotechnol.*, 2014, **9**, 198–203.

Jens Michaelis queried: It is not clear to me why you need a 10nm accuracy if you then co-localise the particle you are tracking to positions/structures that are determined with a resolution of about 200nm. For what questions do you require this accuracy? Couldn't you alter your setup to decrease tracking accuracy but at the same time increase speed?

Haw Yang replied: As we hopefully stated clearly in our earlier publication, for example,[1] the dynamic slowing-down of a virus-like particle approaching a live-cell membrane is but one of the many exciting new possibilities that can be delivered using the 3D multi-resolution imaging idea. We do not believe that colocalization-type microscopy could deliver such dynamical information. Also, when integrated over time, the 3D membrane protrusion on live-cell membranes could be vividly imaged at a resolution that rivals 2D electron microscopy on a fixed specimen. For this kind of data, and also potentially for molecule-level information, one will have to push hard on both the localization precision as well as time resolution, as illustrated in Fig. 1 of our Faraday Discussion paper (DOI: 10.1039/c5fd00090d). We thank Professor Michaelis for the opportunity to clarify our points.

1 K. Welsher *et al.*, *Nat. Nanotechnol.*, 2014, **9**, 198–203.

Fernando Stefani opened a discussion of the paper by Frank Cichos: What is the effect of ionic strength and ion identity on the "swimming" behavior as you move towards smaller nanoparticles?

Frank Cichos replied: Whether or not the ionic strength or the ion valency, for example, modify the swimming behavior depends on the details of the swimming mechanism. If we take for example a charged colloid which is screened by its counter ion cloud in solution, then a temperature gradient across the particle would modify the Debye length of the counter ion cloud and thus also the density. This ion density change across the particle would result in an osmotic pressure difference, which causes the interfacial creep flow and finally the swimming. In this case, the details of the ionic solution would matter. The boundary layer approximation, in which the interfacial creep flow is treated as a hydrodynamic slip boundary condition, will fail once the boundary layer and the swimmer size are on the same order of magnitude (boundary layer ~nm, particles are typically >1nm).

In the experiments presented here, we use a block-copolymer to coat the particle and the glass cover slides. From other experiments we know that the coating gives rise to a non-ionic contribution to thermophoresis, which is much stronger than the electrostatic contributions. Thus we do not observe a dependence on ionic strength.

Himangshu Prabal Goswami enquired: 1) I can visualize the system as a heat engine, where the Janus particle takes heat from the hot side and dumps it in the cold sink. In this process there is both work done and entropy. I am confused about the work done. Normally, I would expect the work done to be the perpetual motion of the Janus swimmer and the entropy to be lost. However it is the entropy that is utilized to create the perpetual motion of the Janus swimmers. In this case, what is the work done and how are you defining the efficiency?

2) What is the physical significance of the geometric factor, $h(R)$, and its connection to temperature? When you multiply its square with R, you get another geometric factor, $k(R)$. How are these two factors different from each other?

Frank Cichos responded: The situation with the forces and the work done in the thermodynamic sense is a bit complicated. Let's say, the swimmer moves in the liquid with a constant velocity due to the interfacial stresses generated by the temperature gradient, then according to Newton's laws the forces exerted by the swimmer on the fluid and the ones exerted by the fluid on the swimmer must add up to zero. Nevertheless, the swimmer does work on the fluid to propel itself, but the work is completely dissipated into heat due to viscous dissipation. As the thermodynamic efficiency is now referring to the external work done, it would be zero as there is no external work done since the fluid belongs to the system. A way of defining the efficiency in this case has been pioneered by Lighthill and picked up later by others. Lighthill[1] has been using the so-called thrust generated which is balanced by the liquid. This thrust multiplied by the propulsion velocity is considered as the work done and is entering an efficiency definition known as the Froude efficiency, which is coming from applied propeller theory. The way we define the efficiency of swimming is to consider a thought experiment, where the swimmer is doing actual external work by coupling it for example to an external weight that the swimmer has to lift against gravity. This work is actually considered by Childress.[2] The force which has to be applied to stop the swimmer completely is then the force lifting the weight, or when switching off the propulsion, the gravitational force of the weight would move the passive swimmer against the viscous drag with a velocity which is the same as the swimming velocity. This force multiplied by the velocity then gives the propulsion power, which we divide by the total absorbed optical power that is used to heat the gold cap. Overall this gives a useful form of the efficiency, which we can exploit experimentally by measuring the propulsion velocity and estimating the absorbed laser power.

The propulsion velocity is determined by the temperature gradient as we deal with self-thermophoretic swimmers. The temperature gradient can be split into two contributions. One is coming from the amount of material that is absorbing the laser radiation to be converted into heat. This delivers the absorbed power. The same absorbed power may, however, generate different temperature gradients. For example, a complete gold shell may absorb the same amount of optical power as a thicker hemisphere but there is no interfacial temperature gradient in the case of the full gold shell and thus no propulsion. This geometry effect, *i.e.* how the power is converted into an interfacial temperature gradient, is measured by the geometry factor $h(R)$. The factor $k(R)$ now summarizes the same effects for the efficiency. It includes, however, also the size dependence of the absorption cross-section and is thus proportional to the square of $h(R)$ and the particle size R. In the case of the Janus swimmers considered here, the efficiency at constant incident heating power will therefore scale with the geometry factor $k(R)$ which is proportional to the inverse volume of the swimmer.

1 M. J. Lighthill, *Commun. Pur. Appl. Math.*, 1952, **5**, 109–118.
2 S. Childress, *J. Fluid Mech.*, 2012, **705**, 77–97.

Michel Orrit asked: The orientation of the large Janus particles you used in your experiment was easy to determine optically from anisotropic scattering spectra. If you want to use smaller particles, they will respond as dipoles, and although the orientation of the dipole axis can be determined from the scattering pattern (a similar effect is at work in nanorods), the polarity of your asymmetric Janus particle will not show up on the scattering diagram. How would you determine the pointing direction of the metal in that case?

Frank Cichos answered: Yes, a measurement, for example, of the polarization of the scattered light for smaller Janus swimmers would deliver only the symmetry axis orientation but not the actual propulsion direction. This is one of the current problems where we do not yet have a working solution we can report on.

Francisco Balzarotti enquired: The direction of the propulsion is determined by rotational diffusion, of which in this context there is no control. Would it be useful to also have control of the rotational diffusion of the swimmers by, for example, using mixed materials like gold and silver and alternating their heating?

Frank Cichos responded: Yes. The intention of the photon nudging is to harness the orientational fluctuations to control the direction of motion by switching the propulsion on and off based on an optical feedback. A control of the orientation of the swimmer is, in general, possible and is an interesting task. One may then consider the system as a model for the swarming and flocking of birds or fish, which actively control their direction. Adding orientation control would bring us closer to several applications. Such studies are currently under way and already exist but are unpublished. One way of doing that could involve a spherical object where an asymmetry of the slip flow in the symmetry plane is induced by a special coating such as an alternating silver and gold coating. Actually, all larger spherical swimmers which have a very slow rotational diffusion move in circles due to slight asymmetries in the light absorbing capping. This is blurred by the rotational diffusion for smaller particles. Another possible method for controlling orientation which we are pursuing is to use non-spherical shapes with multiple heat sources.

Maxime Dahan asked: How do you synthesise the Janus nanoparticles?

Frank Cichos answered: The Janus particles are prepared from colloidal polystyrene or silicon dioxide particles. The particle suspension is spin-casted onto a glass cover slide. Low concentrations are used to avoid clustering of colloids. A thin chromium adhesion layer (typically 3 nm) and a gold layer of 50 nm is then evaporated on the cover slides including the colloids. The particles are removed from the substrate using an ultrasonic bath.

Brahim Lounis enquired: The coating of the particle generally shows imperfections. The quality of the metal cap can affect the movement of the particles. Can you comment on this? Can we vary the viscosity of the medium to improve the energy exchange?

Frank Cichos responded: The local temperature gradient is responsible for the strength of the local thermo-osmotic flow on the Janus particle surface.

Imperfections will modify this boundary flow and typically destroy the symmetry of the flow profile around the particle orientation axis. For larger particles (>2 μm) this often results in a motion which is, not straight along the particle symmetry axis, but circular. This is, however, only visible if the rotational diffusion time is very large. So, in general, asymmetries in the temperature gradient will show up in a more complex motion. One can exploit that by preparing swimmers with a lower degree of symmetry.

The swimmer's velocity is in general proportional to the inverse viscosity in the phoretic mechanisms. This means that a high viscosity will slow down the swimmer. We define the efficiency as the power required to drag a particle of the same size as the swimmer through the liquid. This power is the product of the swimmer's velocity and the Stokes friction force. The Stokes force contains the product of the viscosity of the liquid and the velocity of the particle. Therefore the output power depends on the viscosity times the velocity of the undisturbed swimmer (without weight) squared. Together with the viscosity dependence of the swimming velocity, this results in an efficiency which should depend on the inverse viscosity. Thus, a lower viscosity should increase the swimming efficiency according to our treatment.

Sanli Faez asked: It is mentioned that for a particle diameter smaller than 250 nm, it is not possible to detect the orientation of the Janus particle based on the microscopic image. To determine the thermophoretic velocity, the auto-correlation function of velocity is used in the paper (DOI: 10.1039/c5fd00111k). Is this method fast enough to be used for targeted actuation by photon-nudging?

Frank Cichos replied: Yes, below 250 nm the dark field images do not allow the determination of the orientation from the images by considering their asymmetry and total intensity. To determine the velocity, different methods can be used. One method is the use of the velocity auto-correlation. This is a vector correlation, which decays exponentially due to the randomization of the direction by rotational diffusion. Its value at time zero allows us to obtain the magnitude of the velocity, not its direction. Its decay gives the rotational diffusion time. If the rotational diffusion is so fast that the exponential decay cannot be resolved then this method is also not suitable for the measurement of the propulsion velocity. In this case a different method has to be used. This is coming from the mean squared displacement of the particles. The mean squared displacement (msd) of the swimmer depends linearly on time for times much larger than the rotational diffusion time of the swimmer. The slope of this linear increase of the msd with time yields the effective diffusion coefficient. This effective diffusion coefficient is proportional to the diffusion coefficient of the un-propelled particle plus a term, which depends on the square of the propulsion velocity. Measuring the effective diffusion coefficient, being the sum of these two terms, allows again the measurement of the magnitude of the velocity even though rotational diffusion is much faster than the time resolution of the detection system. However, both methods do not allow the determination of the direction of the velocity and are thus not suitable for photon-nudging, where the orientation of the particle (or the orientation of the velocity) is required for the feedback.

Arindam Chowdhury asked: Why do the particles move predominantly in the *xy* plane and rarely show movement or re-orientation in the z-direction?

Frank Cichos replied: The experiments are typically carried out in a slit pore or, in other words, between two glass slides. The presence of boundaries will influence, on one side, the hydrodynamic flow field that is generated by the swimmer itself, and on the side, the temperature field around the swimmer. Both allow the swimmer to interact with such walls. In the case of a single wall, a particular orientation of the swimmer which is not parallel to the wall is coming from the hydrodynamic interaction with the wall. In the case of a slit pore causing strong confinement (swimmer diameter slightly smaller than the slit width) as in the experiments presented here, the swimmer will stabilize with its orientation parallel to the wall due to the hydrodynamics. The change in the heat conductivity between the solvent and the substrates can also cause interactions of the swimmer, which may be attractive or repulsive. The main contributor to the effect you have enquired about is however the hydrodynamics.

Himangshu Prabal Goswami remarked: I think this experiment can be used to realize Crooks' fluctuation theorem or Jarzynski's equality if the entropy and work can be calculated for forward and backward processes.

If one can think of the energy levels in the Janus particle to be coupled differently to two hot and cold reservoirs, it can represent a quantum junction. Then, by monitoring the quanta of heat transferred between the particle and the reservoirs, the heat fluctuation theorem or the Gallovotti–Cohen symmetry can be looked at. However, I would add that there will be more specific concepts to address before this can be realized.

Frank Cichos commented: Yes, the experimental validation of fluctuation theorems, as well as the theoretical development and verification of fluctuation dissipation theorems for the thermal non-equilibrium of the studied swimmers here is the scope of current research.

Himangshu Prabal Goswami opened a general discussion of the paper by Maxime Dahan by enquiring: Typically, I would expect kinetic distributions to decay exponentially in time. Under what conditions do you observe these power law distributions in time? Is it safe to think that modeling rates as a direct function of energy instead of temperature may generate power laws?

Maxime Dahan replied: You are right. For a simple dissociation process, I expect exponential distribution. In our *in vivo* experiments, we think we are probing many different dissociation processes and that is why we see a broad distribution. The fact that it is a power law might reflect a specific distribution of binding energies but note that it is not sufficient to model rates as a function of energy to generate power laws.

Jens Michaelis asked: What is the shortest binding time that you could determine with your technique? How does this time compare with estimates of truly unspecific binding times? If I understood correctly, then the model you presented which led to the power law dependence of observed binding times is based on binding to sequences which are related to the specific recognition sequence. Thus, I would be interested in what the expected binding time to a completely unrelated sequence is.

Maxime Dahan answered: The shortest time we can access is around 20–30 ms. It is thus possible that the time taken for binding to a completely unrelated sequence is much shorter and we therefore do not observe it in our experiments. That is why we are now trying to do single molecule *in vitro* experiments to access these short events. Yet, one might still argue that the environment in the nucleus is still very different to what can be achieved *in vitro* and that might affect the binding times.

Sanli Faez queried: In comparing the experimental results with the model, there was no indication of the other parts of the cell interior. Is it possible to check the validity of the model by carrying out *in vitro* experiments?

Maxime Dahan responded: I am not sure what you mean by 'other parts of the cell interior'. We are trying to use *in vitro* experiments but it is difficult to test a large variety of binding sequences, compared to what can be directly encountered in the nucleus.

Michel Orrit remarked: I am really fascinated by your results and I'm trying to form a simple picture of that protein looking for the needle-sequence in the haystack-chromatin. It's clear that it's an advantage to switch from 1D to 3D diffusion because 1D diffusion is compact (that is, it explores the same sites many times). Therefore, on the one hand, pure 1D diffusion is very inefficient. To improve the chances of finding the right sequence, you have to release the DNA, fly in 3D to another strand and explore again. On the other hand, pure 3D diffusion is also useless, as you must land often onto the DNA in order to find the right sequence.

From this simple reasoning, I deduce that the sequence isn't likely to be the cause of the dispersion in binding times, as the protein must perform some 1D diffusion, thereby sampling many sequences with different binding efficiencies. Diffusion will average out those differences, which is not what you observe. Therefore, an alternative explanation is that the heterogeneity stems from the protein itself, rather than from the sequence.

This idea is not new: Frauenfelder was first with research on ensembles, then more and more researchers studying single molecules have found that proteins have complex conformational landscapes. Sunney Xie and his collaborators (Peter Lu and Haw Yang, amongst others), Hofkens and others have shown that enzymes undergo an alternation of busy and lazy periods, attributable to a random walk in the space of conformations having different turnover rates. Why not extend this idea to your LaqI protein, which may bind more or less strongly to the DNA chain while exploring, leading to a very broad distribution of binding times?

This brings me to another question: Is there a functional advantage of having a broad distribution of binding times? Investigation of Lévy flights and other stretched kinetics has suggested that this type of search may be advantageous and is indeed used by animals foraging for food or in search of a rare partner.

Maxime Dahan responded: I think your suggestion is extremely valid and needs to be investigated carefully. I think the best way (and right now, for me, the only way) is to do it *in vitro* by measuring the interaction kinetics for different sequences. If your hypothesis is right, we should not have many differences between sequences (fully random or quasi-specific). We are starting these experiments and I'll let you know the results. Regarding the 1D sliding process, I am not sure I agree with you that it will average out all the differences. I could envision a competition between two time scales: one in which the protein diffuses along random DNA (1D binding time) and the dissociation times from quasi-specific sequences.

Thorben Cordes asked: Have you or others tested the dependence of buffer or salt on the time that the protein is bound to DNA using *in vitro* assays? Could a buffer mimicking eukaryotic *vs.* bacterial cells already account for the increase in binding time (comparing the same DNA sequence)?

Maxime Dahan answered: This is a great question but we have not tried any *in vitro* assays with the protein and I am not aware of studies in which buffers have been carefully examined to mimic eukaryotic *vs.* bacterial cells.

Frank Cichos enquired: The existence of a power law distributed non-specific binding time suggests that there might be a biological relevance to this. Can you comment on that? Is a power law distribution of binding times more suitable than an exponential distribution, as it would be then more robust against temperature variations? Are there other examples of power law distributed processes in biology that follow this concept of a broad distribution of timescales as well?

Michel Orrit responded to this question: The question of the relevance of power law distributions in biophysics is a difficult one. Single-enzyme kinetics, for example, have been found to be stretched over several decades of times, thereby deviating from the conventional Michaelis–Menten model.[1] It is not yet clear whether this is just a consequence of the intricacy of the free-energy landscape of proteins (itself required for flexibility and function), or whether the power law in itself presents functional advantages, such as extended times in possibly favorable but hard-to-reach conformations. Similarly, in the present case, the power law behavior of the binding times might be just a consequence of the conformational space of the reaction complex, or have deeper functional implications. Further experiments may help answer this deep-reaching question.

1 H. Yang *et al.*, *Science*, 2003, **302**(5643), 262–266.

Wolfgang Langbein queried: You are reporting average binding times to characterize a process (see for example Table 1 in DOI: 10.1039/c5fd00112a), but in reality a wide distribution of times is found, and a single time does not represent these results well. Given the interpretation with binding affinities, would a better representation of the experimental data not be the distribution of binding affinities which result in the measured times? This would also relate directly to the physical interpretation. Is the transcription assumed to be commencing after a certain attachment time, so that the specificity is regulated by the binding affinity?

Maxime Dahan responded: You are right that we extracted a single number (the average) for the distribution and it does not represent the results well. We could represent the distribution of binding affinities but I don't think it will fundamentally contain more information than the distribution of binding events (which is already provided).

Regarding the specificity of attachment and its relationship to transcription; it is a complicated process in eukaryotic cells. It is widely believed that binding specificity comes from the fact that endogenous transcription factors (not LacI as in our model experiments) operate in a combinatorial way, meaning that the specificity is increased by the fact that several of them are required at the same position. However, the accurate kinetic measurements of all this remain to be performed (in my opinion).

Mark Leake commented: You could adapt Bayesian inference algorithms[1] that analyse tracked particle mobility using mean squared displacement traces to determine if there are binding events in individual tracks.

1 A. Robson, K. Burrage and M. C. Leake, *Philos. Trans. R. Soc., B*, 2013, **368**(1611): 20120029.

Jacek Mika opened a general discussion of the paper by Mark Leake by asking: What is your view on how the maturation of different fluorescent proteins will influence the concentration measurement? mCherry is expected to mature less than GFP; do you see a difference in concentration determination when you use mCherry as compared to GFP? How would you correct for the (under)estimation of protein concentration due to FP tagging of proteins of interest?

Mark Leake answered: It is essential to select fluorescent proteins that mature quickly. Ultimately, maturation effects are inevitable, even in the fast maturing FPs, but what we can do is characterize this effect, for example, by treating bacteria cells with the antibiotic chloramphenicol to suppress further FP expression, then photobleaching all FPs present, and then looking at any subsequent recovery of fluorescence which is indicative of FP maturation. This typically indicates with both GFP and mCherry that up to 10% of the FPs, during the time scale of the experimental observation, are potentially dark, and therefore the concentration estimates may be underestimates by up to 10% under these conditions.

Thorben Cordes asked: What is the dynamic range of your approach and how high or low are the concentrations that you can determine?

Mark Leake answered: The dynamic range is likely to be limited by the pixel depth of the camera used, so the working dynamic range may potentially be *ca.* nM up to *ca.* tens of micromoles.

Himangshu Prabal Goswami asked: What is the source of the second hump in the probability density function with concentration? Are there two distributions which are followed based on the range of concentration?

Mark Leake responded: Replication of DNA in these bacteria precedes cell elongation and division and so at certain stages in the cell cycle there will be a doubling of the number of genes and so a doubling in concentration of the expressed protein (*i.e.* the second hump in the graph of DnaQ).

Maxime Dahan enquired: Until we discover the hypothetical perfect dye that would be small, targetable and non-bleaching, we must live with the fluorescent proteins. Is there a study on the difference of the properties in terms of maturation efficiency in eukaryotic *vs.* bacteria cells?

Mark Leake replied: There are multiple biochemical studies that measure the maturation rate of *ex vivo* fluorescent proteins expressed from different eukaryote and prokaryote sources (*e.g.* research by Roger Tsien *et al.*),[1,2] however the information for *in vivo* rates is more limited currently. What we can do in bacteria (see Mark Leake *et al.*, *Nature*, 2006[3] and later pertaining publications) is incubate cells in the antibiotic chloramphenicol to suppress further fluorescent protein expression, then photobleach all cellular fluorescent proteins and then look at the fluorescence intensity of the cells as a function of time following this; this is a metric for the maturation of the fluorescent proteins. The maturation rate depends on the growth medium quite sensitively, and there is merit in having minimal growth conditions (*e.g.* using minimal media at room temperature instead of at 40 °C *etc.*) since this suppresses expression rates and allows the pool of already expressed fluorescent protein more time to mature before newly expressed fluorescent protein is generated. We have observed that this results in a recovery of ~10% of fluorescence based on GFP and mCherry. We are in the process of obtaining similar data from eukaryotic systems (budding yeast).

1 R. Y. Tsien, *Annu. Rev. Biochem.*, 1998, **67**, 509–544.
2 N. C. Shaner, P. A. Steinbach and R. Y. Tsien, *Nat. Methods*, 2005, **2**, 905–909.
3 M. C. Leake, J. H. Chandler, G. H. Wadhams, F. Bai, R. M. Berry and J. P. Armitage, *Nature*, 2006, **443**, 355–358.

Sandra Mayr opened a general discussion of the paper by Jacek Mika by asking: Do you have an explanation for your observation of cells without large SeqA foci? Can you explain the difference in the physicochemical properties between eYFP and PAmCherry, since this is claimed to be the reason for the missing observation of SeqA localized near the cell membranes?

Jacek Mika replied: We cannot pick out a single explanation for the existence of cells without large SeqA foci. Cells without clearly defined SeqA foci have been observed previously with conventional fluorescence microscopy in cells grown in minimal growth media (slow growth, one chromosome copy).[1,2] In these studies however, the cells were grown on a rich medium (fast growth, LB). Their physiology could have been altered by serine hydroxymate (SHX) treatment and paraformaldehyde fixation (see the main text of these publications for details).[1,2] It should be stressed that the distinction between large and small foci (or SeqA–PAmCherry molecules outside of large foci) was made arbitrarily upon visual inspection (*i.e.*, these "small" foci appeared different in PALM reconstructions as compared to "large" foci, as shown in Fig. 2 in DOI: 10.1039/c5fd00058k). It is therefore possible that the "small" foci observed in the PALM images would appear as "large" foci in conventional fluorescence microscopy due to the fact that a certain detail of the SeqA structures can be obscured by the diffraction limit (see Fig. 1 in DOI: 10.1039/c5fd00058k). It is also possible that, within the "small" SeqA–PAmCherry foci, some of the PAmCherry molecules go undetected (not matured or not photoactivated efficiently; note that the photoactivation efficiency of PAmCherry is only 45%, as determined in *Xenopus* oocytes).[3] It remains possible that a portion of the small spots ("small" foci) are free PAmCherry molecules that were cleaved off from the SeqA–PAmCherry fusion protein.

PAmCherry is photoactivated with 405 nm illumination from a dark state to a bright state that can be excited with 561 nm illumination and detected, after which it soon photobleaches. In some cases, the molecule might return to the bright state, as was observed and quantified in the main text of our manuscript (DOI: 10.1039/c5fd00058k).

eYFP is initially in a bright state and transforms to a dark state upon high intensity illumination at 514 nm. Some of these molecules will become permanently dark (photobleached) while some will return from the dark state as bright molecules. In order to perform the single molecule fitting required in PALM, a vast majority of the initially bright eYFP molecules needs to be in the dark state and subsequently, single, well spatially resolved molecules should be recorded as they return from the dark state. We do not have full control over how many eYFP molecules return from the dark state to the bright state and we did not determine how many of the total eYFP molecules did actually return as bright molecules (blink) and did not become permanently photobleached. We assume that only a small fraction of the actual total eYFP molecules returns from the dark state and we visualize identical emitters several times due to blinking (*i.e.*, multiple returns of the same molecule from the dark state), which leads us to conclude that the fraction of the total eYFP molecules we visualize is rather small (and smaller than that of PAmCherry) as compared to the total amount of eYFP tagged molecules present in the sample.

1 T. Brendler, J. Sawitzke, K. Sergueev and S. Austin, *EMBO J.*, 2000, **19**, 6249–6258.
2 S. Adachi, T. Fukushima and S. Hiraga, *Genes Cells*, 2008, **13**, 181–197.
3 N. Durisic, L. Laparra-Cuervo, A. Sandoval-Álvarez, J. S. Borbely, and M. Lakadamyali, *Nat. Methods*, 2014, **11**, 156–162.

Daniel Cross asked: Could you elaborate on why you chose to use a supervised clustering method to obtain the parameters for the clusters listed?

In order to perform an unsupervised clustering, one would use a compactness-based algorithm in this instance, *e.g.* K-means or agglomerative hierarchical, and then validate using an internal validation measure, or construct an artificial ground truth with assigned cluster labels and directly compare class assignments from the results of the clustering.

Jacek Mika answered: We simply did not consider an automated/unsupervised method for analysing the clusters. Indeed it does seem it would be a good approach to analyze those clusters.

Thorben Cordes asked: How do you explain the big variation in the protein number? Can the copy number effect the function of the protein? Do you find specific phenotypes for high or low copy numbers, *i.e.*, does the length of the bacteria correlate with this?

Jacek Mika replied: There are multiple possible explanations for the big variation in observed protein copy numbers. Partially, the variation can be explained by imperfection of the experimental set up. There is a contribution to the experimental error; *i.e.*, not all PAmCherry-tagged molecules are photoconverted efficiently (although we did continue image acquisition until no new photoactivation events were observed upon visual inspection) and there might be differences in the photoconversion efficiencies between individual cells. Another factor will be the real biological variation resulting from cell-to-cell heterogeneity. High variations in protein copy numbers between cells were observed in other studies where PALM was used to count protein copy numbers in *E. coli.* Endesfelder *et al.*[1] have counted the copy number of the *E. coli* RNA Polymerase tagged with PAmCherry and also found a high standard deviation (about 35%, compared to 50–60% in our study) and explain that this phenomenon reflects cell-to-cell heterogeneity.

Indeed, it has been reported previously that isogenic bacterial cultures generally display heterogeneity at the individual cell level for multiple physiological parameters.[2,3] Elowitz *et al.*[4] report that the amount of protein produced by a particular gene varies from cell to cell. Additionally, in a study by the group of S. Xie, where a library of over 1000 *E. coli* proteins C-terminally tagged with a fluorescent protein, Venus, was constructed and protein copy numbers were counted on a single cell level, also broad distributions of protein copy numbers in cells were reported.[5]

Under the growth conditions used in this study (rich medium, LB) the bacteria cells divided rapidly and had several copies of the bacterial chromosome, as seen by the presence of multiple SeqA foci (see Fig. 5 in DOI: 10.1039/c5fd00058k). This

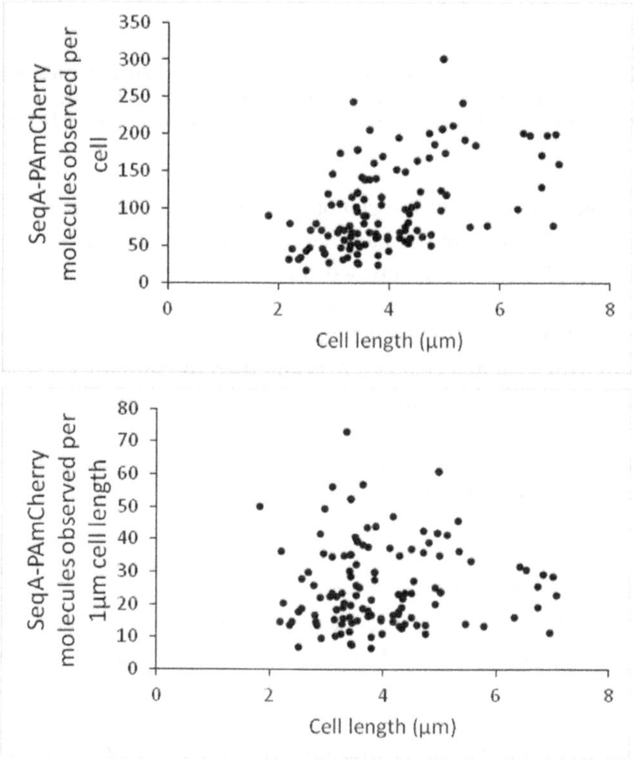

Fig. 1 (Top) SeqA–PAmCherry molecules observed per cell, (Bottom) SeqA–PAmCherry molecules observed per 1 μm cell length.

is in contrast to the slow growth conditions (minimal media) where there is typically 1–2 and a maximum of 4 foci per cell (see a review by Waldminghaus and references therein).[6] If you disregard the experimental error in determining the number of SeqA molecules per cell, we would expect the cell-to-cell variation in SeqA–PAmCherry protein copy numbers to, in part, reflect the cell-to-cell variation in copy numbers of chromosomes. We are not sure what the biological significance of lower copy numbers is, but we believe that the lower copy numbers reflect the lower amount of chromosomes. We did not find any striking phenotypes of bacteria with either high or low copy numbers of SeqA–PAmCherry but we also did not mine the data for this feature.

We plotted the SeqA–PAmCherry protein numbers per cell as a function of cell length (for the data set of cells grown at 37 °C and synchronized in the cell cycle using SHX (see DOI: 10.1039/c5fd00058k for details)), as well as the same data but with the number of molecules normalized by dividing by the cell length (Fig. 1). From the graphs it is clear that the number of observed molecules per cell correlates with the cell length. However, the number of molecules per cell length, a proxy for the concentration, does not correlate strongly with cell length. In other words, the "concentration" of observed SeqA–PAmCherry molecules is most likely independent from the cell length.

1 U. Endesfelder, *et al.*, *Biophys. J.*, 2013, **105**, 172–181.
2 I. Golding, J. Paulsson, S. M. Zawilski and E. C. Cox, *Cell*, 2005, **123**, 1025–1036.
3 M. E. Lindstrom and M. C. Konopka, *Nat. Chem. Biol.*, 2010, **6**, 705–712.
4 M. B. Elowitz, A. J. Levine, E. D. Siggia, P. S. Swain, *Science*, 2002, **297**, 1183–1186.
5 Y. Taniguchi, P. J. Choi, G. W. Li, H. Chen, M. Babu, J. Hearn, A. Emili and X. S. Xie, *Science*, 2010, **329**, 533–538.
6 T. Waldminghaus and K. Skarstad, *Plasmid*, 2009, **61**, 141–150.

Faraday Discussions

PAPER

Single molecule microscopy and spectroscopy: concluding remarks

Niek F. van Hulst

Received 28th October 2015, Accepted 6th November 2015
DOI: 10.1039/c5fd00147a

Chemistry is all about molecules: control, synthesis, interaction and reaction of molecules. All too easily on a blackboard, one draws molecules, their structures and dynamics, to create an insightful picture. The dream is to see these molecules in reality. This is exactly what "Single Molecule Detection" provides: a look at molecules in action at ambient conditions; a breakthrough technology in chemistry, physics and biology. Within the realms of the Royal Society of Chemistry, the Faraday Discussion on "Single Molecule Microscopy and Spectroscopy" was a very appropriate topic for presentation, deliberation and debate. Undoubtedly, the Faraday Discussions have a splendid reputation in stimulating scientific debates along the traditions set by Michael Faraday. Interestingly, back in the 1830's, Faraday himself pursued an experiment that led to the idea that atoms in a compound were joined by an electrical component. He placed two opposite electrodes in a solution of water containing a dissolved compound, and observed that one of the elements of the compound accumulated on one electrode, while the other was deposited on the opposite electrode. Although Faraday was deeply

Fig. 1 The Royal Society of Chemistry at Burlington House in London. The entrance and the stained glass window on the stairway towards the first floor corridor where one finds the bronze bust representing Michael Faraday, protagonist of the early-day lively scientific discussions, which have inspired the "Faraday Discussions".

ICFO – Institut de Ciencies Fotoniques, The Barcelona Institute of Science and Technology, Barcelona and ICREA – Institució Catalana de Recerca i Estudis Avançats, Barcelona, Spain. E-mail: niek.vanhulst@icfo.eu; Web: http://www.icfo.eu

opposed to atomism, he had to recognize that electrical forces were responsible for the joining of atoms. Probably a direct view on the atoms or molecules in his experiment would have convinced him. As such, Michael Faraday might have liked the gathering at Burlington House in September 2015 (Fig. 1). Surely, with the questioning eyes of his bust on the 1st floor corridor, the non-believer Michael Faraday has incited each passer-by to enter into discussion and search for deeper answers at the level of single molecules. In these concluding remarks, highlights of the presented papers and discussions are summarized, complemented by a conclusion on future perspectives.

Introduction

The detection of atoms and molecules has long been restricted to the gas phase under ultrahigh vacuum conditions. Yet in 1989, now over 25 years ago, the first detection of the absorption of a single molecule in the condensed phase was realized by the group of W. E. Moerner, in a highly engineered frequency and Stark modulated scheme.[1] Soon after, in 1990, Michel Orrit and his team also observed the absorption of a single molecule by detecting the background-free fluorescence.[2] The early experiments by Moerner and Orrit all took place under cryogenic conditions, with narrow lines and large absorption cross-sections. In fact, the molecules could be separated spectrally, even without microscopic selection. In 1993 Eric Betzig entered into the world of "single molecule detection" using near-field fluorescence microscopy to image the first individual molecules at room temperature.[3] His near-field approach offered a tiny sub-wavelength excitation volume, which was advantageous to discriminate the individual molecules against their surroundings. Soon it was shown that, also with regular confocal microscopy, and even wide-field microscopy and photon counting CCDs, single molecules can readily be traced. Through these early pioneering actions, Moerner, Orrit and Betzig have laid the ground work for a highly active research field, "single molecule detection", which has led to so many discoveries and applications. Today, single molecule fluorescence detection has evolved to a new frontier in science with high impact in a wide range of disciplines, such as materials research, analytical chemistry, and the biological sciences. Single-molecule fluorescence is particularly a subject of intense interest for biological imaging, through the labeling of biomolecules, such as proteins and nucleotides, to study enzymatic functions, which cannot easily be studied on the bulk scale. The motion and action of several types of motor proteins (myosin/actin) can be followed at the single protein level. More and more, we are getting closer to following the action of single molecules directly in living cells.

Many of these topics were discussed at the Faraday Discussion. Asking around in the Burlington House corridors, some said it was quite late to have a Faraday Discussion on single molecule detection, after 25 years. Others found the meeting very timely as the field is very active and rapidly spreading out in many techniques and applications in both chemistry and biology. Again, others pointed out the crucial role of single molecule detection in the current developments of nano-plasmonics and super-resolution "nanoscopy".

Indeed, it is exactly the capacity to detect and image single molecules at room temperature that triggered the original ideas towards "super-resolution" optical imaging. Essentially, the position of a single isolated molecule can be determined with very high accuracy, limited only by the amount of detected signal: ideally if N photons are detected, Poisson statistics tells us that the location precision improves with $\sqrt{N}$. It was Eric Betzig who in 1994 stated that if each molecule could be discriminated by "some unique optical feature", then a map of their positions could be reconstructed, allowing "molecular resolution imaging".[4] In 2006 Betzig introduced time dependent photoactivation as a discriminator: with an optical stimulus, a tiny fraction of molecules is activated, imaged until photobleaching, and the exact positions determined. By successive iteration of activation–mapping–bleaching, the position map gets filled: Photo-Activation Localization Microscopy (PALM) with nanometric localization accuracy.[5] STORM is an alternative approach, switching molecules temporarily without bleaching.[6] PALM and STORM are commonly termed super-resolution microscopies, yet it should be noted that the diffraction limit is not broken: it is the combination of a well-defined point spread function and a high signal that yields super localisation, all following the rules of diffraction and photon statistics.

Interestingly, Stefan Hell,[7] also in 1994, demonstrated that a fluorescent sample can be switched off by depleting the excited state through stimulated emission, where the reduction of fluorescence shows a highly non-linear response to the depletion power. This non-linearity now is the key to gaining spatial accuracy, allowing one to leave a narrow non-depleted spot at the centre of an excitation profile. Thus, Stefan Hell made STED, stimulated emission depletion microscopy, to work as a super-resolution method.[8] Again, STED perfectly obeys the diffraction limit, but it is the non-linear response of the fluorescent molecule that squeezes the effective spatial imaging response to far below the diffraction limit. It is important to notice that STED does not need the detection of single molecules.

The 2014 Nobel Prize in Chemistry was awarded jointly to William E. Moerner, Eric Betzig and Stefan W. Hell, for "the development of super-resolved fluorescence microscopy". Extraordinarily, at this Faraday Discussion, chaired by Michel Orrit, both William E. Moerner and Eric Betzig presented work. All three pioneers, Orrit, Moerner and Betzig, played a crucial role in the early development and first demonstrations of single molecule detection, thus bringing a great standing to the meeting at the Royal Society of Chemistry.

To capture new developments, the Faraday Discussion focused on four discussion themes:

- *Super-resolution and imaging of soft and biological matter.*

The revolution of super-resolution imaging and several nonlinear optical imaging modalities, to be discussed in the frame of single molecules and single objects.

- *Quantum optics and plasmonics.*

Single molecules combined with plasmonics open up new routes for the enhancement of both excitation and emission, with connections to quantum optics.

- *Probes and sensors for molecular biophysics.*

Plasmonic structures as bright and stable labels, as rulers to probe dynamics, and as antennas to efficiently extract information from the nanoscale; all particularly attractive for biophysical applications

- *Nonlinear optics and coherence in biophysics.*

Tailored light pulses open up a way to manipulate quantum states of single molecules, and to explore coherent effects in biological processes.

Here let's revisit and discuss particular highlights of the various themed sessions.

Super-resolution and imaging of soft and biological matter

The introductory lecture was presented by W. E. Moerner of Stanford University, and it covered single-molecule active control microscopy (SMACM) (DOI: 10.1039/C5FD00149H). He gave a superb introduction into the history and current highlights of single molecule detection in general, and into super-resolution microscopy in particular, thus nicely kicking-off the topic. Beyond a wealth of high-resolution images of new biological applications, such as labelled free amines on living bacterial surfaces, he expressed the continuing need for new robust photoswitchable molecules to pursue the technology further. Next, he presented various solutions to bring super-resolution microscopy from 2D towards 3D. Building on his previously developed double helix illumination, he showed the use of saddle point masks and azimuthal polarisation detection. It should be noted that his student Maurice Lee received the conference poster award for a poster on this topic. Finally, W. E. Moerner made a fine distinction between the localization accuracy of light spots and the determination of real positions of molecules. Due to field dependent aberrations, positions can easily be off by more than 20 nm and up to 100 nm. Using arrays of sub-diffraction limited nanoholes, filled with fluorophores, he presented corrections of the absolute positions to better than 25 nm, thus substantially improving the 3D imaging and tracking. He presented an application of the aberration correction in tracking diffusing molecules simultaneously confined in an ABEL-trap.

The complementary technique to SMACM (PALM and STORM) is STED, which does not require photo-switchability, but uses an engineered two colour Gaussian/donut excitation. As such, STED is a single point method, which requires serial point-by-point scanning and relatively slow imaging. The use of an array of illumination spots is a solution put forward by the Hell group. The array of donut spots for state depletion needs only to be scanned to cover the distance between adjacent donuts, which enhances the frame rate. Yet the single point detection needs to be replaced by a CCD camera; one needs to avoid cross-talk between the various donuts; and polarisation becomes critical. Brahim Lounis from the University of Bordeaux pushed the parallel "lattice" STED to a high frame/rate using four incident beams to create a 2D interference pattern of dips for state depletion, together with a rapid CMOS CCD camera (DOI: 10.1039/C5FD00092K). Mutual polarization of the four beams appeared critical to avoid long depletion lines. He presented STED images with 75 nm resolution and 80 ms frame time.

Scanning probe microscopy and near-field optical microscopy are the predecessors to the current far-field super-resolution microscopies. Sergey Sekatskii from EPFLausanne presented a challenging idea to use a FRET (Fluorescence Resonance Energy Transfer) pair for imaging at molecular resolution: scanning the acceptor molecule on a local probe *versus* the donor molecules in the sample (DOI: 10.1039/C5FD00097A). To avoid premature bleaching of the probe, he designed a single diamond NV centre at the end of the scanning probe. He presented a scanning probe with force control and nice high resolution force microscopy images of ssDNA, yet the optical FRET contrast, so far, did not yield results.

The use of FRET at the level of single donor and acceptor is generally very powerful. The sixth power distance dependence allows distance determination at the sub-nanometer scale and has been used extensively to monitor protein dynamics and folding. Jens Michaelis of Ulm University presented an approach to retrieve quantitative structural information from a single protein using single-molecule FRET (DOI: 10.1039/C5FD00110B). Labelling a protein at many different sites with FRET pairs and measuring all FRET responses for each labelling case, he developed a nano-positioning system (NPS). Interpretation of the data appeared very challenging due to the length of the linkers and mobility of the donor/acceptor. As such, he could determine the most credible volume for each position and thus work his way towards the full protein structure. Clearly an interesting approach, making progress and hopefully delivering the first protein structure one day.

Connected to single molecule FRET probing, Victoria Birkedal of Aarhus University showed methods to retrieve quantitative single molecule FRET efficiencies for use in TIRF microscopy (DOI: 10.1039/C5FD00100E). A particular problem in FRET experiments is the cross-talk, in bleed-through of the donor when detecting the acceptor and in direct excitation of the acceptor. She showed practical approaches and checks for correction and thus achieved quantitative FRET.

High resolution microscopy technology was being pushed firmly by many presenters at the Faraday Discussion. Haw Yang of Princeton University argued that the sequential localization in PALM/STORM is a severe limit on the speed and thus it is better to localize only the object of study and follow that quickly (DOI: 10.1039/C5FD00090D). He presented localization microscopy with 3D feedback of a single molecule, acting as a fast 3D particle tracking system with nanometric resolution. The 3D speed barrier was not the detected signal level any more, instead, the piezo-electric feedback of the various high frequency position controls was. Particle tracking was also shown by Frank Cichos of Chemnitz University (DOI: 10.1039/C5FD00111K). Instead of fluorescence, he directly detected the scattering of metal-coated polymer particles of different sizes, from the micron to nanometer scale. The particles were Janus particles with metal hemispheres, which absorb the incident laser power and convert the energy into heat, thus acting as self-propelled photophoretic swimmers. Tracking the particles, the local heating and speed appeared to depend strongly on particle size.

A powerful alternative for the efficient detection of single molecules, particularly under conditions with a certain background, is the nanoscale reduction of the detection volume, very analogous to the original near-field work of Eric Betzig in 1993. A more practical approach is the detection in a limited volume, such as "zero-mode" waveguides and nanopores, with characteristic dimensions far below 100 nm. Xin Shi of East China University presented the fabrication of nanopores for simultaneous local optical and electrical detection (DOI: 10.1039/C5FD00060B). However, the presented pores were still quite large, ~100 nm, compared to the biomolecular dimensions. Paul Bohn of the University of Notre Dame presented ultra-small zero-dimensional nanophotonic structures to detect a freely diffusing single flavin mononucleotide and detected both spectral and electrochemical responses (DOI: 10.1039/C5FD00072F). Interestingly, he could control the local oxidation/reduction conditions, making the nucleotides go bright or dim, and reveal bimodal distributions at certain pH values. These nanovolume approaches are definitely simple, sensitive and versatile, allowing relatively high background concentrations up to biologically relevant micromolar concentration levels. Their scalability is very promising for applications in biosensing arrays.

Besides pushing the technology, the applications of super-resolution microscopy are rapidly advancing towards nanoscale imaging in living cells. Maxime Dahan of the Institut Curie in Paris (DOI: 10.1039/C5FD00112A), Mark Leake of the University of York (DOI: 10.1039/C5FD00077G) and Jacek Mika of Leuven University (DOI: 10.1039/C5FD00058K) applied localization microscopy to show (de)attachment of protein binding, local distributions of protein concentrations, and quantification and stoichiometry of proteins, all on single cells and bacteria.

The last word on super-resolution was by Eric Betzig, presenting a perspective lecture. In his typical persuasive self-critical style, Eric Betzig made it clear that he had moved on beyond the topic that had earned him the Nobel Prize in 2014. No word was said on his single molecule near-field work in his past. Some hard and critical words were said on STED, due to the high power levels needed for the depletion. Next, he scrutinized the resolution limits of localization microscopy: within the Nyquist criterion, the claimed 10–20 nm accuracy is not realistic; label-linker-length can shift actual positions; local index variations create aberrations; and the labelling density has limits. Altogether he reasoned that ~50 nm accuracy was a more realistic number for super-resolution microscopy. Based on that, he advertised structured illumination microscopy (SIM), which also reaches a resolution below 100 nm, yet does not need photoswitchable labels and allows much faster frame rates. He presented a dashing amount of high resolution images of living cells using SIM. At the same time, he advocated light-sheet microscopy to address the 3rd dimension. Combining light-sheet excitation and structured illumination, Eric Betzig presented a wide variety of detailed full 3D images of cells.

Quantum optics and plasmonics

The detection of a single molecule implies the detection of a single photon emission with fundamentally non-classical photon statistics at room temperature. Together with the observation of discrete quantum jumps and discrete blinking, the detection of single photon emitters is tightly connected to the field of quantum optics. Most quantum optical experiments on single emitters are carried out under cryogenic conditions as they require pure states with very narrow lines. However, at room temperature there is room for certain quantized experiments.

Jörg Wrachtrup of Stuttgart University presented advances in characterizing very weakly coupled spins using diamond NV centres (DOI: 10.1039/C5FD00116A). Diamond NV centres are relatively pure quantum systems with microsecond coherence time, even at room temperature, and stable luminescence. Most importantly, the NV system has an electronic spin, which can be used for local field sensing up to about the 20 nm range. Moreover, effects on nuclear spin can be felt up to 2 nm. The idea of Wrachtrup is to use the NV centre as a local spin sensor for a protein (or other biomolecule) in proximity, ideally towards single molecular NMR sensors. For nuclear spin sensing, he proposes to couple the electron and nuclear spin, such that the nuclear spin range can be enhanced towards 10–15 nm. He presented results on the specific sensing of local isotopes deposited on the diamond surface, an encouraging step towards the projected biological applications.

Pushing the quantum optics even further, Tristan Farrow of Oxford University proposed to directly witness quantum effects in complex molecules, specifically based on a theoretical protocol for entangling two organic molecules by interfering their fluorescence spectra (DOI: 10.1039/C5FD00101C). The authors are working on a set-up to achieve spectral overlap of both molecules under cryogenic conditions to explore their entanglement predictions.

Over the last decade, the fields of nanoscale optics, plasmonics and optical antennas have found a direct connection to the detection of individual nanoparticles and molecules. Nanoplasmonic particles display highly localized fields with a strong mode density, ideal for nanofocusing and enhancement of excitation and emission rates. Thus, when engineered well, the plasmonic particles can be of great advantage to enhance the single molecule response and control the emission characteristics. Enhancement of the fluorescence and Raman response are well-known examples. Several examples in this direction were presented at the Faraday Discussion.

Lukas Novotny of ETH Zurich gave a nice display of the enhanced Raman response of carbon nanotubes and graphene in the proximity of a sharp metallic tip (DOI: 10.1039/C5FD00050E). Super-resolution to 10 nm detail was resolved, while at each position the full Raman spectrum was retrieved, showing effects of strain, defects, and doping. It should be noted that the Raman response of nanocarbon materials is particularly strong, orders of magnitude above the response of organic molecules. Paola Borri of Cardiff University presented transient resonant four-wave mixing on individual metallic nanoparticles, where the phase controlled experiment allowed one to retrieve the spectral phase development for both single and coupled plasmonic particles (DOI: 10.1039/C5FD00079C). Alfred Meixner of Tübingen University showed the enhancement of single quantum dot luminescence when coupled to plasmonic nanocones (DOI: 10.1039/C5FD00074B). Lifetime experiments revealed that both excitation, radiative and non-radiative rates were enhanced. Controlling the excitation polarization, he could choose to enhance the top or bottom side of the cone. Finally Patrick Z. El-Khoury showed the enhanced Raman scattering from aromatic dithiols electrosprayed into plasmonic nanojunctions (DOI: 10.1039/C5FD00036J). Clearly, in all these examples, the control of the near-field interaction between metallic particles and single emitters is of large benefit to enhance the detection sensitivity of a single unit.

Probes and sensors for molecular biophysics

Since the early days of single molecule detection, it has been clear that single molecules are ideal probes for their local nano-environments. Spectral shifts, blinking dynamics, lifetime variations, and polarization changes are all direct observables of fluctuations in the local conformation, chemistry, field, *etc.* In this context, Lukasz Piatkowski of ICFO in Barcelona showed a multicolour excitation approach to reveal large spectral shifts in the emission spectra for molecules embedded in PMMA, even up to almost 200 nm (DOI: 10.1039/C5FD00107B). Such dramatic shifts hamper the detection of all molecules and require changing excitation wavelengths and detection bands. As a result, one can easily be biased by failing to detect a substantial percentage of the molecules. As a solution, he presented first broadband Fourier transform excitation spectroscopy of single molecules, which seems promising to capture the full variation in the excitation spectra. Also, Yuri G. Vainer of the Russian Academy of Sciences presented spectral changes of dyes in a host matrix, yet this time at cryogenic temperature, where the linewidth can be lifetime limited (DOI: 10.1039/C5FD00055F). In practice, he showed that the spectral linewidths would change as a function of the nanometric diffusion depths in the polymer, thus revealing subsurface layer dynamics of polymers with nanometer resolution. Finally, Maxim G. Gladush, also of the Russian Academy of Sciences, showed how single molecule spectro-microscopy can even be used for sub-wavelength refractometry (DOI: 10.1039/C5FD00086F).

Discrete photodissociation is the most limiting effect when using single molecules, both for probing local properties and in super-resolution applications. Enhancing the total number of emitted photons is a permanent wish for the whole community. One of the main routes for photodissociation is through the triplet state. Thorben Cordes of Groningen University presented an effective solution towards photostabilization by quenching of the triplet-state (DOI: 10.1039/C5FD00114E). Attaching a triplet quencher to the Cy5 dye, he proved intramolecular healing and bleach rate reduction with even 2 orders of magnitude. The enhanced photon number is impressive and surely will prove important in both imaging and sensing applications.

Sanli Faez of Utrecht University proposed an original idea to use single molecules for the local sensing of electric fields on the nanoscale (DOI: 10.1039/C5FD00065C). To make an aromatic molecule more field-sensitive, he identified 3-fluoroterrylene with a strongly different LUMO and HOMO. Based on first estimates, a sufficient Stark shift in the ZPL could be expected in a polymer host with an external field applied. Synthesis and experiments are in progress.

Nonlinear optics and coherence in biophysics

During the meeting, the non-linear response of nanoparticles and coherences in the emission of single molecules and NV centres were touched on in the discussions. However, on the whole, the topic appeared rather under-represented compared to the dominance of super-resolution contributions.

Topics beyond the discussion

The collective set of papers presented at the Faraday Discussion reflects the wide range of advances in the field in recent years. The field is gradually coming of age and, in particular, the growing diversity of the applications should be noted. However, the field is constantly looking for new directions. Let's reflect on some very recent developments.

Detection beyond fluorescence

The vast majority of successes in the imaging and spectroscopy of individual molecules at room temperature rely on fluorescence. Yet most relevant molecules do not fluoresce or only weakly fluoresce. Thus, the field has been limited to efficient fluorophores, often used as biolabels. As such, alternative contrast methods have always been high on the agenda. Direct detection of a molecule in absorption has been shown, yet remains very challenging.[9] Reduction of the excitation volume towards dimensions of the molecular cross-section is needed for effective absorption detection; a way to go is nanofocusing using plasmonic particles. Besides absorption, scattering is also a valid alternative. Through clever interferometric detection schemes, scattering detection is now reaching the level of individual proteins.[10] Raman spectroscopy is a valid alternative to fluorescence detection, as also presented during the discussion. Yet enhancements of 10^{10} are needed to compete with the spontaneous emission, making single molecule Raman challenging, and, moreover, harder to interpret than enhanced spectroscopy. Stimulated emission is another way to go; effective stimulated emission imaging has been shown,[11] yet the single molecule limit is still challenging. The photothermal detection of nanoparticles has reached the limits of particles only a few nanometers in size, and the photothermal approach even allows one to detect a single molecule in absorption.[12,13] Electrical read-out of single molecules in tunnel junctions has been shown, but it needs scaling up to ambient conditions for applications.

Ultrafast and coherent control

Ultrafast femtosecond and picosecond spectroscopy is an important topic in spectroscopy and physical chemistry as it reveals the dynamics on the molecular time scale. Single molecule detection based on fluorescence detection only yields photons and typically with a nanosecond lifetime. As a result, the single molecule detection is fundamentally too slow for ultrafast spectroscopy. However, the stimulated excitation is instantaneous and can be used to encode fs effects, while still detecting the fluorescence response. Through pump-probe and coherent control schemes, ultrafast signatures of single molecules have been observed.[14] The route is open to detect coherences, fs–ps energy transfers, charge transfers, *etc.* Recently, the vibration of a single molecule was reported using surface enhanced single molecule CARS.[15]

New single emitters

The detection of single entities started with atoms and molecules and has extended to single quantum dots and single color centres, such as NV centres in diamond. New emitters are fanatically pursued. Detection of single lanthanides has been reported,[16] while systems such as perovskites are explored. Beyond the conventional emitters based on a transition dipole moment, one could also attempt to detect the weaker non-dipolar transitions or even magnetic transitions. Single molecules are detected in the very weak field linear regime; yet even at the few photon level, operating at the nanoscale, a nonlinear single molecule response could be expected provided one couples efficiently to the molecule. Finally, all users of super-resolution microscopy are constantly craving new bright and stable photoswitchable molecules, preferentially small ones with short linkers to allow a higher labelling density.

Acknowledgements

Niek van Hulst gratefully acknowledges the support from the European Commission (ERC Advanced Grant, no. 247330 NanoAntennas); the European

Commission 7th Framework Programme project NanoVista (no. 288263) and Laser Lab Europe (no. 284464); the ESF (European Science Foundation) Research Network Program Plasmon-BioNanosense; the Catalan AGAUR (Agència de Gestió d'Ajuts Universitaris i de Recerca) Project 2014 SGR 01540; the Spanish Ministry of Science MICINN program CSD2007-046 "NanoLight" and Ministry of Economy MINECO projects FIS2012-35527 and FIS2014-55563-REDC; and the Fundació Privada CELLEX (Barcelona) for generous endowment.

References

1 W. E. Moerner and L. Kador, Optical detection and spectroscopy of single molecules in a solid, *Phys. Rev. Lett.*, 1989, **62**, 2535–2538.

2 M. Orrit and J. Bernard, Single pentacene molecules detected by fluorescence excitation in a *p*-terphenyl crystal, *Phys. Rev. Lett.*, 1990, **65**, 2716–2719.

3 E. Betzig and R. J. Chichester, Single Molecules Observed by Near-Field Scanning Optical Microscopy, *Science*, 1993, **262**, 1422.

4 E. Betzig, Proposed Method for Molecular Optical Imaging, *Opt. Lett.*, 1995, **20**, 237.

5 E. Betzig, *et al.*, Imaging intracellular fluorescent proteins at nanometer resolution, *Science*, 2006, **313**, 1642.

6 M. J. Rust, M. Bates and X. Zhuang, Sub-diffraction limit imaging by stochastic optical reconstruction microscopy (STORM), *Nat. Methods*, 2006, **3**, 793.

7 S. W. Hell and J. Wichmann, Breaking the diffraction resolution limit by stimulated emission; stimulated emission depletion microscopy, *Opt. Lett.*, 1994, **19**, 154.

8 V. Westphal and S. W. Hell, Nanoscale resolution in the focal plane of an optical microscope, *Phys. Rev. Lett.*, 2005, **94**, 143903.

9 P. Kukura, M. Celebrano, A. Renn and V. Sandoghdar, Single-molecule sensitivity in optical absorption at room temperature, *J. Phys. Chem. Lett.*, 2010, **1**, 3323–3327.

10 J. Ortega-Arroyo and P. Kukura, Interferometric scattering microscopy (iSCAT): new frontiers in ultrafast and ultrasensitive optical microscopy, *Phys. Chem. Chem. Phys.*, 2012, **14**, 15625–15636.

11 W. Min, S. Lu, S. Chong, R. Roy, G. R. Holtom and X. S. Xie, Imaging Chromophores with Undetectable Fluorescence, *Nature*, 2009, **461**, 1105.

12 D. Boyer, P. Tamarat, A. Maali, B. Lounis and M. Orrit, Photothermal imaging of nanometer-sized metal particles among scatterers, *Science*, 2002, **297**(5584), 1160–1163.

13 A. Gaiduk, M. Yorulmaz, P. V. Ruijgrok and M. Orrit, Room-Temperature Detection of a Single Molecule's Absorption by Photothermal Contrast, *Science*, 2010, **330**, 353–356.

14 D. Brinks, F. D. Stefani, F. Kulzer, R. Hildner, T. H. Taminiau, Y. Avlasevich, K. Müllen and N. F. van Hulst, Visualizing and controlling vibrational wavepackets of single molecules, *Nature*, 2010, **465**, 905–908.

15 S. Yampolsky, D. A. Fishman, S. Dey, E. Hulkko, M. Banik, E. O. Potma and V. A. Apkarian, Seeing a single molecule vibrate through time-resolved coherent anti-Stokes Raman scattering, *Nat. Photonics*, 2014, **8**, 650–656.

16 T. Utikal, E. Eichhammer, L. Petersen, A. Renn, S. Götzinger and V. Sandoghdar, Spectroscopic detection and state preparation of a single praseodymium ion in a crystal, *Nat. Commun.*, 2014, **5**, 3627.

Poster titles

Shedding new light on ancient drugs: Subcellular localisation of novel fluorescent analogues of, *Cinchona* alkaloids in intraerythrocytic *Plasmodium falciparum*, **J. G. Woodland, R. Hunter and T. J. Egan**, *Department of Chemistry, University of Cape Town, South Africa*

Single metallic nanoparticle measure thermal diffusivities using photothermal microscopy, **A. Heber, M. Selmke and F. Cichos**, *Leipzig University, Germany*

Individual gold nanorods: Optical cross-sections and ultrafast dynamics, **A. Zilli, G. Zoriniants, S. Großmann, B. Hecht, W. Langbein and P. Borri**, *Cardiff University, UK*

Advanced pulse pattern generation and fine tuning for single molecule based STED microscopy, **R. Dowler, M. Koenig, P. Reisch, A. Glatz, S. Tannert, T. Schoenau, R. Haertel, T. Roehlicke, M. Sackrow, M. Patting, F. Koberling and R. Erdmann**, *PicoQuant GmbH, Germany*

Asymmetry detection in megamolecules by second-order nonlinear optical microscopy, **Y. Zhao, K. T. T. Hien, G. Mizutani, H. Rutt, K. Amornwachirabodee, M. Okajima and T. Kaneko**, *Japan Advanced Institute of Science and Technology (JAIST), Japan*

Intrinsic light emission from a plasmonic nanoparticle, **G. Lu, Y. Cheng, Y. He, H. Shen, K. Xia and Q. Gong**, *Peking University, China*

Synthesis of labelled heparan sulfate oligosaccharides and proteins, **C. E. Dalton, R. Jeanneret, S. D. Quinn, S. W. Magennis and J. M. Gardiner**, *University of Manchester, UK*

Stimulated emission spectroscopy of single quantum emitters, **L. Piatkowski, N. Accanto, S. Christodoulou, G. Calbris, I. Moreels and N. van Hulst**, *ICFO - The Institute of Photonic Sciences, Spain*

Interferometric broadband excitation spectroscopy of single molecules at room temperature, **E. Gellings, L. Piatkowski and N. van Hulst**, *ICFO The Institute of Photonic Sciences, Spain*

Role of P4 stem modulating functional dynamics of SAM-I riboswitch, **D. C. Perez-Gonzalez, J. Boudreault, D. A. Lafontaine and J. C. Penedo**, *University of St Andrews, UK*

Dimensionality effects on fluorescence resonance energy transfer between single semiconductor nanocrystals and multiple dye acceptors, **I. Hadar and U. Banin**, *The Hebrew University of Jerusalem, Israel*

3D super-resolution fluorescence microscopy with the corkscrew point spread function, **M. Lee, M. Lew, Y. Shechtman, A. Diezmann and W. Moerner**, *Stanford University, USA*

Single-molecule tracer diffusion dynamics in heterogeneous systems: Hop, stop and jump?, **J. Mahato, S. Bhattacharya, D. Sharma and A. Chowdhury**, *Indian Institute of Technology, Bombay, India*

Functionalizing large gold nanoparticles beyond 100 nm for small gaps in dimer nanoantennas, **C. Vietz, B. Lalkens, G. P Acuna and P. Tinnefeld**, *Technische Universität Braunschweig, Germany*

Single-molecule study of interactions between gold nanostructures and fluorescent dyes on DNA origami, **B. Wünsch, P. Holzmeister, C. Vietz, G. Acuna and P. Tinnefeld**, *Technical University of Braunschweig, Germany*

Fluorescence enhancement on tip-specific functionalized gold nanorods, **P. M. R. Paulo, S. Martins, E. Garcia-Fernandez and S. M. B. Costa**, *Centro de Química Estrutural, Instituto Superior Técnico, Portugal*

Automated classification of Rhesus D antigen expression on erythrocytes at single molecule level, **S. Mayr, D. Borgmann, H. Polin, S. Schaller, L. Obritzberger, V. Dorfer, C. Gabriel, S. Winkler and J. Jacek**, *University of Applied Sciences, Austria*

Super resolution imaging *via* spectral separation of quantum dots, **K. Keseroglu, M. Song and A. V. Sapelkin**, *Queen Mary, University of London, UK*

Growing dibenzoterrylene doped anthracene crystals for optical microcavity experiments, **K. D. Major, Y.-H. Lien, C. Polisseni, S. Grandi, A. Clark, J. Hwang and E. A. Hinds**, *Imperial College London, UK*

Coupling single dye molecules to nanophotonic waveguides, **A. S. Clark, Sebastien Boissier, Kyle Major, Samuele Grandi, Claudio Polisseni, Jaesuk Hwang and E. A. Hinds**, *Imperial College London, UK*

Novel method for sub-wavelength thin film growth for single photon emission from dye molecules, **C. Polisseni, K. W. Kho, Kyle Major, A. S. Clark, J. Hwang and E. Hinds**, *Imperial College London, UK*

Solution single-molecule microscopy of the conjugated polymer P3HT and the influence of solvent, **F. Tenopala-Carmona, P. A. Dalgarno, S. Fronk, J. C. Penedo, G. C. Bazan and I. D. W. Samuel**, *University of St Andrews, UK*

Area-selective electroless deposition of gold on AFM tip apexes for probes in tip-enhanced Raman spectroscopy, **H. Itasaka, M. Nishi, M. Shimizu and K. Hirao**, *Kyoto university, Japan*

Dark-field assisted raman spectroscopy system for *in situ* monitoring of nanowire catalysed reaction at molecule level, **H.-W. Li, X. Shi and Y.-T. Long**, *East China University of Science and Technology, China*

Charge-tunnelling and self-trapping dependent quantum dot photodynamics, **A. Fisher and M. Osborne**, *University of Sussex, UK*

Molecular dynamics with non-adiabatic and spin-orbit effects applied to deactivation of thiophene and selenophene, **M. Pederzoli and J. Pittner**, *J. Heyrovsky Institute of Physical Chemistry, Czech Republic*

Synthesis of biocompatible Au-ZnTe core–shell and ZnTe nanoparticles for bioimaging and biosafety studies, **N. Revaprasadu, S. Mlowe and R. Dunpall**, *University of Zululand, South Africa*

Electroluminescence and coherence assisted luminescence in single molecule junctions: a diagrammatic approach, **H. P. Goswami, W. Hua, Y. Zhang, S. Mukamel and U. Harbola**, *Indian Institute of Science, India*

Intersystem crossing rates of single perylene molecules, **N. R. Verhart, P. Navarro, S. Faez and M. Orrit**, *Universiteit Leiden, The Netherlands*

Optical excitation of magnetic dipole transitions in Eu^{3+}-doped nanoparticles, **M. Kasperczyk, S. Person, D. Ananias, L. D. Carlos and L. Novotny**, *ETH Zürich, Switzerland*

The Skinner Prize for the best poster was awarded to Mr Maurice Lee of Stanford University, USA, for his poster on 3D super-resolution fluorescence microscopy with the corkscrew point spread function.

List of participants

Dr Patricia Abellan, *SuperSTEM, United Kingdom*
Professor Mona Arsanious, *National Research Centre, Egypt*
Mr Ruediger Bader, *Hamamatsu Photonics UK Limited, United Kingdom*
Dr Francisco Balzarotti, *Department of Nanobiophotonics, Max Planck Insitute for Biophysical Chemistry, Germany*
Dr Yossef Bar, *SRC, Israel*
Dr Justine Bentley, *Laser Quantum Ltd, United Kingdom*
Dr Eric Betzig, *Howard Hughes Medical Institute, United States*
Mrs Victoria Birkedal, *Aarhus University, iNANO center, Denmark*
Dr Paul Bohn, *University of Notre Dame, USA*
Professor Paola Borri, *Cardiff University, United Kingdom*
Dr Michelle Carey, *Royal Society of Chemistry, United Kingdom*
Mr Alexander Carr, *University of Cambridge, United Kingdom*
Mr Pavan Kumar Challa, *University of Cambridge, United Kingdom*
Dr Arindam Chowdhury, *Indian Institute of Technology Bombay, India*
Prof Dr Frank Cichos, *Molecular Nanophotonics Group, Germany*
Dr Alex Clark, *Imperial College London, United Kingdom*
Dr Thorben Cordes, *University of Groningen, Netherlands*
Dr Marie Cote, *Royal Society of Chemistry, United Kingdom*
Mr Hugh Cowley, *RSC, United Kingdom*
Mr Daniel Cross, *University of Birmingham, United Kingdom*
Dr Maxime Dahan, *Institut Curie, France*
Miss Charlotte Dalton, *University of Manchester, United Kingdom*
Miss Katie Dryden-Holt, *Royal Society of Chemistry, United Kingdom*
Prof Dr Jiangfeng Du, *University of Science and Technology of China, China*
Mr Jamie Durrani, *Royal Society of Chemistry, United Kingdom*
Mr Yvan Eilers, *Max-Planck-Institut für biophysikalische Chemie, Germany*
Dr Patrick El-Khoury, *Pacific Northwest National Laboratory, United States*
Dr Geraint Evans, *Illumina, United Kingdom*
Mr Sanli Faez, *Leiden University, Netherlands*
Mr Thomas Farrell, *IOP Publishing, United Kingdom*
Dr Tristan Farrow, *University of Oxford, United Kingdom*
Mr Aidan Fisher, *University of Sussex, United Kingdom*
Dr Paul Fitzsimons, *Fianium Ltd, United Kingdom*
Ms Esther Gellings, *ICFO- The Institute of Photonic Sciences, Spain*
Dr Maxim Gladush, *Institute of Spectroscopy RAS; Moscow State Pedagogical University, Russia*
Mr Ido Hadar, *The Hebrew University of Jerusalem, Israel*
Dr Mike Hann, *GSK, United Kingdom*
Mr Andre Heber, *Leipzig University, Molecular Nanophotonics Group, Germany*
Dr Suzanne Howson, *Royal Society of Chemistry, United Kingdom*
Mr Hiroki Itasaka, *Kyoto University, Japan*
Ms Ija Jusuk, *TU Braunschweig / Nanobiosciences, Germany*

Mr Mark Kasperczyk, *ETH Zurich*, *Switzerland*
Mr Hakan Keles, *GSK*, *United Kingdom*
Mr Kemal Keseroglu, *Queen Mary, University of London*, *United Kingdom*
Dr Felix Koberling, *PicoQuant GmbH*, *Germany*
Professor Dame Irena Kratochvilova, *Institute of Physivs AS CR*, *Czech Republic*
Professor Wolfgang Langbein, *Cardiff University*, *United Kingdom*
Ms Alice Laures, *GSK*, *United Kingdom*
Dr Katherine Lawrence, *Wiley VCH Verlag GmbH & Co. KGaA*, *Germany*
Professor Mark Leake, *Biological Physical Sciences Institute (BPSI)*, *United Kingdom*
Mr Maurice Lee, *Stanford University*, *United States*
Mr Jemin Justin Lee, *Yonsei University*, *South Korea*
Mr Yuandi Li, *Royal Society of Chemistry*, *United Kingdom*
Mr Yitao Long, *East China University of Science and Technology*, *China*
Professor Brahim Lounis, *Univesity of Bordeaux*, *France*
Dr Guowei Lyu, *Peking University*, *China*
Mr Kyle Major, *Imperial College London*, *United Kingdom*
Dr Sofia Alexandra Martins, *Instituto Superior Técnico- Universidade de Lisboa*, *Portugal*
Miss Sandra Mayr, *University of Applied Sciences*, *Austria*
Prof Dr Alfred J. Meixner, *University of Tübingen*, *Germany*
Prof Dr Jens Michaelis, *University of Ulm*, *Germany*
Dr Jacek Mika, *Katholieke Universiteit Leuven*, *Belgium*
Professor William Moerner, *Stanford University*, *United States*
Miss Jessie-May Morgan, *Royal Society of Chemistry*, *England*
Professor Andrew Mount, *The University of Edinburgh*, *United Kingdom*
Miss Fiona Nalden, *Royal Society of Chemistry*, *United Kingdom*
Mrs Lisa-Maria Needham, *University of Cambridge*, *United Kingdom*
Mr Lukas Novotny, *ETH Zürich*, *Switzerland*
Prof Dr Michel Orrit, *Leiden University*, *Netherlands*
Dr Mark Osborne, *University of Sussex*, *United Kingdom*
Mr James Owens, *Hamamatsu Photonics UK Limited*, *United Kingdom*
Dr Pedro M. R. Paulo, *Centro de Química Estrutural, Instituto Superior Técnico*, *Portugal*
Mr Marek Pederzoli, *J. Heyrovsky Institute of Physical Chemistry*, *Czech Republic*
Mr D. Cibran Perez-Gonzalez, *SUPA School of Physics and Astronomy, University of St. Andrews, St. Andrews, Scotland, KY16 9SS, United Kingdom*, *United Kingdom*
Dr Lukasz Piatkowski, *ICFO- The Institute of Photonic Sciences*, *Spain*
Mr Claudio Polisseni, *Imperial College London*, *United Kingdom*
Mr Himangshu Prabal Goswami, *Indian Institute of Science*, *India*
Professor Neerish Revaprasadu, *University of Zululand*, *South Africa*
Mr Christopher Robinson, *Hamamatsu Photonics UK Limited*, *United Kingdom*
Professor Vahid Sandoghdar, *Max Planck Institute for the Science of Light*, *Germany*
Dr Sergey Sekatskii, *EPFL*, *Switzerland*
Mr Xin Shi, *East China University of Science and Technology*, *China*
Dr Fazhan Shi, *University of Science and Technology of China*, *China*
Dr Gary Skinner, *Illumina*, *United Kingdom*
Ms Claire Springett, *Royal Society of Chemistry*, *United Kingdom*
Dr Fernando Stefani, *Max Planck Insitute for Biophysical Chemistry*, *Germany*
Professor Robert Taylor, *University of Oxford*, *United Kingdom*
Mr Francisco Tenopala Carmona, *Organic Semiconductor Centre, SUPA, School of Physics and Astronomy and Laboratory for Biophysics and Biomolecular Dynamics, Biomedical Science Research Complex, University of St Andrews*, *United Kingdom*
Mr Yuri Vainer, *Institute for Spectroscopy RAS*, *Russia*

Prof Dr Niek van Hulst, *ICFO- The Institute of Photonics Sciences*, *Spain*
Professor Vlatko Vedral, *University of Oxford*, *United Kingdom*
Mr Nico Verhart, *Leiden University*, *Netherlands*
Ms Carolin Vietz, *Technische Universität Braunschweig*, *Germany*
Mr Nathaniel Wand, *University of Birmingham*, *United Kingdom*
Mr Charles Williams, *Imperial College London*, *United Kingdom*
Professor Phil Williams, *University of Nottingham*, *United Kingdom*
Mr John Woodland, *University of Cape Town*, *South Africa*
Prof Dr Joerg Wrachtrup, *3rd Institute of Physics, University of Stuttgart*, *Germany*
Mrs Bettina Wünsch, *Technische Universität Braunschweig / Institut für Physikalische und Theoretische Chemie / NanoBioSciences*, *Germany*
Professor Haw Yang, *Princeton University*, *United States*
Mr Piotr Zdankowski, *University of Dundee*, *United Kingdom*
Mr Yue Zhao, *Japan Advanced Institute of Science and Technology*, *Japan*
Mr Attilio Zilli, *Cardiff University*, *United Kingdom*